STUDY GUIDE

to the third editions of

CHEMISTRY

Bailar, Moeller, Kleinberg, Guss, Castellion, Metz

and

CHEMISTRY

with Inorganic Qualitative Analysis

Moeller, Bailar, Kleinberg, Guss, Castellion, Metz

Clyde Metz
College of Charleston

HBJ

Harcourt Brace Jovanovich, Publishers
and its subsidiary, Academic Press

San Diego New York Chicago Austin Washington, D.C.
London Sydney Tokyo Toronto

Front Cover: Paul Silverman, Test Tube Abstract. © Fundamental Photographs.

Requests for permission to make copies of any part of the work should be mailed to: Permissions, Harcourt Brace Jovanovich, Publishers, Orlando, Florida 32887

ISBN: 0-15-506458-4

Printed in the United States of America

TABLE OF CONTENTS

INTRODUCTION

YOUR STUDY GUIDE

Understanding chemistry is not easy for many students, even though most claim to put in many long hours of study. How much time is needed? Most educational experts suggest a minimum of three hours of study outside the classroom for each hour in the classroom. This means that for a typical general chemistry class that meets each week for three lectures, a recitation, and a laboratory, at least 18 hours of outside study each week would be required.

Activities such as writing chapter outlines, rewriting lectures notes, reviewing old examinations and quizzes, writing laboratory reports, doing homework assignments, and cramming for tests are essential parts of a typical plan of study, but before any of these learning techniques can be beneficial, you should understand the material--and *that* is the purpose of this study guide.

Each chapter of this study guide is divided into several major sections. The purpose and use of each section are summarized below.

CHAPTER OVERVIEW

This section presents a very brief survey of the major topics presented in the textbook chapter. When studying a chapter for the first time, you should make a quick survey of the chapter to identify the major topics so that you can see where the chapter is leading.

COMPETENCIES

This section lists in greater detail the major topics that should be learned in the chapter. These topics are categorized into lists of important terms for which definitions should be learned, general concepts and skills that should be mastered, and numerical exercises that should be understood. As you read the chapter for the first time, you should refer to these lists to help you identify the sections that will require extra study.

QUESTIONS

This section presents a series of multiple-choice, matching, fill-in-the-blank, true-false, and short-answer questions designed to check your understanding of the new terms and of the general concepts presented in the chapter. Answers are provided so that you can check your progress.

SKILLS

This section presents in-depth studies of certain concepts that must be mastered. Following the explanation and examples, skill exercises with answers is provided to test your competency.

NUMERICAL EXERCISES

This section is designed to help you with the quantitative material presented in the chapter. Although the textbook contains numerous example problems and exercises, additional examples and exercises with answers are presented in this section.

PRACTICE TEST

Here is your chance to check your overall understanding of the material in the chapter. The practice test consists of a mixture of multiple-choice and matching questions as well as a few problems. Most of the questions have been used on actual examinations. Answers are provided so that you can check your progress.

ADDITIONAL NOTES

Because each course, instructor, and student is different, this section is left for you to include additional material presented in class, notes from outside reading assignments, doodles, and the like.

CHAPTER 1

THE NATURE OF CHEMISTRY

CHAPTER OVERVIEW

Several important definitions are given in this chapter, and the various subdivisions of chemistry are described. A short discussion about the role of chemistry in society concludes the chapter.

COMPETENCIES

Definitions for the following terms should be learned:

- biological sciences
- chemical change
- chemistry
- descriptive chemistry
- inorganic chemistry
- matter
- organic chemistry
- physical change
- physical sciences
- principles of chemistry
- qualitative analysis
- quantitative analysis
- theory

General concepts and skills that should be learned include

- the scope of the study of chemistry
- the subdivisions of chemistry

QUESTIONS

1. Match the correct term to each definition.

(a) the study of living matter
(b) the study of natural laws and processes other than those peculiar to living matter
(c) the study of matter, the changes that matter can undergo, and the laws that describe these changes

(i) biological sciences
(ii) chemistry
(iii) descriptive chemistry
(iv) inorganic chemistry
(v) organic chemistry
(vi) physical sciences

(d) the explanation of chemical facts, for example, by theories and mathematics (vii) principles of chemistry

(e) the description of the elements and their compounds, their physical states, and how they behave

(f) the chemistry of compounds of carbon with hydrogen and of their derivatives

(g) the chemistry of all the elements and their compounds, with the exception of the compounds of carbon with hydrogen and their derivatives

2. Which of the following statements are true? Rewrite each false statement so that it is correct.

(a) A physical change involves a change in the composition of matter.

(b) Quantitative analysis involves the determination of the amounts of substances in a sample.

(c) Organic chemistry is the chemistry of compounds of carbon with hydrogen and of their derivatives.

(d) Matter is the unifying principle or group of principles that explains a body of facts or phenomena.

3. Distinguish clearly between (a) descriptive chemistry and principles of chemistry, (b) a chemical change and a physical change, and (c) inorganic chemistry and organic chemistry.

4. Choose the response that best completes each statement.

(a) Something that has mass is

(i) a physical science (ii) matter

(iii) a theory (iv) a physical change

(b) The identification of substances is

(i) qualitative analysis (ii) chemical engineering

(iii) quantitative analysis (iv) physical change

(c) A chemical change results in

(i) no change in the composition of the matter involved

(ii) a unifying principle

(iii) the study of inorganic matter

(iv) a change in the composition of matter

(d) Descriptive chemistry

(i) explains chemical facts

(ii) is the study of the composition of matter

(iii) is the description of the elements and their compounds, their physical states, and how they behave

(iv) none of these answers

Answers to questions

1. (a) i, (b) vi, (c) ii, (d) vii, (e) iii, (f) v, (g) iv.
2. (a) F, replace *physical* with *chemical*; (b) T; (c) T; (d) F, replace *Matter* with *A theory*.
3. (a) *Descriptive chemistry* deals with the description of the elements and their compounds, their physical states, and how they behave; *principles of chemistry* are explanations of chemical facts. (b) In a *chemical change* a change in the composition of matter occurs; in a *physical change* there is no change in the composition of matter. (c) *Inorganic chemistry* deals with the chemistry of all of the elements and their compounds except for compounds of carbon with hydrogen and their derivatives; *organic chemistry* is the study of compounds of carbon with hydrogen and their derivatives.
4. (a) ii, (b) i, (c) iv, (d) iii.

PRACTICE TEST

1 (50 points). Choose the response that best completes each statement.

(a) The category of science that includes botany, physiology, ecology, and genetics is

(i) chemical engineering (ii) the physical sciences

(iii) the biological sciences (iv) none of these answers

(b) A change in which the composition of matter changes is

(i) a physical change (ii) quantitative analysis

(iii) a chemical change (iv) none of these answers

(c) The chemistry of chemical compounds of carbon with hydrogen and of their derivatives, whether from living or nonliving matter, is

(i) inorganic chemistry (ii) organic chemistry

(iii) physical chemistry (iv) a biological science

(d) Physical chemistry is the

(i) application of the methods and theories of physics to the study of chemical changes and the properties of matter

(ii) application of chemistry on a large scale

(iii) study of qualitative and quantitative analysis
(iv) study of noncarbon compounds

(e) Which branch of chemistry would not involve the study of carbon-hydrogen compounds?
(i) bioinorganic chemistry (ii) clinical chemistry
(iii) pharmaceutical chemistry (iv) inorganic chemistry

2 (50 points). Briefly discuss the concept that chemistry is an experimental science.

Answers to practice test

1. (a) iii, (b) iii, (c) ii, (d) i, (e) iv.
2. Some of the points that could be mentioned are that reproducible data are used to verify or to help formulate a theory, chemical and physical properties can be observed, experiments can be costly and time-consuming, and there is always an opportunity for the unexpected.

ADDITIONAL NOTES

CHAPTER 2

UNITS, MEASUREMENTS, AND NUMBERS

CHAPTER OVERVIEW

In this chapter two important aspects of chemical arithmetic are discussed--significant figures and scientific notation. The modern units of measurement are reviewed, with the emphasis being placed on the SI system. The dimensional method that is used in problem solving is described and a comprehensive problem-solving method is introduced.

COMPETENCIES

Definitions for the following terms should be learned:

- density
- exact numbers
- force
- mass
- pressure
- scientific notation
- significant figures
- weight

General concepts and skills that should be learned include

- using numbers and units to express the measurement of a physical property
- expressing the uncertainty in a measurement
- determining the number of significant figures in a number
- expressing large and small values of measurements by using prefixes for units and scientific notation
- expressing the units of the seven base physical quantities and the units of derived physical quantities by using the SI system
- using the dimensional method in problem solving
- deriving conversion factors from equalities
- using the comprehensive method introduced in this chapter for problem solving

Numerical exercises that should be understood include

- calculating a numerical answer and expressing the answer to the proper number of significant figures
- calculating a numerical answer involving factors written in scientific notation
- converting values of heat and energy, length, volume, mass, density, force, pressure, and temperature given in one system of units to another system by using conversion factors (dimensional method)
- calculating density, mass, or volume when two of these variables are given

QUESTIONS

1. Complete the following sentences by filling in the various blanks with these terms: density, mass, pressure, and significant figures.

The (a)______ of a substance can be calculated by dividing its mass by its volume. The calculated value should be expressed to the correct number of (b)______ as determined by the uncertainties in the values of (c)______ and volume. If the substance is a solid or a liquid, the volume varies only slightly with (d)______ and temperature. However, the volume of a gaseous substance varies considerably with these physical quantities.

2. Distinguish clearly between (a) base and derived physical quantities, (b) mass and weight, and (c) force and pressure.

3. Which of the following statements are true? Rewrite each false statement so that it is correct.

(a) The expresssion for the measurement of a physical quantity consists of a numerical value and a unit.

(b) Energy is a base physical quantity.

(c) The density of a substance is dependent on the total amount of substance present.

(d) Zeros in a number are not significant.

(e) In addition or subtraction, the mathematical operation is performed and the answer is rounded to the digit that has the greatest amount of uncertainty.

(f) The dimensional method of calculation treats units in the same manner as numbers.

(g) The magnitude of a physical quantity is unchanged by converting the units from one system to another.

4. Write three quantitative statements concerning the items on your desk (for example: 6 pens; thickness of book is 7/8 inch). What two parts to each statement have you written in each of your statements?

5. How can the uncertainty in a number be expressed? If the uncertainty is not expressed for a given number, what do we normally assume?

6. How can the number of significant figures in a given number be determined? When is a zero considered to be a significant digit and when is it not?

7. State the rules for determining the number of significant figures in the answers for calculations involving (a) addition or subtraction and (b) multiplication or division.

8. What is an exact number? How are the rules of significant figures applied to calculations involving exact numbers?

9. What is the procedure for changing a number greater than unity to standard scientific notation? What is the procedure for a number less than unity?

10. How is the number of significant figures in a number that is expressed in scientific notation determined?

11. State the rule for adding or subtracting numbers expressed in scientific notation. What is the rule for multiplication or division?

12. Complete the following table describing the seven base physical quantities in the SI system:

Quantity	Symbol	Name of unit	Unit symbol
(a)	T		
(b)			s
(c) mass			
(d)		mole	
(e) electrical current			
(f)	I		
(g)			cd

13. Several units defined by the SI system are too large or too small for some uses. How are multiples or fractions of a given base or derived unit expressed?

14. How are the units of a physical quantity treated in a numerical calculation?

15. What is a conversion factor? How are these factors derived?

16. To how many significant figures should conversion factors, physical constants, or other numbers (such as π), be expressed in a calculation?

17. List the four steps given in the general problem-solving method discussed in the text. Identify in which step the following questions might be asked: (a) What is the connection between the known and the unknown? (b) Did the answer have in the correct units? (c) What is known? (d) Are significant figures used correctly? (e) What is unknown? (f) Is the answer reasonable? (g) What is necessary to make the connection?

Answers to questions

1. (a) density, (b) significant figures, (c) mass, (d) pressure.
2. (a) Seven *base physical quantities* and units have been chosen to describe all other physical quantities; *derived physical quantities* are those that are derived from the seven base quantities, and the units for the derived quantities are derived from the base units. (b) *Mass* is a representation of the quantity of matter; *weight* is the force a body exerts because of the pull of gravity on the mass of that body. (c) A *force* is an interaction that can cause a change in the motion or state of rest of a body; *pressure* is a force exerted per unit area.
3. (a) T; (b) F, change *base* to *derived*; (c) F, change *dependent on* to *independent of*; (d) F, insert *that simply locate the decimal point* between *Zeros* and *in*; (e) T; (f) T; (g) T.
4. Typical answers could be 4 paper clips, 40 buttons on calculator, 150 watt light bulb; or, pen is 17 cm long, textbook is 9.5 inch x 1.5 inch x 6.5 inch. Each statement contains a numerical value and a unit.
5. The uncertainty in a number can be expressed by using a ± symbol and a value that includes the range of the uncertainty. If the uncertainty is not expressed, we assume that the uncertainty lies in the last significant digit to the right and that it is ±1.
6. The number of significant figures is found by counting from left to right, beginning with the first nonzero digit and ending with the digit that has the uncertain value. Zeros are significant when they appear in the middle of a number or at the end of any number that includes a decimal point.

Zeros are not significant when they precede the first nonzero digit or usually are not considered significant at the end of a number given without a decimal point.

7. (a) To maintain the correct number of significant figures in addition or subtraction, round the answer to the place (before or after the decimal point) with the greatest uncertainty. (b) To maintain the correct number of significant figures in multiplication or division, round the answer to the same number of significant figures as in the number with the fewest significant figures.
8. An exact number is one with no uncertainty; it arises by directly counting whole items or by definition. Because exact numbers do not have any uncertainty, they do not limit the number of significant figures in calculations.
9. Move the decimal point to the left until a number from 1 to 9.99... is obtained, then write x 10^n where n is the number of digits that the decimal point was moved; move the decimal point to the right until a number from 1 to 9.99... is obtained, then write x 10^{-n} where n is the number of digits that the decimal point was moved.
10. All of the digits in the numerical factor are significant.
11. To add or subtract numbers expressed in scientific notation, the numbers must all be given with the same exponent. The factors are added or subtracted, with the rules governing significant figures applied to the factor exactly as for any number. The power of 10 is then the same as in the numbers added or subtracted.

    To multiply or divide numbers expressed in scientific notation, multiply or divide the factors, obeying the rules of significant figures. Then find the power of 10 for the answer by adding powers of 10 for multiplication or subtracting them for division.
12. (a) temperature, kelvin, K; (b) time, t, second; (c) m, kilogram, kg; (d) amount of substance, n, mol; (e) I, ampere, A; (f) length, meter, m; (g) luminous intensity, I_V, candela.
13. Multiples or fractions of a given base or derived unit are indicated by using standard prefixes before the unit.
14. In a numerical problem, units are included in setting up the calculation and are treated exactly as numbers would be.

15. A conversion factor is based on the relationship between two units. Conversion factors are derived from equalities.
16. Conversion factors, physical constants, and similar numbers should contain at least the same number of significant figures as the other factors in multiplication or division and should be expressed to at least the same decimal place as the number with the fewest decimal places in addition and subtraction.
17. The four steps are (1) study the problem and be sure you understand it, (2) decide how to solve the problem, (3) set up the problem and solve it, and (4) check the result. Question (a) will be asked in step 2, (b) 4, (c) 1, (d) 4, (e) 1, (f) 4, and (g) 2.

SKILLS

Skill 2.1 Systems of units

All measurements of physical quantities are expressed with a numerical value and a unit (or dimension). In the SI system, there are seven base physical quantities and units from which all other physical quantities and units can be derived. The units commonly used in chemistry for some of the base and derived physical quantities are given in the table on the next page. The SI units are given in boldface type. Some English units are given in parentheses.

Skill Exercise: Identify which of the following measurements are expressed in correct units: the (a) length of a string in cm, (b) volume in dm^3, (c) energy in kJ, (d) density in g/mm, (e) temperature in K, (f) mass in m, (g) length in nm, and (h) pressure in kg/m^2. *Answer*: a, b, c, e, g.

Skill 2.2 Using significant figures

Numbers that result from measurements are never exact. There is always some uncertainty as a result of random errors in the measurement. This uncertainty is reflected in the number of significant figures to which the number is reported--the number contains all of the digits that are known with certainty plus the first digit to the right that has an uncertain value. The uncertainty in this digit is usually assumed to be ±1 unless otherwise stated. For example, the uncertainty in 432 is ±1, in 0.2645 is ±0.0001, and in 5260 is probably ±10.

Skill Exercise: Express the uncertainty in the following numbers: (a) 6.027, (b) 14,000, (c) 0.026, (d) 428.3, (e) 0.000293, and (f) 47. *Answers*: (a) ±0.001, (b) probably ±1,000, (c) ±0.001, (d) ±0.1, (e) ±0.000001, (f) ±1.

physical quantity/unit		unit symbol	approximate conversion factor
length			
	meter	**m**	
	centimeter	cm	1 cm = 1×10^{-2} m
	nanometer	nm	1 nm = 1×10^{-9} m
	angstrom	Å	1 Å = 1×10^{-10} m
	(inch	inch	1 inch = 2.54 cm)
mass			
	kilogram	**kg**	
	gram	g	1 g = 1×10^{-3} kg
	atomic mass unit	u	1 u = 1.6606×10^{-27} kg
	(pound	lb	1 lb = 454 g)
time			
	second	**s**	
electrical current			
	ampere	**A**	
temperature			
	kelvin	**K**	
	degree Celsius	°C	T(K) = T(°C) + 273.15
amount of substance			
	mole	**mol**	
quantity of electricity			
	Coulomb	**C** = A s	
electromotive force			
	volt	**V** = kg m^2/A s^3	
energy			
	joule	**J** = kg m^2/s^2	
	calorie	cal	1 cal = 4.184 J
	liter atmosphere	L atm	1 L atm = 101.325 J
	wave number	cm^{-1}	1 cm^{-1} = 1.986×10^{-23} J
	atomic mass unit	u	1 u = 1.492×10^{-10} J
frequency			
	hertz	**Hz** = s^{-1}	
pressure			
	pascal	**Pa** = kg/m s^2	
	atmosphere	atm	1 atm = 101,325 Pa
	bar	bar	1 bar = 1×10^5 Pa
	torr	Torr	1 Torr = 133.3 Pa
	millimeter of mercury	mm Hg, mmHg	1 mm Hg = 133.3 Pa
volume			
	cubic meter	**m^3**	
	liter	L	1 L = 1×10^{-3} m^3
	milliliter	mL	1 mL = 1×10^{-6} m^3
	cubic centimeter	cm^3, cc	1 cm^3 = 1×10^{-6} m^3
	(quart	qt	1 qt = 0.946 L)

The number of significant figures in a number is found by counting the number of digits in that number, beginning on the left with the first nonzero digit and ending on the right with the digit that has the uncertain value. Zeros that are part of the number or are at the end of a number and follow a decimal point are significant. Zeros that are used to locate the decimal point in a number less than 1 are not considered significant and those at the end of a number with no decimal point are ambiguous. For example, each of the following numbers has three significant figures

123	1 23	123	123	123
432	4.32	0.432	0.0432	0.000432

Each of these numbers has four significant figures:

1234	1 234	1234
1036	1.036	0.01036

Each of these numbers has two significant figures:

12	12	1 2
78	0.0078	7.8

1 2	12	12	12
9.0	0.90	0.00090	90.

The decimal point in 90. indicates that the uncertainty is ±1 and that the zero is significant. Because of the absence of a decimal point, the number of significant figures in each of the following numbers is not clear:

123?	123 ???
4320	432,000

Skill Exercise: Determine the number of significant figures in each of the following numbers: (a) 1492, (b) 106,000, (c) 0.2080, (d) 3.5070, (e) 100., (f) 0.0003, and (g) 1000.10. *Answers*: (a) 4, (b) 3 (or more), (c) 4, (d) 5, (e) 3, (f) 1, (g) 6.

The number of significant figures in the answer to a calculation is determined by the uncertainty in the numbers used in the calculation. For addition or subtraction, the answer is rounded off to the place with the greatest uncertainty. An easy way to apply this rule is to draw a vertical line after the digit of the greatest uncertainty in the numbers being added or subtracted and round off the answer to the same place. For example:

$$\begin{array}{r|l} 0 & .62 \\ 273 & \\ \hline [273 & .62] \end{array} \qquad \begin{array}{r|l} 0.29 & 6 \\ 4.41 & \\ 8.92 & 73 \\ \hline [13.63 & 33] \end{array} \qquad \begin{array}{r|l} 0.0526 & \\ -0.0553 & 4 \\ \hline [-0.0027 & 4] \end{array}$$

$$274 \qquad 13.63 \qquad -0.0027$$

For multiplication or division, the answer is rounded off to the same number of significant figures as in the factor with the fewest significant figures. For example

$$(26.5)(4.3) = [113.95] = 110 \qquad \frac{2.996}{0.432} = [6.9351852] = 6.94$$

$$(0.13)(0.02697) = [0.0035061] = 0.0035 \qquad \frac{6.034}{349.6} = [0.0172597] = 0.01726$$

For a more complicated calculation involving addition and/or subtraction and multiplication and/or division, solve the problem by using all numbers as given, inspect the problem to determine which factor has the largest uncertainty, and round the answer off to the proper number of significant figures as determined by the least certain factor. For example

$$C = (0.103) - (0.062)(425) = [-26.247] = -26$$

To see this, do the calculation one step at a time:

$$C = (0.103) - (0.062)(425) = (0.103) - [26.35]$$
$$= (0.103) - (26) = -26$$

Skill Exercise: Perform the following calculations and express the answers to the proper number of significant figures:

(a) $0.0036 + 9000$ (b) $9.26 - 0.16$

(c) $(0.0036)(9000)$ (d) $(9.26)/(0.16)$

(e) $\dfrac{(52.9)(1467)}{(8126)}$ (f) $\dfrac{(0.0026)(1855)}{(0.3)}$

(g) $C = \dfrac{1}{0.25} + (0.062)(7300)$ (h) $F = \dfrac{(12.3) - (11.9)}{(11.9)}$

Answers: (a) 9000, (b) 9.10, (c) 30 (or 32), (d) 58, (e) 9.55, (f) 20, (g) C = 460, (h) F = 0.03.

If the calculation involves an exact number, the correct number of significant figures in the answer is determined by applying the usual rules to the factors that are uncertain. For example, in the calculations below the numbers 273.15, 1000, and 24 are exact and do not limit the number of significant figures in the answer

$$\begin{array}{r} 3.652 \\ \underline{273.15} \\ 276.802 \end{array} \qquad \frac{4.3}{1000} = 0.0043 \qquad (6.28)(24) = 151$$

When supplying numerical values for physical contents (e.g., $\pi = 3.14159265...$, $R = 0.0820568$, and $m_e = 0.0005485802$) be sure to express the value of the constant to enough significant figures so that it does not limit the number of significant figures in the answer. A good rule to follow is to express the constant to at least one more place than is really needed. For example

$$A = \pi(0.32)^2 = (3.14)(0.32)^2 = 32$$

$$V = (4/3)\pi(5.629)^3 = (4/3)(3.1416)(5.629)^3 = 747.1$$

$$V = \frac{(2.0)R(310.5)}{1.6345} = \frac{(2.0)(0.0821)(310.5)}{1.6345} = 31$$

$$m = 1.0072764 + m_e = 1.0072764 + 0.00054858 = 1.0078250$$

Skill Exercise: Perform the following calculations and express the answers to the proper number of significant figures (assume the factors 1000, 60, and 273.15 to be exact):

(a) $(10.63)(60)$ (b) $273.15 - 2.4$ (c) $\frac{(1.36)(8.364)(273.15)}{1000}$

and express the value of π or R, respectively, that should be used in the following calculations:

(d) $(14.35)\pi$ (e) $(3.2)R$ (f) $(5.1) + \pi$ (g) $(6.2) - R$

Answers: (a) 637.8, (b) 270.8, (c) 3.11, (d) 3.1416, (e) 0.0821, (f) 3.14, (g) 0.08.

Skill 2.3: Using scientific notation for large and small numbers

Scientific notation is the representation of a number in which the significant figures of the number are retained usually in a factor from 1 to 9.99... and the decimal point is given by a power of 10. Briefly, for numbers greater than unity, the decimal point is moved to the left in the number until a number that is from 1 to 9.99... is reached and the power of 10 is equal to the number of digits that the decimal point was moved. For numbers less than unity, the decimal point is moved to the right in the number until a number that is from 1 to 9.99... is reached, and the power of 10 is equal to the number of digits that the decimal point was moved except that a negative sign appears in the power. For example

$$1700 = 1.7 \times 10^{3} \qquad 0.07230 = 7.230 \times 10^{-2}$$
$$\leftarrow\leftarrow\leftarrow \qquad\qquad\qquad \rightarrow\rightarrow$$

Skill Exercise: Express the following numbers in scientific notation: (a) 460; (b) 7,294,000; (c) 1.6; (d) 0.0002; and (e) 0.0486. *Answers*: (a) 4.6×10^{2}, (b) 7.294×10^{6}, (c) 1.6×10^{0}, (d) 2×10^{-4}, (e) 4.86×10^{-2}.

To convert a number written in scientific notation to a number written in the usual form, simply reverse the foregoing steps. The power of 10 indicates the number of places to move the decimal point--to the left for negative values and to the right for positive values. For example

$$2 \times 10^{5} = 200{,}000 \qquad 4.3 \times 10^{-3} = 0.0043$$
$$\rightarrow\rightarrow\ \rightarrow\rightarrow\rightarrow \qquad\qquad\qquad \leftarrow\leftarrow\leftarrow$$

Skill Exercise: Write the following numbers in the usual arithmetic form: (a) 1.72×10^{3}, (b) 6.35×10^{-6}, (c) 3.0×10^{-3}, (d) 5×10^{7}, and (e) 5×10^{-7}. *Answers*: (a) 1720; (b) 0.00000635; (c) 0.0030; (d) 50,000,000; (e) 0.0000005.

Many engineering students use another system of exponential notation known as engineering notation. In this system of representing numbers, the factor is a number from 1 to 999.99... (rather than from 1 to 9.99...) and the exponent is always a positive or negative multiple of 3 (rather than being any integer). For example

$20{,}000{,}000 = 20 \times 10^{6}$	$2{,}000 = 2 \times 10^{3}$	$0.2 = 200 \times 10^{-3}$
$2{,}000{,}000 = 2 \times 10^{6}$	$200 = 200 \times 10^{0}$	$0.02 = 20 \times 10^{-3}$
$200{,}000 = 200 \times 10^{3}$	$20 = 20 \times 10^{0}$	$0.002 = 2 \times 10^{-3}$
$20{,}000 = 20 \times 10^{3}$	$2 = 2 \times 10^{0}$	$0.0002 = 200 \times 10^{-6}$

The use of numbers expressed in engineering notation does not differ from that for scientific notation.

When performing calculations with numbers expressed in scientific notation, the rules governing significant figures are applied to the numerical factors exactly as for any number. Most electronic calculators will give the answer to a calculation involving scientific notation directly if the range of the answer is between $\pm 1 \times 10^{-99}$ to $\pm 9.999... \times 10^{99}$.

If addition or subtraction is to be done "by hand," all numbers must be changed so that they all have the same power of 10, the arithmetic is performed for the factors, the power of 10 for the answer is the same as that for the numbers used, and the answer is restated to the proper number of significant figures and power of 10 (if necessary). For example, to subtract 6.2×10^3 from 3.21×10^5, using either

$$\begin{array}{r|l} 3.21 & \quad \times 10^5 \\ -0.06 & 2 \times 10^5 \\ \hline [3.14 & 8 \times 10^5] \\ \end{array} \quad \text{or} \quad \begin{array}{r|l} 321 & \quad \times 10^3 \\ -\ \ 6 & .2 \times 10^3 \\ \hline [314 & .8 \times 10^3] \\ \end{array}$$

$$3.15 \times 10^5 \qquad\qquad 315 \times 10^3 = 3.15 \times 10^5$$

gives the correct answer.

If multiplication or division is to be done "by hand," the arithmetic is performed for the factors, the power of 10 is found by adding the powers of 10 for multiplication and subtracting them for division, and the answer is restated to the proper number of significant figures and power of 10 (if necessary). For example

$$\frac{(6.022 \times 10^{23})(1.66 \times 10^{-10})}{(3.65 \times 10^{2})} = [2.7387726 \times 10^{11}] = 2.74 \times 10^{11}$$

The power in the above answer is given by (23) + (-10) - (2) = 11.

Skill Exercise: Perform the following calculations and express the answers in scientific notation to the proper number of significant figures:

(a) $3.2 \times 10^5 + 6.8 \times 10^3$ (b) $5.2 \times 10^7 + 4.9 \times 10^7$

(c) $7.0 \times 10^{-2} - 3.1 \times 10^{-1}$ (d) $3.462 \times 10^3 - 1.2 \times 10^2$

(e) $\dfrac{(6.656 \times 10^{-34})^3(3.2 \times 10^{-8})}{4.26 \times 10^{-17}}$ (f) $\dfrac{(4.26 \times 10^{-10})(8.2 \times 10^{12})}{(135)}$

Answers: (a) 3.3×10^5, (b) 1.01×10^8, (c) -2.4×10^{-1}, (d) 3.34×10^3, (e) 2.2×10^{-91}, (f) 2.6×10^1.

Skill 2.4 Using prefixes for large and small numbers

There is an alternative method to scientific notation for expressing large and small numbers. Instead of writing 10 to a power, the decimal point information is expressed as a prefix, the symbol for which is placed before the unit on the number. The common prefixes are

$$10^6 = \text{M} \quad 10^3 = \text{k} \quad 10^{-2} = \text{c} \quad 10^{-3} = \text{m} \quad 10^{-6} = \mu \quad 10^{-9} = \text{n}$$

Skill Exercise: Select from among these equivalencies the ones that are correct: (a) 0.15 m = 15 cm, (b) 1 L = 100 mL, (c) 1000 nm = 1 μm, (d) 4.3×10^3 J = 4.3 kJ, and (e) 0.1056 g = 105.6 mg. *Answer*: a, c, d, e.

Skill 2.5 Using the dimensional method in problem solving

The units on numbers and conversion factors can be cancelled just like numbers. Many times an incorrect solution to a problem can be spotted by observing that the wrong units appear in the answer. For example, a student incorrectly used the equation for density, $d = m/V$, to solve for volume and obtained the answer

$$V = md = (16.3\ \text{g})(0.892\ \text{g/mL}) = 14.5\ \text{g}^2/\text{mL}$$

This answer is obviously wrong because the units for volume are not g²/mL. The student then solved the equation for volume again and correctly obtained

$$V = \frac{m}{d} = \frac{16.3\ \text{g}}{0.892\ \text{g/mL}} = 18.3\ \text{mL}$$

Skill Exercise: Determine the units on the answers for each of the following expressions: (a) (g)/(g/mol), (b) (L)(mol/L), (c) (g/mol)/(molecules/mol), (d) (mol)(g/mol), and (e) (mol)/(mol/L). *Answers*: (a) mol, (b) mol, (c) g/molecule, (d) g, (e) L.

Skill 2.6 Deriving conversion factors

Conversion factors for use in the dimensional method in problem solving are derived from equalities such as

$$1\ \text{atm} = 760\ \text{Torr}$$

This equality yields two conversion factors,

$$1\ \text{atm} = 760\ \text{Torr} \qquad\qquad 1\ \text{atm} = 760\ \text{Torr}$$

$$\frac{1\ \text{atm}}{1\ \text{atm}} = \frac{760\ \text{Torr}}{1\ \text{atm}} \qquad\qquad \frac{1\ \text{atm}}{760\ \text{Torr}} = \frac{760\ \text{Torr}}{760\ \text{Torr}}$$

$$1 = \frac{760\ \text{Torr}}{1\ \text{atm}} \qquad\qquad \frac{1\ \text{atm}}{760\ \text{Torr}} = 1$$

which allows the conversion of atmospheres to Torr or vice versa. For example

$$(745\ \text{Torr})\left[\frac{1\ \text{atm}}{760\ \text{Torr}}\right] = 0.980\ \text{atm}$$

$$(15.2\ \text{atm})\left[\frac{760\ \text{Torr}}{1\ \text{atm}}\right] = 11{,}600\ \text{Torr}$$

Skill Exercise: Derive two conversion factors from each of the following equalities: (a) 1 nm = 1×10^{-9} m, (b) 1 cm³ = 1 mL, and (c) 11.3 g Pb = 1 cm³ Pb. *Answers*: (a) (1 nm/1×10^{-9} m), (1×10^{-9} m/1 nm); (b) (1 cm³/1 mL), (1 mL/1 cm³); (c) (1 cm³ Pb/11.3 g Pb), (11.3 g Pb/1 cm³ Pb).

Skill 2.7 Solving problems

The series of steps listed below can often help in solving numerical problems:

1. Study the problem and be sure you understand it.
   Answer the questions: (a) What is unknown? (b) What is known?
2. Decide how to solve the problem.
   Answer the questions: (a) What is the connection between the known and the unknown? (b) What is necessary to make the connection?
3. Set up the problem and solve it.
   Utilize the dimensional method in the solution.
4. Check the result.
   Answer the questions: (a) Are significant figures and the location of the decimal point correct? (b) Did the answer come out in the correct units? (c) Is the answer reasonable?

Even though each step and each question may not actually be written out as the problem is solved, the steps and questions are a guide to a logical approach to thinking about the problem. Example problems marked with an asterisk (*) throughout this book are solved by this method.

NUMERICAL EXERCISES

Example 2.1 Counting significant figures

How many significant figures are in the following numbers: (a) 1.01506, (b) 1.500, (c) 0.013900, (d) 45.6702, (e) 1500, and (f) 60.?

The rules for determining the number of significant figures in a number are presented in *Skill 2.2 Using significant figures*. Application of these rules involves marking the significant figures in each number with a small number:

1 23456 (a) 1.01506	1 234 (b) 1.500	12345 (c) 0.013900
12 3456 (d) 45.6702	12?? (e) 1500	12 (f) 60.

The number of significant figures in each number is (a) 6, (b) 4, (c) 5, (d) 6, (e) 2 (or more), and (f) 2.

Exercise 2.1 Indicate the number of significant figures in the following numbers: (a) 3.1415, (b) 0.0372, (c) 1000., (d) 36.0, and (e) 0.002070.
Answers: (a) 5, (b) 3, (c) 4, (d) 3, (e) 4.

Example 2.2 Significant figures--addition/subtraction

Perform the following addition/subtraction problems and express the answers to the proper number of significant figures:

(a) 703 + 7 + 0.66

(b) 7.26 - 0.2

(c)
$$\begin{array}{r} 6.0235 \\ 0.1257 \\ 19.23 \\ 4 \\ \hline \end{array}$$

The rules for determining the number of significant figures in the answer for an addition or subtraction problem are presented in *Skill 2.2 Using significant figures*. Upon application of these rules, with the aid of the vertical line marking the digit with the greatest uncertainty, the answers are

(a)
$$\begin{array}{r|l} 703 & \\ 7 & \\ 0 & .66 \\ \hline [710 & .66] \end{array}$$
711

(b)
$$\begin{array}{r|l} 7.2 & 6 \\ -0.2 & \\ \hline [7.0 & 6] \end{array}$$
7.1

(c)
$$\begin{array}{r|l} 6 & .0235 \\ 0 & .1257 \\ 19 & .23 \\ 4 & \\ \hline [29 & .3792] \end{array}$$
29

Exercise 2.2 Perform the following addition/subtraction problems and express the answers to the proper number of significant figures: (a) 0.306 - 11.29, (b) 10 + 0.0036 + 4.733, and (c) 302.65 + 1.27 + 0.036. *Answers*: (a) -10.98, (b) 10 (or 15), (c) 303.96.

Example 2.3 Significant figures--multiplication/division

Perform the following multiplication/division problems and express the answers to the proper number of significant figures:

(a) (0.306)(11.29)

(b) (220)(35,000)

(c) (2.21)(0.3)/(10.000)

(d) $\dfrac{(16{,}000)(0.0002)(1.2)}{(2000)(0.006)(0.00032)}$

The rules for determining the number of significant figures in the answer for multiplication/division problems are presented in *Skill 2.2 Using significant figures*. Application of these rules and rounding off of the answers to the same number of significant figures as in the factor with the fewest such figures yield these answers:

(a) (0.306)(11.29) = [3.45474] = 3.45

(b) (220)(35,000) = 7,700,000 (or 7.70×10^6)

(c) $\dfrac{(2.21)(0.3)}{(10.000)} = [0.0663] = 0.07$

(d) $\dfrac{(16{,}000)(0.0002)(1.2)}{(2000)(0.006)(0.00032)} = 1000$

Exercise 2.3 Perform the following multiplication/division problems and express the answers to the proper number of significant figures:

(a) (10.)(3.625)/(14.97) (b) (10,000)(3.14)

(c) (1.632)(9.624) (d) (0.002)(1.357)/(0.248)

Answers: (a) 2.4, (b) 30,000 (or 31,400), (c) 15.71, (d) 0.01.

Example 2.4 Significant figures--arithmetic

Perform the following calculation and express the answer to the proper number of significant figures:

$$\frac{(42.892) - (41.75)}{3.257}$$

The rules for determining the number of significant figures in a multistep calculation involving addition/subtraction and multiplication/division are presented in *Skill 2.2 Using significant figures*. In this case the answer should be rounded to three significant figures because the numerator (42.892 - 41.75) = [1.142] = 1.14 is the term containing the fewest significant figures:

$$\frac{(42.892) - (41.75)}{3.257} = [0.3506294] = 0.351$$

Exercise 2.4 Perform the following calculation and express the answer to the proper number of significant figures:

$$\frac{2.72}{9.68} + 2.0$$

Answer: 2.3.

Example 2.5 Scientific notation

Express each of the following numbers in scientific notation: (a) 0.000002; (b) 0.831; (c) 36,000,000; (d) 1006; (e) 32,600; and (f) 0.000706.

The rules for converting numbers into scientific notation are presented in *Skill 2.3 Using scientific notation for large and small numbers*. In applying these rules, the number of decimal places that the decimal point has been moved is marked by arrows:

(a) $0.000002 = 2 \times 10^{-6}$
→→→→→→

(b) $0.831 = 8.31 \times 10^{-1}$
→

(c) $36,000,000 = 3.6 \times 10^{7}$
← ←←← ←←←

(d) $1006 = 1.006 \times 10^{3}$
←←←

(e) $32,600 = 3.26 \times 10^{4}$
← ←←←

(f) $0.000706 = 7.06 \times 10^{-4}$
→→→→

Exercise 2.5 Express each of the following numbers in scientific notation: (a) 0.0063, (b) 1038, (c) 100., (d) 1030, (e) 0.6×10^{23}, and (f) 1.030. *Answers*: (a) 6.3×10^{-3}, (b) 1.038×10^{3}, (c) 1.00×10^{2}, (d) 1.03×10^{3}, (e) 6×10^{22}, (f) 1.030×10^{0}.

Example 2.6 Scientific notation

Express each of the following numbers in conventional notation: (a) 3.0×10^{-4}, (b) 1.03×10^{5}, (c) 2.162×10^{-5}, (d) -3.2×10^{-1}, and (e) 1.17×10^{-13}.

The rules for converting numbers from scientific to conventional notation are given in *Skill 2.3 Using scientific notation for large and small numbers*. In applying these rules, the number of decimal places that the decimal point has been moved is marked by arrows:

(a) $3.0 \times 10^{-4} = \underset{\leftarrow\leftarrow\leftarrow\leftarrow}{0.00030}$ (b) $1.03 \times 10^{5} = \underset{\rightarrow\rightarrow\;\rightarrow\rightarrow\rightarrow}{103{,}000}$

(c) $2.162 \times 10^{-5} = \underset{\leftarrow\leftarrow\leftarrow\leftarrow\leftarrow}{0.00002162}$ (d) $-3.2 \times 10^{-1} = \underset{\leftarrow}{-0.32}$

(e) $1.17 \times 10^{-13} = \underset{\leftarrow\leftarrow\leftarrow\leftarrow\leftarrow\leftarrow\leftarrow\leftarrow\leftarrow\leftarrow\leftarrow\leftarrow\leftarrow}{0.000000000000117}$

Exercise 2.6 Express each of the following numbers in conventional notation: (a) 1.78×10^{4}, (b) 1.09×10^{3}, (c) -9.70×10^{-1}, (d) 8.5×10^{8}, and (e) 6.19×10^{0}. *Answers*: (a) 17,800; (b) 1,090; (c) -0.970; (d) 850,000,000; (e) 6.19.

Example 2.7 Scientific notation--arithmetic

Perform the following calculations and express the answers in scientific notation to the proper number of significant figures:

(a) $(1 \times 10^{-3})(2 \times 10^{5})^2$ (b) $(47.34) + (1.729 \times 10^{3}) - (4 \times 10^{-3})$

(c) $(6.35 \times 10^{-4}) - (6.35 \times 10^{-6})$ (d) $\dfrac{(6.0235 \times 10^{21})(32.0)}{(1.0)(0.00472)}$

The rules for performing calculations with numbers written in scientific notation are presented in *Skill 2.3 Using scientific notation for large and small numbers*. Application of the rules of significant figures gives

(a) $(1 \times 10^{-3})(2 \times 10^{5})^2 = 4 \times 10^{7}$

(b)

$$\begin{array}{r|l} 47 & .34 \\ 1729 & \\ -\ 0 & .004 \\ \hline [1776 & .336] \end{array}$$

$1776 = 1.776 \times 10^3$

←←←

(c)

$$\begin{array}{r|l} 635 & \quad \times 10^{-6} \\ -\ 6 & .35 \times 10^{-6} \\ \hline [628 & .65 \times 10^{-6}] \end{array}$$

$629 \times 10^{-6} = 6.29 \times 10^{-4}$

←←

(d) $\dfrac{(6.0235 \times 10^{21})(32.0)}{(1.0)(0.00472)} = [40837.288 \times 10^{21}] = 4.1 \times 10^{25}$

Exercise 2.7 Perform the following calculations and express the answers in scientific notation to the proper number of significant figures: (a) $(8)(2 \times 10^{-2})^3$, (b) $(7.762 \times 10^{-4}) - (9.11 \times 10^{-6})$, (c) $(5.61 \times 10^6) + (2.2 \times 10^4)$, (d) $(6.283 \times 10^{-27})/(6.63)$, and (e) $[(3.15 \times 10^{-6}) + (2.9397 \times 10^{-7})]/(6.022 \times 10^{23})$. *Answers*: (a) 6×10^{-5}, (b) 7.671×10^{-4}, (c) 5.63×10^6, (d) 9.48×10^{-28}, (e) 5.72×10^{-30}.

Example 2.8 Temperature conversion*

Cesium metal melts at 28.4 °C. Convert this temperature to the Fahrenheit and Kelvin scales.

This problem is analyzed using the comprehensive problem-solving approach presented in *Skill 2.7 Solving problems.*

1. <u>Study the problem and be sure you understand it</u>.
   (a) What is unknown?
   The Fahrenheit and Kelvin temperatures.
   (b) What is known?
   The Celsius temperature: 28.4 °C.
2. <u>Decide how to solve the problem</u>.
   (a) What is the connection between the known and the unknown?
   The connection between the Celsius temperature and the Fahrenheit temperature is based on the relative sizes of the degrees (9 F degrees = 5 C degrees) and the freezing point of water (0 °C = 32 °F). The connection between the Celsius temperature and the Kelvin temperature is based only on the different zero points of the two scales (0 K = -273.15 °C) because the sizes of the degrees are identical (1 C degree = 1 K).
   (b) What is necessary to make the connection?
   Use the relationships

$$(1.8\ F°/1\ C°)T(°C) = T(°F) - 32$$

$$T(K) = T(°C) + (273.15)$$

3. Set up the problem and solve it.

$$T(°F) = (1.8\ F°/1\ C°)(28.4\ °C) + 32 = 83.1\ °F$$

$$T(K) = (28.4\ °C) + (273.15) = 301.6\ K$$

4. Check the result.
   (a) Are significant figures and the location of the decimal point correct?
   Yes. The factors 1.8, 32, and 273.15 are exact, so the number of significant figures in each answer is determined by the uncertainty in the original temperature (28.4 °C). The calculation of T(°F) is a case in which several operations are combined. Solving the equation as written in one step on a calculator gives 83.12 as the answer, but not all of these digits are significant. Inspection shows that (1.8)(28.4) = 51.1 and that the addition step (51.1 + 32) retains the three significant figures. In the calculation of T(K), the answer should be expressed to the nearest 0.1 K because of the uncertainty in the tenths place in 28.4 °C.
   (b) Did the answer come out in the correct units?
   Yes.
   (c) Is the answer reasonable?
   Yes. Meteorologists are now giving temperatures in both Celsius and Fahrenheit units, and because 25 °C is roughly 78 °F, it makes sense that 28.4 °C would be equal to 83.1 °F.

Exercise 2.8 Nitrogen boils at -195.8 °C. Convert this temperature to the Fahrenheit and Kelvin scales. *Answers*: -320.4 °F, 77.4 K.

Example 2.9 Dimensional method*

A bit is the smallest piece of binary information used by a computer. A byte consists of 8 bits. A standard 8 inch memory diskette stores 128 bytes of information in a sector and there are 26 sectors on a track. If 14 tracks on a diskette are filled, how many bits are stored on this diskette?

This problem is analyzed using the comprehensive problem-solving approach presented in *Skill 2.7 Solving problems*.

1. Study the problem and be sure you understand it.
   (a) What is unknown?
   The number of bits on the partially filled diskette.
   (b) What is known?
   The number of tracks that are filled on the diskette: 14.
2. Decide how to solve the problem.
   (a) What is the connection between the known and the unknown?
   The connection is provided by conversion factors based on the known relationships between bits, bytes, sectors, and tracks: 1 byte = 8 bits, 1 sector = 128 bytes, and 1 track = 26 sectors.
   (b) What is necessary to make the connection?
   Set up a dimensional calculation that includes conversion factors that allow conversion of tracks to bits.
3. Set up the problem and solve it.
   The solution can be set up in one step:

$$(14 \text{ tracks})\left[\frac{26 \text{ sectors}}{1 \text{ track}}\right]\left[\frac{128 \text{ bytes}}{1 \text{ sector}}\right]\left[\frac{8 \text{ bits}}{1 \text{ byte}}\right] = 372{,}736 \text{ bits}$$

4. Check the result.
   (a) Are significant figures and the location of the decimal point correct?
   Yes. All numbers are exact, so all digits in the answer are significant.
   (b) Did the answer come out in the correct units?
   Yes.
   (c) Is the answer reasonable?
   Yes. A memory diskette should contain a large number of the smallest pieces of information.

Exercise 2.9 An athlete runs the 100.00 yd dash in 10.1 s. If this velocity is maintained, how long will it take the athlete to run 100.00 m? *Answer*: 11.0 s.

Example 2.10 Unit conversion*

Engine capacities in American cars are usually given in cubic-inch displacement; for European cars they are usually given in cubic-centimeter or liter displacement. (Displacement is the total volume swept out by the pistons in one stroke.) A typical European luxury sports car might have an engine displacement of 2.30 L. Express this displacement in units of cubic inches. How does this compare with the displacement of a typical American family car, which is about 400 cubic inches?

This problem is analyzed using the comprehensive problem-solving approach presented in *Skill 2.7 Solving problems*.

1. <u>Study the problem and be sure you understand it.</u>
   (a) What is unknown?
   The displacement of the European car engines in units of cubic inches.
   (b) What is known?
   The displacement in units of liters.
2. <u>Decide how to solve the problem.</u>
   (a) What is the connection between the known and the unknown?
   The connection is the conversion factor between liters and cubic inches.
   (b) What is necessary to make the connection?
   The desired conversion factor can be found by using conversion factors that can be derived from the following equalities: (see *Skill 2.1 Systems of units*): 2.54 cm = 1 inch, 1 mL = 1 cm^3, and 1 L = 1000 mL.
3. <u>Set up the problem and solve it.</u>
   The equality between liters and cubic inches can be found by using the dimensional method:

$$1\ \text{L} = (1\ \text{L})\left[\frac{1000\ \text{mL}}{1\ \text{L}}\right]\left[\frac{1\ \text{cm}^3}{1\ \text{mL}}\right]\left[\frac{1\ \text{inch}}{2.54\ \text{cm}}\right]^3 = 61.0\ \text{inch}^3$$

   The conversion factor for volume conversion is (61.0 $inch^3$/1 L).
   The displacement of the European car is

$$(2.30\ \text{L})\left[\frac{61.0\ \text{inch}^3}{1\ \text{L}}\right] = 140.\ \text{inch}^3$$

   The displacement of the European car is roughly one-third (140. $inch^3$/400 $inch^3$) that of the American car.
4. <u>Check the result.</u>
   (a) Are significant figures and the location of the decimal point correct?
   Yes. Both the numbers for the conversion factor and the volume have three significant figures and so the answer would have three significant figures.
   (b) Did the answer come out in the correct units?
   Yes.
   (c) Is the answer reasonable?
   Yes. The engine displacement for both types of cars should be approximately the same.

Exercise 2.10 Calculate the number of cubic centimeters in one cubic foot. *Answer*: 2.83×10^4 cm^3/ft^3.

Example 2.11 Unit conversion

The British thermal unit (Btu) is the common unit for rating home air conditioners and furnaces. It is defined as the quantity of heat required to raise the temperature of one pound of water by one degree Fahrenheit. A calorie is the quantity of heat required to raise the temperature of one gram of water by one degree Celsius. How many calories are equivalent to one Btu?

The equalities needed to make this conversion are 5 C degrees = 9 F degrees and 1 lb = 454 g.

$$\left[\frac{1 \text{ lb (F degree)}}{1 \text{ Btu}}\right]\left[\frac{5 \text{ C degrees}}{9 \text{ F degrees}}\right]\left[\frac{454 \text{ g}}{1 \text{ lb}}\right]\left[\frac{1 \text{ cal}}{1 \text{ g (C degree)}}\right] = 252 \text{ cal/Btu}$$

One Btu is equal to 252 calories.

Exercise 2.11 To raise the temperature of 100 mL of water from room temperature to the boiling point requires about 7.5 kcal of heat. Using the conversion factor 1 cal = 4.184 J, express this measurement in MJ. *Answer*: 0.031 MJ.

Example 2.12 Unit conversion

The density of metallic silver is 10.5 g/mL. Express this density in units of kg/m^3.

The mass conversion is based on 1 kg = 1000 g:

$$\left[10.5 \ \frac{\text{g}}{\text{mL}}\right]\left[\frac{1 \text{ kg}}{1000 \text{ g}}\right] = 0.0105 \text{ kg/mL}$$

The volume conversion is made by converting mL to cm^3 and converting cm^3 to m^3. The equality for this second step is found by cubing the factor for conversion from centimeters to meters.

$$\left[0.0105 \ \frac{\text{kg}}{\text{mL}}\right]\left[\frac{1 \text{ mL}}{1 \text{ cm}^3}\right]\left[\frac{10^2 \text{ cm}}{1 \text{ m}}\right]^3 = 1.05 \times 10^4 \text{ kg/m}^3$$

The density of silver is 1.05×10^4 kg/m^3.

Exercise 2.12 The density of air is 0.95 g/L at 1 atm and 100 °C. Express this density in units of mg/mL. *Answer*: 0.95 mg/mL.

Example 2.13 Density*

Calculate the density of a body that weighs 320.0 g and has a volume of 45 cm^3.

This problem is analyzed using the comprehensive problem-solving approach presented in *Skill 2.7 Solving problems*.

1. Study the problem and be sure you understand it.
   (a) What is unknown?
   The density.
   (b) What is known?
   The mass and the volume.
2. Decide how to solve the problem.
   (a) What is the connection between the known and the unknown?
   Density is mass per unit volume.
   (b) What is necessary to make the connection?
   Divide the mass by the volume: $d = m/V$.
3. Set up the problem and solve it.

$$d = \frac{m}{V} = \frac{320.0\ \text{g}}{45\ \text{cm}^3} = 7.1\ \text{g/cm}^3$$

4. Check the result.
   (a) Are significant figures and the location of the decimal point correct?
   Yes. The number of significant figures in the answer is limited by the two significant figures in the volume measurement.
   (b) Did the answer come out in the correct units?
   Yes. The units of g/cm^3 are acceptable units for density.
   (c) Is the answer reasonable?
   Without knowledge of the physical state of the substance, this is difficult to answer. Typical solids and liquids have densities in the range of a few g/cm^3 and gases have densities in the range of a few g/L.

Exercise 2.13 The density of water is 1.00 g/cm^3. What is the mass of a gallon of water in pounds? *Answer*: 8.33 lb.

PRACTICE TEST

1 (20 points). Choose the response that best completes each statement.

(a) The number of significant figures in 0.03040 is

(i) 3 (ii) 4

(iii) 5 (iv) 6

(b) The name of the physical quantity which is defined as the mass divided by the volume is the

(i) weight (ii) pressure

(iii) density (iv) energy

(c) The uncertainty in 932.3 is

(i) ±0.1 (ii) ±100.0

(iii) ±0.01 (iv) none of these answers

(d) The number 1×10^{-16} is a

(i) small negative number (ii) large negative number

(iii) small positive number (iv) large positive number

(e) The number 52.0 in scientific notation is

(i) 5.2×10^{1} (ii) 5.20×10^{-1}

(iii) 5.20×10^{2} (iv) none of these answers

(f) The set of units equivalent to $\frac{\text{m/s}}{\text{L/kg}}$ is

(i) kg m/L s (ii) m/L kg s

(iii) m L/s kg (iv) none of these answers

(g) A commonly used unit in chemistry to express volume is the

(i) cubic foot (ii) quart

(iii) erg (iv) milliliter

(h) A student is asked to determine how many revolutions the second hand on a clock will make in two weeks. Which is the correct dimensional analysis setup to solve this problem?

(i) $(2\text{ wks})\left[\frac{1\text{ wk}}{7\text{ days}}\right]\left[\frac{1\text{ day}}{24\text{ h}}\right]\left[\frac{1\text{ h}}{60\text{ min}}\right]\left[\frac{1\text{ min}}{1\text{ revolution}}\right]$

(ii) $(2\text{ wks})\left[\frac{7\text{ days}}{1\text{ wk}}\right]\left[\frac{24\text{ h}}{1\text{ day}}\right]\left[\frac{60\text{ min}}{1\text{ h}}\right]\left[\frac{1\text{ revolution}}{1\text{ min}}\right]$

(iii) $\left[\frac{1}{2\text{ wks}}\right]\left[\frac{1\text{ wk}}{7\text{ days}}\right]\left[\frac{1\text{ day}}{24\text{ h}}\right]\left[\frac{1\text{ h}}{60\text{ min}}\right]\left[\frac{1\text{ revolution}}{1\text{ min}}\right]$

(iv) none of these answers

(i) Which statement is *not* true concerning the Kelvin temperature scale?
   (i) The freezing point of water is 273 K.
   (ii) The scale is defined as 0° and 100° for the freezing and boiling points of water.
   (iii) The kelvin is the same size as the Celsius degree.
   (iv) A Fahrenheit temperature can be calculated from a Kelvin temperature.

(j) Which is the correct answer for (0.6238)(6.6)?
   (i) 4.11708   (ii) 4
   (iii) 7.2   (iv) 4.1

2 (30 points). Choose the best answer for each statement.

(a) The correct answer to $(4.56 \times 10^3)/(1.22 \times 10^{-2})$ is
   (i) 3.74×10^5 (ii) 37.37
   (iii) 3.73×10^5 (iv) 5.56×10^1

(b) The correct answer to $(3.25 \times 10^{-5}) + (2.16 \times 10^{-7})$ is
   (i) 3.2716×10^{-5} (ii) 3.27×10^{-5}
   (iii) 2.19×10^{-7} (iv) 1.505×10^2

(c) The correct answer to $\dfrac{(6.022 \times 10^{24})}{(130.6) + (2.5 \times 10^4) - (6.3)}$ is
   (i) 4.6×10^{22} (ii) 4.84×10^{22}
   (iii) 2.397×10^{20} (iv) 2.4×10^{20}

(d) The approximate radius of a hydrogen atom is 5.29×10^{-11} m. Convert this radius to nanometers.
   (i) 0.529 nm   (ii) 0.0529 nm
   (iii) 5.29 nm   (iv) 5.29×10^{-8} nm

(e) The heat of combustion of one mole of naphthalene is 5.15×10^3 kJ. Express this heat of combustion in units of joules.
   (i) 5.15 J   (ii) 5.15×10^6 J
   (iii) 5150 J   (iv) 5.15×10^9 J

(f) Given: 3 blocks = 2 rods, 1 mile = 80 chains, 1 chain = 4 rods, and 100 links = 1 chain. Calculate the number of links in 3 miles.
   (i) 2.4×10^4 (ii) 3.75×10^{-4} links
   (iii) 1800 links   (iv) none of these answers

(g) Given: 1 mol U = 6.022×10^{23} atoms, 238 g U = 1 mol U, and 238 u U = 1 atom U. Determine the conversion factor between u and g.
   (i) 1 g/u   (ii) 6.022×10^{23} g/u
   (iii) 1.66×10^{-24} g/u   (iv) 1 g/u mol

(h) Calculate the volume occupied by 27.0 g of mercury. The density of mercury is 13.5 g/cm^3.

(i) 365 cm^3 (ii) 2×10^{-2} L

(iii) 0.500 cm^3 (iv) 2.00 cm^3

(i) If 4.33×10^{13} helium atoms were laid next to each other in a straight line, they would cover a distance of 5.00 miles. What is the radius of a helium atom?

(i) 0.186 nm (ii) 93 Å

(iii) 0.0929 nm (iv) 18.6 Å

(j) A stockroom attendant needed a minimum of 1000 perfect test tubes. He knew from previous experience that 6 out of every 100 test tubes stored in cartons would be broken or chipped. Each carton contained 25 test tubes. How many cartons would he need to transport to the stockroom from storage?

(i) 43 cartons (ii) 42.4 cartons

(iii) 40 cartons (iv) 3 cartons

3 (5 points). Determine the number of significant figures in each of the following numbers: (a) 0.0309; (b) 209; (c) 0.290; (d) 7421.0; and (e) 4.26×10^3.

4 (10 points). Express 72 °F in units of (a) °C and (b) K.

5 (10 points). Express 60.9 kg in units of (a) g, (b) mg, and (c) lb.

6 (10 points). A parsec is a unit of length used in astronomy. Given 1 parsec = 1.92×10^{13} miles, calculate the number of meters in a parsec by using the conversion factor 1 inch = 2.54 cm.

7 (15 points). A flask that has a volume of 5.00 mL and a mass of 24.921 g was used to determine the density of a liquid and a solid.

(a) The combined mass of the flask and liquid is 32.254 g. Calculate the density of the liquid.

(b) A 3.456 g sample of the solid was placed in the empty flask and enough of the liquid was added to fill the flask to the 5.00 mL volume. The combined mass of the flask, liquid, and solid was 33.101 g. Calculate the density of the solid. Hint: First find the volume of the liquid by using the density from part (a) and then find the volume of the solid.

Answers to practice test

1. (a) ii, (b) iii, (c) i, (d) iii, (e) iv, (f) i, (g) iv, (h) ii, (i) ii, (j) iv.
2. (a) i, (b) ii, (c) iv, (d) ii, (e) ii, (f) i, (g) iii, (h) iv, (i) iii, (j) i.
3. (a) 3, (b) 3, (c) 3, (d) 5, (e) 3.
4. (a) 22.2 °C, (b) 295.4 K.
5. (a) 6.09×10^4 g, (b) 6.09×10^7 mg, (c) 134 lb.
6. 3.09×10^{16} m/parsec.
7. (a) 1.47 g/mL, (b) 1.93 g/mL.

ADDITIONAL NOTES

CHAPTER 3

CHEMISTRY: THE SCIENCE OF MATTER

CHAPTER OVERVIEW

Various classifications of matter that are based on the composition and on the chemical and physical properties of the matter are discussed. States of matter and changes in these states are introduced.

An introduction to modern atomic theory is presented in this chapter. The basis of this introduction is the discussion of five classic sets of experiments--cathode ray experiments, canal ray experiments, α-particle scattering experiments, the discovery of the neutron, and x-ray spectra experiments. Atomic symbols are introduced and nuclear arithmetic--calculations involving the atomic number, the neutron number, the mass number, and the atomic mass--is discussed.

COMPETENCIES

Definitions for the following terms should be learned:

- α-particle
- aqueous solution
- atom
- atomic mass
- atomic mass unit
- atomic number
- canal rays
- cathode rays
- changes of state
- chemical compound
- chemical properties
- chemical reaction
- Coulomb force
- deuterium
- dissolution
- electrodes
- electron
- elements
- fluorescent
- fundamental particle
- gas-discharge tubes
- heterogeneous mixture
- homogeneous mixture
- ions

- isotopes
- mass number
- mixture
- neutron
- neutron number
- nucleus
- phase
- physical properties
- potential energy
- proton
- pure substance
- solute
- solution
- solvent
- spectrum
- states of matter
- subatomic particle
- tritium

General concepts and skills that should be learned include

- using symbols and names to identify the various elements
- the fundamental ideas of Dalton's atomic theory
- the kinds of matter
- identifying chemical and physical properties
- the components of a solution
- the various states of matter and their general properties
- briefly describing and interpreting the results for each of the five classic experiments that led to the modern atomic theory
- the relationship among atomic number, mass number, and neutron number
- the relationship among the masses of the isotopes, their natural abundance, and the atomic mass of an element

Numerical exercises that should be understood include

- calculating the atomic number, neutron number, or mass number when two of these variables are given
- calculating the mass of an atom in atomic mass units or in grams when one of these variables is given
- calculating the atomic mass, the isotopic composition, or the atomic mass of an isotope when two of these variables are given

QUESTIONS

1. Complete the following sentences by filling in the various blanks with these terms: α-particle, atomic number, isotopes, mass number, mass spectrometer, neutrons, neutron number, nucleus, protons, and subatomic particle.

An instrument used to measure the actual masses and relative abundances of (a)_______ is known as a (b)_______. For a given element, the (c)______ represents the number of protons in the (d)______ and is always the same, but the number of (e)______ given by the (f)______ will be different for one isotope compared to another. The (g)______ represents the sum of the number of neutrons and (h)______ in that isotope. Although an (i)______ is not a (j)______, it played an important part in the experiments that helped in understanding the structure of an atom.

2. Complete the following sentences by filling in the various blanks with these terms: chemical compound, chemical properties, chemical reactions, heterogeneous mixture, homogeneous mixture, phases, physical properties, and pure substance.

Consider a rain puddle with a drop of motor oil floating on top of it. This is an example of a (a)______ in which the two (b)______ are easily seen. The reason for the layer formation is that the two materials are not soluble in each other and have different densities--examples of the (c)______ of the materials. The layers can be separated into water, which is an example of a (d)______, and a (e)______ of hydrocarbons (compounds containing carbon and hydrogen), which makes up the motor oil. Each (f)______ in the motor oil has a set of unique (g)______ that can be observed by a series of (h)______.

3. Match the correct term to each definition.

(a) a homogeneous part of a system in contact with, but separate from, the other parts of the system
(b) the process of one substance dissolving in another
(c) the condition or form in which matter exists (e.g., gas, solid, or liquid)
(d) a homogeneous mixture of two or more substances
(e) the component of a solution usually present in the smaller amount
(f) the component of a solution usually present in the larger amount
(g) a combination (of variable composition) of two or more substances in which the substances combined retain their identity
(h) a solution of any substance in water

(i) aqueous solution
(ii) dissolution
(iii) mixture
(iv) phase
(v) solute
(vi) solution
(vii) solvent
(viii) states of matter

4. Which of the following statements are true? Rewrite each false statement so that it is correct.
   (a) The symbols for the elements contain two letters.
   (b) Dalton's atomic theory states that chemical combination is the union of atoms of different elements such that the ratios of the masses of the combined atoms are simple whole numbers.
   (c) Cathode rays are streams of electrons flowing from the cathode toward the anode in a gas-discharge tube.
   (d) The atomic number gives the position of an element in the periodic table and is equal to the positive charge on the nucleus of each atom of that element.
   (e) The neutron number is very close to the atomic mass expressed in atomic mass units.
   (f) The elements in a compound can be separated from each other by physical means.
   (g) The individual components in a homogeneous mixture remain physically separate and can be seen as separate components.
   (h) An element is a pure substance that contains only atoms of the same atomic number.

5. Distinguish clearly between (a) cathode rays and canal rays, (b) atomic mass and atomic number, (c) physical and chemical properties, (d) mixtures and pure substances, (e) elements and chemical compounds, (f) homogeneous and heterogeneous mixtures, and (g) atoms and ions.

6. Choose the best answer for each statement.
   (a) The detailed study of what phenomenon in a gas-discharge tube resulted in the discovery and characterization of the properties of electrons?
      (i) cathode rays (ii) canal rays
      (iii) wavelengths (iv) line spectrum
   (b) A conductor through which electrical current enters or leaves a conducting medium is an
      (i) absorption spectrum (ii) electrode
      (iii) excited state (iv) atomic isotope
   (c) Select the one statement that is *not* true of cathode rays:
      (i) They travel in straight lines from anode to cathode.
      (ii) They produce fluorescence.

(iii) They expose photographic film.

(iv) They interact with applied electrical and magnetic fields.

(d) Select the one statement that is true of canal rays:

(i) They do not produce fluorescence.

(ii) They expose photographic film.

(iii) They are deflected more than cathode rays by magnetic and electrical fields.

(iv) They are deflected in the same direction as cathode rays by magnetic and electric fields.

(e) A fundamental, subatomic, negatively charged particle is the

(i) proton (ii) neutron

(iii) electron (iv) isotope

(f) Which is not a subatomic particle?

(i) neutron (ii) proton

(iii) electron (iv) electrode

(g) Hydrogen that consists of atoms containing nuclei having one proton and two neutrons is called

(i) fluorescent (ii) deuterium

(iii) tritium (iv) protium

(h) Which is a fundamental particle?

(i) α-particle (ii) nucleus

(iii) isotope (iv) proton

7. The symbols for the elements fall into three classifications based on the number of letters in the symbol. Describe how a symbol in each classification is written. Be sure to include a symbol for an element in your description of each classification.

8. List and briefly discuss the points of the atomic theory as formulated by John Dalton.

9. What do we mean by a "pure substance"? Is an element a pure substance? Is a chemical compound a pure substance?

10. What is the smallest particle of an element?

11. What is a chemical compound? What type of process will separate the elements combined in a chemical compound?

12. Name the three states of matter. Can all substances exist in all of these states?

13. Explain how cathode rays and canal rays are produced from atoms in a gas-discharge tube.

14. Is an α-particle a subatomic particle? What is the relationship of an α-particle to a helium atom?

15. Each of the following scientists made a major contribution to the understanding of atomic structure. For each, briefly describe the experiment or theory that he contributed and briefly interpret the significance of the results: (a) William Crookes, (b) Eugen Goldstein, (c) Joseph J. Thomson, (d) Robert A. Millikan, (e) Ernest Rutherford, (f) Henry G. J. Moseley, (g) James Chadwick, (h) Julius Plücker, (i) George J. Stoney, (j) Wilhelm Wien, (k) John Dalton, and (l) Hans Geiger and Ernest Marsden.

16. Compare the atomic models of Dalton, Thomson, and Rutherford.

17. Prepare a table showing (i) the symbol, (ii) the mass in grams, (iii) the mass in atomic mass units, (iv) the relative electrical charge, and (v) the electrical charge (in Coulombs) for (a) an electron, (b) a proton, (c) a neutron, and (d) an α-particle.

18. What is the name for the force of electrical attraction or repulsion between charged particles? Write the equation for this force and define the symbols used.

19. What is the relationship among the atomic number (Z), mass number (A), and neutron number (N)?

20. What does the term "atomic mass" mean? Given the fraction of each naturally occurring isotope and its mass, how is the atomic mass for an element determined?

21. What is the relationship between the mass number of an isotope and the mass of the isotope expressed in atomic mass units?

22. What instrument is used to determine the relative masses of the isotopes of an element? How can this instrument also indicate the relative abundance of each isotope?

23. What is the definition of an atomic mass unit?

24. What are physical properties? Write a few physical properties for water.

25. Describe each of the following by placing *x*'s in as many boxes as are appropriate:

	(i) Matter	(ii) Mixture	(iii) Homogeneous mixture	(iv) Heterogeneous mixture	(v) Pure substance	(vi) Compound	(vii) Element
(a) Sugar							
(b) Brine*							
(c) Air							
(d) Pencil lead*							
(e) Steam							
(f) Diamond							
(g) Gasoline*							
(h) Dry Ice*							
(i) Oil on a rain puddle							

*Brine is salt water, pencil lead is a clay-graphite mixture, gasoline contains many chemical compounds, and dry ice is solid carbon dioxide.

26. What are interconversions between states of matter called? Illustrate three such interconversions by using ice, water, and steam.

Answers to questions

1. (a) isotopes, (b) mass spectrometer, (c) atomic number, (d) nucleus, (e) neutrons, (f) neutron number, (g) mass number, (h) protons, (i) α-particle, (j) subatomic particle.
2. (a) heterogeneous mixture, (b) phases, (c) physical properties, (d) chemical compound or pure substance, (e) homogeneous mixture, (f) chemical compound or pure substance, (g) chemical properties, (h) chemical reactions.
3. (a) iv, (b) ii, (c) viii, (d) vi, (e) v, (f) vii, (g) iii, (h) i.
4. (a) F, change *two* to *one, two, or three*; (b) T; (c) T; (d) T; (e) F, change *neutron* to *mass*; (f) F, change *physical* to *chemical*; (g) F, change *homogeneous* to *heterogeneous*; (h) T.
5. (a) *Cathode rays* are negatively charged rays that travel from the cathode to the anode in a gas-discharge tube and are identical for all substances; *canal rays* are positively charged rays that travel from the anode to the cathode and are different for different substances. (b) The *atomic mass* is the mass of the atom determined by the sum of the masses of the constituent

atomic particles; the *atomic number* of an element is the number of protons present in the nucleus of an atom of that element. (c) *Physical properties* are properties that are observed or measured without changing the identity of a substance; *chemical properties* are properties that are observed or measured which result in a change of identity of a substance. (d) *Mixtures* are combinations of two or more pure substances that can be separated by physical processes because the substances are not chemically united; *pure substances* are forms of matter that have identical chemical and physical properties no matter what the source of the substance. (e) *Elements* are pure substances that contain only one kind of atom; *chemical compounds* are pure substances of definite composition in which atoms of two or more elements are chemically combined. (f) *Homogeneous mixtures* contain components that are thoroughly intermingled; *heterogeneous mixtures* contain components that remain physically separate. (g) An *atom* is the smallest particle of an element and will be electrically neutral because of equal numbers of protons and electrons; an *ion* is an atom or group of atoms that has an electrical charge because of the gain or loss of electrons.

6. (a) i, (b) ii, (c) i, (d) ii, (e) iii, (f) iv, (g) iii, (h) iv.
7. One-letter symbols consist of a capital letter, e.g., N; two-letter symbols consist of a capital letter followed by a lowercase (small) letter, e.g., Na; three-letter symbols consist of a capital letter and two lowercase letters, e.g., Unq.
8. The points of Dalton's atomic theory are
   (a) all matter consists of tiny particles called atoms
   (b) atoms of one element can neither be subdivided nor changed into atoms of any other element
   (c) atoms can neither be created nor destroyed
   (d) all atoms of the same element are identical
   (e) atoms of one element differ from atoms of other elements
   (f) chemical combination is the union of atoms in simple, whole-number ratios
9. A pure substance is a form of matter than has the same physical and chemical properties, no matter what its source. Yes, each element is a pure substance. Yes, every compound is a pure substance.
10. The smallest particle of an element is an atom.

11. A chemical compound is a substance of definite, fixed composition in which atoms of two or more elements are chemically combined. The elements can be separated only by a chemical change.
12. The three states of matter are solid, liquid, and gas. No, not all substances can exist in all of these states.
13. Application of an electrical potential between the electrodes in a gas-discharge tube causes electrons to be released from the cathode. They collide with gaseous atoms in the tube, knocking out of each of these atoms one or more additional electrons, which form the cathode rays. The positive ions resulting from electron loss constitute the canal rays.
14. No, an α-particle is not a subatomic particle. An α-particle is a helium ion with a charge of +2 produced by the loss of two electrons from the atom.
15. (a) *Crookes* performed gas-discharge tube experiments, which led to an understanding of atomic structure; (b) *Goldstein* performed canal ray experiments, which led to an understanding about the remainder of the atom once electrons were removed; (c) *Thomson* measured the charge-to-mass ratio for electrons, which made it possible to know either the charge or the mass of an electron once the other was known; (d) *Millikan* measured the charge on electrons, which made possible the calculation of the mass; (e) *Rutherford* performed radioactivity experiments, which discovered the α-particle; (f) *Moseley* studied the x-ray spectra of elements, which determined the atomic number of the elements; (g) *Chadwick* discovered the neutron, which was needed to explain the atomic mass of an element; (h) *Plücker* observed that cathode rays were influenced by magnetic fields, which proved that electrons were electrically charged; (i) *Stoney* gave a role to electrons by stating that electricity is carried by individual negatively charged particles; (j) *Wien* measured the charge-to mass ratio for canal rays, which gave the mass of the canal ray, showing that they were the remainder of the atom once electrons were removed; (k) *Dalton* stated an early atomic theory that helped to formalize the properties of atoms; (l) *Geiger* and *Marsden* performed α-particle scattering experiments, which proved the existence of the nucleus.
16. Dalton's atoms were hard and indivisible; Thompson's atoms contained positively and negatively charged particles uniformly distributed inside a sphere; Rutherford's atoms contained a very small nucleus that contained nearly all of the mass and all of the positive charge of the atom.

17. (a) e^-, 9.11 x 10^{-28} g, 1/1837 u, -1, -1.602 x 10^{-19} C; (b) p^+, 1.673 x 10^{-24} g, 1 u, +1, 1.602 x 10^{-19} C; (c) *n*, 1.675 x 10^{-24} g, 1 u, 0, 0; (d) α, 6.647 x 10^{-24} g, 4 u, +2, 3.204 x 10^{-19} C.
18. The force is known as the Coulomb force; force = kq_1q_2/r^2 where *k* is a proportionality constant, q_1 and q_2 are the charges on the particle, and *r* is the distance between the particles.
19. $A = Z + N$.
20. Atomic mass is the average mass of atoms of the naturally occurring element relative to (1/12)th the mass of ^{12}C. The atomic mass is calculated by multiplying each isotope fraction by the mass of that isotope and adding the results for all isotopes.
21. The mass number is equal to the nearest whole integer of the mass expressed in atomic mass units.
22. The relative masses of isotopes of an element are determined with a mass spectrometer. It will also indicate the relative abundances of each isotope.
23. An atomic mass unit is defined as 1/12 of the mass of one carbon-12 atom. relative mass in u is 1.6605655 x 10^{-24} g = 1 u.
24. Physical properties can be measured or observed without changing the composition and identity of a substance; e.g., melting point is 0 °C, boiling point is 100 °C, density is 1 g/cm³, colorless and odorless.
25. (a) i, v, vi; (b) i, ii, iii; (c) i, ii, iii; (d) i, ii, iii; (e) i, v, vi; (f) i, v, vii,; (g) i, ii, iii; (h) i, v, vi; (i) i, ii, iv.
26. These interconversions are called changes of state. Ice melts to water or water freezes to ice; water evaporates to steam or steam condenses to water; and ice sublimes to steam or steam deposits to ice.

SKILLS

Skill 3.1 Identifying chemical and physical properties

Physical properties can be exhibited, measured, or observed without resulting in a change in the composition and identity of the substance in question. On the other hand, chemical properties can be observed only in a chemical reaction in which at least one substance is changed in composition and identity. For example, the boiling point of water is a physical property because there has been no change in the composition of water (it is still H_2O), whereas the tendency of water to undergo reaction with metallic sodium is a chemical property because the combination of the two substances results in a change in composition (the formation of hydrogen gas and a solution of sodium hydroxide).

Skill Exercise: Identify which of the following properties are classified as (i) physical and (ii) chemical: (a) the ability of a piece of wood to burn, (b) the ability of an acid to undergo reaction with a base to form water and a salt, (c) the melting point of metallic lead at 328 °C, (d) the yellowish color of metallic copper, (e) the density of liquid water is 1.00 g/mL, and (f) the ability of metallic aluminum to form an oxide coating. *Answers*: (a) ii, (b) ii, (c) i, (d) i, (e) i, (f) ii.

Skill 3.2 Describing matter

Matter can be classified as either a pure substance or a mixture, based on whether it can have a variable composition or not. A pure substance may be classified as an element or a compound, depending on whether or not it can be converted to a simpler form of matter by chemical methods. A mixture may be classified as homogeneous or heterogeneous, based on whether or not the components of the mixture remain physically separate and can be seen as separate components. For example, sugar is considered to be a pure substance and a compound, while seawater is considered to be a mixture that is homogeneous.

Skill Exercise: Describe the following items: (a) a steel nail, (b) a piece of copper wire, (c) a piece of bread, (d) water, (e) sand, (f) toothpaste, and (g) limestone. *Answers*: (a) homogeneous mixture, (b) elemental substance, (c) heterogeneous mixture, (d) pure substance and compound, (e) pure substance and compound, (f) homogeneous mixture, (g) pure substance and compound.

NUMERICAL EXERCISES

Example 3.1 Nuclear arithmetic*

What is the composition of $^{90}_{40}Zr$ nuclei? How does this composition differ from that of $^{94}_{40}Zr$ nuclei?

This problem is analyzed using the comprehensive problem-solving approach presented in *Skill 2.7 Solving problems*.

1. Study the problem and be sure you understand it.
   (a) What is unknown?
   The nuclear composition of each of the isotopes.

(b) What is known?

The atomic number of Zr is 40, as given by the leading subscript on the symbol, and the respective mass numbers of the isotopes are 90 and 94, as given by the leading superscript on the symbol.

2. Decide how to solve the problem.

(a) What is the connection between the known and the unknown?

The number of protons in a nucleus is equal to the atomic number. The mass number represents the total number of protons and neutrons in the nucleus.

(b) What is necessary to make the connection?

The atomic number, and therefore the number of protons is known. Subtracting the number of protons from the mass number gives the number of neutrons in the nucleus ($N = A - Z$, where A is the mass number, Z is the atomic number, and N is the neutron number).

3. Set up the problem and solve it.

In $^{90}_{40}Zr$, the number of protons is 40 and the number of neutrons is

$$N = A - Z = (90) - (40) = 50$$

In $^{94}_{40}Zr$, the number of protons is 40 and the number of neutrons is

$$N = (94) - (40) = 54$$

The two nuclei have the same number of protons, but $^{94}_{40}Zr$ has four more neutrons in each nucleus than does $^{90}_{40}Zr$.

4. Check the result.

(a) Are the significant figures and the location of the chemical point correct?

Yes. All of these numbers are exact.

(b) Is the answer reasonable?

Yes. Most nuclei have roughly the same numbers of protons as neutrons.

Exercise 3.1 What is the composition of a nucleus of $^{149}_{61}Pm$? *Answer*: 61 protons, 88 neutrons.

Example 3.2 Nuclear arithmetic

What is the mass number of the isotope of lawrencium that contains 154 neutrons?

The periodic table shows that the atomic number of lawrencium (Lr) is 103. Using the relationship among mass number, atomic number, and neutron number gives

$$A = Z + N = 103 + 154 = 257$$

The mass number of this isotope is 257.

Exercise 3.2 A certain nucleus contains 145 neutrons and has a mass number of 239. Identify the element. *Answer*: Pu.

Example 3.3 Atomic mass*

What is the mass in grams of a selenium-74 atom given that its mass is 73.9225 u?

This problem is analyzed using the comprehensive problem-solving approach presented in *Skill 2.7 Solving problems.*

1. Study the problem and be sure you understand it.
   (a) What is unknown?
   The mass of a ^{74}Se atom in grams.
   (b) What is known?
   The mass of a ^{74}Se atom in atomic mass units.
2. Decide how to solve the problem.
   (a) What is the connection between the known and the unknown?
   The connection is the conversion factor between atomic mass units and grams, which is based on the relative atomic mass scale and the definition of an atomic mass unit (1 u = (1/12)th the mass of an atom of ^{12}C).
   (b) What is necessary to make the connection?
   $1 \text{ u} = 1.6605655 \times 10^{-24} \text{ g}$
3. Set up the problem and solve it.

$$(73.9225 \text{ u})\left[\frac{1.66057 \times 10^{-24} \text{ g}}{1 \text{ u}}\right] = 1.22753 \times 10^{-22} \text{ g}$$

4. Check the result.
   (a) Are significant figures and the location of the decimal point correct?
   Yes. The conversion factor is given to six significant figures because the mass data contained six significant figures.
   (b) Did the answer come out in the correct units?
   Yes.
   (c) Is the answer reasonable?
   Yes. The mass of an atom would be expected to be very small.

Exercise 3.3 What is the mass of an atom of germanium-70 in atomic mass units if the mass is 1.16114×10^{-22} g? *Answer*: 69.9244 u.

Example 3.4 Atomic mass

Natural selenium consists of 0.9% Se atoms weighing 73.9225 u, 9.0% weighing 75.9192 u, 7.6% weighing 76.9199 u, 23.5% weighing 77.9173 u, 49.8% weighing 79.9165 u, and 9.2% weighing 81.9167 u. Calculate the atomic mass for Se.

The atomic mass can be found by multiplying each percentage composition expressed as a decimal fraction by the respective mass and adding these results. The result is

$$\begin{aligned} \text{atomic mass} &= (0.009)(73.9225\ \text{u}) + (0.090)(75.9192\ \text{u}) \\ &\quad + (0.076)(76.9199\ \text{u}) + (0.235)(77.9173\ \text{u}) \\ &\quad + (0.498)(79.9165\ \text{u}) + (0.092)(81.9167\ \text{u}) \\ &= (0.7 + 6.8 + 5.8 + 18.3 + 39.8 + 7.5)\ \text{u} \\ &= 78.9\ \text{u} \end{aligned}$$

The atomic mass of Se is 78.9 u.

Exercise 3.4 Determine the atomic mass of Ge from the following isotopic composition: 20.5% of 69.9243 u, 27.4% of 71.9217 u, 7.8% of 72.9234 u, 36.5% of 73.9219 u, and 7.8% of 75.9214 u. *Answer*: 72.6 u.

Example 3.5 Atomic mass

The atomic mass of lithium is 6.941 u. Natural lithium consists of two isotopes, ^{6}Li (mass = 6.01512 u) and ^{7}Li (mass = 7.01600 u). What is the isotopic composition of naturally occurring lithium?

The atomic mass can be found by multiplying each fraction by the respective mass and adding these results. Letting x be the fraction of ^{6}Li, the following equation can be written and solved giving

$$\begin{aligned} 6.941\ \text{u} &= (x)(6.01512\ \text{u}) + (1 - x)(7.01600\ \text{u}) \\ 6.941 &= 7.01600 - (1.00088)x \\ (1.00088)x &= 0.075 \\ x &= 0.075 \end{aligned}$$

Naturally occurring lithium consists of 7.5% ^{6}Li and 92.5% ^{7}Li.

Exercise 3.5 The atomic mass of vanadium is 50.9415 u. Natural vanadium consists of two isotopes, ^{50}V (mass = 49.9472 u) and ^{51}V (mass = 50.9440 u).

Determine the isotopic composition of naturally occurring vanadium. *Answer*: 0.25% ^{50}V and 99.75% ^{51}V.

PRACTICE TEST

1 (10 points). Match the correct term to each definition.

(a) homogeneous matter having a set of distinctive properties and fixed composition no matter what its source

(b) a combination of two or more substances with variable composition in which the substances retain their identities

(c) a pure substance formed by the chemical combination of two or more elements

(d) a pure substance that cannot be converted into a simpler form of matter by chemical means

(e) something that has mass

(i) compound
(ii) element
(iii) matter
(iv) mixture
(v) pure substance

2 (10 points). Describe the following materials in terms of (i) a homogeneous mixture, (ii) a heterogeneous mixture, (iii) a compound, or (iv) an element: (a) air, (b) water, (c) diamond, (d) chalk, and (e) a pencil.

3 (60 points). Choose the response that best completes each statement.

(a) The symbol for strontium is
(i) S (ii) Si
(iii) Sr (iv) Sn

(b) The symbol for tungsten is
(i) W (ii) Ti
(iii) Th (iv) Tc

(c) The symbol for potassium is
(i) K (ii) P
(iii) Pt (iv) Ag

(d) The symbol Br represents
(i) boron (ii) barium
(iii) beryllium (iv) bromine

(e) The symbol Th represents
(i) thallium (ii) thorium
(iii) tantalum (iv) titanium

(f) The person who is credited with the discovery of the electron is

(i) Bohr (ii) Thomson

(iii) Dalton (iv) Rutherford

(g) The experiment in which cathode rays were deflected by an electrical field demonstrated that

(i) cathode rays are negatively charged

(ii) cathode rays are α-particles

(iii) cathode rays have mass

(iv) none of these answers

(h) Experiments by Rutherford showed

(i) that the mass of the electron is equal to the mass of a neutron

(ii) that the charge-to-mass ratio is identical for all cathode ray particles

(iii) that most of the mass of an atom occupies a very small portion of the volume of the atom

(iv) the existence of the neutron

(i) The results of Moseley's experiments made it possible to assign values to

(i) the number of neutrons in the nucleus

(ii) the number of protons in the nucleus

(iii) the atomic mass of the element

(iv) none of these answers

(j) Which particle is the heaviest?

(i) proton (ii) neutron

(iii) β-particle (iv) α-particle

(k) The respective electrical charges on the α-particle, β ray, and γ ray are

(i) +2, -1, +1 (ii) 0, -1, +2

(iii) -1, +2, 0 (iv) +2, -1, 0

(l) The Coulomb force between a proton and a neutron is

(i) repulsive when they are near

(ii) large when they are far apart

(iii) zero

(iv) attractive

(m) Identify the respective atomic number, neutron number, and number of electrons in $^{206}_{82}Pb$.

(i) 82, 124, 82 (ii) 124, 82, 82

(iii) 82, 82, 124 (iv) none of these answers

(n) A typical atom of boron will contain
(i) 5 protons, 5 neutrons, and 10 electrons
(ii) 5 protons, 5 neutrons, and 5 electrons
(iii) 10 protons, 5 neutrons, and 5 electrons
(iv) 5 protons, 10 neutrons, and 5 electrons

(o) Which pair of symbols represent isotopes?
(i) $^{14}_{6}C$ and $^{14}_{7}N$ (ii) $^{25}_{12}Mg$ and $^{12}_{6}C$
(iii) $^{12}_{6}C$ and $^{14}_{6}C$ (iv) $^{10}_{5}B$ and $^{11}_{6}B$

(p) Which is not a chemical reaction?
(i) iron rusting (ii) milk souring
(iii) coal burning (iv) liquid air boiling

(q) Which is not a physical state of matter?
(i) solid (ii) gas
(iii) compound (iv) liquid

(r) A solution of sugar in hot tea is an example of a
(i) mixture (ii) compound
(iii) element (iv) heterogeneous system

(s) In an aqueous solution, water is the
(i) solute (ii) solvent
(iii) fundamental particle (iv) subatomic particle

(t) An array of radiation or particles spread out according to the increasing or decreasing magnitude of some physical property is
(i) the Coulombic force (ii) a canal ray
(iii) fluorescence (iv) a spectrum

4 (10 points). The mass of ^{6}Li is 6.01512 u. Calculate the atomic mass in grams for this isotope.

5 (10 points). Calculate the atomic mass of naturally occurring sulfur. The percentage composition is 95.0% ^{32}S (atomic mass = 31.97207 u), 0.76% ^{33}S (atomic mass = 32.97146 u), 4.22% ^{34}S (atomic mass = 33.96786 u), and 0.02% ^{36}S (atomic mass = 35.96709 u).

Answers to practice test

1. (a) v, (b) iv, (c) i, (d) ii, (e) iii.
2. (a) i, (b) iii, (c) iv, (d) iii, (e) ii.
3. (a) iii, (b) i, (c) i, (d) iv, (e) ii, (f) ii, (g) i, (h) iii, (i) ii, (j)

iv, (k) iv, (l) iii, (m) i, (n) ii, (o) iii, (p) iv, (q) iii, (r) i, (s) ii, (t) iv.

4. 9.98850×10^{-24} g.
5. 32.1 u.

ADDITIONAL NOTES

CHAPTER 4

ATOMS, MOLECULES, AND IONS

CHAPTER OVERVIEW

The concepts of molecular and ionic compounds are introduced. The writing of formulas and the naming of chemical compounds are discussed. An introduction to writing and balancing chemical equations to describe chemical reactions is given. The expressing of the concentration of a solution in terms of molarity is briefly discussed. Finally, molecular masses and the mass relationships that can be derived from chemical formulas are presented--with emphasis on Avogadro's number and the concept of molar mass and the mole.

COMPETENCIES

Definitions for the following terms should be learned:

- anions
- Avogadro's number
- binary compound
- catalyst
- cations
- chemical equation
- chemical formula
- chemical nomenclature
- concentration
- diatomic molecules
- dissociation of an ionic compound
- empirical formula
- empirical relationship
- formula unit
- ionization
- molar mass
- molarity
- mole
- molecular formula
- molecular mass
- molecule
- monatomic ions
- percentage composition
- polyatomic ions
- polyatomic molecule
- polymers
- products
- reactants
- salts
- simplest formula

General concepts and skills that should be learned include

- the relationships among atoms, ions, molecules, and ionic compounds
- the nomenclature (naming and writing of symbols or formulas) of cations, anions, ionic compounds, acids, and binary molecular compounds
- writing, balancing, and interpreting chemical equations to describe chemical reactions
- the meaning and use of the mole
- the relationship between molecular mass and molar mass
- the relationship among the solution concentration expressed in molarity, the volume of the solution, and the number of moles of solute
- the relationship between the composition of a compound on a mass basis and its chemical formula

Numerical exercises that should be understood include

- calculating the molecular mass and molar mass of a substance by using atomic masses and the chemical formula
- calculating the number of moles or the number of molecules by using Avogadro's number when one of these variables is given
- calculating the number of moles, mass, or molar mass when two of these variables are given
- calculating the concentration of a solute in molarity, the number of moles of solute, or the volume of the solution when two of these variables are given
- calculating the percentage composition for a substance from the chemical formula, or finding the empirical formula of a compound from the percentage composition
- determining the molecular formula of a compound from the empirical formula and molecular mass

QUESTIONS

1. Complete the following sentences by filling in the various blanks with these terms: anion, binary compound, cation, monatomic ion, polyatomic ions, and polyatomic molecules.

   If an atom has an electrical charge because it has gained or lost electrons, it is known as a (a)________. If the electrical charge is positive, the ion is called a (b)________ and if negative, an (c)________.

The combination of monatomic cations and monatomic anions results in the formation of a (d)_______. The ions formed from (e)_______ are known as (f)_______.

2. Complete the following sentences by filling in the various blanks with these terms: catalyst, chemical equation, chemical formulas, products, and reactants.

   A (a)_______ is a shorthand notation describing a chemical process that takes place. The (b)_______ of the (c)_______ are written on the left side of an arrow that is read as "yields," and the formulas of the (d)_______ are written on the right side of the arrow. Reaction conditions, such as the presence of a (e)_______, are often written over the arrow.

3. Complete the following sentences by filling in the various blanks with these terms: atomic mass, atomic mass units, diatomic molecule, empirical formula, molar mass, molecular formula, and molecular mass.

   The simplest formula or (a)_______ for a gaseous compound is F. Additional experiments showed that the (b)_______ is F_2, which means that the compound exists as a (c)_______. The (d)_______ is found by multiplying the (e)_______ by 2 and if this is expressed in units of g/mol instead of (f)_______, it is known as the (g)_______.

4. Match the correct term to each definition.

(a) the formation of ions from a nonionic substance	(i) catalysts
(b) the number of moles of solute per liter of solution	(ii) chemical equation
(c) substances that affect the rate of a reaction, but can be recovered from the reaction unchanged	(iii) concentration
(d) a quantitative statement of the amount of solute in a given amount of solvent or solution	(iv) dissociation of an ionic compound
(e) ionic compounds formed between cations and and anions	(v) ionization
(f) the transformation of a neutral ionic compound into positive and negative ions in solution	(vi) molarity
	(vii) polymer
	(viii) products
	(ix) reactants
	(x) salts

(g) the new substances that are produced in a chemical reaction
(h) very large molecules formed by linking smaller molecules together
(i) the symbols and formulas representing the total chemical change that occurs in a chemical reaction
(j) the substances that are changed in a chemical reaction

5. Which of the following statements are true? Rewrite each false statement so that is it correct.
(a) The molar mass is the same as the molecular mass.
(b) The empirical formula represents the actual number of atoms that are combined in each molecule of the compound.
(c) The law of definite proportions states that if two elements combine to form more than one compound, the masses of one element that combine with a fixed mass of the other will be related to each other as ratios of small whole numbers.
(d) A molecule of oxygen, O_2, has a mass of 32.0 g.
(e) A chemical equation is a statement of the law of conservation of mass.

6. Distinguish clearly between/among (a) atoms, molecules, and ions; (b) cations and anions; (c) molecules and formula units; (d) molecular mass and molar mass; (e) molecules and moles; (f) simplest and molecular formulas; (g) a chemical reaction and a chemical equation; and (h) reactants and products.

7. Choose the response that best completes each statement.
(a) The number of molecules in a mole is given by
(i) the chemical formula (ii) the percentage composition
(iii) Avogadro's number (iv) the empirical formula
(b) Dividing the molar mass of a monatomic element by Avogadro's number gives the
(i) actual mass of an atom (ii) percentage composition
(iii) molecular mass (iv) simplest formula
(c) The number of atoms of the various elements making up a compound are given in the
(i) empirical formula (ii) chemical formula
(iii) simplest formula (iv) percentage composition
(d) The information for an ionic compound given by the formula unit is the same as for a molecular compound given by the

(i) molecule (ii) mole
(iii) anion (iv) concentration

(e) The collective term for the rules and regulations that govern the naming of chemical compounds is
(i) the law of definite proportions (ii) chemical nomenclature
(iii) the standard solution (iv) catalysis

(f) The amount of substance of a system that contains as many elementary entities as there are atoms in 12 g of carbon-12 is a
(i) molecule (ii) atom
(iii) mole (iv) formula unit

(g) The fraction by mass of each element in a compound can be expressed as
(i) the percentage composition (ii) the mass % for solutions
(iii) the law of multiple proportions (iv) the simplest formula

8. Of what type of particles is a molecular compound composed? Of what type of particles is an ionic compound composed? What is the difference between these types of particles?

9. Define the terms (a) "monatomic molecule," (b) "diatomic molecule," and (c) "polyatomic molecule." Give two examples of each.

10. How are ions formed from atoms? What happens to the nucleus of an atom as this process occurs?

11. Define the terms "cation" and "anion." Give two examples of each type of ion.

12. What information is given by a chemical formula? What additional information is given by a structural formula?

13. Briefly discuss the difference between/among (a) 2H and H_2; (b) C_2H_2 and C_6H_6; (c) Hg_2Cl_2 and $HgCl_2$; and (d) O, O^{2-}, O_2, O_2^{2-}, and O_3.

14. What determines the ratio of cations to anions in an ionic compound? What would be the cation-to-anion ratio in ionic compounds containing a cation with a +2 charge and anions with (a) a -1 charge, (b) a -2 charge, and (c) a -3 charge?

15. Write the formulas for the ionic compounds formed between Ca^{2+} and (a) Cl^-, (b) O^{2-}, (c) N^{3-}, (d) CN^-, and (e) HSO_4^-.

16. Write the formulas for the ionic compounds formed between SO_4^{2-} and (a) Na^+, (b) Zn^{2+}, and (c) Al^{3+}.

17. Define the term "chemical nomenclature."

18. State the rules for naming (a) monatomic cations (b) monatomic anions (c) binary ionic compounds, (d) binary molecular compounds, and (e) salts containing polyatomic ions.

19. Name the following cations: (a) K^+, (b) Fe^{2+}, (c) Fe^{3+}, (d) Cu^+, (e) Cu^{2+}, (f) Ag^+, and (g) NH_4^+.

20. Write the formula for each of the following cations: (a) calcium ion, (b) cobalt(II) ion, (c) tin(II) ion, (d) tin(IV) ion, (e) mercuric or mercury(II) ion, (f) mercurous or mercury(I) ion, and (g) magnesium ion.

21. Name the following anions: (a) Cl^-, (b) O^{2-}, (c) N^{3-}, (d) F^-, (e) OH^-, (f) SO_3^{2-}, and (g) NO_3^-.

22. Write the formula for each of the following anions: (a) iodide ion, (b) phosphide ion, (c) sulfide ion, (d) bromide ion, (e) peroxide ion, (f) acetate ion, and (g) chlorate ion.

23. Name the following compounds: (a) $(NH_4)_2Cr_2O_7$, (b) $Ba(ClO_3)_2$, (c) $Cd(NO_3)_2$, (d) $PbCl_2$, (e) KCN, (f) H_2CO_3, (g) $Cr(OH)_3$, (h) $MgHPO_4$, (i) Na_2S, (j) N_2O_5, and (k) PCl_5.

24. Write the formula for each of the following compounds: (a) ammonium chloride, (b) copper(II) bromide, (c) bismuth nitrate, (d) silver phosphate, (e) perchloric acid, (f) potassium permanganate, (g) mercury(II) acetate, (h) aluminum dihydrogen phosphate, (i) bromine trichloride, and (j) sulfur hexafluoride.

25. What is a chemical reaction? How is a chemical reaction expressed in written form? What does it mean to have a "balanced" chemical equation?

26. How is the physical state of each of the reactants and products represented in a chemical equation? List the usual symbols used and indicate their meaning.

27. Using the format

$$A \xrightarrow{B} C$$

indicate by the respective letter(s) where each of these parts of a chemical equation would appear: (a) reactants, (b) products, (c) reaction conditions, (d) physical states of substances, (e) coefficients, and (f) chemical formulas and symbols.

28. A student wrote the following equation to describe the decomposition of hydrogen peroxide, H_2O_2, into oxygen and water:

$$H_4O_4 \rightarrow 2H_2O + O_2$$

What is wrong with this equation?

29. Write a sentence describing in words each of the chemical reactions represented by the following chemical equations:

(a) $2S(s) + 2H_2O(g) + 3O_2(g) \rightarrow 4H^+(aq) + 2SO_4^{2-}(aq)$

(b) $2N_2O(g) \xrightarrow{\Delta} 2N_2 + O_2(g)$

(c) $CO(g) + 2H_2(g) \xrightarrow{400\ °C,\ 200\ atm} CH_3OH(g)$

(d) $Mg(s) + Cu^{2+}(aq) \rightarrow Cu(s) + Mg^{2+}(aq)$

(e) $2MnO_4^-(aq) + 3H_2S(g) \rightarrow 2MnO_2(s) + 3S(s) + 2OH^-(aq) + 2H_2O(l)$

(f) $H_2O(s) \xrightarrow{0\ °C,\ 1\ atm} H_2O(l)$

30. Choose the chemical equations that are correctly balanced:

(a) $AgCl(s) \rightarrow Ag^+ + Cl_2(g)$

(b) $FeAsS(s) \xrightarrow{\Delta} FeS(s) + As(g)$

(c) $2Sb_2S_3(s) + 9O_2(g) \xrightarrow{\Delta} Sb_4O_6(s) + 6SO_2(g)$

(d) $B_2H_6(g) + 3H_2O(l) \rightarrow 2H_3BO_3(ether) + 6H_2(g)$

(e) $2NaOH(aq) + Cl_2(g) \rightarrow NaOCl(aq) + NaCl(aq) + H_2O(l)$

31. Balance the following chemical equations:

(a) $NH_4Cl(s) + CaO(s) \rightarrow CaCl_2(s) + NH_3(g) + H_2O(g)$

(b) $Al(s) + I_2(\text{in } CS_2) \rightarrow AlI_3(s)$

(c) $H_2SO_4(aq) + HBr(g) \rightarrow Br_2(g) + SO_2(g) + H_2O(l)$

(d) $Li_3N(s) + H_2O(l) \rightarrow NH_3(g) + LiOH(aq)$

(e) $HNO_3(l) + P_4O_{10}(s) \rightarrow N_2O_5(s) + HPO_3(s)$

32. Write and balance the chemical equations describing the following reactions:

(a) Solid arsenic undergoes reaction with gaseous chlorine to produce solid arsenic trichloride.

(b) Liquid nitric acid decomposes to form gaseous nitrogen dioxide, gaseous oxygen, and liquid water.

(c) The mineral chalcocite, Cu_2S, is heated in oxygen, forming copper(I) oxide, Cu_2O, and sulfur dioxide, SO_2.

(d) Upon gentle heating, ammonium nitrate, NH_4NO_3, decomposes to form gaseous dinitrogen monoxide, N_2O, and steam and upon strong heating, ammonium nitrate decomposes explosively to form nitrogen, steam, and oxygen.

(e) Butane, $C_4H_{10}(g)$, burns in air by reacting with the oxygen to produce carbon dioxide and water.

33. What is the numerical value of Avogadro's number? What is the relationship of Avogadro's number to the mole?

34. How do we calculate the molecular mass of a chemical compound from its formula? What are the units that are used to express molecular mass and molar mass?

35. What is the relationship between the molecular mass and the molar mass of a substance?

36. What is the commonly used unit for expressing concentration that involves the number of moles of solute and the volume of the solution?

37. How is the volume of solution that contains a given number of moles of solute found from the molarity? How is the number of moles of solute in a given volume of solution found from the molarity?

38. What is the relationship between the empirical formula of a compound and its molecular formula? What information about the compound is usually used to find the molecular formula from the empirical formula?

Answers to questions

1. (a) monatomic ion, (b) cation, (c) anion, (d) binary compound, (e) polyatomic molecules, (f) polyatomic ions.
2. (a) chemical equation, (b) chemical formulas, (c) reactants, (d) products, (e) catalyst.
3. (a) empirical formula, (b) molecular formula, (c) diatomic molecule, (d) molecular mass, (e) atomic mass, (f) atomic mass units, (g) molar mass.
4. (a) v, (b) vi, (c) i, (d) iii, (e) x, (f) iv, (g) viii, (h) vii, (i) ii, (j) ix.

5. (a) F, change *is* to *has*, insert *numerical value* between *same* and *as*, insert *and is expressed in g/mol rather than atomic mass units* after *molecular mass*; (b) F, change *empirical* to *molecular*; (c) F, replace *if ... numbers* with *in pure compounds the elements are always combined in the same definite proportions by mass*; (d) F, replace *g* by *u*; (e) T.
6. (a) *Atoms* are the smallest particles of an element; *molecules* are the smallest particles of a pure substance; *ions* are positively or negatively charged atoms or groups of atoms. (b) *Cations* are positively charged ions; *anions* are negatively charged ions. (c) *Molecules* are the smallest particles of a molecular compound; *formula units* represent the simplest unit of an ionic compound. (d) The *molecular mass* is the mass of one molecule or formula unit of a substance; the *molar mass* is the mass of a mole of that substance. (e) *Molecules* are the smallest particles of a molecular substance; *moles* represent an Avogadro's number of particles. (f) The *simplest formula* represents the simplest whole-number ratio of atoms in a compound; the *molecular formula* is the actual numbers of each element in a molecule. (g) A *chemical reaction* is a process in which the identity and composition of at least one substance changes; a *chemical equation* is the set of symbols and formulas that represents the chemical change that occurs in a chemical reaction. (h) *Reactants* are substances that are changed in a chemical reaction; *products* are the new substances that are produced in a chemical reaction.
7. (a) iii, (b) i, (c) ii, (d) i, (e) ii, (f) iii, (g) i.
8. Molecules make up molecular compounds. Ions make up ionic compounds. Ions have electrical charges and molecules are neutral.
9. (a) A monatomic molecule contains only one atom, e.g., Ne and Ar; (b) A diatomic molecule contains two atoms, e.g., O_2 and HF; (c) A polyatomic molecule contains more than two atoms, e.g., CO_2 and H_2SO_4.
10. Ions are formed by the gain or loss of electrons; the nucleus does not change during this process.
11. A cation is a positively charged ion formed by the loss of one or more electrons, e.g., Na^+ and Al^{3+}; an anion is a negatively charged ion formed by the gain of one or more electrons, e.g., Cl^- and SO_4^{2-}.
12. A chemical formula gives the symbols for the elements present in the substance and uses subscripts to indicate how many atoms of each element are

included; a structural formula also indicates how the atoms are connected together.

13. (a) 2 H represents two atoms of hydrogen; H_2 represents a diatomic molecule of hydrogen.
    (b) C_2H_2 represents a molecule containing two atoms of carbon and two atoms of hydrogen; C_6H_6 represents a molecule containing six atoms of carbon and six atoms of hydrogen.
    (c) Hg_2Cl_2 represents one formula unit of two atoms of mercury as the Hg_2^{2+} ion and two chloride ions, Cl^-. $HgCl_2$ represents one formula unit of one mercury atom as the Hg^{2+} ion and two chloride ions, Cl^-.
    (d) O represents one atom of oxygen; O^{2-} represents one oxide ion; O_2 represents a diatomic molecule of oxygen; O_2^{2-} represents the peroxide ion of two oxygen atoms plus two electrons; O_3 represents a triatomic molecule of oxygen.
14. The ratio is determined by the requirement of electrical neutrality; (a) 1 cation with 2 anions, (b) 1 cation with 1 anion, (c) 3 cations with 2 anions.
15. (a) $CaCl_2$, (b) CaO, (c) Ca_3N_2, (d) $Ca(CN)_2$, (e) $Ca(HSO_4)_2$.
16. (a) Na_2SO_4, (b) $ZnSO_4$, (c) $Al_2(SO_4)_3$.
17. Chemical nomenclature refers to the rules and regulations that govern the naming of chemical compounds.
18. (a) The name of a monatomic cation consists of the name of the element, the charge on the ion given inside parentheses as a Roman numeral, and the word "ion."
    (b) Monatomic anions are named by writing the root of the name of the element modified by the ending "ide," followed by the work "ion."
    (c) A binary ionic compound is named by giving the cation name first, followed by the anion name.
    (d) The name of a binary molecular compound consists of the name of the first element modified with a prefix identifying the number of atoms and the name of the second element modified with both a similar prefix and the ending "ide."
    (e) The name of a salt containing a polyatomic ion consists of the name of the cation followed by the name of the anion.
19. (a) potassium ion, (b) iron(II) ion, (c) iron(III) ion, (d) copper(I) ion, (e) copper(II) ion, (f) silver ion, (g) ammonium ion.
20. (a) Ca^{2+}, (b) Co^{2+}, (c) Sn^{2+}, (d) Sn^{4+}, (e) Hg^{2+}, (f) Hg_2^{2+}, (g) Mg^{2+}.

21. (a) chloride ion, (b) oxide ion, (c) nitride ion, (d) fluoride ion, (e) hydroxide ion, (f) sulfite ion, (g) nitrate ion.
22. (a) I^-, (b) P^{3-}, (c) S^{2-}, (d) Br^-, (e) O_2^{2-}, (f) CH_3COO^-, (g) ClO_3^-.
23. (a) ammonium dichromate, (b) barium chlorate, (c) cadmium nitrate, (d) lead chloride, (e) potassium cyanide, (f) carbonic acid, (g) chromium(III) hydroxide, (h) magnesium monohydrogen phosphate, (i) sodium sulfide, (j) dinitrogen pentoxide, (k) phosphorus pentachloride.
24. (a) NH_4Cl, (b) $CuBr_2$, (c) $Bi(NO_3)_3$, (d) Ag_3PO_4, (e) $HClO_4$, (f) $KMnO_4$, (g) $Hg(CH_3COO)_2$, (h) $Al(H_2PO_4)_3$, (i) $BrCl_3$, (j) SF_6.
25. A chemical reaction is any process in which at least one substance is changed in composition and identity to become a chemically different substance; a chemical equation is written to express a reaction; a balanced chemical equation has the same numbers of each kind of atom on both the reactants side and the products side.
26. An abbreviation for the physical state of the substance is placed in parentheses and appears after the respective formula; g = gas, l = liquid, s = solid, aq = aqueous solution.
27. (a) A; (b) C; (c) B; (d) A, B, C; (e) A, C; (f) A, B, C.
28. The student incorrectly changed the subscripts of hydrogen peroxide, making the equation wrong. Use of the coefficient 2 for H_2O_2 correctly balances the equation.
29. (a) Solid sulfur undergoes reaction with steam and gaseous oxygen to form aqueous hydrogen ion and aqueous sulfate ion. (b) Heating gaseous dinitrogen monoxide produces gaseous nitrogen and gaseous oxygen. (c) Gaseous carbon monoxide undergoes reaction with gaseous hydrogen at 400 °C and 200 atm pressure to produce gaseous methanol. (d) Solid magnesium undergoes reaction with an aqueous solution of copper(II) ion to give solid copper and an aqueous solution of magnesium ion. (e) An aqueous solution of permanganate ion undergoes reaction with gaseous hydrogen sulfide to give solid manganese dioxide, solid sulfur, aqueous hydroxide ion, and liquid water. (f) Ice melts at 0 °C and 1 atm pressure to give liquid water.
30 b, c, e.
31. (a) $2NH_4Cl(s) + CaO(s) \rightarrow CaCl_2(s) + 2NH_3(g) + H_2O(g)$, (b) $2Al(s) + 3I_2(\text{in } CS_2) \rightarrow 2AlI_3(s)$, (c) $H_2SO_4(aq) + 2HBr(g) \rightarrow Br_2(g) + SO_2(g) + 2H_2O(l)$, (d) $Li_3N(s) + 3H_2O(l) \rightarrow NH_3(g) + 3LiOH(aq)$, (e) $4HNO_3(l) + P_4O_{10}(s) \rightarrow 2N_2O_5(s) + 4HPO_3(s)$.

32. (a) $2As(s) + 3Cl_2(g) \rightarrow 2AsCl_3(s)$, (b) $4HNO_3(l) \rightarrow 4NO_2(g) + O_2(g) + 2H_2O(l)$, (c) $2Cu_2S(s) + 3O_2(g) \xrightarrow{\Delta} 2Cu_2O(s) + 2SO_2(g)$, (d) $NH_4NO_3(s) \xrightarrow{\Delta} N_2O(g) + 2H_2O(g)$, $2NH_4NO_3(s) \xrightarrow{\Delta} 2N_2(g) + 4H_2O(g) + O_2(g)$, (e) $2C_4H_{10}(g) + 13O_2(g) \rightarrow 8CO_2(g) + 10H_2O(l)$.
33. The value of Avogadro's number is 6.022×10^{23} entities/mol. The mole is a counting unit, numerically equal to Avogadro's number.
34. The molecular mass of a chemical compound is calculated by summing the atomic masses of the total number of atoms in the formula of the compound. Molecular mass is expressed in atomic mass units and molar mass is expressed in grams.
35. For a compound, the molar mass is the mass in grams numerically equal to the molecular mass in atomic mass units.
36. The commonly used unit for expressing concentration is molarity.
37. The volume of solution which contains a given number of moles of solute is calculated by multiplying the number of moles by (1/molarity). The number of moles of solute in a given volume of solution is calculated by multiplying the volume by the molarity.
38. The molecular formula of a compound is an integer multiple of the empirical formula. This relationship is usually determined from the molecular or molar mass of the compound.

SKILLS

Skill 4.1 Interpreting chemical formulas

The chemical formula for a compound identifies what elements are present in the compound by the use of the appropriate chemical symbols and states how many atoms of each element are present by the use of subscripts following the respective elemental symbol. For example, the formula for hydrogen peroxide, H_2O_2, shows that the compound contains hydrogen and oxygen and that the formula unit (in this case a molecule) contains two atoms of each element. The formula for water, H_2O, shows that this compound also contains the same elements, but in different amounts--two atoms of hydrogen and one atom of oxygen.

A structural formula will show the number of atoms and the arrangement of the atoms in the compound. For example, the structural formulas for hydrogen peroxide and water are

H - O - O - H H - O - H

The first formula shows that each hydrogen atom is attached to an oxygen atom and that the oxygen atoms are connected to each other. The second formula shows that each hydrogen atom is bonded to the oxygen atom.

Skill Exercise: (a) How many carbon atoms are in a molecule of C_6H_6, benzene? (b) What is the total number of atoms in a formula unit of $Mg(OH)_2$? (c) How many water molecules are in a formula unit of $CuSO_4 \cdot 5H_2O$? (d) What elements make up NH_4NO_3? In what proportion do they do so? (e) What is the ratio of S atoms to O atoms in the sulfate ion, SO_4^{2-}? (f) How many and what type(s) of bonds are in

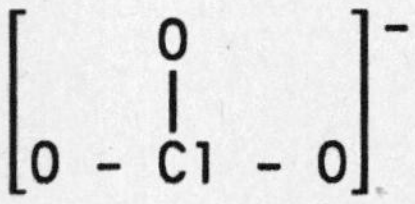

Answers: (a) 6; (b) 5; (c) 5; (d) N, H, and O, 2 N atoms to 4 H atoms to 3 O atoms; (e) 1 S atom to 4 O atoms; (f) three Cl-O bonds.

Skill 4.2 Naming simple inorganic compounds

Several substances have common names. For example,

H_2O	water	NH_3	ammonia

These names are relatively easy to remember and systematic names for these substances are very seldom used.

Each element has a distinctive name and symbol. For example

Zr	zirconium	Zn	zinc	Y	yttrium
Yb	ytterbium	Xe	xenon	V	vanadium
U	uranium	Unh	unnilhexium	W	tungsten

Some of the elements naturally occur as molecules--not as atoms; for example

N_2	molecular nitrogen	P_4	molecular phosphorus
	dinitrogen		tetraphosphorus
	"nitrogen"		"phosphorus"

The systematic name is given by adding a prefix to the elemental name to show the number of atoms that are combined in the molecule. Very few chemists actually call these elements by these systematic names, and usually the simple name of the element represents the molecular form. Occasionally, the distinction between the atomic and molecular forms is made as follows:

N	atomic nitrogen	N_2	molecular nitrogen

In order to specify a certain isotope of an element by name, the name of the element is given, followed by the mass number. For example

^{14}C	carbon-14	^{238}U	uranium-238

The isotopes of hydrogen usually carry their traditional distinctive names:

^{1}H	protium	^{2}H or D	deuterium	^{3}H or T	tritium
	hydrogen-1		hydrogen-2		hydrogen-3

Skill Exercise: Name the following elements: (a) Cl_2, (b) Na, (c) ^{33}S, (d) He, and (e) D. Write the symbols for (f) molecular oxygen, (g) atomic fluorine, (h) potassium, (i) disodium, and (j) chlorine-37. *Answers*: (a) chlorine, dichlorine, or molecular chlorine; (b) sodium; (c) sulfur-33; (d) helium; (e) deuterium; (f) O_2; (g) F; (h) K; (i) Na_2; (j) ^{37}Cl.

Simple cations are formed by the removal of one or more electrons from an atom. The name of a simple cation consists of the name of the element, the ionic charge given as a Roman numeral in parentheses, and the word "ion." For example

Al^{3+}	aluminum(III) ion	Ba^{2+}	barium(II) ion
	aluminum ion		barium ion
Cr^{2+}	chromium(II) ion	Cr^{3+}	chromium(III) ion
Fe^{2+}	iron(II) ion	Fe^{3+}	iron(III) iron
	ferrous ion		ferric ion

Note that for elements that commonly form only one cation, the Roman numeral is omitted. An older system of naming cations is sometimes used in which the names of many of the elements revert to the original Latin or historical names and the ionic charge is indicated by the suffix "ous" for the ion with the lower charge and the suffix "ic" for the ion with the higher charge.

Skill Exercise: Name the following cations: (a) Na^+, (b) Sn^{2+}, (c) Sn^{4+}, (d) Mn^{2+}, and (e) Ca^{2+}. Write the symbols for (f) vanadium(III) ion, (g) silver ion, (h) zinc ion, (i) cuprous ion, and (j) copper(I) ion. *Answers*: (a) sodium ion, (b) tin(II) ion or stannous ion, (c) tin(IV) ion or stannic ion, (d) manganese(II) ion or manganous ion, (e) calcium ion, (f) V^{3+}, (g) Ag^+, (h) Zn^{2+}, (i) Cu^+, (j) Cu^+.

Simple anions are formed by the addition of one or more electrons to an atom. The name of a simple anion consists of the root of the name of the element modified by the suffix "ide" and the word "ion." For example

Cl^-	chloride ion	O^{2-}	oxide ion
C^{4-}	carbide ion	P^{3-}	phosphide ion

Skill Exercise: Name the following anions: (a) F^-, (b) S^{2-}, and (c) N^{3-}. Write the symbols for (d) bromide ion, (e) selenide ion, and (f) iodide ion. *Answers*: (a) fluoride ion, (b) sulfide ion, (c) nitride ion, (d) Br^-, (e) Se^{2-}, (f) I^-.

The name of a binary ionic compound consists of the name of the cation followed by the name of the anion. For example

Al_2O_3	aluminum oxide	$MgCl_2$	magnesium chloride
NaF	sodium fluoride	KI	potassium iodide
$FeCl_2$	iron(II) chloride ferrous chloride	$FeCl_3$	iron(III) chloride ferric chloride
CrO	chromium(II) oxide	Cr_2O_3	chromium(III) oxide
LiH	lithium hydride	BaO_2	barium peroxide

Some binary molecular compounds are named in a similar fashion. For example

BN	boron nitride	H_2S	hydrogen sulfide

If an element forms more than one binary molecular compound with another element, prefixes are used to indicate the number of atoms. For example

NO	nitrogen monoxide [nitrogen(II) oxide]	NO_2	nitrogen dioxide [nitrogen(IV) oxide]
N_2O	dinitrogen monoxide [nitrogen(I) oxide]	N_2O_3	dinitrogen trioxide [nitrogen(III) oxide]
N_2O_4	dinitrogen tetroxide [nitrogen(IV) oxide dimer]	N_2O_5	dinitrogen pentoxide [nitrogen(V) oxide]

A second method that is sometimes used for indicating the relative numbers of atoms in these compounds is shown above by the names in brackets. The numbers in parentheses are the oxidation numbers of the nitrogen atom; see *Skill 10.2 Determining oxidation numbers*.

Skill Exercise: Name the following binary compounds: (a) BaS, (b) AuBr, (c) Cu_2O, (d) MnS_2, (e) ICl_3, (f) SO_3, (g) CaI_2, and (h) ZnF_2. Write the formulas for (i) gold(III) bromide, (j) manganese(II) sulfide, (k) cupric oxide, (l) silver oxide, (m) nitrogen trichloride, and (n) dialuminum hexachloride. *Answers*: (a) barium sulfide, (b) gold(I) bromide, (c) copper(I) oxide or cuprous oxide, (d) manganese(IV) sulfide, (e) iodine trichloride or iodine(III) chloride, (f) sulfur trioxide or sulfur(VI) oxide, (g) calcium iodide, (h) zinc fluoride, (i) $AuBr_3$, (j) MnS, (k) CuO, (l) Ag_2O, (m) NCl_3, (n) Al_2Cl_6.

Bases contain the hydroxide ion, OH^-. The nomenclature of bases is just like that of binary compounds. For example

NaOH	sodium hydroxide	$Mn(OH)_2$	manganese(II) hydroxide

Aqueous solutions of many binary inorganic compounds containing hydrogen are acidic. They are named with the prefix "hydro" before the element name, the suffix "ic" after the root of the element name, and the word "acid." For example

$H_2S(aq)$	hydrosulfuric acid	HCl(aq)	hydrochloric acid
HF(aq)	hydrofluoric acid	HBr(aq)	hydrobromic acid

HI(aq) hydroiodic acid

The following table lists some of the common oxygen-containing acids and the anions that are formed from them:

H_2CO_3	carbonic acid	HCO_3^-	hydrogen carbonate ion
		CO_3^{2-}	carbonate ion
HNO_3	nitric acid	NO_3^-	nitrate ion
HNO_2	nitrous acid	NO_2^-	nitrite ion
$HClO_4$	perchloric acid	ClO_4^-	perchlorate ion
H_3PO_4	phosphoric acid	$H_2PO_4^-$	dihydrogen phosphate ion
		HPO_4^{2-}	hydrogen phosphate ion
		PO_4^{3-}	phosphate ion
H_2SO_4	sulfuric acid	HSO_4^-	hydrogen sulfate ion
		SO_4^{2-}	sulfate ion
H_2SO_3	sulfurous acid	HSO_3^-	hydrogen sulfite ion
		SO_3^{2-}	sulfite ion
CH_3COOH	acetic acid	CH_3COO^-	acetate ion

The nomenclature of salts of these acids consists of the cation name followed by the anion name. For example

Na_2CO_3	sodium carbonate	$KHSO_4$	potassium hydrogen sulfate
NH_4NO_3	ammonium nitrate	$PbCO_3$	lead carbonate

Skill Exercise: Name the following compounds: (a) $Ca(CH_3COO)_2$, (b) $Ba(NO_2)_2$, (c) $(NH_4)_2CO_3$, (d) $Ca(OH)_2$, and (e) $NaClO_4$. Write the formulas for (f) lithium hydroxide, (g) sulfuric acid, (h) sodium hydrogen carbonate, (i) ammonium sulfate, and (j) silver nitrate. *Answers*: (a) calcium acetate, (b) barium nitrite, (c) ammonium carbonate, (d) calcium hydroxide, (e) sodium perchlorate, (f) LiOH, (g) H_2SO_4, (h) $NaHCO_3$, (i) $(NH_4)_2SO_4$, (j) $AgNO_3$.

Additional nomenclature of inorganic compounds is discussed in *Skill 31.1 Naming complexes* and in *Skill 20.3 Naming oxoacids and anions of oxoacids*.

Skill 4.3 Writing and balancing chemical equations

Each year, in general chemistry courses in high schools and colleges, thousands of students prepare small samples of oxygen gas by heating potassium chlorate in the presence of a catalyst, manganese dioxide. It can be shown that potassium chloride is also a product of this reaction. The chemical equation describing this reaction is written by placing the formula of the reactant ($KClO_3$) on the left side of the arrow used to separate the reactants from the products and the formulas of the products (O_2 and KCl) on the right side of the arrow, to give the following unbalanced chemical equation:

$$KClO_3 \rightarrow KCl + O_2$$

Because a chemical equation is a statement of the law of conservation of mass, the equation must be balanced by putting coefficients in front of the various formulas so that the total number of atoms of each kind on the reactants side of the arrow is equal to the total number of atoms of each kind on the products side. For the reaction above the balanced equation is

$$2KClO_3 \rightarrow 2KCl + 3O_2$$

because there are two K atoms on each side, two Cl atoms on each side, and six O atoms on each side.

To complete the equation, a description of the physical state of each substance is added and the additional information about the reaction is indicated over the arrow. The complete equation for this reaction is

$$2KClO_3(s) \xrightarrow[\Delta]{MnO_2} 2KCl(s) + 3O_2(g)$$

Such a chemical equation can be read as a "sentence": Solid potassium chlorate, when heated in the presence of a catalyst--manganese dioxide, yields solid potassium chloride and gaseous oxygen.

Skill Exercise: Write a balanced chemical equation for each of the following reactions: (a) an aqueous solution of silver nitrate reacts with an aqueous solution of calcium chloride to form a precipitate of silver chloride and an aqueous solution of calcium nitrate and (b) an aqueous solution of sodium hydroxide reacts with an aqueous solution of sulfuric acid to form water and an aqueous solution of sodium sulfate. Write the "sentence" form for each of the following equations: (c) $Zn(s) + CuSO_4(aq) \rightarrow ZnSO_4(aq) + Cu(s)$ and (d) $2NaI(aq) + Br_2(l) \rightarrow 2NaBr(aq) + I_2(s)$. *Answers*: (a) $2AgNO_3(aq) + CaCl_2(aq) \rightarrow 2AgCl(s) + Ca(NO_3)_2(aq)$, (b) $2NaOH(aq) + H_2SO_4(aq) \rightarrow 2H_2O(l) + Na_2SO_4(aq)$, (c) solid zinc plus an aqueous solution of copper(II) sulfate yields an aqueous solution of zinc sulfate plus copper metal, (d) an aqueous solution of sodium iodide plus liquid bromine yields an aqueous solution of sodium bromide plus solid iodine.

It is important to always work with a balanced chemical equation. Until more specialized methods for balancing chemical equations are introduced, the "inspection method" will be used. For example, to balance the chemical equation for the burning of propane, C_3H_8, in air to give water and carbon dioxide

$$C_3H_8(g) + O_2(g) \xrightarrow{\text{not balanced}} CO_2(g) + H_2O$$

we begin with the most complicated formula, C_3H_8, on the left, and add coefficients so that there are three C atoms and eight H atoms on the right.

$$C_3H_8(g) + O_2(g) \xrightarrow{\text{not balanced}} 3CO_2(g) + 4H_2O(l)$$

There are ten O atoms on the right. We balance the number of oxygen atoms by using 5 in front of the O_2 giving

$$C_3H_8(g) + 5O_2(g) \rightarrow 3CO_2(g) + 4H_2O(l)$$

The equation is balanced because

on the left side:	3 C atoms	8 H atoms	(5)(2) = 10 O atoms
on the right side:	3 C atoms	(4)(2) = 8 H atoms	(3)(2) + 4 = 10 O atoms

Skill Exercise: Balance the following chemical equations: (a) $N_2(g) + H_2(g) \rightarrow NH_3(g)$, (b) $C_2H_5OH(l) + O_2(g) \rightarrow CO_2(g) + H_2O(l)$, (c) $C_{12}H_{22}O_{11}(s) + O_2(g) \rightarrow CO_2(g) + H_2O(l)$, (d) $Al_2O_3(s) + C(s) + Cl_2(g) \rightarrow AlCl_3(s) + CO(g)$, (e) $MgO(s) + Si(s) \rightarrow Mg(s) + SiO_2(s)$, (f) $H_3PO_3(l) \rightarrow H_3PO_4(l) + PH_3(g)$, and (g) $Pb(s) + H_2O(l) + O_2(g) \rightarrow Pb(OH)_2(aq)$. *Answers*: (a) $N_2 + 3H_2(g) \rightarrow 2NH_3(g)$, (b) $C_2H_5OH(l) + 3O_2(g) \rightarrow 2CO_2(g) + 3H_2O(l)$, (c) $C_{12}H_{22}O_{11}(s) + 12O_2(g) \rightarrow 12CO_2(g) + 11H_2O(l)$, (d) $Al_2O_3(s) + 3C(s) + 3Cl_2(g) \rightarrow 2AlCl_3(s) + 3CO(g)$, (e) $2MgO(s) + Si(s) \rightarrow 2Mg(s) + SiO_2(s)$, (f) $4H_3PO_3(l) \rightarrow 3H_3PO_4(l) + PH_3(g)$, (g) $2Pb(s) + 2H_2O(l) + O_2(g) \rightarrow 2Pb(OH)_2(aq)$.

Further discussion of the interpretation of chemical equations is presented in *Skill 6.2 Interpreting chemical equations involving gases* and *Skill 5.1 Interpreting chemical equations*; balancing chemical equations is discussed in *Skill 16.2 Balancing redox equations: oxidation number method* and *Skill 24.1 Balancing redox equations: half-reaction method.*

Skill 4.4 Finding the number of moles

The number of moles (*n*) of a substance is related to the number of particles (*N*) by Avogadro's number (6.022×10^{23} entities/mol):

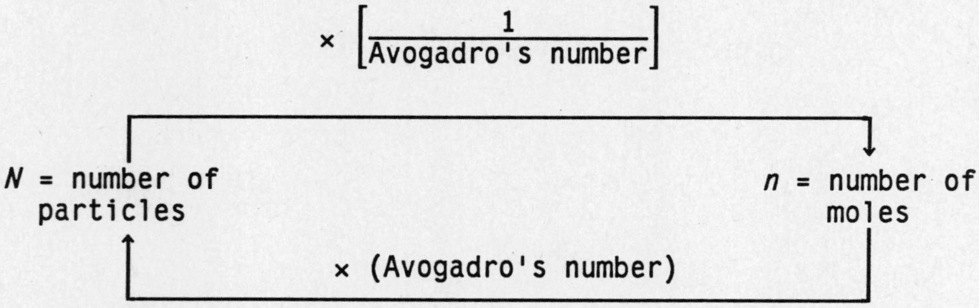

Each of these relationships can be written in mathematical equation form; a sample calculation follows for each conversion:

$$n = N\left[\frac{1}{\text{Avogadro's number}}\right]$$

$$= (5 \times 10^{25}\ CO_2\ \text{molecules})\left[\frac{1\ \text{mol}\ CO_2}{6.022 \times 10^{23}\ CO_2\ \text{molecules}}\right]$$

$$= 80\ \text{mol}\ CO_2$$

$$N = n(\text{Avogadro's number}) = (3.0\ \text{mol He})\left[\frac{6.022 \times 10^{23}\ \text{He atoms}}{1\ \text{mol He}}\right]$$

$$= 1.8 \times 10^{24}\ \text{He atoms}$$

The number of moles in a given mass of a substance is related to the mass (*m*) by the molar mass of the substance:

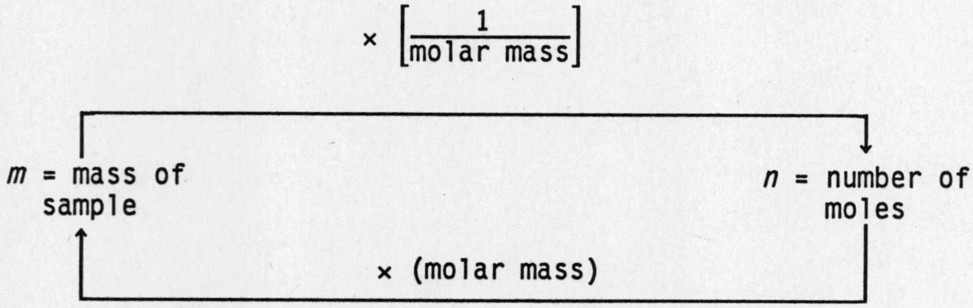

Each of these relationships can be written in mathematical equation form; a sample calculation follows for each conversion:

$$n = m\left[\frac{1}{\text{molar mass}}\right] = (15.3\ \text{g S})\left[\frac{1\ \text{mol S}}{32.06\ \text{g S}}\right] = 0.477\ \text{mol S}$$

$$m = n(\text{molar mass}) = (0.39\ \text{mol NaCl})\left[\frac{58.44\ \text{g NaCl}}{1\ \text{mol NaCl}}\right] = 23\ \text{g NaCl}$$

The number of moles of solute in a solution is related to the volume of the solution (V) by the molarity (number of moles of solute/volume of solution in liters):

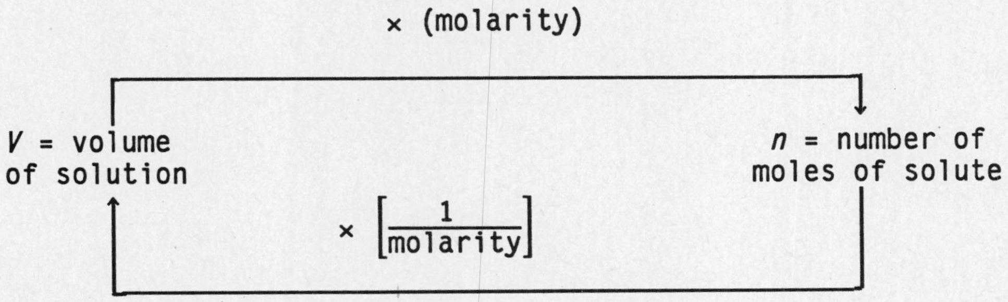

Each of these relationships can be written in mathematical equation form; a sample calculation follows for each conversion:

$$n = V(\text{molarity}) = (0.50\ \text{L soln})\left[\frac{3.0\ \text{mol}\ H_2SO_4}{1\ \text{L soln}}\right] = 1.5\ \text{mol}\ H_2SO_4$$

$$V = n\left[\frac{1}{\text{molarity}}\right] = (0.25\ \text{mol NaOH})\left[\frac{1\ \text{L soln}}{0.105\ \text{mol NaOH}}\right] = 2.4\ \text{L soln}$$

For ideal gases (see Chapter 6), the number of moles of gas in a given volume (V) measured at a known pressure (P) and absolute temperature (T) is given by the ideal gas law,

$$n = \frac{PV}{RT}$$

where R is the ideal gas constant (R = 0.0821 L atm/K mol). Sample calculations using the ideal gas law are presented in *Example 6.10* and *Example 6.11*.

For electrochemical reactions (see Chapter 24), the number of moles of electrons involved in an oxidation-reduction reaction is related to the amount of electrical charge by the faraday (96,484.56 C/mol):

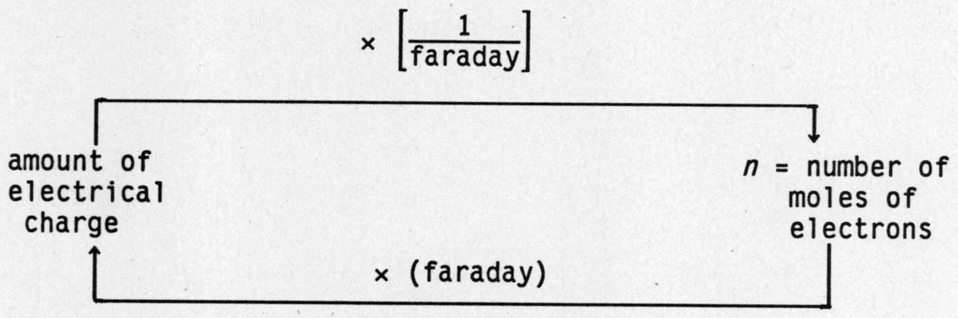

All of these relationships are summarized in the following sketch:

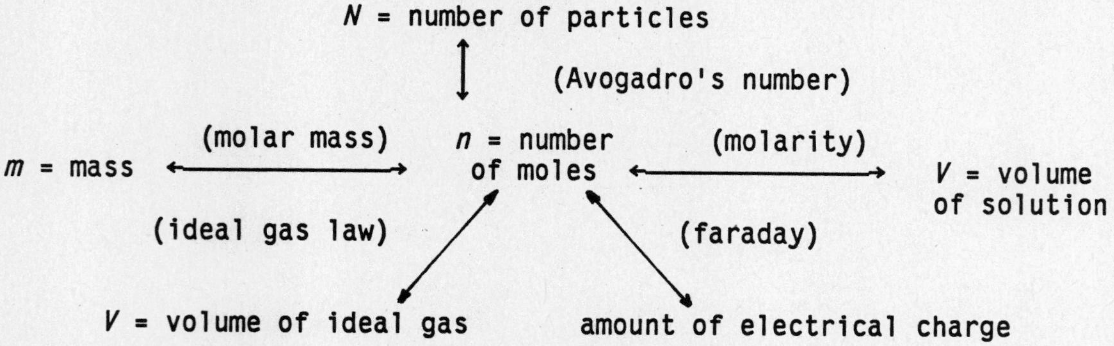

Skill Exercise: Find the number of moles of substance in each of the following samples: (a) 1.0 g H_2, (b) 4.26 x 10^{25} molecules of N_2, (c) 2.63 kg O_3, and (d) 37 mL of 0.517 M HCl. *Answers*: (a) 0.50 mol H_2, (b) 70.7 mol N_2, (c) 54.8 mol O_3, (d) 0.019 mol HCl.

NUMERICAL EXERCISES

Example 4.1 Molecular mass and molar mass

Find the molecular mass and the molar mass for $Fe(NO_3)_3 \cdot 9H_2O$.

The chemical formula indicates that there are 1 Fe atom, 3 N atoms, (3)(3) + 9 = 18 O atoms, and (9)(2) = 18 H atoms in the compound. The molecular mass is

	number of atoms		atomic mass (u/atom)		mass (u)
Fe	1	x	55.85	=	55.85
N	3	x	14.01	=	42.03
O	18	x	16.00	=	288.0
H	18	x	1.01	=	18.2
molecular mass of $Fe(NO_3)_3 \cdot 9H_2O$				=	404.1 u

The molar mass is the mass in grams numerically equal to the molecular mass in atomic mass units. Thus the molar mass of $Fe(NO_3)_3 \cdot 9H_2O$ is 404.1 g (often written 404.1 g/mol).

Exercise 4.1 Find the molecular mass and the molar mass of $Pb(BF_4)_2$.
Answers: 380.8 u, 380.8 g/mol.

Example 4.2 Avogadro's number and moles

The molar mass of gold is 196.97 g. What is the average mass (in grams) of one atom?

The average mass is determined by dividing the molar mass by the number of particles in the sample (Avogadro's number):

$$\frac{(196.97 \text{ g Au}/1 \text{ mol Au})}{(6.022 \times 10^{23} \text{ Au atoms}/1 \text{ mol Au})} = 3.2708 \times 10^{-22} \text{ g/atom}$$

The average mass of a gold atom is 3.2708×10^{-22} g. Note that the value of Avogadro's number is given to five significant figures because of the number of significant figures in the mass value.

Exercise 4.2 What is the average mass (in grams) of one atom of neon? *Answer*: 3.3509×10^{-23} g.

*Example 4.3 Avogadro's number and moles**

A sample of $CaCl_2$ contains 4×10^{30} formula units. How many moles of $CaCl_2$ are in the sample?

This problem is analyzed using the comprehensive problem-solving approach presented in *Skill 2.7 Solving problems*.

1. Study the problem and be sure you understand it.
   (a) What is unknown?
   The number of moles of $CaCl_2$.
   (b) What is known?
   The number of formula units in the sample.
2. Decide how to solve the problem.
   (a) What is the connection between the known and the unknown?
   The connection between the number of moles and the number of particles is Avogadro's number; see *Skill 4.4 Finding the number of moles*.
   (b) What is necessary to make the connection?
   Multiplication by (1/Avogadro's number) will convert the number of particles into the number of moles.

3. Set up the problem and solve it.

The number of moles is

$$n = N\left[\frac{1}{\text{Avogadro's number}}\right]$$

$$= (4 \times 10^{30} \text{ formula units})\left[\frac{1 \text{ mol}}{6.022 \times 10^{23} \text{ formula units}}\right]$$

$$= 7 \times 10^{6} \text{ mol}$$

4. Check the result.

(a) Are significant figures and the location of the decimal point correct?

Yes.

(b) Did the answer come out in the correct units?

Yes.

(c) Is the answer reasonable?

Yes. Because there were more than Avogadro's number of formula units, there is more than one mole of substance.

Exercise 4.3 How many molecules of H_2SO_4 are in 1.5 µmol of H_2SO_4? *Answer*: 9.0×10^{17} molecules.

Example 4.4 Molar mass and moles*

A chemical reaction requires 4.25 mol of potassium chlorate, $KClO_3$. What mass of $KClO_3$ is needed?

This problem is analyzed using the comprehensive problem-solving approach presented in *Skill 2.7 Solving problems*.

1. Study the problem and be sure you understand it.

(a) What is unknown?

The mass of $KClO_3$.

(b) What is known?

The number of moles.

2. Decide how to solve the problem.

(a) What is the connection between the known and the unknown?

The connection between the number of moles and the mass is the molar mass; see *Skill 4.4 Finding the number of moles*.

(b) What is necessary to make the connection?

Multiplication by (molar mass) will convert the number of moles into mass.

3. Set up the problem and solve it.

The mass of $KClO_3$ needed is

$$m = (4.25 \text{ mol } KClO_3)\left[\frac{122.55 \text{ g } KClO_3}{1 \text{ mol } KClO_3}\right] = 521 \text{ g } KClO_3$$

4. Check the result.
   (a) Are significant figures and the location of the decimal point correct?
   Yes.
   (b) Did the answer come out in the correct units?
   Yes.
   (c) Is the answer reasonable?
   Yes. If one mole of $KClO_3$ weighs roughly 120 g, then a little over four moles should weigh about 500 g.

Exercise 4.4 How many moles are contained in 10.0 g of FeS_2? *Answer*: 0.0834 mol FeS_2.

Example 4.5 Molar mass, Avogadro's number, and moles

A sealed flask is marked "3.0 g NO_2." How many (a) moles, (b) molecules, (c) N atoms, (d) O atoms, and (e) total atoms are present in the flask?

(a) The number of moles is found by using the mass and molar mass:

$$n = m\left[\frac{1}{\text{molar mass}}\right] = (3.0 \text{ g } NO_2)\left[\frac{1 \text{ mol } NO_2}{46.01 \text{ g } NO_2}\right] = 0.065 \text{ mol } NO_2$$

(b) The number of molecules in the sample is found by multiplying the number of moles by Avogadro's number:

$$N = n(\text{Avogadro's number}) = (0.065 \text{ mol})\left[\frac{6.022 \times 10^{23} \text{ molecules}}{1 \text{ mol}}\right]$$

$$= 3.9 \times 10^{22} \text{ molecules}$$

(c) Each NO_2 molecule contains 1 atom of N, so in the sample

$$(3.9 \times 10^{22} \text{ molecules } NO_2)\left[\frac{1 \text{ N atom}}{1\ NO_2 \text{ molecule}}\right] = 3.9 \times 10^{22} \text{ atoms of N}$$

(d) Each NO_2 molecule contains 2 atoms of O, so in the sample

$$(3.9 \times 10^{22} \text{ molecules } NO_2)\left[\frac{2 \text{ O atoms}}{1 \text{ NO molecule}}\right] = 7.8 \times 10^{22} \text{ atoms of O}$$

(e) Each NO_2 molecule contains a total of 3 atoms, so in the sample

$$(3.9 \times 10^{22} \text{ molecules } NO_2)\left[\frac{3 \text{ atoms}}{1 \; NO_2 \text{ molecule}}\right] = 1.2 \times 10^{23} \text{ atoms}$$

Exercise 4.5 A reagent bottle contains 2.5 mol of NaCl. How many (a) grams, (b) formula units, (c) sodium ions, (d) chloride ions, and (e) total ions are present? *Answers*: (a) 150 g, (b) 1.5×10^{24} formula units, (c) 1.5×10^{24} sodium ions, (d) 1.5×10^{24} chloride ions, (e) 3.0×10^{24} ions.

Example 4.6 Molarity

A lead nitrate solution was prepared by mixing 173 g $Pb(NO_3)_2$ with 999 g of water to produce 871 mL of solution. Calculate the molarity of this solution.

The number of moles of solute--the $Pb(NO_3)_2$--is

$$n = m\left[\frac{1}{\text{molar mass}}\right] = (173 \text{ g } Pb(NO_3)_2)\left[\frac{1 \text{ mol } Pb(NO_3)_2}{331.23 \text{ g } Pb(NO_3)_2}\right]$$

$$= 0.522 \text{ mol } Pb(NO_3)_2$$

Molarity is found by dividing the number of moles of solute by the solution volume in liters:

$$\text{molarity} = \frac{\text{number of moles of solute}}{\text{volume of solution in liters}}$$

$$= \frac{0.522 \text{ mol } Pb(NO_3)_2}{(871 \text{ mL})(1 \text{ L}/1000 \text{ mL})} = 0.599 \text{ mol } Pb(NO_3)_2/\text{L}$$

The solution is 0.599 M.

Exercise 4.6 Exactly two liters of an aqueous solution contained 75.0 g of ethyl alcohol, CH_3CH_2OH. What is the concentration of the solution expressed in molarity? *Answer*: 0.815 M.

Example 4.7 Molarity and moles*

A bottle in the laboratory is marked "0.15 M $AgNO_3$." How many moles of silver nitrate are in 250 mL of this solution?

This problem is analyzed using the comprehensive problem-solving approach presented in *Skill 2.7 Solving problems*.

1. Study the problem and be sure you understand it.
   (a) What is unknown?
   The number of moles of $AgNO_3$ in the sample.
   (b) What is known?
   The concentration of the solution and the volume of the sample.

2. Decide how to solve the problem.
   (a) What is the connection between the known and the unknown?
   The connection between the number of moles of solute and the volume of solution is the molarity; see *Skill 4.4 Finding the number of moles*.
   (b) What is necessary to make the connection?
   Multiplication by the molarity will give the number of moles of solute in the given volume of solution.
3. Set up the problem and solve it.
   The number of moles of $AgNO_3$ is

$$n = V(\text{molarity}) = (250 \text{ mL soln})\left[\frac{1 \text{ L}}{1000 \text{ mL}}\right]\left[\frac{0.15 \text{ mol } AgNO_3}{1 \text{ L soln}}\right]$$

$$= 0.038 \text{ mol } AgNO_3$$

4. Check the result.
   (a) Are significant figures and the location of the decimal point correct?
   Yes.
   (b) Did the answer come out in the correct units?
   Yes.
   (c) Is the answer reasonable?
   Yes. One liter of a 1 M solution would contain one mole. Because there is less than one liter and the concentration is less than 1 M, there should be less than one mole.

Exercise 4.7 Find the volume of a 0.664 M solution of KBr needed to contain exactly 0.01 mol of KBr. *Answer*: 15.1 mL.

Example 4.8 Percentage composition

Determine the percentage composition of the iron compound given in *Example 4.1*.

The percentage composition is found by dividing the mass of each element in one formula unit (or in one mole) of the compound by the molecular mass (or molar mass) of the compound and multiplying by 100:

$$\%\text{ Fe} = \left[\frac{55.85 \text{ u}}{404.00 \text{ u}}\right](100) = 13.82\% \text{ Fe} \qquad \%\text{ O} = \left[\frac{287.99 \text{ u}}{404.00 \text{ u}}\right](100) = 71.285\% \text{ O}$$

$$\%\text{ N} = \left[\frac{42.02 \text{ u}}{404.00 \text{ u}}\right](100) = 10.40\% \text{ N} \qquad \%\text{ H} = \left[\frac{18.14 \text{ u}}{404.00 \text{ u}}\right](100) = 4.490\% \text{ H}$$

The sum of the four percentages equal 100.00%.

Exercise 4.8 Determine the percentage composition of the lead compound given in Exercise 4.1. *Answer*: 54.41 mass % Pb, 5.678 mass % B, 39.91 mass % F.

Example 4.9 Percentage composition

Calculate the percentage of phosphorus in $Ca_3(PO_4)_2$.

The molar mass of $Ca_3(PO_4)_2$ is 310.18 g and one mole of $Ca_3(PO_4)_2$ contains 61.94 g of phosphorus. The percentage of phosphorus is

$$\% \text{ P} = \left[\frac{61.94 \text{ g}}{310.18 \text{ g}}\right](100) = 19.97\% \text{ P}$$

Exercise 4.9 Determine the percentage of phosphorus in $Ca(H_2PO_4)_2$. *Answer*: 26.46% P.

Example 4.10 Empirical formula

A hydrocarbon contains 80.1 mass % C and 19.9 mass % H. What is the empirical formula for this compound?

This problem is best solved by setting up a table in which the number of moles of each type of atom is calculated on the basis of 100.0 g of the compound, the mole ratio is calculated, and the mole ratio is scaled to integer values if necessary. The entries for the mole ratio in the table are found by dividing the respective values for number of moles by the smallest value of number of moles.

	C	H
mass	80.1 g	19.9 g
number of moles	$(80.1 \text{ g C})\left[\frac{1 \text{ mol C}}{12.0 \text{ g C}}\right]$	$(19.9 \text{ g H})\left[\frac{1 \text{ mol H}}{1.01 \text{ g H}}\right]$
	= 6.68 mol C	= 19.7 mol H
mole ratio n/n_C	$\frac{6.68}{6.68} = 1.00$	$\frac{19.7}{6.68} = 2.95$
relative number of moles	1	3

The empirical formula is CH_3.

Exercise 4.10 Find the empirical formula of a hydrocarbon containing 82.7 mass % C and 17.3 mass % H. *Answer*: C_2H_5.

Example 4.11 Molecular formula

The approximate molar mass of the hydrocarbon considered in *Example 4.10* is 29 g/mol. What is the molecular formula and exact molar mass of this compound?

The molecular mass for the empirical formula is 12.011 + 3(1.0080) = 15.035 u. A comparison of this value with the approximate molecular mass of 29 u shows that the molecular formula is double the empirical formula:

$$\frac{29\ \text{u}}{15.035\ \text{u}} = 1.9$$

so the true formula is C_2H_6 and the exact molar mass is

$$(2)(15.035\ \text{g/mol}) = 30.070\ \text{g/mol}$$

Exercise 4.11 Suppose the approximate molar mass of the hydrocarbon considered in Exercise 4.10 is 60 g/mol. What is the molecular formula and exact molar mass of this compound? *Answer*: C_4H_{10}, 58.123 g/mol.

PRACTICE TEST

1 (5 points). Match the correct term to each definition.

(a) the amount of substance of a system which contains as many elementary entities as there are atoms in 12 g of carbon-12
(b) the smallest particle of an element that can participate in a chemical reaction
(c) the simplest ratio in which the atoms or ions are combined in an ionic compound
(d) formed from a neutral atom by the loss or addition of one or more electrons
(e) the smallest particle of a pure substance that has the composition and chemical properties of that substance and is capable of independent existence

(i) atom
(ii) formula unit
(iii) ion
(iv) mole
(v) molecule

2 (35 points). Choose the response that best completes each statement.

(a) A binary compound
  (i) always has a cation and an anion
  (ii) always exists as a diatomic molecule
  (iii) has a molar mass based on its true formula
  (iv) always dissociates upon dissolution

(b) The component in solution that is present in the largest amount is usually the
  (i) solute (ii) solvent
  (iii) polyatomic molecule (iv) catalyst

(c) An ionic compound dissolves in water by the process of
(i) ionization (ii) dissociation
(iii) forming a formula unit (iv) catalysis

(d) The chemical formula that represents the actual number of atoms in the formula unit or molecule is the
(i) molecular formula (ii) empirical formula
(iii) simplest formula (iv) stoichiometric coefficient

(e) A substance that affects the rate of a reaction, but is recovered from the reaction unchanged, is
(i) an oxidizing agent (ii) a reducing agent
(iii) a catalyst (iv) a polymer

(f) The elements present in baking soda, sodium hydrogen carbonate, are
(i) K, C, O (ii) H, C, O
(iii) Na, H, C (iv) Na, O, C, H

(g) The molecular mass of H_3PO_4 is
(i) 48 u (ii) 98 g
(iii) 49 u (iv) 98 u

(h) The mass of one mole of $Fe(OH)_3$ is
(i) 73 g (ii) 75 g
(iii) 107 g (iv) none of these answers

(i) One millimole of HNO_3 weighs
(i) 6.3×10^{-2} g (ii) 6.3×10^{4} g
(iii) 1.26 g (iv) 6.3 mg

(j) The molarity of a solution containing 8.00 g methyl alcohol, CH_3OH, in 500. mL solution is
(i) 4.00 M (ii) 0.400 M
(iii) 0.250 M (iv) 0.500 M

(k) The molecular masses of chlorine and oxygen are 71 u and 32 u, respectively. Which of the following statements is true?
(i) A mole of chlorine contains more molecules than a mole of oxygen.
(ii) A mole of chlorine is heavier than a mole of oxygen.
(iii) A mole of oxygen contains more molecules than a mole of chlorine.
(iv) A mole of oxygen is heavier than a mole of chlorine.

(l) In 1 g of nitrogen, 1 g of oxygen, and 1 g of sulfur there are
(i) the same number of atoms of each element
(ii) more atoms of nitrogen than of sulfur or oxygen

(iii) fewer atoms of nitrogen than of sulfur or oxygen
(iv) more atoms of sulfur than of oxygen and nitrogen

(m) The number of molecules in 0.020 g of SO_3 is
(i) 0.00025 molecules (ii) 4.2×10^{-28} molecules
(iii) 1.2×10^{22} molecules (iv) 1.5×10^{20} molecules

(n) How many atoms are in 0.10 mol of SO_2?
(i) 6.0×10^{23} (ii) 18×10^{-23}
(iii) 6.0×10^{22} (iv) 1.8×10^{23}

(o) The compound having the greatest percentage by mass of phosphorus is
(i) H_3PO_2 (ii) H_3PO_3
(iii) H_3PO_4 (iv) PCl_5

(p) An organic molecule has the empirical formula C_2H_4O. Its molecular mass is about 90 u. Its molecular formula is
(i) $CH_2O_{0.5}$ (ii) C_2H_4O
(iii) $C_4H_8O_2$ (iv) $C_6H_{12}O_3$

(q) A pure oxide of chlorine was found to have a molecular mass of 67.5 u. The mass % of oxygen in it must be
(i) 47% (ii) 67%
(iii) 51% (iv) none of these answers

(r) The coefficients for H^+ and CO_2 that balance the equation

$$C_2O_4^{2-} + MnO_4^- + H^+ \rightarrow Mn^{2+} + CO_2(g) + H_2O(l)$$

are
(i) 8, 2 (ii) 16, 10
(iii) 2, 2 (iv) none of these answers

3 (10 points). Write the chemical formula for each of the following substances:
(a) calcium perchlorate (b) boron trifluoride
(c) barium hydroxide (d) oxide ion
(e) gold(III) bromide (f) potassium
(g) antimony (h) sodium sulfide
(i) hydrosulfuric acid (j) lithium sulfite

4 (10 points). Name each of the following substances:
(a) W (b) Mg (c) ICl (d) Bi_2O_3
(e) KCN (f) $BaCr_2O_7$ (g) SF_6 (h) $HNO_2(aq)$
(i) NH_4NO_3 (j) Ag_2HPO_4

5 (5 points). Balance the following chemical equation:

$MnO_2(s) + HCl(aq) \rightarrow MnCl_2(aq) + Cl_2(g) + H_2O(l)$

Write a sentence describing this reaction in words.

6 (5 points). What would be the cation-to-anion ratios in ionic compounds containing an anion with a -2 charge and cations with (a) a +1 charge, (b) a +2 charge, and (c) a +3 charge?

7 (10 points). The formula for the mineral albite is $NaAlSi_3O_8$. Determine the percentage composition.

8 (10 points). Topaz contains 30.0 mass % Al, 15.6 mass % Si, 53.3 mass % O, and 1.1 mass % H. Find the empirical formula for this substance.

9 (10 points). A volumetric flask contains 100.00 mL of 0.250 M C_2H_5OH. Considering only the solute, how many (a) moles of C_2H_5OH, (b) molecules of C_2H_5OH, (c) atoms of H, (d) atoms of C, and (e) total atoms are present?

Answers to practice test

1. (a) iv, (b) i, (c) ii, (d) iii, (e) v.
2. (a) iii, (b) ii, (c) ii, (d) i, (e) iii, (f) iv, (g) iv, (h) iii, (i) i, (j) iv, (k) ii, (l) ii, (m) iv, (n) iv, (o) i, (p) iii, (q) i, (r) iii.
3. (a) $Ca(ClO_4)_2$, (b) BF_3, (c) $Ba(OH)_2$, (d) O^{2-}, (e) $AuBr_3$, (f) K, (g) Sb, (h) Na_2S, (i) $H_2S(aq)$, (j) Li_2SO_3.
4. (a) tungsten, (b) magnesium, (c) iodine monochloride, (d) bismuth(III) oxide, (e) potassium cyanide, (f) barium dichromate, (g) sulfur hexafluoride, (h) nitrous acid, (i) ammonium nitrate, (j) silver monohydrogen phosphate.
5. $MnO_2(s) + 4HCl(aq) \rightarrow MnCl_2(aq) + Cl_2(g) + 2H_2O(l)$; solid manganese dioxide undergoes reaction with hydrochloric acid to give an aqueous solution of manganese(II) chloride, gaseous chlorine, and liquid water.
6. (a) 2 cations to 1 anion, (b) 1 cation to 1 anion, (c) 2 cations to 3 anions.
7. 8.767 mass % Na, 10.29 mass % Al, 32.13 mass % Si, 48.81 mass % O.
8. $Al_2SiO_6H_2$.
9. (a) 0.0250 mol, (b) 1.51×10^{22} molecules, (c) 9.06×10^{22} H atoms, (d) 3.02×10^{22} C atoms, (e) 1.36×10^{23} atoms.

ADDITIONAL NOTES

CHAPTER 5

CHEMICAL REACTIONS AND STOICHIOMETRY

CHAPTER OVERVIEW

A method of classifying chemical reactions is introduced. A discussion of chemical stoichiometry (the calculation of quantitative relationships for reactants and products involved in chemical changes) is given.

COMPETENCIES

Definitions for the following terms should be learned:

- actual yield
- chemical reactivity
- combination reaction
- decomposition reaction
- displacement reaction
- limiting reactant
- net ionic equation
- partner-exchange reaction
- percent yield
- precipitate
- precipitation
- spectator ions
- stoichiometric amount
- stoichiometry
- theoretical yield

General concepts and skills that should be learned include

- classifying chemical reactions
- writing and using net ionic equations
- interpreting a chemical equation on a mass basis
- interpreting a chemical equation on a molar basis
- determining mole ratios
- using the four-step method for solving stoichiometry problems

Numerical exercises that should be understood include

- showing that a balanced chemical equation obeys the law of conservation of mass
- calculating the amounts of reactants and products involved in a chemical reaction

- calculating the percent yield of a reaction
- calculating the amount of products formed in the presence of a limiting reactant
- calculating the amount of a desired product formed in a series of reactions

QUESTIONS

1. Complete the following sentences by filling in the various blanks with these terms: limiting reactant, net ionic equation, percent yield, precipitate, spectator ions, stoichiometric amounts, and theoretical yield.

The amounts of reactants in a chemical reaction may be adjusted to be the (a)________--the exact amounts of substances required according to the equation--or so that one reactant is the (b)________--the single reactant that determines the maximum amount of product formed because it is in the smallest amount. The (c)________ is the maximum amount of product that can be obtained from the known amounts of reactants, whereas the (d)________ is the actual yield obtained divided by the theoretical yield converted to a percentage. Often, when writing chemical equations involving the formation of a (e)________, the (f)________ is written, in which only the ions actually involved in the reaction are included and the (g)________ are omitted.

2. Match the correct term to each definition.

(a) a chemical reaction in which two compounds interact as follows: $AC + BD \rightarrow AD + BC$

(b) a chemical reaction in which a single compound breaks down to give two or more other substances: $C \rightarrow A + B$

(c) a chemical reaction in which two reactants combine to give a single product: $A + B \rightarrow C$

(d) a chemical reaction in which one substance takes the place of part of a compound: $A + BC \rightarrow AC + B$

(i) combination reaction
(ii) decomposition reaction
(iii) displacement reaction
(iv) partner-exchange reaction

3. Which of the following statements are true? Rewrite each false statement so that it is correct.

(a) A limiting reactant is present in the largest stoichiometric amount and determines the maximum amount of products that can be formed.

(b) Only two products can be formed in a decomposition reaction.
(c) The theoretical yield of a reaction is found by dividing the actual yield of the product by the maximum amount possible and multiplying by 100.
(d) The mass relationships of reactants and products in a chemical equation are true no matter what units are used to express the mass because molecular masses are derived from a relative scale.
(e) A valid assumption is that mass is conserved in "ordinary" chemical reactions.
(f) Spectator ions have mass and charge and participate in a reaction.
(g) The net ionic equation shows only the species involved in the chemical change and not all of the ions present.

4. Distinguish clearly between (a) limiting and excess reactants, (b) theoretical and actual yields, and (c) overall and net ionic chemical equations.

5. State the law of conservation of mass. Is this law obeyed by "ordinary" chemical reactions? Explain.

6. List the names of the four types of chemical reactions. Give a definition for each type.

7. Write balanced chemical equations to describe each of the following reactions:
(a) the partner-exchange reaction between barium oxide and sulfuric acid
(b) the combination reaction between barium oxide and sulfur trioxide
(c) the partner-exchange reaction between sodium carbonate and calcium hydroxide
(d) the displacement reaction between magnesium and hydrochloric acid
(e) the electrolytic decomposition of water

8. What is an activity series of metals? Using chemical equations, compare the information summarized in the activity series given in Table 5.2 of the text about (a) potassium (an active metal) and (b) copper (a rather inactive metal).

9. How does a net ionic equation differ from the overall chemical equation? In what way other than a mass balance, must a net ionic equation be balanced?

10. What is the basis for writing the formula of a substance as separate ions or not?

11. Look up the definition of stoichiometry in a dictionary. In what ways is this definition similar to or different from the one used in this book? What is the historical basis of this word?

12. How does a chemical equation provide the connection between the known and unknown facts in a stoichiometry problem?

13. How are mole ratios determined from a chemical equation? Write as many mole ratios as you can for the following equation:

$$2Al(OH)_3(s) + 3H_2SO_4(aq) \rightarrow Al_2(SO_4)_3(aq) + 6H_2O(l)$$

14. List the four steps that can be used to solve most stoichiometry problems.

Answers to questions

1. (a) stoichiometric amounts, (b) limiting reactant, (c) theoretical yield, (d) percent yield, (e) precipitate, (f) net ionic equation, (g) spectator ions.
2. (a) iv, (b) ii, (c) i, (d) iii.
3. (a) F, change *largest* to *smallest*; (b) F, change *Only two* to *Two or more*; (c) F, change *theoretical* to *percent*; (d) T; (e) T; (f) F, insert *do not* between *and participate*; (g) T.
4. (a) The *limiting reactant* determines the maximum amount of products that will be produced as well as the actual amounts of the other reactants that will react; an *excess reactant* is a reactant that is present in a larger amount than the stoichiometric amount as determined by the limiting reactant and will not completely react. (b) The *theoretical yield* is the predicted yield of product based on 100% completion of the reaction; the *actual yield* is the yield of product that is obtained. (c) An *overall chemical equation* shows the formulas of all chemical species in a chemical reaction; a *net ionic* equation shows only the species involved in the chemical change and excludes spectator ions.
5. The mass of the universe is constant; yes; mass-to-energy conversions in ordinary chemical reactions are so small that they cannot be detected with current instrumentation.
6. The four types of chemical reactions are combination, decomposition, displacement, and partner exchange. In combination reactions, two reactants combine to give a single product. In a decomposition reaction, a single compound breaks down to give two or more other substances. A displacement reaction is one in which the atoms or ions of one substance take the place

of other atoms or ions in a compound. In a partner-exchange reaction, atoms, or ions or two compounds exchange to form two different compounds.

7. (a) $BaO(s) + H_2SO_4(aq) \rightarrow BaSO_4(s) + H_2O(l)$
   (b) $BaO(s) + SO_3(g) \rightarrow BaSO_4(s)$
   (c) $Na_2CO_3(aq) + Ca(OH)_2(aq) \rightarrow CaCO_3(s) + 2NaOH(aq)$
   (d) $Mg(s) + 2HCl(aq) \rightarrow H_2(g) + MgCl_2(aq)$
   (e) $2H_2O(l) \xrightarrow{\text{electricity}} 2H_2(g) + O_2(g)$
8. An activity series of metals is a list of metals in the order of their tendency to react with acids, water, and each other. (a) $2K(s) + 2H_2O(l) \rightarrow 2KOH(aq) + H_2(g)$, $2K(s) + 2H_2O(g) \rightarrow 2KOH(s) + H_2(g)$, $2K(s) + 2H^+ \rightarrow 2K^+ + H_2(g)$, $K_2O(s) + H_2(g) \rightarrow$ no reaction; (b) $Cu(s) + H_2O(l \text{ or } g) \rightarrow$ no reaction, $Cu(s) + H^+ \rightarrow$ no reaction, $CuO(s) + H_2(g) \xrightarrow{\Delta} Cu(s) + H_2O(g)$, $CuO(s) \xrightarrow{\Delta}$ no reaction.
9. A net ionic equation shows only the species involved in the chemical change; an overall equation includes all of the species present in the reaction mixture. A net ionic equation must be balanced with respect to the charge on each side of the arrow.
10. Ionic substances that are considered soluble (see Table 5.3 of the text) are written in the ionic form; the formulas for all insoluble ionic substances, molecular substances, gases, etc. are written in their usual form.
11. The dictionary definition, "the branch of science that deals with the application of the laws of definite proportions and conservation of matter and energy to chemical reactions," includes the topics discussed in Chapter 4 as well as in the present chapter and is more comprehensive because it includes energy. The root words are from the Greek for element and measurement.
12. A chemical equation provides the connection between the known and unknown facts in a stoichiometry problem via the coefficients, which represent the mole ratios.
13. Mole ratios are determined by the stoichiometric coefficients of the species involved. For the given chemical equation, the mole ratios are (2 mol $Al(OH)_3$)/(3 mol H_2SO_4), (2 mol $Al(OH)_3$)/(1 mol $Al_2(SO_4)_3$), (2 mol $Al(OH)_3$)/(6 mol H_2O), (3 mol H_2SO_4)/(1 mol $Al_2(SO_4)_3$), (3 mol H_2SO_4)/(6 mol H_2O), (1 mol $Al_2(SO_4)_3$)/(6 mol H_2O), and the inverse of these six.
14. Write the balanced chemical equation, convert the known information to number of moles, use mole ratios to find the unknown in terms of moles, convert from number of moles to desired unknown quantity.

SKILLS

Skill 5.1 Interpreting chemical equations

The basis of writing and balancing chemical equations is presented in *Skill 4.3 Writing and balancing chemical equations*. For example, the balanced chemical equation describing the reaction of methane, ammonia, and oxygen,

$$2CH_4(g) + 2NH_3(g) + 3O_2(g) \xrightarrow{800\ °C,\ catalyst} 2HCN(g) + 6H_2O(g)$$

shows that there are 2 C atoms, 14 H atoms, 2 N atoms, and 6 O atoms on each side of the reaction. On a mass basis, this means that (2 C atoms)(12 u/1 C atom) + (2 N atoms)(14 u/1 N atom) + (14 H atoms)(1 u/1 H atom) + (6 O atoms)(16 u/1 O atom) = 162 u of reactants are undergoing chemical change to form 162 u of products, and mass has been conserved. This chemical equation can also be interpreted as stating that 162 g of reactants are undergoing chemical change to form 162 g of products. In fact, *any* mass unit can be used because atomic and molar masses are relative numbers: 162 ton of reactants are undergoing chemical change to form 162 ton of products.

Skill exercise: Interpret the chemical equation

$$2NH_4NO_3(s) \xrightarrow{\Delta} 2N_2(g) + 4H_2O(g) + O_2(g)$$

on (a) an atomic basis (numbers of atoms), (b) a mass basis using atomic mass units, (c) a mass basis using grams, and (d) a mass basis using pounds. *Answers*: (a) 4 N atoms, 8 H atoms, and 6 O atoms on each side; (b) (4 N atoms)(14 u/1 N atom) + (8 H atoms)(1 u/1 H atom) + (6 O atoms)(16 u/1 O atom) = 160. u reactants and 160. u products; (c) 160. g reactants and 160. g products; (d) 160. lb reactants and 160. lb products.

A molecular interpretation of chemical equations can also be given. For example, the above reaction involving methane, ammonia, and oxygen shows that 2 molecules of CH_4 undergo reaction with 2 molecules of NH_3 and 3 molecules of O_2 to produce 2 molecules of HCN and 6 molecules of H_2O.

Skill exercise: Interpret the chemical equation given above for the decomposition of NH_4NO_3 on a molecular basis. *Answer*: 2 NH_4NO_3 formula units undergo reaction to form 2 N_2 molecules, 4 H_2O molecules, and 1 O_2 molecule.

The usual interpretation of a chemical equation is done on a molar basis. For example, the above reaction involving methane, ammonia, and oxygen shows that 2 mol CH_4 undergo reaction with 2 mol NH_3 and 3 mol O_2 to produce 2 mol HCN and 6 mol H_2O.

Skill exercise: Interpret the chemical equation involving NH_4NO_3 given above on a molar basis. *Answer*: 2 mol NH_4NO_3 undergo reaction to give 2 mol N_2, 4 mol H_2O, and 1 mol O_2.

Skill 5.2 Types of chemical reactions

Most of the time, a chemical reaction can be classified as either a combination reaction, a decomposition reaction, a displacement reaction, or a partner-exchange reaction.

In combination reactions, two reactants combine to give a single product:

$$A + B \rightarrow C$$

for example

$$4Fe(s) + 3O_2(g) \rightarrow 2Fe_2O_3(s)$$
$$3Mg(s) + N_2(g) \rightarrow Mg_3N_2(s)$$
$$CaO(s) + H_2O(l) \rightarrow Ca(OH)_2(s)$$
$$6Li_2O(s) + P_4O_{10}(g) \rightarrow 2Li_3PO_4(s)$$

In decomposition reactions, a single compound breaks down to give two or more substances:

$$C \rightarrow A + B$$

Decomposition is essentially the reverse of combination. For example

$$2H_2O(l) \xrightarrow{\text{electricity}} 2H_2(g) + O_2(g)$$
$$NiCO_3(s) \xrightarrow{\Delta} NiO(s) + CO_2(g)$$
$$(NH_4)_2Cr_2O_7(s) \rightarrow N_2(g) + Cr_2O_3(s) + 4H_2O(g)$$
$$CuSO_4\cdot 5H_2O(s) \xrightarrow{\Delta} CuSO_4(s) + 5H_2O(g)$$
$$SnCl_4\cdot 6H_2O(s) \xrightarrow{\Delta} SnO_2(s) + 4HCl(g) + 4H_2O(g)$$

In displacement reactions, the atoms or ions of one substance take the place of other atoms or ions in a compound:

$$A + BC \rightarrow AC + B$$

Many displacement reactions involve a metal displacing another metal from a compound, hydrogen from water, or hydrogen from an acid. The following are examples of displacement reactions:

$$Zn(s) + 2HCl(aq) \rightarrow ZnCl_2(aq) + H_2(g)$$
$$2Al(s) + Fe_2O_3(s) \rightarrow 2Fe(s) + Al_2O_3(s)$$
$$2NaI(aq) + Br_2(aq) \rightarrow 2NaBr(aq) + I_2(aq)$$

Displacement reactions are also called replacement or single-replacement reactions.

In partner-exchange reactions, two compounds interact as follows:

$$AC + BD \rightarrow AD + BC$$

The following are examples of partner-exchange reactions:

$$CaCO_3(s) + 2HCl(aq) \rightarrow CaCl_2(aq) + H_2O(l) + CO_2(g)$$
$$KOH(aq) + HI(aq) \rightarrow KI(aq) + H_2O(l)$$
$$H_2SO_4(aq) + Ba(OH)_2(aq) \rightarrow BaSO_4(s) + 2H_2O(l)$$

Partner-exchange reactions are also called double decomposition, double displacement, or metathesis reactions.

Skill exercise: Classify each of the following reactions as one of the four types:

(a) $F_2(g) + H_2(g) \rightarrow 2HF(g)$

(b) $F_2(g) + 2NaCl(aq) \rightarrow 2NaF(aq) + Cl_2(g)$

(c) $XeF_6(s) \rightarrow XeF_4(s) + F_2(g)$

(d) $CaF_2(aq) + 2HCl(aq) \rightarrow CaCl_2(aq) + 2HF(g)$

(e) $2CO_2(g) \rightarrow 2CO(g) + O_2(g)$

(f) $2Cs(s) + I_2(g) \rightarrow 2CsI(s)$

(g) $Zn(s) + 2AgNO_3(aq) \rightarrow Zn(NO_3)_2(aq) + Ag(s)$

(h) $AgNO_3(aq) + HCl(aq) \rightarrow AgCl(s) + HNO_3(aq)$

(i) $XeF_6(s) + RbF(s) \rightarrow RbXeF_7(s)$

Answers: (a) combination, (b) displacement, (c) decomposition, (d) partner exchange, (e) decomposition, (f) combination, (g) displacement, (h) partner exchange, (i) combination.

Skill 5.3 Writing net ionic equations

Often the complete chemical equation is not written to describe a reaction, but rather a chemical equation is used that involves only the ions, atoms, and molecules that are directly involved in the reaction. In other words, the spectator ions are omitted. For example, if an acid such as $HCl(aq)$ or $HNO_3(aq)$ is added to an aqueous solution of K_2CrO_4, the yellow solution changes to an orange solution, indicating the presence of $K_2Cr_2O_7(aq)$. The two reactions can be represented by the following equations:

$$2K_2CrO_4(aq) + 2HCl(aq) \rightarrow K_2Cr_2O_7(aq) + 2KCl(aq) + H_2O(l)$$
$$2K_2CrO_4(aq) + 2HNO_3(aq) \rightarrow K_2Cr_2O_7(aq) + 2KNO_3(aq) + H_2O(l)$$

However, K_2CrO_4, HCl, HNO_3, $K_2Cr_2O_7$, KCl, and KNO_3 all exist in the form of dissociated ions in aqueous solutions (see Table 5.3 of the text). Thus these equations can be rewritten

$$4K^+ + 2CrO_4^{2-} + 2H^+ + 2Cl^- \rightarrow 4K^+ + Cr_2O_7^{2-} + 2Cl^- + H_2O(l)$$
$$4K^+ + 2CrO_4^{2-} + 2H^+ + 2NO_3^- \rightarrow 4K^+ + Cr_2O_7^{2-} + 2NO_3^- + H_2O(l)$$

And after elimination of the spectator ions, both equations can be represented by the same net ionic equation:

$$2CrO_4^{2-} + 2H^+ \rightarrow Cr_2O_7^{2-} + H_2O(l)$$

In fact, the net ionic equation tells us that any chromate that is dissociated in aqueous solution (e.g., K_2CrO_4, Na_2CrO_4, $(NH_4)_2CrO_4$) will undergo reaction with any strong acid (e.g., HCl, HI, HNO_3) to form the dichromate ion and water. Note that electrical charge on each side of the equation is balanced,

$$(1)(-2) + (1)(+1) = -2 = (1)(-2)$$

as is the number of atoms.

Skill exercise: Write the net ionic equations for the following reactions:

(a) $2AgNO_3(aq) + CaCl_2(aq) \rightarrow Ca(NO_3)_2(aq) + 2AgCl(s)$

(b) $2NaOH(aq) + 2HBr(aq) \rightarrow 2H_2O(l) + 2NaBr(aq)$

(c) $Zn(s) + CuSO_4(aq) \rightarrow ZnSO_4(aq) + Cu(s)$

(d) $2NaI(aq) + Br_2(l) \rightarrow 2NaBr(aq) + I_2(s)$

Use the following information: $AgNO_3$, $CaCl_2$, $Ca(NO_3)_2$, NaOH, HBr, NaBr, $CuSO_4$, $ZnSO_4$, and NaI react as dissociated ions in aqueous solution. *Answers*: (a) $Ag^+ + Cl^- \rightarrow AgCl(s)$, (b) $OH^- + H^+ \rightarrow H_2O(l)$, (c) $Zn(s) + Cu^{2+} \rightarrow Cu(s) + Zn^{2+}$, (d) $2I^- + Br_2(l) \rightarrow 2Br^- + I_2(s)$.

A list of solubilities of ionic substances in water (see Table 5.3 of the text) often helps in deciding how to write the formulas for these substances in a net ionic equation. For example, the formula for $AgNO_3$ would be written as $Ag^+ + NO_3^-$ because it is soluble and the formula for AgCl would be written as AgCl(s) because it is insoluble. Likewise the formula for Na_2S would be written as $2Na^+ + S^{2-}$ because it is soluble and the formula for CuS would be written as CuS(s) because it is insoluble.

Skill exercise: Write the formulas for the following ionic substances as they should appear when writing a net ionic equation occurring in an aqueous solution:

(a) NH_4NO_2, (b) $KClO_3$, (c) AgI, (d) $KMnO_4$, and (e) $Zn_3(PO_4)_2$. *Answers*: (a) $NH_4^+ + NO_2^-$, (b) $K^+ + ClO_3^-$, (c) AgI(s), (d) $K^+ + MnO_4^-$, (e) $Zn_3(PO_4)_2(s)$.

The formation of a product that is a precipitate, a gas that leaves the reaction mixture, or a molecular substance often causes partner-exchange reactions to occur. For example, the net ionic equation for the reaction

$$KOH(aq) + HI(aq) \rightarrow KI(aq) + H_2O(l)$$

is simply the formation of a molecular substance:

$$OH^- + H^+ \rightarrow H_2O(l)$$

The two favorable changes in the reaction

$$H_2SO_4(aq) + Ba(OH)_2(aq) \rightarrow BaSO_4(s) + 2H_2O(l)$$

are the formation of a precipitate and the formation of a molecular product.

Skill exercise: Determine whether or not the following partner-exchange reactions will occur.

(a) $ZnS(s) + K(CH_3COO)(aq) \rightarrow$

(b) $ZnCl_2(aq) + NaOH(aq) \rightarrow$

(c) $NH_4Cl(aq) + NaNO_3(aq) \rightarrow$

(d) $Pb(NO_3)_2(aq) + Na_2SO_4(aq) \rightarrow$

(e) $MnS(s) + HCl(aq) \rightarrow$

Write net ionic equations for those reactions that occur. *Answer*: (a) no reaction; (b) $Zn^{2+} + 2OH^- \rightarrow Zn(OH)_2(s)$, (c) no reaction, (d) $Pb^{2+} + SO_4^{2-} \rightarrow PbSO_4(s)$, (e) $MnS(s) + 2H^+ \rightarrow Mn^{2+} + H_2S(g)$.

Skill 5.4 Using the activity series

The activity series given in Table 5.2 of the text is a list of metals in order of decreasing reactivity to form aqueous cations by the loss of electrons. For example, because Zn is listed above Cu in the activity series, Zn will undergo a displacement reaction with Cu^{2+} by transferring electrons to Cu^{2+} according to the equation

$$Zn(s) + Cu^{2+} \rightarrow Zn^{2+} + Cu(s)$$

The reverse of this reaction will not occur; that is,

$$Cu(s) + Zn^{2+} \rightarrow \text{no reaction}$$

because the position of Cu is below that of Zn in the activity series.

Although the activity series is useful for predicting this type of displacement reaction, it cannot be used to predict reactions between elements, e.g.,

$$Na(s) + K(s) \rightarrow \qquad K(s) + H_2(g) \rightarrow$$

nor between ions, e.g.,

$$Fe^{2+} + Cu^{2+} \rightarrow \qquad Sn^{2+} + Fe^{3+} \rightarrow$$

Information concerning the activity of the metals in displacing hydrogen from water and acids and concerning reactions of the oxides of the metals is also summarized along the sides of the list of metals in Table 5.2 of the text. For example, the table shows that the reaction between CaO and H_2

$$CaO_2 + H_2(g) \rightarrow Ca(s) + H_2O(l)$$

would not take place.

Skill exercise: Predict whether the following reactions will occur or not and complete the equation for those that will occur: (a) $Cu(s) + H^+ \rightarrow$, (b) $Fe^{2+} + Pb(s) \rightarrow$, (c) $Ca(s) + Al^{3+} \rightarrow$, (d) $H_2(g) + Cu^{2+} \rightarrow$, (e) $Na(s) + H^+ \rightarrow$, (f) $H_2(g) + Zn^{2+} \rightarrow$, (g) $Ni(s) + Pb^{2+} \rightarrow$, (h) $Ag(s) + Au^+ \rightarrow$, (i) $Fe(s) + H_2O(l) \rightarrow$, (j) $CuO(s) \xrightarrow{\Delta}$, (k) $Zn(s) + H_2O(l) \rightarrow$, (l) $Na(s) + H_2O(g) \rightarrow$, (m) $Zn^{2+} + Ag^+ \rightarrow$, and (n) $Al(s) + Pb(s) \rightarrow$. *Answers*: (a) no reaction, (b) no reaction, (c) $3Ca(s) + 2Al^{3+} \rightarrow 3Ca^{2+} + 2Al(s)$, (d) $H_2(g) + Cu^{2+} \rightarrow 2H^+ + Cu(s)$, (e) $2Na(s) + 2H^+ \rightarrow 2Na^+ + H_2(g)$, (f) no reaction, (g) $Ni(s) + Pb^{2+} \rightarrow Ni^{2+} + Pb(s)$, (h) $Ag(s) + Au^+ \rightarrow Ag^+ + Au(s)$, (i) no reaction, (j) no reaction, (k) no reaction, (l) $2Na(s) + 2H_2O(g) \rightarrow H_2(g) + 2NaOH(aq)$, (m) no reaction, (n) no reaction.

Skill 5.5 Stoichiometry

The calculation of the quantitative relationships in chemical change is called stoichiometry. Usually these quantitative relationships involve the relative amounts of reactants and products: masses; volumes of gases; concentrations or volumes of solutions; numbers of atoms, ions, molecules, and moles; and amounts of electrical charge.

The complete method for solving many stoichiometric problems consists of four steps:

1. Write the balanced chemical equation.
2. Convert the known information to number of moles.
3. Use mole ratios from the chemical equation to find the unknown in terms of number of moles.

4. Convert from number of moles to the unknown quantity that is desired.

The "heart" of the method of solving stoichiometric problems is the use of the mole ratio in Step (3). This mole ratio is simply a conversion factor (see *Skill 2.6 Deriving conversion factors*) derived from the molar interpretation of the chemical equation (see *Skill 5.1 Interpreting chemical equations*). For example, the mole ratio that would be used to find the relationship between Na and H_2 in the reaction

$$2Na(s) + 2H_2O(g) \rightarrow H_2(g) + 2NaOH(aq)$$

is (1 mol H_2)/(2 mol Na), and the mole ratio that would be used to find the tionship between Na and H_2O is (2 mol Na)/(2 mol H_2O) [or (1 mol Na)/(1 mol H_2O)]. As with all conversion factors, the inverse of the factor is also true; for example, (2 mol Na)/(1 mol H_2) is also a relation between Na and H_2 in this reaction.

Skill exercise: Determine the mole ratio that would be used to find the relationship between (a) $K_2C_2O_4$ and CaC_2O_4, (b) $Ca(NO_3)_2$ and $K_2C_2O_4$, (c) KNO_3 and CaC_2O_4, (d) $Ca(NO_3)_2$ and KNO_3, (e) $K_2C_2O_4$ and KNO_3, and (f) $Ca(NO_3)_2$ and CaC_2O_4 based on the following equation:

$$Ca(NO_3)_2(aq) + K_2C_2O_4(aq) \rightarrow CaC_2O_4 + 2KNO_3(aq)$$

Answers: (a) (1 mol CaC_2O_4)/(1 mol $K_2C_2O_4$), (b) (1 mol $Ca(NO_3)_2$)/(1 mol $K_2C_2O_4$), (c) (2 mol KNO_3)/(1 mol CaC_2O_4), (d) (2 mol KNO_3)/(1 mol $Ca(NO_3)_2$), (e) (2 mol KNO_3)/(1 mol $K_2C_2O_4$), (f) (1 mol CaC_2O_4)/(1 mol $Ca(NO_3)_2$). [Any of these factors could be written in the inverse form.]

Steps (2) and (4) are essentially the calculations presented in *Skill 4.4 Finding the number of moles* and Step (1) is covered in *Skill 4.3 Writing and balancing chemical equations*. The last three steps can be expressed in a flow chart:

known information

Step (2) *Skill 4.4*

↓

number of moles of known

Step (3) mole ratio

↓

number of moles of unknown

Step (4) *Skill 4.4*

↓

desired quantity

For example, for the reaction

$$CaCO_3(s) + 2HCl(aq) \rightarrow CaCl_2(aq) + H_2O(l) + CO_2(g)$$

(a) the volume of HCl (of a known concentration) needed for a given mass of $CaCO_3$ might be the unknown or (b) the concentration of $CaCl_2$ produced for a given number of HCl molecules might be the unknown. [These are only two of over a dozen types of problems that could be solved for this system.] The flow charts illustrating Steps (2) to (4) for these problems would be

(a)

$$m_{CaCO_3} \xrightarrow{\left[\frac{1}{\text{molar mass}}\right]} n_{CaCO_3} \xrightarrow{\left[\frac{2\text{ mol HCl}}{1\text{ mol }CaCO_3}\right]} n_{HCl} \xrightarrow{\left[\frac{1}{\text{molarity}}\right]} V_{HCl}$$

(b)

$$N_{HCl} \xrightarrow{\left[\frac{1}{\text{Avogadro's number}}\right]} n_{HCl} \xrightarrow{\left[\frac{1\text{ mol }CaCl_2}{2\text{ mol HCl}}\right]} n_{CaCl_2} \xrightarrow{\left[\frac{1}{\text{volume}}\right]} M_{CaCl_2}$$

Skill exercise: Write the flow charts illustrating Steps (2) to (4) for the equation

$$MnO_2(s) + 4HCl(aq) \rightarrow MnCl_2(aq) + 2H_2O(l) + Cl_2(g)$$

for (a) the mass of Cl_2 produced from a given mass of MnO_2 and (b) the volume of HCl (at a known concentration) needed to undergo reaction with a given mass of MnO_2. *Answers*:

(a)

$$m_{MnO_2} \xrightarrow{\left[\frac{1}{\text{molar mass}}\right]} n_{MnO_2} \xrightarrow{\left[\frac{1\text{ mol }Cl_2}{1\text{ mol }MnO_2}\right]} n_{Cl_2} \xrightarrow{[\text{molar mass}]} m_{Cl_2}$$

(b)

$$m_{MnO_2} \xrightarrow{\left[\frac{1}{\text{molar mass}}\right]} n_{MnO_2} \xrightarrow{\left[\frac{4\text{ mol HCl}}{1\text{ mol }MnO_2}\right]} n_{HCl} \xrightarrow{\left[\frac{1}{\text{molarity}}\right]} V_{HCl}$$

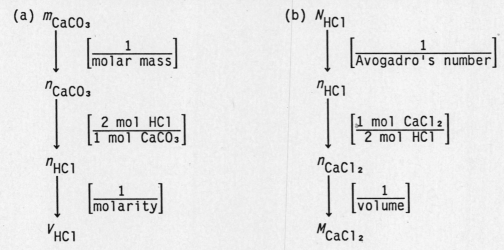

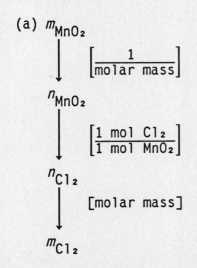

NUMERICAL EXERCISES

Example 5.1 Interpreting chemical equations

Small quantities of ammonia gas can be produced by allowing an ionic nitride such as Mg_3N_2 to undergo reaction with water:

$$Mg_3N_2(s) + 6H_2O(l) \rightarrow 3Mg(OH)_2(s) + 2NH_3(g)$$

Show that this equation is in agreement with the law of conservation of mass.

The chemical equation states that one mole of Mg_3N_2, or 100.95 g, undergoes reaction with six moles of H_2O, or (6 mol H_2O)(18.02 g/mol H_2O) = 108.12 g, to give three moles of $Mg(OH)_2$, or (3 mol $Mg(OH)_2$)(58.33 g/mol $Mg(OH)_2$) = 174.99 g, and two moles of NH_3, or (2 mol NH_3)(17.03 g/mol NH_3) = 34.06 g. Thus the mass of reactants, 100.95 + 108.12 = 209.07 g, equals the mass of products, 174.99 + 34.06 = 209.05 g. [The small difference, 0.02 g, results from the rounding off of molar mass values to the nearest 0.01 g.]

Exercise 5.1 Elemental silicon can be produced by heating SiO_2 with C:

$$SiO_2(s) + 2C(s) \rightarrow Si(s) + 2CO(g)$$

Show that this equation is in agreement with the law of conservation of mass.
Answer: 60.09 g SiO_2 + 24.02 g C = 28.09 g Si + 56.02 g CO.

Example 5.2 Interpreting chemical equations

Tungsten, a metal of intermediate activity, is produced by the following displacement reaction.

$$3H_2(g) + WO_3(s) \xrightarrow{\Delta} W(s) + 3H_2O(g)$$

Interpret this equation on a molecular, molar, and mass basis.

	$3H_2(g)$ +	$WO_3(s)$ →	$W(s)$ +	$3H_2O(g)$
molecular basis:	3 molecules	1 molecule	1 atom	3 molecules
molar basis:	3 mol	1 mol	1 mol	3 mol
mass basis:	$(3\ \text{mol})\left[\frac{2.02\ \text{g}}{1\ \text{mol}}\right]$ = 6.06 g	$(1\ \text{mol})\left[\frac{231.85\ \text{g}}{1\ \text{mol}}\right]$ = 231.85 g	$(1\ \text{mol})\left[\frac{183.85\ \text{g}}{1\ \text{mol}}\right]$ = 183.85 g	$(3\ \text{mol})\left[\frac{18.02\ \text{g}}{1\ \text{mol}}\right]$ = 54.06 g

Exercise 5.2 The chemical equation describing the decomposition of nitroglycerin, a powerful explosive, is

$$4\ \begin{array}{l} CH_2(NO_3) \\ | \\ CH(NO_3)\ (l) \\ | \\ CH_2(NO_3) \end{array} \rightarrow 6N_2(g) + 12CO_2(g) + 10H_2O(g) + O_2(g)$$

Interpret this equation on a molecular, molar, and mass basis. *Answer*:

molecular basis:	4 molecules	6 molecules	12 molecules	10 molecules	1 molecule
molar basis:	4 mol	6 mol	12 mol	10 mol	1 mol
mass basis:	908.4 g	168.1 g	528.1 g	180.2 g	32.0 g

Example 5.3 Stoichiometry

What mass of H_2O will undergo reaction with 1.00 g Mg_3N_2 in the reaction given in *Example 5.1*?

The four-step procedure used for solving stoichiometric problems begins with writing the balanced chemical equation:

$$Mg_3N_2(s) + 6H_2O(l) \rightarrow 3Mg(OH)_2(s) + 2NH_3(g)$$

The second step is to convert the given information to number of moles.

$$(1.00\text{ g }Mg_3N_2)\left[\frac{1\text{ mol }Mg_3N_2}{100.95\text{ g }Mg_3N_2}\right] = 0.00991\text{ mol }Mg_3N_2$$

The third step is to inspect the chemical equation to find the mole ratio between the known and unknown quantities. In this case, the chemical equation tells us that 6 moles of H_2O are needed to undergo reaction with 1 mol of Mg_3N_2, so that

$$(0.00991\text{ mol }Mg_3N_2)\left[\frac{6\text{ mol }H_2O}{1\text{ mol }Mg_3N_2}\right] = 0.0595\text{ mol }H_2O$$

are needed to undergo reaction. Converting back to mass (the fourth step of the procedure) gives

$$\text{mass} = (0.0595\text{ mol }H_2O)\left[\frac{18.02\text{ g }H_2O}{1\text{ mol }H_2O}\right] = 1.07\text{ g }H_2O$$

The required amount of water is 1.07 g.

Exercise 5.3 What mass of carbon is needed to undergo reaction with 15.2 g of SiO_2 in the reaction given in *Exercise 5.1*? *Answer*: 6.08 g.

Example 5.4 Stoichiometry*

What mass of $Mg(OH)_2$ is produced for every 10.0 g Mg_3N_2 that undergoes reaction according to the equation given in *Example 5.1*?

This problem is analyzed using the comprehensive problem-solving approach presented in *Skill 2.7 Solving problems.*

1. Study the problem and be sure you understand it.
   (a) What is unknown?
   The mass of $Mg(OH)_2$ produced.
   (b) What is known?
   The mass of Mg_3N_2 that undergoes reaction.
2. Decide how to solve the problem.
   (a) What is the connection between the known and the unknown?
   The connection is provided by the molar interpretation of the chemical equation

$$Mg_3N_2(s) + 6H_2O(l) \rightarrow 3Mg(OH)_2(s) + 2NH_3(g)$$

   (b) What is necessary to make the connection?
   The standard four-step method for solving stoichiometry problems--see *Skill 5.5 Stoichiometry*--can be used to calculate the amount of product produced by the reaction of the known amount of reactant. Step (1) is to write the balanced chemical equation--given above. The remaining three steps are outlined in the flow chart

$$m_{Mg_3N_2}$$

Step 2 ↓ $\left[\frac{1}{\text{molar mass}}\right]$

$$n_{Mg_3N_2}$$

Step 3 ↓ $\left[\frac{3\text{ mol }Mg(OH)_2}{1\text{ mol }Mg_3N_2}\right]$

$$n_{Mg(OH)_2}$$

Step 4 ↓ [molar mass]

$$m_{Mg(OH)_2}$$

3. Set up the problem and solve it.
   Steps (2) to (4) can be done separately or combined into one calculation.

$$(10.0\text{ g }Mg_3N_2)\left[\frac{1\text{ mol }Mg_3N_2}{100.95\text{ g }Mg_3N_3}\right]\left[\frac{3\text{ mol }Mg(OH)_2}{1\text{ mol }Mg_3N_2}\right]\left[\frac{58.33\text{ g }Mg(OH)_2}{1\text{ mol }Mg(OH)_2}\right]$$

$$= 17.3\text{ g }Mg(OH)_2$$

4. Check the result.
   (a) Are significant figures and the location of the decimal point correct?
   Yes. The molar masses are given to more than enough significant figures.

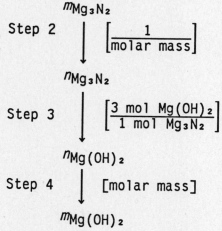

(b) Did the answer come out in the correct units?

Yes. Note that by identifying "g Mg_3N_2," "mol $Mg(OH)_2$," etc., it is easy to see exactly what quantity is being dealt with at each step of the calculation.

(c) Is the answer reasonable?

Yes. The answer is neither too large nor too small for 10.0 g of a reactant.

Exercise 5.4 What mass of Si is produced for each 5.25 g of SiO_2 that undergoes reaction in the equation given in Exercise 5.1? *Answer*: 2.45 g.

Example 5.5 Stoichiometry

Water gas is the mixture of CO and H_2 produced by the reaction

$$H_2O(g) + C(s) \xrightarrow{\Delta} \underbrace{CO(g) + H_2(g)}_{\text{water gas}}$$

What mass in pounds of water gas is formed by passing 1.00 lb of steam over white-hot carbon? Assume that 45% of the steam undergoes reaction.

Note that in this problem the mass in pounds could be converted to grams, the problem solved in the usual way, and the answer converted back to pounds. There is really no reason for the two mass conversions because they cancel each other. Because all of the mass-to-mass relationships are relative, it is common to use other types of "moles," e.g., kg-mol, lb-mol, and ton-mol, in solving problems involving masses of reagents given in these units.

Step 2: $$(1.00 \text{ lb } H_2O)\left[\frac{1 \text{ lb-mol } H_2O}{18.02 \text{ lb } H_2O}\right] = 0.0555 \text{ lb-mol } H_2O$$

Step 3: $$(0.0555 \text{ lb-mol } H_2O)\left[\frac{1 \text{ lb-mol CO}}{1 \text{ lb-mol } H_2O}\right] = 0.0555 \text{ lb-mol CO}$$

$$(0.0555 \text{ lb-mol } H_2O)\left[\frac{1 \text{ lb-mol } H_2}{1 \text{ lb-mol } H_2O}\right] = 0.0555 \text{ lb-mol } H_2$$

Step 4: $$(0.0555 \text{ lb-mol CO})\left[\frac{28.01 \text{ lb CO}}{1 \text{ lb-mol CO}}\right] + (0.0555 \text{ lb-mol } H_2)\left[\frac{2.02 \text{ lb } H_2}{1 \text{ lb-mol } H_2}\right]$$
$$= 1.67 \text{ lb water gas}$$

The yield is

$$(1.67 \text{ lb water gas})(0.45) = 0.75 \text{ lb water gas}$$

Exercise 5.5 Chlorine is prepared by the electrical decomposition of a concentrated brine solution:

$$2NaCl(aq) + 2H_2O(l) \xrightarrow{\text{electricity}} Cl_2(g) + H_2(g) + 2NaOH(aq)$$

What is the theoretical mass of chlorine that can be obtained from 1.00 ton of NaCl? *Answer*: 0.607 ton Cl_2.

Example 5.6 Stoichiometry

Consider the following equation for photosynthesis.

$$CO_2(g) + H_2O(l) \xrightarrow{\text{unbalanced}} C_6H_{12}O_6(aq) + O_2(g)$$

(a) What mass of CO_2 is necessary to undergo reaction with 1.00 g of water? (b) How many O_2 molecules will be produced? (c) The hexose, $C_6H_{12}O_6$, that is produced is dissolved in 0.500 L of solution; what is the molarity of the solution?

The balanced chemical equation is

$$6CO_2(g) + 6H_2O(l) \rightarrow C_6H_{12}O_6(aq) + 6O_2(g)$$

The number of moles of water is

$$(1.00 \text{ g } H_2O)\left[\frac{1 \text{ mol } H_2O}{18.02 \text{ g } H_2O}\right] = 0.0555 \text{ mol } H_2O$$

(a) The mass of CO_2 needed is

$$(0.0555 \text{ mol } H_2O)\left[\frac{6 \text{ mol } CO_2}{6 \text{ mol } H_2O}\right]\left[\frac{44.01 \text{ g } CO_2}{1 \text{ mol } CO_2}\right] = 2.44 \text{ g } CO_2$$

(b) The number of molecules of O_2 is

$$(0.0555 \text{ mol } H_2O)\left[\frac{6 \text{ mol } O_2}{6 \text{ mol } H_2O}\right]\left[\frac{6.022 \times 10^{23} \text{ } O_2 \text{ molecules}}{1 \text{ mol } O_2}\right]$$

$$= 3.34 \times 10^{22} \text{ } O_2 \text{ molecules}$$

(c) The concentration of the hexose solution is

$$(0.0555 \text{ mol } H_2O)\left[\frac{1 \text{ mol } C_6H_{12}O_6}{6 \text{ mol } H_2O}\right] = 0.00925 \text{ mol } C_6H_{12}O_6$$

$$\frac{0.00925 \text{ mol } C_6H_{12}O_6}{0.500 \text{ L}} = 0.0185 \text{ mol } C_6H_{12}O_6/\text{L}$$

Exercise 5.6 Calculate the mass of CO_2, required to precipitate 2.00×10^{-2} g of calcium carbonate from excess limewater (aqueous calcium hydroxide). How many $CaCO_3$ formula units are produced?

$$Ca(OH)_2(aq) + CO_2(g) \rightarrow CaCO_3(s) + H_2O(l)$$

Answer: 8.8 mg CO_2, 1.20 x 10^{20} formula units $CaCO_3$.

Example 5.7 Stoichiometry--percent yield

The actual yield of $Mg(OH)_2$ produced in *Example 5.4* was 15.2 g. What is the percent yield?

The percent yield is found by dividing the actual yield, 15.2 g, by the theoretical yield, 17.3 g, and converting the fraction to percentage by multiplying by 100:

$$\text{percent yield} = \left[\frac{15.2\ \text{g}}{17.3\ \text{g}}\right] \times 100 = 87.9\ \%$$

The percent yield is 87.9 %.

Exercise 5.7 What is the percent yield if 0.65 g of Si is formed in the reaction described in Exercise 5.4? *Answer*: 27 %.

Example 5.8 Stoichiometry--limiting reactant*

What mass of $NaNO_2$ will be produced (assuming 100 % yield) from the reaction of 62.3 g $NaNO_3$ with 87.8 g Pb?

$$NaNO_3(aq) + Pb(s) \rightarrow NaNO_2(aq) + PbO(s)$$

This problem is analyzed using the comprehensive problem-solving approach presented in *Skill 2.7 Solving problems*.

1. Study the problem and be sure you understand it.
   (a) What is unknown?
   The mass of $NaNO_2$ produced.
   (b) What is known?
   The masses of $NaNO_3$ and Pb available to undergo reaction.
2. Decide how to solve the problem.
   (a) What is the connection between the known and the unknown?
   The molar interpretation of the chemical equation.
   (b) What is necessary to make the connection?
   The masses of both reactants are given. Unless these are stoichiometric amounts, one of the reactants will be the limiting reactant and one will be in excess. The number of moles of each reactant available must be calculated and compared with the required amounts in order to determine which is the limiting reactant. Then the calculation of the unknown is completed based on the limiting reactant.

3. Set up the problem and solve it.

The number of moles of each reactant is

$$(62.3\ g\ NaNO_3)\left[\frac{1\ mol\ NaNO_3}{84.99\ g\ NaNO_3}\right] = 0.733\ mol\ NaNO_3$$

$$(87.8\ g\ Pb)\left[\frac{1\ mol\ Pb}{207.2\ g\ Pb}\right] = 0.424\ mol\ Pb$$

If all of the $NaNO_3$ undergoes reaction, the amount of Pb needed is

$$(0.733\ NaNO_3)\left[\frac{1\ mol\ Pb}{1\ mol\ NaNO_3}\right] = 0.733\ mol\ Pb$$

The reaction mixture contains only 0.424 mol Pb, so the lead is the limiting reactant and $NaNO_3$ is in excess. The mass of $NaNO_2$ produced is determined by the amount of Pb:

$$(0.424\ mol\ Pb)\left[\frac{1\ mol\ NaNO_2}{1\ mol\ Pb}\right]\left[\frac{69.00\ g\ NaNO_2}{1\ mol\ NaNO_2}\right] = 29.3\ g\ NaNO_2$$

There will be

$$(62.3\ g\ NaNO_3) - (0.424\ mol\ Pb)\left[\frac{mol\ NaNO_3}{1\ mol\ Pb}\right]\left[\frac{84.99\ g\ NaNO_3}{1\ mol\ NaNO_3}\right]$$
$$= 26.3\ g\ NaNO_3$$

in excess.

4. Check the result.

(a) Are significant figures and the location of the decimal point correct?

Yes.

(b) Did the answer come out in the correct units?

Yes.

(c) Is the answer reasonable?

Yes.

Exercise 5.8 What mass of Hg_2Cl_2 can be produced by the reaction

$$HgSO_4(s) + Hg(l) + 2NaCl(s) \rightarrow Hg_2Cl_2 + Na_2SO_4(s)$$

if 100.0 g $HgSO_4$ are mixed with 100.0 g of Hg and 100.0 g of NaCl? *Answer*: 159.1 g.

Example 5.9 Stoichiometry--consecutive reactions*

In the Solvay process for making Na_2CO_3 (at one time widely used, but now obsolete in the United States), calcium carbonate is heated to give CaO and CO_2

with an average yield of 95%. The CO_2 is passed into an aqueous solution of NaCl and NH_3 to give NH_4HCO_3 with an average yield of 95%. The NH_4HCO_3 undergoes reaction immediately with the NaCl to give a precipitate of $NaHCO_3$ with an average yield of 85%. The $NaHCO_3$ is filtered out, dried, and heated to give Na_2CO_3 and H_2O quantitatively. [The ammonium chloride in the solution is heated with calcium oxide to give calcium chloride, water, and ammonia (which is recycled).] What mass of sodium carbonate will be obtained for each 1.00 kg of $CaCO_3$ used?

This problem is analyzed using the comprehensive problem-solving approach presented in *Skill 2.7 Solving problems*.

1. Study the problem and be sure you understand it.
   (a) What is unknown?
   The mass of Na_2CO_3 produced.
   (b) What is known?
   The mass of $CaCO_3$ that undergoes reaction in the first of a series of reactions, and the reactants and products in each reaction in the series of reactions.
2. Decide how to solve the problem.
   (a) What is the connection between the known and the unknown?
   The molar interpretation of the following chemical equations:

$$CaCO_3(s) \xrightarrow{\Delta} CaO(s) + CO_2(g)$$
$$CO_2(g) + NH_3(aq) + H_2O(l) \rightarrow NH_4HCO_3(aq)$$
$$NH_4HCO_3(aq) + NaCl(aq) \rightarrow NaHCO_3(s) + NH_4Cl(aq)$$
$$2NaHCO_3(s) \xrightarrow{\Delta} Na_2CO_3(s) + CO_2(s) + H_2O(g)$$

   (b) What is necessary to make the connection?
   The equations for the individual steps must be added together so that the intermediate products (NH_4HCO_3 and $NaHCO_3$) cancel. The mole ratio needed to connect the amounts of $CaCO_3$ and Na_2CO_3 is determined from the overall equation.

$$CaCO_3 \rightarrow CaO + CO_2$$
$$(2)[CO_2 + NH_3 + H_2O \rightarrow NH_4HCO_3]$$
$$(2)[NH_4HCO_3 + NaCl \rightarrow NaHCO_3 + NH_4Cl]$$
$$2NaHCO_3 \rightarrow Na_2CO_3 + CO_2 + H_2O$$

$$CaCO_3 + 2NH_3 + H_2O + 2NaCl \rightarrow CaO + 2NH_4Cl + Na_2CO_3$$

   The mole ratio is (1 mol Na_2CO_3)/(1 mol $CaCO_3$).

3. Set up the problem and solve it.
   Once the mole ratio is known, this problem is solved by using Steps (2) to (4) of the usual method.

$$(1.00\ \text{kg CaCO}_3)\left[\frac{1000\ \text{g}}{1\ \text{kg}}\right]\left[\frac{1\ \text{mol CaCO}_3}{100.09\ \text{g CaCO}_3}\right]\left[\frac{1\ \text{mol Na}_2\text{CO}_3}{1\ \text{mol CaCO}_3}\right]$$

$$\times \left[\frac{105.99\ \text{g Na}_2\text{CO}_3}{1\ \text{mol Na}_2\text{CO}_3}\right] = 1060\ \text{g Na}_2\text{CO}_3$$

   Because the reactions do not give quantitative yields, the actual amount will be

$$(1060\ \text{g Na}_2\text{CO}_3)(0.95)(0.95)(0.85)(1.00) = 810\ \text{g Na}_2\text{CO}_3$$

4. Check the result.
   (a) Are significant figures and the location of the decimal point correct?
       Yes.
   (b) Did the answer come out in the correct units?
       Yes.
   (c) Is the answer reasonable?
       Yes.

Exercise 5.9 The Ostwald process for preparing nitric acid is

$$4NH_3(g) + 5O_2(g) \xrightarrow{\text{Pt-Rh, 600 °C - 1000 °C}} 4NO(g) + 6H_2O(g)$$

$$2NO(g) + O_2(g) \rightarrow 2NO_2(aq)$$

$$3NO_2(g) + H_2O(l) \rightarrow 2HNO_3(aq) + NO(g)$$

Write the overall equation for this process and calculate the percent yield if 2.00 lb of pure NH_3 gave 2.80 lb HNO_3. *Answer*: $NH_3(g) + 2O_2(g) \rightarrow HNO_3(aq) + H_2O(l)$, 37.8 %.

Example 5.10 Stoichiometry

A 5.6 g sample of pure iron was dissolved in excess dilute hydrochloric acid. A gas identified as hydrogen was released and collected. The mass of hydrogen was 0.20 g. Write the chemical equation describing this reaction.

The reactants and one of the products are known, so the following partial chemical equation can be written

$$Fe(s) + HCl(aq) \xrightarrow{\text{not balanced}} H_2(g) + X$$

where X represents some substance containing Fe and Cl. The balanced equation can be found by first finding the mole ratio of hydrogen to iron, which will give the coefficients for Fe and H_2. These will allow the coefficient for HCl to be determined, which in turn should make it possible to determine the formula of X.

The number of moles of Fe that undergoes reaction is

$$(5.6 \text{ g Fe})\left[\frac{1 \text{ mol Fe}}{55.85 \text{ g Fe}}\right] = 0.10 \text{ mol Fe}$$

and the number of moles of H_2 produced is

$$(0.20 \text{ g } H_2)\left[\frac{1 \text{ mol } H_2}{2.02 \text{ g } H_2}\right] = 0.10 \text{ mol } H_2$$

For every 0.10 mol of Fe that undergoes reaction, 0.10 mol of H_2 is produced. In other words, the mole ratio is (1 mol H_2)/(1 mol Fe) and the chemical equation must reflect this ratio. Going back to the unbalanced equation above, the coefficient 2 is placed before the HCl to achieve a balance of hydrogen atoms.

$$Fe(s) + 2HCl(aq) \rightarrow H_2(g) + X$$

On the reactants side there are 1 Fe atom and 2 Cl atoms that are not accounted for on the products side. Assigning X = $FeCl_2$, which is the formula for one of the known chlorides of iron, the balanced equation can be written

$$Fe(s) + 2HCl(aq) \rightarrow H_2(g) + FeCl_2(aq)$$

Exercise 5.11 A mixture of $CaCO_3$ and SiO_2 was treated with hydrochloric acid.

$$CaCO_3(s) + 2HCl(aq) \rightarrow CaCl_2(aq) + H_2O(l) + CO_2(g)$$
$$SiO_2(s) + HCl(aq) \rightarrow \text{no reaction}$$

A 1.027 g sample of the mixture gave 0.274 g of CO_2. What is the percent composition of the mixture? *Answer*: 60.7 mass % $CaCO_3$, 39.3 mass % SiO_2.

PRACTICE TEST

1 (20 points). Choose the response that best completes each statement.

(a) The single reagent that determines the maximum amount of product because it is present in the smallest stoichiometric amount is the
(i) limiting reactant (ii) spectator ion
(iii) precipitate (iv) catalyst

(b) The maximum amount of a product that can be obtained from a known amount of reactants is the

(i) percent yield (ii) actual yield
(iii) theoretical yield (iv) normality

(c) Use the reactions

$$Zn + Cu^{2+} \rightarrow Cu + Zn^{2+}$$
$$Ag + Cu^{2+} \rightarrow \text{no reaction}$$

to determine the correct order of Zn, Cu, and Ag in the activity series.

(i) Ag > Cu > Zn (ii) Cu > Ag > Zn
(iii) Cu > Zn > Ag (iv) Zn > Cu > Ag

(d) Which partner-exchange reaction would be expected to occur in aqueous solution?

(i) $NaCl + KBr \rightarrow KCl + NaBr$
(ii) $Pb(NO_3)_2 + K_2SO_4 \rightarrow 2KNO_3 + PbSO_4$
(iii) $NH_4NO_3 + KI \rightarrow NH_4I + KNO_3$
(iv) $LiOH + NaCl \rightarrow LiCl + NaOH$

(e) The mole ratio of Cl^- to ClO_3^- is

$$6Cl_2(g) + 12OH^- \rightarrow 10Cl^- + 2ClO_3^- + 6H_2O(l)$$

(i) 2/10 (ii) 1/2
(iii) 10/6 (iv) 10/2

2 (10 points). A confirmatory test in many qualitative analysis schemes for the presence of Co^{2+} involves the equations

$$CoCl_2(aq) + 6KNO_2(aq) \rightarrow K_4[Co(NO_2)_6](aq) + 2KCl(aq)$$
$$K_4[Co(NO_2)_6](aq) + KNO_2(aq) + 2HCl(aq) \rightarrow K_3[Co(NO_2)_6](s) + NO(g) + H_2O(l) + 2KCl(aq)$$

Given the information that all soluble potassium salts, $CoCl_2$, and HCl exist in the ionic form in aqueous solution, write the net ionic equations for the two reactions.

3 (35 points). Choose the response that best completes each statement.

(a) How many moles of Fe_2O_3 can be produced from 56 g of iron and 16 g of oxygen?

(i) 1.00 (ii) 0.33
(iii) 0.50 (iv) 0.17

(b) What volume of 0.136 M HCl is required to undergo reaction with 3.62 g of KOH?

(i) 474 mL (ii) 237 mL
(iii) 26.6 mL (iv) none of these answers

(c) What mass of zinc sulfide can be produced by the reaction of 85.0 g of Zn and 48.0 g of S?

(i) 146 g (ii) 133 g

(iii) 127 g (iv) 97.44 g

(d) What mass of carbon dioxide is produced for 1.00 g of ethane, C_2H_6, that is burned?

$$2C_2H_6(g) + 7O_2(g) \rightarrow 4CO_2(g) + 6H_2O(g)$$

(i) 2.93 g (ii) 5.87 g

(iii) 1.47 g (iv) 44.01 g

(e) Exactly 5 g Mg is added to 25.0 mL of 6.0 M HCl. What mass of H_2 will be produced?

(i) 0.412 g (ii) 0.30 g

(iii) 0.15 g (iv) none of these answers

4 (10 points). A rather patriotic set of beakers can be prepared by adding concentrated aqueous NH_3 to solutions of phenolphthalein (represented by H_2In), $Pb(NO_3)_2$, and very dilute $CuSO_4$. The net ionic equations are

$$\underset{\text{colorless}}{H_2In(aq)} + 2OH^- \rightarrow \underset{\text{red}}{In^{2-}} + 2H_2O(l)$$

$$Pb^{2+} + 2OH^- \rightarrow \underset{\text{white}}{Pb(OH)_2(s)}$$

$$Cu^{2+} + 4NH_3(aq) \rightarrow \underset{\text{blue}}{[Cu(NH_3)_4]^{2+}}$$

Classify each reaction as a combination, decomposition, displacement, or partner-exchange reaction.

5 (10 points). Sodium hypochlorite undergoes decomposition

$$NaOCl(s) \xrightarrow{\text{unbalanced}} NaClO_3(s) + NaCl(s)$$

(a) What mass of $NaClO_3$ is produced by 10.0 g of NaOCl decomposing?

(b) If the actual yield is 4.2 g, what is the percent yield?

6 (10 points). A very dilute solution of acetic acid, CH_3COOH, was neutralized by a solution of NaOH.

$$CH_3COOH(aq) + NaOH(aq) \rightarrow Na(CH_3COO)(aq) + H_2O(l)$$

If 25.0 mL of the acetic acid solution required 1.6 mL of 0.025 M NaOH for the reaction, what is the concentration of the acid solution?

7 (5 points). The formation of sulfur trioxide is a two-step process.

$$S(s) + O_2(g) \rightarrow SO_2(g)$$
$$2SO_2(g) + O_2(g) \rightarrow 2SO_3(s)$$

What mass of SO_3 will be produced by burning 100.0 g of S in 100.0 g of O_2?

Answers to practice test

1. (a) i, (b) iii, (c) iv, (d) ii, (e) iv.
2. $Co^{2+} + 6NO_2^- \rightarrow [Co(NO_2)_6]^{4-}$, $[Co(NO_2)_6]^{4-} + NO_2^- + 2H^+ \rightarrow [Co(NO_2)_6]^{3-} + NO(g) + H_2O(l)$.
3. (a) ii, (b) i, (c) iii, (d) i, (e) iii.
4. partner exchange, partner exchange, combination.
5. (a) 4.77 g, (b) 88 %.
6. 0.0016 mol/L.
7. 168 g.

ADDITIONAL NOTES

CHAPTER 6

THE GASEOUS STATE

CHAPTER OVERVIEW

In this chapter the general properties of gases are reviewed and some practical aspects of measuring pressure are discussed. Next, the kinetic-molecular theory of gases is presented. Boyle's law and Charles' law, which deal with changes in pressure, volume, and temperature for a fixed quantity of gas, are presented. The amount of gas involved is introduced into the pressure-volume-temperature relationships through Avogadro's law, Gay-Lussac's law, and the ideal gas law. Finally, Dalton's law of partial pressures, stoichiometry of reactions involving gases, Graham's laws of effusion and diffusion, and deviations from the ideal gas law are discussed.

COMPETENCIES

Definitions for the following terms should be learned:

- absolute temperature scale
- absolute zero
- Avogadro's law
- barometer
- Boyle's law
- Charles' law
- combined gas law
- constant
- Dalton's law of partial pressures
- diffusion
- dynamic equilibrium
- effusion
- equilibrium vapor pressure
- Gay-Lussac's law of combining volumes
- Graham's law
- ideal gas
- ideal gas law
- Kelvin temperature scale
- kinetic energy
- manometer
- mole fraction
- partial pressure
- standard molar volume
- standard temperature and pressure (STP)
- variables
- volatile

General concepts and skills that should be learned include

- the general properties of the gaseous state
- the five statements summarizing the kinetic-molecular theory of gases and their description of the behavior of ideal gases
- the relationships among pressure, volume, temperature, and number of moles of a gas
- the definition of standard temperature and pressure (STP) for a gas
- the relationship given by the chemical equation for the relative volumes of gases involved in a chemical reaction
- the contribution of each individual gas to the properties of a gaseous mixture
- the differences between ideal and real gases

Numerical exercises that should be understood include

- finding the final pressure, volume, or temperature of an ideal gas by using Boyle's law, Charles' law, or the combined gas law when the initial conditions and the other final variable(s) are given
- determining the respective volumes of ideal gases involved in a chemical reaction by using Gay-Lussac's law
- calculating the number of moles or the volume of an ideal gas by using Avogadro's law when one of these variables is given
- calculating the pressure, volume, temperature, or the number of moles of an ideal gas by using the ideal gas law when three of these variables are given
- calculating density or molar mass for an ideal gas by using the ideal gas law
- calculating the total pressure of a mixture of ideal gases, the partial pressure of a gas in a mixture, or the mole fraction of a gas in a mixture by using Dalton's law when two of these variables are given
- calculating the amounts of reactants and products involved in a chemical reaction involving gases
- finding relative rates of effusion of gases from densities or molar masses by using Graham's law
- calculating the pressure or temperature of a real gas by using the van der Waals equation

QUESTIONS

1. Complete the following sentences by filling in the various blanks with these terms: absolute temperature scale, atmosphere, barometers, ideal gas, manometers, pressure, and Torr.

The various equations describing the (a) ________-volume-temperature behavior of an (b) ________ require the temperature to be expressed by using an (c) ________. Common units for expessing pressure are the pascal, (d) ________, and (e) ________. These pressures are measured by using (f) ________ for atmospheric pressures and (g) ________ for other pressures.

2. Match the correct term to each definition.

(a) PV = constant for a fixed amount of gas at a fixed temperature
(b) $rate_1/rate_2 = \sqrt{d_2/d_1}$ for two gases at the same temperature and pressure
(c) $P_1V_1/T_1 = P_2V_2/T_2$ for a fixed amount of gas
(d) $P_{total} = P_1 + P_2 + \ldots$
(e) V/T = constant for a fixed amount of gas at a fixed pressure
(f) $PV = nRT$
(g) $n_1/V_1 = n_2/V_2$ for a fixed pressure and temperature
(h) the ratios of volumes of gases involved in a chemical reaction, under fixed pressure and temperature conditions, are small whole numbers
(i) $(P + an^2/V^2)(V - nb) = nRT$

(i) Avogadro's law
(ii) Boyle's law
(iii) Charles' law
(iv) combined gas law
(v) Dalton's law of partial pressure
(vi) Gay-Lassac's law of combining volumes
(vii) Graham's law
(viii) ideal gas law
(ix) van der Waals' equation

3. Which of the following statements are true? Rewrite each false statement so that it is correct.

(a) A gauge on a cylinder of compressed gas reads zero pressure. This means the tank is empty.
(b) An increase from 10 °C to 20 °C will approximately double the volume of an ideal gas.
(c) Pressure is a force exerted per unit volume.
(d) Avogadro's law states that equal volumes of gases, measured under the same conditions of temperature and pressure, contain equal numbers of atoms.

(e) At the same temperature, the molecules of different gases, whether they are light or heavy, have the same average kinetic energy.

(f) Total energy is conserved during collisions between molecules although it can be transferred from one molecule to another.

(g) The average speed of heavier gaseous molecules is greater than that of lighter molecules at the same temperature.

(h) The molar volumes of all ideal gases at the same temperature and pressure are the same.

(i) Dalton's law of partial pressures allows the finding of the relative ratios of volumes of gases reacting and being produced in a chemical reaction.

4. Distinguish clearly between (a) a barometer and a manometer, (b) an ideal gas and a real gas, (c) the actual molecular speed and the average molecular speed, (d) a direct proportionality and an inverse proportionality, (e) effusion and diffusion, and (f) a gauge pressure and an absolute pressure.

5. Choose the response that best completes each statement.

(a) For an ideal gas, the physical quantity 22.4 L represents

(i) the molar volume (ii) the standard molar volume

(iii) absolute zero (iv) the rate of effusion

(b) Charles' law implies that for an ideal gas, the volume will become zero at

(i) absolute zero

(ii) the liquefaction point

(iii) the point at which the rate of effusion equals the rate of diffusion

(iv) 0 °C

(c) The partial pressure for a substance in a gaseous mixture can be found by multiplying the total pressure by

(i) the rate of effusion (ii) the equilibrium vapor pressure

(iii) the mole fraction (iv) absolute zero

(d) The escaping of molecules in the gaseous state, one by one, without collisions, through a hole of molecular dimensions is

(i) diffusion (ii) effusion

(iii) the equilibrium vapor pressure (iv) sublimation

(e) The mixing of molecules by random motion and collisions until the mixture becomes homogeneous is
(i) the partial pressure (ii) effusion
(iii) diffusion (iv) vapor pressure

(f) The equilibrium vapor pressure above a liquid is an example of
(i) effusion (ii) a dynamic equilibrium
(iii) the ideal gas law (iv) the momentum

6. Briefly describe your mental picture of an ideal gas.

7. Choose the properties that are characteristic of a gas:
(a) has negligible compressibility
(b) has infinite expandability
(c) assumes shape of container but with a flat surface and fixed volume
(d) flows rapidly because of very small viscosity
(e) consists of molecules that are moving with complete disorder
(f) forms either a liquid or a solid upon cooling
(g) expands as the temperature is decreased
(h) has a density similar to that of a liquid, but less than that of a solid

8. Each of the five statements at the right summarizing the kinetic-molecular theory is general enough for each to include several facts about gases. For example, that the volume of an atom of He is roughly 3.4×10^{-24} mL and the molar volume occupied by the gas is 22.4 L at STP illustrates that the space between molecules is large compared to the molecular size (the first statement). Match the correct statement to each observable fact.

(a) The pressure of a gas results from impacts with the walls of the container.
(b) Gases are easily compressible.
(c) Energy is conserved during molecular collisions.
(d) Temperature is a measure of the average kinetic energy of molecules.
(e) Pressure is directly proportional to the temperature under fixed volume conditions.

(i) Molecules are relatively far apart, and the spaces between the molecules are large compared to their size.
(ii) Molecules are in constant motion.
(iii) The average speed of molecules is proportional to the temperature.
(iv) Molecules of different gases have the same average kinetic energy at the same temperature.

(f) All molecules in a sample of gas do not have the same speed.

(g) Gases with lower molar mass move faster, on the average, than gases with higher molar mass at the same temperature.

(h) The volume actually occupied by molecules is negligible.

(v) Molecular collisions are elastic.

9. Two different gases have the same average kinetic energy at the same temperature. How do the average speeds of the molecules differ?

10. What causes the pressure on the walls of a container holding a gas?

11. Explain, in terms of the kinetic-molecular theory, the direct relationship between the pressure and the temperature of a gas in a fixed volume.

12. In sentence and equation forms, state (a) Boyle's law and (b) Charles' law.

13. What are the conditions known as standard temperature and pressure (STP)?

14. State Gay-Lussac's law of combining volumes. What is the relationship between the volumes of gases and the coefficients used to balance a chemical equation?

15. State Avogadro's law. How does this explain that any ideal gas will have the same molar volume under the same pressure and temperature conditions?

16. What is the numerical value of the standard molar volume of an ideal gas?

17. State the ideal gas law in mathematical form. Identify the variables to which the number of moles is directly proportional.

18. Show that (a) Boyle's law, (b) Charles' law, and (c) Avogadro's law are special cases of the ideal gas law.

19. What type of relationship exists between the molar mass and the density of an ideal gas?

20. State Dalton's law of partial pressures. What is the relationship between the partial pressures and the composition of a gaseous mixture?

21. How are the stoichiometric coefficients from a balanced chemical equation related to the relative volumes of gaseous reactants and products involved in the reaction?

22. What is the relationship between the rate of effusion of a gas and its density? What is the relationship between the rate of effusion and the molar mass of the gas?

23. How is the total pressure of a gaseous mixture related to the partial pressures of the components? Briefly explain why this is true.

24. What two properties of real gas molecules cause deviations from ideal gas behavior?

25. Under what experimental conditions will real gases most resemble ideal gases?

26. The volume of a real gas is not equal to zero at 0 K as predicted by Charles' law. Why?

27. None of the statements given in Question 8 that describe an ideal gas are completely valid when applied to real gases. For example, real gases can be liquefied, which means that under certain conditions the forces of attraction between molecules are not negligible. Identify which of the statements (i) to (v) in Question 8 are not valid for each of the following facts: (a) the van der Waals constant *b* is related to the volume actually occupied by a molecule and (b) the van der Waals constant *a* is related to the intermolecular attractions.

28. What types of experiments could be used to prove that molecular chlorine is diatomic?

Answers to questions

1. (a) pressure, (b) ideal gas, (c) absolute temperature scale, (d) torr or atmosphere, (e) atmosphere or torr, (f) barometers, (g) manometers.
2. (a) ii, (b) vii, (c) iv, (d) v, (e) iii, (f) viii, (g) i, (h) vi, (i) ix.
3. (a) F, replace *is empty* with *contains a gas at atmospheric pressure*; (b) F, change *°C* to *K* in both places; (c) F, change *volume* to *area*; (d) F, change *atoms* to *molecules*, (e) T; (f) T; (g) F, replace *greater* with *less*; (h) T; (i) F, change *Dalton's law of partial pressures* to Gay-Lussac's law of combining volumes.
4. (a) A *barometer* is used to measure atmospheric pressure; a *manometer* is used to measure the pressure of the enclosed system to which it is connected. (b) An *ideal gas* obeys the ideal gas law exactly; a *real gas* shows deviations from the ideal gas law. (c) The *actual molecular speed* is the speed at which a given molecule travels; the *average molecular speed* represents the average

speed for the collection of molecules or represents the speed of an average molecule. (d) A variable related to a second variable by a *direct proportionality* will increase as the second increases or decrease as the second decreases; a variable related to a second variable by an *inverse proportionality* will increase as the second decreases or decrease as the second increases. (e) *Effusion* is the escape (one by one) of molecules through a hole of molecular size; *diffusion* is the mixing caused by collisions until a solution is formed. (f) A *gauge pressure* represents the difference between the absolute pressure of a gas and atmospheric pressure; the *absolute pressure* is the pressure that the gas exhibits under the given set of conditions.

5. (a) ii, (b) i, (c) iii, (d) ii, (e) iii, (f) ii.
6. An ideal gas is a set of rapidly, randomly moving particles that are very small compared to the size of their container and that undergo a large number of collisions in a unit time.
7. b, d, e, f.
8. (a) ii, (b) i, (c) v, (d) iv, (e) iii, (f) iii, (g) iv, (h) i.
9. For two different gases with the same average kinetic energy at the same temperature, the average speed of the lighter one will be greater.
10. The pressure on the walls of a container holding a gas is caused by molecular collisions.
11. For a gas in a fixed volume, its pressure is directly proportional to the temperature. The speed of the molecules increases as the temperature increases and so the molecules collide with the walls more often and with more force, increasing the pressure.
12. (a) Boyle's law is stated as follows: at constant temperature, the volume of a given mass of gas is inversely proportional to the pressure upon the gas.

$$PV = \text{constant}$$

    (b) Charles' law is stated as follows: at constant pressure, the volume of a given mass of gas is directly proportional to the absolute temperature.

$$V = (\text{constant})T$$

13. 0 °C or 273.15 K and 1 atm or 760 Torr pressure.
14. Gay-Lussac's law of combining volumes is stated as follows: when gases react or gaseous products are formed, the ratios of the volumes of the gases involved, measured at the same temperature and pressure, are small whole

numbers. The volumes of gases and the coefficients used to balance the chemical equation are the same.

15. Avogadro's law is stated as follows: equal volumes of gases, measured under the same conditions of temperature and pressure, contain equal numbers of molecules. All ideal gases therefore have the same molar volume at STP since equal numbers of molecules correspond to equal numbers of moles of gases.
16. 22.4 L/mol
17. The ideal gas law stated in mathematical form is $PV = nRT$. The number of moles is directly proportional to the pressure and volume.
18. (a) $PV = nRT$ = constant for constant n and T, (b) $V = (nR/P)T$ = (constant)T for constant n and P, (c) $V = (RT/P)n$ = (constant)n for constant T and P.
19. A direct proportion exists.
20. Dalton's law of partial pressures is stated as follows: in a mixture of gases, the total pressure exerted is the sum of the pressures that each gas would exert if it were present alone under the same conditions. The mole fraction of each component of a mixture is the same as the fraction of the total pressure exerted by that component.
21. The volumes are directly proportional to the coefficients.
22. The rate of effusion of a gas is inversely proportional to the square root of its density. The rate of effusion of a gas is also inversely proportional to the square root of its molar mass.
23. The total pressure equals the sum of the partial pressures; each gas acts independently of the others present and so the total effect is the sum of each individual effect.
24. The two properties of real gas molecules that cause deviations from ideal gas behavior are intermolecular forces of attraction and the small, but real, volumes of the molecules.
25. Ideal gas behavior is approached at low pressure and high temperature by real gases.
26. Not only do real gases have actual molecular volumes (which ideal gases do not), they also form liquids and solids (which do not obey Charles' law) before reaching absolute zero.
27. (a) i, (b) v.
28. Experiments that would show molecular chlorine to be diatomic include (a) measurement of pressure, volume, temperature, and mass of a sample and use

of the ideal gas law; (b) reaction with another gas such as hydrogen and measurement of volumes of reactants and product(s) and use of Gay-Lussac's law; and (c) measurement of its effusion and use of Graham's law.

SKILLS

Skill 6.1 Variables and proportionality

A variable is a measurable property. For example, the mass and volume of a substance are two variables that are related by the equation (mass) = (density)(volume) and the speed and time of travel are two variables that are related by the equation (speed)(time) = (distance of trip).

If one variable (y) is directly proportional to a second variable (x), written $y \propto x$, this means that the value of y will increase in the same ratio as the value of x increases and will decrease in the same ratio as the value of x decreases. For example, 20 mL of lead weighs twice as much as 10 mL of lead and 5 mL of lead weighs half as much as 10 mL of lead.

A direct proportion can be written in the form

$$y = kx$$

where k is called the proportionality constant. The mass is directly proportional to the volume of a substance--the proportionality constant relating mass and volume is the density.

If one variable (y) is inversely proportional to a second variable (x), written $y \propto (1/x)$, this means that the value of y will increase in the same ratio as the value of x decreases and the value of y will decrease in the same ratio as the value of x increases. For example, doubling the speed will decrease the time required for a given trip to one-half its original value and slowing down to one-half of the original speed will double the time for the trip.

An inverse proportion can be written in the form

$$y = k(1/x) \qquad \text{or} \qquad (x)(y) = k$$

where k is the proportionality constant. The speed is inversely proportional to the time--the proportionality constant is the length of the trip.

More complicated proportions are often encountered in chemistry. For example, kinetic energy

$$(\text{kinetic energy}) = (1/2)(\text{mass})(\text{speed})^2$$

is directly proportional to the mass and directly proportional to the square of the speed--the proportionality constant is (1/2). The pH of a solution

$$pH = \log\left[\frac{1}{\text{concentration of } H^+}\right]$$

is directly proportional to the logarithm of the inverse of the concentration of H^+.

Skill exercise: Classify the following proportions between variables as either direct or inverse and identify the proportionality constant: (a) $PV = nRT$ at constant n and T, (b) $PV = nRT$ at constant V and n, (c) $PV = nRT$ at constant P and n, (d) $PV = nRT$ at constant P and T, (e) $PV = nRT$ at constant P and V, and (f) $PV = nRT$ at constant V and T. *Answers*: (a) P is inversely proportional to V, (nRT); (b) P is directly proportional to T, (nR/V); (c) V is directly proportional to T, (nR/P); (d) V is directly proportional to n, (RT/P); (e) n is inversely proportional to T, (PV/R); (f) P is directly proportional to n, (RT/V).

Skill 6.2 Interpreting chemical equations involving gases

For a chemical reaction involving gases, Gay-Lussac's law of combining volumes states that the relative gas volumes involved in the reaction (measured at the same temperature and pressure) are the same as the coefficients for the gaseous compounds in the balanced equation. For example, for the reaction

$$3H_2(g) + N_2(g) \rightarrow 2NH_3(g)$$

the chemical equation states that if all volumes of gases are measured under the same temperature and pressure conditions, 3 volumes of hydrogen will react with 1 volume of nitrogen to give 2 volumes of ammonia. The word "volume" means that all volumes are given in the same units (e.g., liters).

A series of "conversion factors" can be derived from Gay-Lussac's law which can be used (see *Skill 2.5 Using the dimensional method in problem solving*) to find the relationships between the various volumes. For example, the conversion factors for the ammonia reaction are (3 volumes H_2)/(2 volumes NH_3), (2 volumes NH_3)/(3 volumes H_2), (3 volumes H_2)/(1 volume N_2), (1 volume N_2)/(3 volumes H_2), (2 volumes NH_3)/(1 volume N_2), and (1 volume N_2)/(2 volumes NH_3).

Note that Gay-Lussac's law is valid only for gaseous substances involved in a chemical reaction. For example, in the reaction

$$CH_4(g) + 2O_2(g) \rightarrow CO_2(g) + 2H_2O(l)$$

only volume relationships between CH_4, O_2, and CO_2 can be derived. Gay-Lussac's law cannot be used to discuss the volume of water because it is a liquid.

Avogadro's law states that under constant pressure and temperature equal volumes of gases contain equal numbers of molecules (hence the same number of moles). Because these are the same conditions as for Gay-Lussac's law, a chemical equation can also be interpreted in terms of number of molecules and number of moles. For example, the equation

$$3H_2(g) + N_2(g) \rightarrow 2NH_3(g)$$

can also be interpreted as stating that 3 molecules of hydrogen react with 1 molecule of nitrogen to give 2 molecules of ammonia, and that as 3 moles of hydrogen react with 1 mole of nitrogen to give 2 moles of ammonia. [Although this skill considers only gaseous reactants and products, the molecular and molar interpretations of a chemical equation are valid for all states of matter (see *Skill 5.1 Interpreting chemical equations*).]

Skill exercise: Write the "conversion factors" that can be derived for the gaseous substances in terms of volumes, number of molecules, and number of moles for each of the following equations: (a) $2CO(g) + O_2(g) \rightarrow 2CO_2(g)$, (b) $C(s) + O_2(g) \rightarrow CO_2(g)$, and (c) $HCl(g) + NH_3(g) \rightarrow NH_4Cl(s)$. *Answers*: (a) (2 volumes CO)/(1 volume O_2), (1 volume O_2)/(2 volumes CO), (2 volumes CO)/(2 volumes CO_2), (2 volumes CO_2)/(2 volumes CO), (1 volume O_2)/(2 volumes CO_2), (2 volumes CO_2)/(1 volume O_2), (2 molecules CO)/(1 molecule O_2), (1 molecule O_2)/(2 molecules CO), (2 molecules CO)/(2 molecules CO_2), (2 molecules CO_2)/(2 molecules CO), (1 molecule O_2)/(2 molecules CO_2), (2 molecules CO_2)/(1 molecule O_2), (2 mol CO)/(1 mol O_2), (1 mol O_2)/(2 mol CO), (2 mol CO)/(2 mol CO_2), (2 mol CO_2)/(2 mol CO), (1 mol O_2)/(2 mol CO_2), (2 mol CO_2)/(1 mol O_2); (b) (1 volume O_2)/(1 volume CO_2), (1 volume CO_2)/(1 volume O_2), (1 molecule O_2)/(1 molecule CO_2), (1 molecule CO_2)/(1 molecule O_2), (1 mol O_2)/(1 mol CO_2), (1 mol CO_2)/(1 mol O_2); (c) (1 volume HCl)/(1 volume NH_3), (1 volume NH_3)/(1 volume HCl), (1 molecule HCl)/(1 molecule NH_3), (1 molecule NH_3)/(1 molecule HCl), (1 mol HCl)/(1 mol NH_3), (1 mol NH_3)/(1 mol HCl).

Skill 6.3 Mixtures of gases

In a mixture of ideal gases that do not undergo reaction with each other, there is no limit to how much of one ideal gas will dissolve in another. Because of this wide range of composition, a convenient concentration unit to use is the mole fraction,

$$X_i = \frac{n_i}{n_1 + n_2 + \dots}$$

where X_i is the mole fraction of component i, n_i is the number of moles of component i in the mixture, and $n_1 + n_2 + \dots$ represents the total number of moles of all of the gases making up the mixture. [The mole fraction can be used to express concentrations of components in any solution, not just mixtures of gases.]

Dalton's law of partial pressures states that the total pressure exerted by a mixture of gases is equal to the sum of these partial pressures:

$$P_{\text{total}} = P_1 + P_2 + \dots$$

and because the volume and temperature are the same for each gas in the mixture, the pressure that each gas exerts is directly proportional to the number of moles of that gas present in the mixture.

$$P_i = X_i P_{\text{total}}$$

For example, in a mixture containing 1 mol of nitrogen and 1 mol of argon at 1 atm, these equations show that $X_{N_2} = X_{Ar} = 0.5$ and $P_{N_2} = P_{Ar} = 0.5$ atm.

Skill exercise: Consider a mixture containing 0.1 mol He and 0.4 mol Ne at a total pressure of 5 atm. Calculate the mole fractions and partial pressures of each gas. *Answers*: $X_{He} = 0.2$, $X_{Ne} = 0.8$; $P_{He} = 1$ atm, $P_{Ne} = 4$ atm.

NUMERICAL EXERCISES

Example 6.1 P-V-T-n relationships: V and P

A 3.0 g sample of gaseous Ar originally at 14.2 Torr and 57 L is compressed at constant temperature so that the final pressure is 712 Torr. What is the final volume?

Solving Boyle's law for V_2 and substituting $P_1 = 14.2$ Torr, $V_1 = 57$ L, and $P_2 = 712$ Torr gives

$$V_2 = V_1\left[\frac{P_1}{P_2}\right] = (57 \text{ L})\left[\frac{14.2 \text{ Torr}}{712 \text{ Torr}}\right] = 1.14 \text{ L}$$

The final volume of the gas is 1.14 L. Applying common sense to the problem indicates that the final volume should be less than the original volume because the pressure was increased.

Exercise 6.1 Under constant temperature conditions, the volume of 0.12 g of gaseous ozone is increased from 1.00 L to 10.0 L. If the original pressure of the ozone was 0.075 atm, find the final pressure. *Answer*: 0.0075 atm.

Example 6.2 P-V-T-n relationships: V and T

The temperature of a 0.102 mol sample of gaseous O_2 is changed from 25 °C to 125 °C under constant pressure conditions. If the original volume of the sample was 6.2 L, find the final volume.

The temperatures in all gas law problems must be expressed on an absolute temperature scale:

$$T_1 = 25\ °C + 273 = 298\ K \qquad T_2 = 125\ °C + 273 = 398\ K$$

Solving Charles' law for V_2 and substituting $V_1 = 6.2$ L and the temperatures above gives

$$V_2 = V_1\left[\frac{T_2}{T_1}\right] = (6.2\ L)\left[\frac{398\ K}{298\ K}\right] = 8.3\ L$$

The final volume of the O_2 is 8.3 L. Applying common sense to the problem indicates that the final volume should be greater than the original volume because the temperature was increased.

Exercise 6.2 A balloon decreased in size from 0.831 ft^3 to 0.790 ft^3 as it was taken from a room at 23 °C into a cold garage. Assuming that the pressure in both places was the same, calculate the temperature in the garage. *Answer*: 8 °C.

Example 6.3 P-V-T-n relationships: P and T

The temperature of an ideal gas was increased from 125 °C to 250. °C. What adjustment must be made in the pressure so that the volume of the gas is unchanged?

For a fixed amount of gas held at constant volume, pressure and temperature are inversely related.

$$\frac{P_1}{T_1} = \frac{P_2}{T_2} \qquad \text{or} \qquad P_2 = P_1\left[\frac{T_2}{T_1}\right]$$

The unknown in this problem is not P_2, the final pressure. It is how the initial pressure must be changed in order to maintain a constant volume.

The initial and final temperatures are

T_1 = 125 °C + 273 = 398 K $\qquad$ T_2 = 250. °C + 273 = 523 K

Therefore

$$P_2 = P_1\left[\frac{523\ \text{K}}{398\ \text{K}}\right] = 1.31\ P_1$$

For the volume to remain constant, the pressure has to be increased so that it is 1.31 times the original pressure.

Exercise 6.3 The pressure of an ideal gas at 25 °C was decreased from 762 Torr to 741 Torr at constant volume. What was the new temperature? *Answer*: 17 °C.

Example 6.4 P-V-T-n relationships: P, V, and T

The temperature of one mole of an ideal gas was changed from -10 °C to 10 °C and the pressure was changed from 105 kPa to 415 kPa. Will the final volume of the gas be more or less than the original volume?

Solving the combined gas for the final volume and substituting the data gives

$$V_2 = V_1\left[\frac{P_1}{P_2}\right]\left[\frac{T_2}{T_1}\right] = V_1\left[\frac{105\ \text{kPa}}{415\ \text{kPa}}\right]\left[\frac{283\ \text{K}}{263\ \text{K}}\right] = 0.272\ V_1$$

The final volume will be a little over one-fourth of the original volume.

Exercise 6.4 What will be the final temperature for a given amount of gas originally at 65 °C, 4.2 L, and 757 Torr if the volume is changed to 7.2 L and the pressure is changed to 755 Torr? *Answer*: 310 °C.

Example 6.5 P-V-T-n relationships: P, V, and T

What will be the overall effect on the volume of an ideal gas of doubling the absolute temperature and then doubling the volume, both processes being done at constant pressure?

The volume is directly proportional to the temperature; thus for the first process

$$V_2 = V_1\left[\frac{T_2}{T_1}\right] = V_1\left[\frac{2T_1}{T_1}\right] = 2V_1$$

The second step simply doubles the volume of V_2, giving $4V_1$. The final volume is four times as large as the original volume.

Exercise 6.5 What will be the overall effect on the volume of an ideal gas of doubling the pressure and then doubling the volume, both processes being done at constant temperature? *Answer*: no change.

Example 6.6 P-V-T-n relationships: P, V, and T*

A 4.2 L sample of air measured at -65 °C and 0.79 atm was adjusted to standard temperature and pressure. What is the volume of this gas at STP?

This problem is analyzed using the comprehensive problem-solving approach presented in *Skill 2.7 Solving problems.*

1. Study the problem and be sure you understand it.
   (a) What is unknown?
   The volume at STP.
   (b) What is known?
   The original volume, temperature, and pressure of the gas and the final temperature (273 K) and pressure (1 atm).
2. Decide how to solve the problem.
   (a) What is the connection between the known and the unknown?
   The combined gas law relates pressures, volumes, and temperatures of ideal gases.
   (b) What is necessary to make the connection?
   The combined gas law must be solved for the volume at STP and the known values of volume, pressure, and temperature used to find V_{STP}.
3. Set up the problem and solve it.
   The volume of the gas at STP is

$$V_{STP} = V_1\left[\frac{P_1}{P_{STP}}\right]\left[\frac{T_{STP}}{T_1}\right] = (4.2\text{ L})\left[\frac{0.79\text{ atm}}{1.00\text{ atm}}\right]\left[\frac{273\text{ K}}{208\text{ K}}\right] = 4.4\text{ L}$$

4. Check the result.
   (a) Are significant figures and the location of the decimal point correct?
   Yes.
   (b) Did the answer come out in the correct units?
   Yes. Any pressure unit--atm, Pa, bar, Torr, etc.-- may be used in the (P_1/P_{STP}) term, just as long as the same unit is used in both the numerator and denominator.
   (c) Is the answer reasonable?
   Yes. The pressure and temperature changes being made are not very large, so the net result is that the final volume should not be drastically different from the original volume. In fact, because both the pressure and temperature are increased, the changes somewhat offset each other.

Exercise 6.6 Exactly 1 L of a gas measured at STP was heated to 750 K and expanded to 1.35 m^3. What is the pressure of this gas under these conditions? *Answer*: 0.0020 atm.

Example 6.7 P-V-T-n relationships: Gay-Lussac's law*

Gaseous water is formed by the reaction of hydrogen with oxygen. Is there a significant volume increase as the reaction takes place at constant temperature and pressure?

This problem is analyzed using the comprehensive problem-solving approach presented in *Skill 2.7 Solving problems*.

1. Study the problem and be sure you understand it.
   (a) What is unknown?
   Whether the gas volume increases, decreases, or remains approximately the same as hydrogen undergoes reaction with oxygen to form steam.
   (b) What is known?
   The reactants are gaseous hydrogen and gaseous oxygen, and that the product is gaseous water.
2. Decide how to solve the problem.
   (a) What is the connection between the known and the unknown?
   The chemical equation gives the connection by showing the relative volumes of gases involved in the reaction.

$$2H_2(g) + O_2(g) \rightarrow 2H_2O(g)$$

   (b) What is necessary to make the connection?
   Application of Gay-Lussac's law of combining volumes (see *Skill 6.2 Interpreting chemical equations involving gases*) to this equation shows that two volumes of hydrogen combine with one volume of oxygen to produce two volumes of steam.
3. Set up the problem and solve it.
   Three volumes of gaseous reactants are forming two volumes of gaseous products and there should be a decrease in volume as the reaction takes place. No, a significant volume increase does not take place.
4. Check the result.
   (a) Is the answer reasonable?
   Yes.

Exercise 6.7 Under a given set of pressure and temperature conditions, what volume of gaseous O_2 is needed to react with 2.0 L of CO(g) to form CO_2(g)? What volume of CO_2(g) will be formed? The equation is

$$2CO(g) + O_2(g) \rightarrow 2CO_2(g)$$

Answer: 1.0 L of O_2, 2.0 L of CO_2.

Example 6.8 P-V-T-n relationships: Avogadro's law

A typical cardboard shoe box is 6.5 inch x 4.0 inch x 12.5 inch. A typical file cabinet measures 26.5 inch x 14.75 inch x 52.5 inch. What is the ratio of the number of "molecules" of air in the cabinet to the number in the shoe box?

The desired ratio can be found by using Avogadro's law as follows:

$$\frac{N_{\text{cabinet}}}{N_{\text{box}}} = \frac{V_{\text{cabinet}}}{V_{\text{box}}} = \frac{(26.5)(14.75)(52.5)\ \text{inch}^3}{(6.5)(4.0)(12.5)\ \text{inch}^3} = 63$$

There are 63 times as many "molecules" in the cabinet.

Exercise 6.8 A lecture bottle of gaseous N_2 is about 36 inch³ in volume and a large gas cylinder is about 5200 inch³ in volume. What is the ratio of the number of moles in the gas cylinder to that in a lecture bottle? *Answer*: 140.

Example 6.9 P-V-T-n relationships: molar volume*

The density of methane at STP is 0.717 g/L. Calculate the molar mass of methane.

This problem is analyzed using the comprehensive problem-solving approach presented in *Skill 2.7 Solving problems.*

1. <u>Study the problem and be sure you understand it</u>.
   (a) What is unknown?
   The molar mass.
   (b) What is known?
   The density at STP (0.717 g/L) and the standard molar volume (22.4 L/mol).
2. <u>Decide how to solve the problem</u>.
   (a) What is the connection between the known and the unknown?
   To calculate the molar mass, the mass and the number of moles must be determined for a sample of the substance. The standard molar volume provides the connection between the volume at STP and the number of moles in that volume. The density provides the connection between the volume and the mass in that volume.

(b) What is necessary to make the connection?

The number of moles of methane in a chosen volume, such as 1.00 L, can be found by using the standard molar volume. The mass of this volume can be found by using the density. The molar mass can then be found by dividing the mass by the number of moles.

3. Set up the problem and solve it.

The mass of 1.00 L of methane at STP is

$$(1.00\text{ L})\left[\frac{0.717\text{ g methane}}{1\text{ L}}\right] = 0.717\text{ g methane}$$

The number of moles in 1.00 L of methane at STP is

$$(1.00\text{ L})\left[\frac{1\text{ mol methane}}{22.4\text{ L}}\right] = 0.0446\text{ mol methane}$$

The molar mass is

$$\frac{0.717\text{ g methane}}{0.0446\text{ mol methane}} = 16.1\text{ g/mol}$$

4. Check the result.

(a) Are significant figures and the location of the decimal point correct?

Yes.

(b) Did the answer come out in the correct units?

Yes.

(c) Is the answer reasonable?

Yes. Molar masses of gaseous materials at STP are usually between 2 g/mol and 200 g/mol.

Exercise 6.9 The molar mass of ethane, C_2H_6, is 30.1 g. What is the density of this gas at STP? *Answer*: 1.34 g/L.

Example 6.10 P-V-T-n relationships: ideal gas law

A flask is marked "250. mL." How many moles of a gas would be present in this flask at typical room conditions of 23 °C and 736 Torr?

Solving the ideal gas law for the number of moles and substituting the data gives

$$n = \frac{PV}{RT} = \frac{(736\text{ Torr})(1\text{ atm}/760\text{ Torr})(250.\text{ mL})(1\text{ L}/1000\text{ mL})}{(0.0821\text{ L atm/K mol})(23 + 273)\text{ K}} = 0.00996\text{ mol}$$

The flask would hold 0.00996 mol of gas.

Exercise 6.10 What pressure is needed to confine 0.35 mol of CO_2(g) at 450 °C in a volume of 1.6 L? *Answer*: 13 atm.

Example 6.11 P-V-T-n relationships: ideal gas law

The mass of an "empty" flask is actually the sum of the masses of the flask and the air in the flask. In some experiments it is necessary to know how much air is in a flask. Calculate the mass of air in an "empty" flask if the flask has a volume of 270 mL and the temperature and pressure of the air in the flask are 22 °C and 755 Torr, respectively. (Air is a mixture of gases that can be treated as having an average molar mass of 29.0 g/mol.)

The number of moles of air in the flask is

$$n = \frac{PV}{RT} = \frac{(755\ \text{Torr})(1\ \text{atm}/760\ \text{Torr})(270\ \text{mL})(1\ \text{L}/1000\ \text{mL})}{(0.0821\ \text{L atm/K mol})(22 + 273)\ \text{K}} = 0.011\ \text{mol}$$

The mass is

$$(0.011\ \text{mol})\left[\frac{29.0\ \text{g}}{1\ \text{mol}}\right] = 0.32\ \text{g}$$

The "empty" flask contains 0.32 g of air.

Exercise 6.11 Most student analytical balances have a precision of ±0.1 mg. To what volume of air measured at 752 Torr and 28 °C does this mass correspond? *Answer*: 0.09 mL.

Example 6.12 Density of ideal gases*

What is the density of ozone, O_3, at 22 °C and 0.935 atm?

This problem is analyzed using the comprehensive problem-solving approach presented in *Skill 2.7 Solving problems*.

1. <u>Study the problem and be sure you understand it</u>.
   (a) What is unknown?
   The density.
   (b) What is known?
   The temperature and pressure of the gas and its molecular formula.
2. <u>Decide how to solve the problem</u>.
   (a) What is the connection between the known and the unknown?
   To calculate the density, the mass and volume of a sample of the substance must be determined. The ideal gas law relates the number of moles to the volume and the molar mass relates the number of moles to the mass.
   (b) What is necessary to make the connection?
   An expression for density in terms of P, T, and molar mass (M) can be derived from the ideal gas law. The mass is

$$m = nM$$

and the volume is

$$V = \frac{nRT}{P}$$

Substituting these equations for mass and volume into the definition of density gives

$$d = \frac{m}{V} = \frac{nM}{nRT/P} = \frac{PM}{RT}$$

This equation relates the density of a gas to pressure, temperature, and molar mass.

3. Set up the problem and solve it.
   The density is

$$d = \frac{PM}{RT} = \frac{(0.935 \text{ atm})(48.0 \text{ g/mol})}{(0.0821 \text{ L atm/mol K})(295 \text{ K})} = 1.85 \text{ g/L}$$

4. Check the result.
   (a) Are significant figures and the location of the decimal point correct?
       Yes.
   (b) Did the answer come out in the correct units?
       Yes.
   (c) Is the answer reasonable?
       Yes. The density of most gases is on the order of a few grams per liter.

Exercise 6.12 The density of nitrosyl fluoride at 20 °C and 1.00 atm is 2.18 g/L. Find the molar mass of this compound assuming ideal gas behavior.
Answer: 52 g/mol.

Example 6.13 Dalton's law

The partial pressures of $H_2(g)$ and $O_2(g)$ in a mixture are 110 Torr and 350 Torr, respectively. What are the respective mole fractions?

Use of Dalton's law gives the following relationship:

$$P_{\text{total}} = P_{H_2} + P_{O_2} = 110 \text{ Torr} + 350 \text{ Torr} = 460 \text{ Torr}$$

The mole fraction of each component is

$$X_{H_2} = \frac{P_{H_2}}{P_{\text{total}}} = \frac{110 \text{ Torr}}{460 \text{ Torr}} = 0.24 \qquad X_{O_2} = \frac{P_{O_2}}{P_{\text{total}}} = \frac{350 \text{ Torr}}{460 \text{ Torr}} = 0.76$$

For this gaseous mixture, $X_{H_2} = 0.24$ and $X_{O_2} = 0.76$. Note that the sum of the mole fractions is unity.

Exercise 6.13 A flask contains a mixture of air at 1.00 atm such that X_{N_2} = 0.80, X_{O_2} = 0.19, and X_{Ar} = 0.01. What are the partial pressures of the gases? *Answer*: 0.80 atm N_2, 0.19 atm O_2, and 0.01 atm Ar.

Example 6.14 Dalton's law

How many moles of oxygen are in a 250 mL sample of "wet" oxygen measured at 765 Torr and 28 °C? The vapor pressure of water at this temperature is 28.349 Torr.

The partial pressure of the oxygen is

$$P_{O_2} = P_{total} - P_{H_2O} = 765 \text{ Torr} - 28.349 \text{ Torr} = 737 \text{ Torr}$$

Using the ideal gas law gives

$$n = \frac{PV}{RT} = \frac{(737 \text{ Torr})(1 \text{ atm}/760 \text{ Torr})(250 \text{ mL})(1 \text{ L}/1000 \text{ mL})}{(0.0821 \text{ L atm/K mol})(28 + 273) \text{ K}} = 0.0098 \text{ mol } O_2$$

The sample contains 0.0098 mol O_2.

Exercise 6.14 The composition of a mixture of helium and neon is X_{He} = 0.25. How many moles of helium are in 1.5 L of the mixture measured at 1.01 atm and 675 K? *Answer*: 0.0068 mol He.

Example 6.15 Stoichiometry

What volume of NH_3 measured at 735 Torr and 45 °C will be produced for every 10.0 g Mg_3N_2 that undergoes the following reaction?

$$Mg_3N_2(s) + 6H_2O(l) \rightarrow 3Mg(OH)_2(s) + 2NH_3(g)$$

Problems involving stoichiometry are solved using the four step process presented in *Skill 5.5 Stoichiometry*. The balanced chemical equation is given above.

$$(10.0 \text{ g } Mg_3N_2)\left[\frac{1 \text{ mol } Mg_3N_2}{100.95 \text{ g } Mg_3N_2}\right] = 0.0991 \text{ mol } Mg_3N_2$$

$$(0.0991 \text{ mol } Mg_3N_2)\left[\frac{2 \text{ mol } NH_3}{1 \text{ mol } Mg_3N_2}\right] = 0.198 \text{ mol } NH_3$$

$$V = \frac{nRT}{P} = \frac{(0.198 \text{ mol } NH_3)(0.0821 \text{ L atm/K mol})(318 \text{ K})}{(735 \text{ Torr})(1 \text{ atm}/760 \text{ Torr})}$$

$$= 5.35 \text{ L}$$

The expected volume would be 5.35 L

Exercise 6.15 What volume of CO is produced at 4.25 atm and 1250 K for each 1.00 g SiO_2 that undergoes reaction?

$$SiO_2(s) + 2C(s) \rightarrow Si(s) + 2CO(g)$$

Answer: 0.804 L.

Example 6.16 Stoichiometry

Slaked lime, $Ca(OH)_2$, is an important industrial chemical produced from limestone, $CaCO_3$, by the following reactions:

$$CaCO_3(s) \xrightarrow{\Delta} CaO(s) + CO_2(g)$$

$$CaO(s) + H_2O(l) \rightarrow Ca(OH)_2$$

The CO_2 obtained is sometimes used to produce dry ice. To study the first reaction in the laboratory, a 25.0 g sample of pure limestone was heated at a moderate temperature for about 30 min and the CO_2 formed as collected at 1.00 atm and 25°C. The volume of gas was 4.20 L. What percentage of the $CaCO_3$ decomposed?

The equation for the reaction is given above. The number of moles of CO_2 is

$$n = \frac{PV}{RT} = \frac{(1.00 \text{ atm})(4.20 \text{ L})}{(0.0821 \text{ L atm/K mol})(298 \text{ K})} = 0.172 \text{ mol } CO_2$$

Using mole ratios to find the mass of $CaCO_3$ that underwent reaction gives and molar mass

$$(0.172 \text{ mol } CO_2)\left[\frac{1 \text{ mol } CaCO_3}{1 \text{ mol } CO_2}\right]\left[\frac{100.0 \text{ g } CaCO_3}{1 \text{ mol } CaCO_3}\right] = 17.2 \text{ g } CaCO_3$$

The percentage of $CaCO_3$ that reacted is

$$\frac{17.2 \text{ g}}{25.0 \text{ g}} \times 100 = 68.8 \%$$

Only 68.8 % of the $CaCO_3$ decomposed in the period of heating.

Exercise 6.16 A 1.16 L sample of H_2 measured at 15 °C and 0.987 atm was mixed with a 1.0 L sample of O_2 measured at 25 °C and 1.028 atm. An electrical spark ignited this mixture to produce water. Calculate the mass of water produced.
Answer: 1.2 g H_2O.

Example 6.17 Graham's law

A rubber balloon contains helium and hydrogen at equal partial pressures. What is the rate of effusion of hydrogen compared to that of helium?

Using Graham's law of effusion gives

$$\frac{\text{rate}_{H_2}}{\text{rate}_{He}} = \sqrt{\frac{\text{molar mass of He}}{\text{molar mass of } H_2}} \quad \sqrt{\frac{4.00}{2.02}} = 1.41$$

Hydrogen will effuse 1.41 times as fast as the helium.

Exercise 6.17 The relative rates of effusion of methane, CH_4, and an unknown gas were 1.58 to 1.00. What is the molecular mass of the unknown gas? *Answer*: 39.9 u.

Example 6.18 Real gases

Assume that the gas in *Example 6.10* is NO_2, which has the following van der Waals coefficients: $a = 5.284\ L^2\ atm/mol^2$ and $b = 0.04424$ L/mol. Calculate the temperature at which 0.00996 mol of NO_2 occupies 250. mL at 736 Torr assuming the van der Waals equation to be valid. What is the difference from 23 °C?

Solving the van der Waals equation for temperature and substituting in the data gives

$$T = \left[P + \frac{an^2}{V_2}\right]\frac{(V - nb)}{nR}$$

$$= \left[(736\ \text{Torr})\left[\frac{1\ \text{atm}}{760\ \text{Torr}}\right] + \frac{(5.284\ L^2\ \text{atm/mol}^2)(0.00996\ \text{mol})^2}{(0.250\ L)^2}\right]$$

$$\times \left[\frac{(0.250\ L) - (0.00996\ \text{mol})(0.04424\ L/\text{mol})}{(0.00996\ \text{mol})(0.0821\ L\ \text{atm/K mol})}\right]$$

$$= 298\ K = 25\ °C$$

There is a 2 °C difference (about 0.7%).

Exercise 6.18 Repeat Exercise 6.10 assuming CO_2 to obey the van der Waals equation: $a = 3.592\ L^2\ atm/mol^2$ and $b = 0.04267$ L/mol. *Answer*: 13 atm, no significant change.

PRACTICE TEST

1 (5 points). Match the correct mathematical statement to the respective gas law.

(a) Boyle's law (i) $P_{total} = P_1 + P_2 + \ldots$

(b) Charles' law (ii) PV = constant for fixed T, n

(c) Dalton's law (iii) $PV = nRT$

(d) Graham's law (iv) V/T = constant for fixed P, n

(e) ideal gas law (v) $rate_1/rate_2 = \sqrt{d_2/d_1}$

2 (24 points). Choose the response that best completes each statement.

(a) One of the causes for major deviations from ideal gas behavior that a real gas shows is the

(i) negligible volume of the molecules

(ii) presence of intermolecular forces

(iii) uncertainty in absolute zero

(iv) infinite expandability

(b) A manometer is used for measuring

(i) relative rates of diffusion (ii) standard molar volumes

(iii) pressure (iv) temperature

(c) The equilibrium vapor pressure above a liquid is an example of

(i) dynamic equilibrium (ii) the combined gas law

(iii) effusion (iv) Dalton's law

(d) Extrapolation of experimental results by Charles gave rise to

(i) an absolute temperature scale

(ii) mole fractions

(iii) Gay-Lussac's law of combining volumes

(iv) a dynamic equilibrium

(e) For an ideal gaseous mixture

(i) the rates of effusion are identical

(ii) the equilibrium vapor pressure is equal to the standard molar volume

(iii) the partial pressure of a component is equal to the total pressure multiplied by the respective mole fraction

(iv) the partial pressures of the gases are directly proportional to the molar mass

(f) Which does *not* describe an ideal gas mixture?

(i) Gaseous molecules are in random, constant motion.

(ii) Gaseous molecules undergo elastic collisions.

(iii) Gaseous molecules of each type have the same average velocity.

(iv) Gaseous molecules have negligible volume compared to the volume of the container.

(g) Which gas has the greatest average kinetic energy at a given temperature?

(i) hydrogen (ii) neon

(iii) carbon dioxide (iv) none of these answers

(h) One mole each of N_2 and O_2 are mixed with two moles of He. The final mixture is at STP. The partial pressure of O_2 is

(i) 4 atm (ii) 1 atm
(iii) 1/4 atm (iv) 1/3 atm

(i) The pressure of an ideal gas will be doubled by

(i) doubling the temperature (at fixed volume and mass)
(ii) halving the temperature (at fixed volume and mass)
(iii) doubling both the temperature and the number of moles (at fixed volume)
(iv) halving the temperature and doubling the number of moles (at fixed volume)

(j) The density of a gas

(i) increases with increasing T (ii) decreases with increasing P
(iii) is a universal constant (iv) none of these answers

(k) Which sample of gas contains more molecules--one that contains one mole of CO_2 at STP or one mole of CH_2CH_2 at 300 K and 760 Torr?

(i) CO_2 (ii) neither, same number
(iii) CH_2CH_2 (iv) insufficient information

(l) At the same temperature and pressure, what is the approximate rate of effusion of Ar to He?

(i) $\sqrt{1/10}$ (ii) $\sqrt{10}$
(iii) 1/10 (iv) 1/100

3 (21 points). Choose the response that best completes each statement.

(a) A gaseous system is at 0 °C and 1.5 atm. What is its pressure at 200. °C?

(i) 2.6 atm (ii) 0.87 atm
(iii) 0.75 atm (iv) 1.0 atm

(b) Starting with 10.0 L of an ideal gas, what is the volume after increasing the temperature from 11 °C to 22 °C and increasing the pressure from 1.2 atm to 2.4 atm?

(i) 10.0 L (ii) 40.0 L
(iii) 5.19 L (iv) 22.4 L

(c) Which sample contains the most molecules?

(i) 10 g hydrogen (ii) 40 g neon
(iii) 4 g helium (iv) 70 g chlorine

(d) The volume of 0.50 mol of gaseous CF_4 at 1.0 atm and 273 K is

(i) 0.5 L (ii) 11 L

(iii) 22.4 L (iv) 34 L

(e) One gram of a gaseous compound of boron and hydrogen occupies 0.821 L at 1.00 atm and 3 °C. The compound is

(i) BH_3 (ii) B_2H_6

(iii) B_3H_{12} (iv) B_4H_{10}

(f) The molar mass of a gas that weighs 0.6376 g and occupies 0.250 L at 373 K and 1.00 atm is

(i) 100.5 g/mol (ii) 1890 g/mol

(iii) 7910 g/mol (iv) 78.1 g/mol

(g) The partial pressure of oxygen in a flask containing 28 g of N_2 and 16 g of O_2 at a total pressure of 6.0 atm is

(i) 1.2 atm (ii) 3.6 atm

(iii) 4.0 atm (iv) none of these answers

4 (10 points). A 273 mL sample of $Cl_2(g)$ was collected at 725 Torr and 21 °C. (a) What will be the volume of the gas at STP? (b) What is the mass of Cl_2 in the sample?

5 (10 points). A porous vessel contains gaseous N_2O, NO, and NO_2 at equal partial pressures and temperature. (a) Based on the postulates of the kinetic-molecular theory, place these gases in order of increasing average velocity. (b) Using Graham's law, calculate the ratio of the rates of effusion for these gases. (c) Briefly discuss your answers for parts (a) and (b).

6 (10 points). A flask designed for collecting gases had a volume of 257 mL. When filled with air, the flask weighed 62.597 g; when filled with a gas containing Ge and H, it weighed 63.959 g. The density of air under the experimental conditions is 1.293 g/L. (a) What is the mass of the empty flask? (b) What is the density of the Ge-H gas? (c) Find the molar mass of the Ge-H gas given that the density was measured at 20 °C and 1.0 atm. (d) Find the molecular formula if the empirical formula of the gas is $(GeH_3)_x$.

7 (10 points). A 62.7 mL sample of hydrogen was collected by water displacement at 18 °C. The pressure of the "wet" hydrogen was 763 Torr and the partial pressure of water was 15 Torr. (a) Calculate the mole fractions of the gases in the mixture. (b) How many moles of H_2 and H_2O are in the gaseous mixture?

(c) What is the volume of "dry" hydrogen reported at STP? (d) What mass of hydrogen was collected? (e) What is the pressure necessary to confine the "wet" hydrogen at 18 °C if the volume of the mixture is halved? Hint: All of the water vapor does not remain in the gaseous form.

8 (10 points). Chlorine may be prepared in the laboratory by the reaction

$2NaCl(aq) + MnO_2(s) + 3H_2SO_4(aq) \rightarrow MnSO_4(aq) + 2NaHSO_4(aq) + Cl_2(g) + H_2O(l)$

What mass of NaCl is needed to prepare 500.0 mL Cl_2 at 730.0 Torr and 25 °C? Will 1.50 g MnO_2 be enough to prepare this amount of Cl_2?

Answers to practice test

1. (a) ii, (b) iv, (c) i, (d) v, (e) iii.
2. (a) ii, (b) iii, (c) i, (d) i, (e) iii, (f) iii, (g) iv, (h) iii, (i) i, (j) iv, (k) ii, (l) i.
3. (a) i, (b) iii, (c) i, (d) ii, (e) ii, (f) iv, (g) iv.
4. (a) 242 mL, (b) 0.766 g.
5. (a) $NO_2 < N_2O < NO$, (b) $rate_{NO}/rate_{N_2O}/rate_{NO_2} = 1.238/1.211/1.000$, (c) effusion deals with molecular velocity.
6. (a) 62.265 g, (b) 6.59 g/L, (c) 150 g/mol, (d) Ge_2H_6.
7. (a) $X_{H_2O} = 0.20$, $X_{H_2} = 0.980$; (b) $n_{H_2O} = 5.2 \times 10^{-5}$ mol, $n_{H_2} = 2.58 \times 10^{-3}$ mol; (c) 57.8 mL; (d) 5.20 mg; (e) $P_{H_2} = 1496$ Torr, $P_{H_2O} = 15$ Torr, $P_{total} = 1511$ Torr.
8. 2.29 g NaCl; 1.70 g MnO_2 required, no.

ADDITIONAL NOTES

CHAPTER 7

THERMOCHEMISTRY

CHAPTER OVERVIEW

The chapter first discusses some aspects of elementary thermodynamics--work, heat, and internal energy change. The concept of enthalpy, the important quantity used to deal with heat in chemistry, is introduced. A discussion of heat capacity and calorimetry is presented. Standard enthalpy changes are defined and the use of enthalpies in calculating heats of reactions, heats of formation of compounds, heat of combustion, and heats of changes of state is explained.

COMPETENCIES

Definitions for the following terms should be learned:

- calorimeter
- combustion
- endothermic
- enthalpy
- exothermic
- first law of thermodynamics
- heat
- heat capacity
- heat of reaction
- Hess's law
- intermolecular forces
- internal energy
- internal energy change
- intramolecular forces
- molar heat capacity
- specific heat
- standard enthalpy changes
- standard enthalpy of combustion
- standard enthalpy of formation
- standard enthalpy of reaction
- standard state
- surroundings
- system
- thermochemical equation
- thermochemistry
- thermodynamics
- work

General concepts and skills that should be learned include

- applying the law of conservation of energy to various chemical and physical processes
- interpreting relative strengths of intermolecular and intramolecular forces as measured by energy changes
- relating the work and heat exchanged between a system and its surroundings to the resulting internal energy change of the system
- predicting whether pressure-volume work is done by a system on the surroundings, is done by the surroundings on the system, or is negligible for a process
- relating the heat transferred in a process and the heat capacity to the temperature change of a system
- writing thermochemical equations--particularly for the formation of a substance, for the combustion of a substance, and for a change of state of a substance
- identifying the standard state of a substance at a given temperature
- predicting the favorability of a process using the energy change as a criterion
- using known thermochemical equations to determine the energy involved for another thermochemical equation

Numerical exercises that should be understood include

- calculating work, heat, or internal energy change for a system when two of these variables are given using the first law of thermodynamics
- calculating the energy involved in a heating or cooling process, the heat capacity, the amount of material, or the temperature change when three of these variables are given
- determining the heat of reaction, a temperature change, or a calorimeter constant when two of these variables are given
- calculating the heat of reaction from a thermochemical equation
- calculating the enthalpy change for a process from known values of ΔH for other processes by using Hess's law
- calculating the heat of reaction from given values of heats of formation of the reactants and products

QUESTIONS

1. Complete the following sentences by filling in the various blanks with these
    terms: changes of state, endothermic process, enthalpy change, exothermic
    process, intermolecular forces, intramolecular forces, and standard enthalpy
    of formation.

    The (a) _enthalpy change_ is the energy transferred under constant pressure
    conditions between the system and the surroundings during a chemical or
    physical change. If heat energy is absorbed by the system, the process is
    an (b) _endothermic process_ and if the sign of ΔH is negative, the process is an
    (c) _exothermic process_. The enthalpy change for (d) _changes of state_ --physical processes that do
    not produce new compounds, but change the form or state of a substance--
    reflects the strength of the (e) _intermolecular forces_ --the various forces of attraction or
    repulsion between individual molecules. The (f) _standard enthalpy of formation_ for a compound is
    the heat gained or lost as one mole of the compound is formed by the direct
    combination of the elements in their standard states at the specified
    temperature and indirectly reflects the strengths of the (g) _intramolecular forces_

2. Match the correct term to each definition.

    (iv) (a) the amount of energy required to raise   (i) calorimeter
           the temperature of a given amount   (ii) calorimeter constant
           of material by 1 °C   (iii) electron affinity

    (vi) (b) the amount of energy required to raise   (iv) heat capacity
           the temperature of a gram of material   (v) ionization energy
           by 1 °C   (vi) specific heat

    (i) (c) a device for measuring the heat given   (vii) standard enthalpy changes
           off or absorbed in a chemical reaction

    (ii) (d) the total heat capacity of a calorimeter

    (vii) (e) enthalpy changes expressed for substances in their standard states

    (v) (f) enthalpy change for the removal of one electron from an atom or an ion
           in the gaseous state

    (iii) (g) energy liberated as heat for the addition of an electron to an atom in
           the gaseous state

3. Which of the following statements are true? Rewrite each false statement so
    that it is correct.

    (a) A positive sign on an enthalpy change means that an endothermic process
         has occurred.

(b) The calorimeter constant is the amount of heat required to raise 1 g of a substance by one degree.

(c) The enthalpy of fusion is numerically equal, but opposite in sign to the enthalpy of crystallization.

(d) The standard state of any substance is the physical state in which it is most stable at 100 kPa atm pressure and a specified temperature.

(e) The heat transferred under constant volume conditions is the enthalpy change.

4. Distinguish clearly between (a) system and surroundings, (b) exothermic and endothermic, (c) intermolecular and intramolecular forces, (d) heat of formation and heat of reaction, (e) internal energy and enthalpy, (f) ΔH and $\Delta H°$, and (g) molar heat capacity and specific heat.

5. Choose the response that best completes each statement.

(a) The study of the thermal energy change associated with the physical or chemical changes of pure substances is

(i) stoichiometry (ii) thermochemistry
(iii) calorimetry (iv) the specific heat

(b) The physical state in which any substance is most stable at 100 kPa pressure and a specified temperature is the

(i) standard enthalpy (ii) enthalpy of formation
(iii) standard state (iv) heat capacity

(c) The standard enthalpy of formation for a compound is the heat of formation of one mole of the compound formed by

(i) direct combination of the elements
(ii) intermolecular forces
(iii) precipitation
(iv) the electron affinities and ionization energies of the elements in their standard states at the specified temperature

(d) If ΔH is negative, the process

(i) will always be spontaneous (ii) is endothermic
(iii) will be a phase change (iv) is exothermic

(e) Which is *not* a property of ΔH for a reaction?

(i) directly proportional to the quantities of reactants or products
(ii) equal in magnitude and opposite in sign to ΔH for the reaction in the reverse direction

(iii) independent of the number and nature of intermediate steps that may be involved in that reaction

(iv) always negative

6. Define the term "thermodynamics." What is the name given to the study of the thermal energy changes that accompany chemical and physical changes?

7. Briefly describe why chemical reactions involve energy.

8. In thermodynamics, how is a "system" defined? What types of interactions can take place between a system and its surroundings?

9. What is internal energy? What changes occur in a system in which an internal energy change is occurring?

10. State the first law of thermodynamics in equation form. Explain this equation in your own words.

11. What will be the sign of the value of the energy absorbed by a system if an endothermic process has occurred? What will be the sign for an exothermic process?

12. What is the sign for work being done on the surroundings by the system? What is the sign for work done by the surroundings on the system?

13. The value of ΔE of a system was negative for a given process. What does this imply?

14. What happens to a system as pressure-volume work is done on it while no heat exchange is permitted between the system and its surroundings?

15. For each of the following chemical and physical changes at constant pressure, is the pressure-volume work (i) done by the system (the substances undergoing the change) on the surroundings, (ii) done by the surroundings on the system, or (iii) negligible?

   (a) $CaCO_3(s) \xrightarrow{\Delta} CaO(s) + CO_2(g)$

   (b) $Cl_2(g) + H_2O(l) \rightarrow HCl(aq) + HClO(aq)$

   (c) $Zn(s) + CuSO_4(aq) \rightarrow ZnSO_4(aq) + Cu(s)$

   (d) $CO_2(s) \rightarrow CO_2(g)$

   (e) $2NO_2(g) \rightleftharpoons N_2O_4(g)$

16. What is the special name given to the heat absorbed or released under constant pressure conditions? What is the symbol used to represent this quantity?

17. Write the equation that relates ΔH to ΔE and interpret the equation in your own words. In what kind of processes is the difference between ΔH and ΔE usually significant?

18. During a chemical process, the volume of a system remained constant. What is the value of the work for the process? Will the heat exchanged between the system and its surroundings be equal to ΔH or ΔE?

19. What does the symbol ΔT represent? Will ΔT be positive or negative for a system for which the temperature is increasing?

20. What does the term "heat capacity" mean? What do we call the heat capacity of a mole of a substance? What do we call the heat capacity of a gram of a substance?

21. What law is the basis for the calculations done in calorimetry? How is this law written in equation form?

22. What is the name of the device used for measuring the heat absorbed or released during a thermochemical process?

23. Define the term "heat capacity of the calorimeter." Why is this quantity usually determined experimentally? Briefly explain two methods for determining the heat capacity of the calorimeter.

24. What is the advantage of designing a calorimeter so that the contribution by the heat capacity of the calorimeter to the calculation of the heat is negligible? What is the advantage if the contribution by the heat capacity of the calorimeter is so large that the heat capacities of the contents of the calorimeter are negligible?

25. What two additional types of information are included in a thermochemical equation besides the stoichiometric chemical equation?

26. Define the term "heat of reaction." If a chemical reaction occurs at constant external pressure, to what is the value of q equal?

27. Identify which of the following reactions are endothermic:

(a) $S(s) + O_2(g) \rightarrow SO_2(g)$ $\Delta H° = -296.830$ kJ

(b) $H_2S(g) \rightarrow H(g) + HS(g)$ $\Delta H° = 381.27$ kJ

(c) $3O_2(g) \rightarrow 2O_2(g)$ $\Delta H° = 285.4$ kJ

(d) $FeCO_2(s,\text{siderite}) + 2HCl(aq) \rightarrow FeCl_2(aq) + H_2O(l) + CO_2(g)$

$\Delta H° = -27.9$ kJ

28. Define the term "standard state" as it refers to a substance. Could the standard state of a substance vary as the temperature is changed?

29. Briefly describe how to write a thermochemical equation for a formation reaction. Write the chemical equations for the formation of (a) $NaCl(s)$, (b) $C_6H_6(l)$, (c) $CaCO_3(s)$, (d) $Cl(g)$, and (e) $Fe_2O_3(s)$ at 25 °C.

30. What is the enthalpy of formation of an element in its standard state? What is the enthalpy of formation of an element in a nonstandard state?

31. Two thermochemical equations are the same except that in the second one the products and reactants are reversed. What is the relationship between the two values of ΔH?

32. Two thermochemical equations are the same except that in the second one the reactants and products have stoichiometric coefficients that are n times as large as those in the other. What is the relationship between the two values of ΔH?

33. Two known thermochemical equations can be added to give a third thermochemical equation. What is the relationship among the three values of $\Delta H°$?

34. State Hess's law. Why is it important in thermochemistry?

35. If the standard heats of formation are known for all reactants and products in a chemical equation, how can the standard heat of reaction be calculated?

36. Briefly describe how to write a thermochemical equation for the combustion of a carbon-hydrogen-oxygen compound. Write the chemical equations for the combustion of (a) $C_6H_6(l)$, (b) $C(s)$, (c) $CH_3OH(l)$, and (d) $CH_3COOH(l)$ at 25 °C.

37. As energy is added or removed during a phase change, what happens to the temperature of the system?

38. What is the source of the energy released during an exothermic change of state? What happens to the energy added during an endothermic change of state?

39. What is the term used for (a) a solid undergoing a change to a liquid and (b) a solid undergoing a change to a gas? (c) What are the terms that are used for the opposite processes?

40. What is the relationship between the heat of vaporization and the heat of condensation (or liquefaction) of a substance?

41. Why is the heat of vaporization usually larger than the heat of fusion for a substance?

42. Suppose it were possible to melt a substance and then vaporize it at the same temperature. What would be the relationship among the heats of fusion, vaporization, and sublimation?

Answers to questions

1. (a) enthalpy change, (b) endothermic process, (c) exothermic process, (d) changes of state, (e) intermolecular forces, (f) standard enthalpy of formation, (g) intramolecular forces.
2. (a) iv, (b) vi, (c) i, (d) ii, (e) vii, (f) v, (g) iii.
3. (a) T; (b) F, replace *calorimeter constant* with *specific heat*; (c) T; (d) T; (e) F, change *enthalpy* to *internal energy*.
4. (a) The *system* is the portion of the universe that is being studied for work and heat transfer; the *surroundings* are the rest of the universe. (b) In an *exothermic* process energy is transferred from the system to the surroundings; in an *endothermic* process energy is transferred from the surroundings to the system. (c) *Intermolecular forces* are forces between molecules; *intra-molecular forces* are forces within a molecule (interatomic forces). (d) The *heat of formation* is the heat of reaction for the formation of one mole of a substance from elements by direct combination; the *heat of reaction* is the heat involved for any chemical process. (e) The *internal energy* is the energy contained within a system; *enthalpy* is defined as $E + PV$. (f) ΔH represents the change in enthalpy for a process; $\Delta H°$ represents the enthalpy change for a process occurring at 100 kPa pressure. (g) The *molar heat capacity* is the heat capacity for one mole of a substance; the *specific heat* is the heat capacity for one gram of a substance.
5. (a) ii, (b) iii, (c) i, (d) iv, (e) iv.
6. Thermodynamics is the study of energy transformations; thermochemistry.
7. Chemical reactions involve the making and breaking of chemical bonds and changes of state--both processes involve energy changes in the form of heat energy.

8. A thermodynamic system is a portion of the universe for which energy transformations are being studied. The types of interactions that take place between a system and its surroundings include work and heat flow.
9. Internal energy is all of the energy contained with a chemical system. A change in the internal energy of a chemical system results in a temperature change, a phase change, or a chemical reaction.
10. The equation representing the first law of thermodynamics is $\Delta E = q + w$. This equation means that the change in internal energy of a system equals the heat flow and the work transferred between the system of its surroundings.
11. The sign on the value of the energy transferred by a system for an endothermic process is positive. The sign for an exothermic process is negative.
12. For work being done on the surroundings by the system, the sign of the value is negative. For work done by the surroundings on the system, the sign of the value is positive.
13. The internal energy of the system decreased during the process.
14. The internal energy of the system must increase, which usually produces a temperature increase, a change of state, or a chemical reaction.
15. (a) i, (b) ii, (c) iii, (d) i, (e) ii.
16. The special name given to the heat exchanged between a system and its surroundings under constant pressure conditions is the enthalpy change. The symbol used to represent enthalpy change is ΔH.
17. $\Delta H = \Delta E + \Delta(PV)$; ΔE represents the change of heat energy for a process under constant pressure conditions; ΔH differs significantly from ΔE only in processes involving gases.
18. During a process in which the volume of a system remained constant, the value of the work is zero. The heat exchanged between the system and its surroundings is equal to ΔE.
19. The symbol ΔT represents the difference between the final temperature and the initial temperature. For a system in which the temperature is increasing, ΔT will be positive.
20. The heat capacity of a substance is the amount of heat required to raise the temperature of a given amount of a substance by one kelvin. The heat capacity of a mole of a substance is called the molar heat capacity. The heat capacity of a gram of a substance is called the specific heat.

21. The law of conservation of energy applies for calculations done in calorimetry; $q + q_{gained} + q_{lost} = 0$.
22. The device used for measuring the heat absorbed or released during a thermochemical process is called a calorimeter.
23. The heat capacity of the calorimeter is the total heat capacity of the components of the calorimeter. The heat capacity of the calorimeter is usually determined experimentally because of the difficulty in calculating the heat capacity of each of the components.

    One method for determining the heat capacity of a calorimeter is to introduce a known amount of heat into the calorimeter with an electric heater and measure the rise in temperature. Another method is to introduce a known amount of a heated metal of known specific heat. A third method is to carry out in the calorimeter a reaction for which the heat of reaction is known.
24. If the calorimeter is designed so that the contribution by the heat capacity of the calorimeter to the calculation of the heat is negligible, the calculations are simplified because the heat will be equal to that calculated for the contents only. If the contribution by the heat capacity of the calorimeter is so large that the heat capacities of the calorimeter contents are negligible, the calculations are simplified because the heat will be equal to that calculated for the calorimeter only.
25. The additional types of information are the physical states of all substances and the energy of process.
26. The heat of reaction is the total amount of heat released or absorbed between the beginning of a reaction and the return of the substances produced to the original temperature. If the chemical reaction occurs at constant pressure, the value of q equals the enthalpy change, ΔH.
27. The endothermic reactions are b and c.
28. The standard state of a substance is the stable physical state at 0.1 MPa pressure and at the specified temperature; yes, the standard state of a substance will change if the substance undergoes a change of state before attaining the new temperature.
29. The equation consists of elements in their standard states as the reactants and one mole of the substance as the product; (a) $Na(s) + \frac{1}{2}Cl_2(g) \rightarrow NaCl(s)$, (b) $6C(s) + 3H_2(g) \rightarrow C_6H_6(l)$, (c) $Ca(s) + C(s) + \frac{3}{2}O_2(g) \rightarrow CaCO_3(s)$, (d) $\frac{1}{2}Cl_2(g) \rightarrow Cl(g)$, (e) $2Fe(s) + \frac{3}{2}O_2(g) \rightarrow Fe_2O_3(s)$.

30. $\Delta H_f^\circ = 0$, $\Delta H_f^\circ \neq 0$.
31. $\Delta H_2 = -\Delta H_1$.
32. $\Delta H_2 = n\Delta H_1$.
33. $\Delta H_3 = \Delta H_1 + \Delta H_2$.
34. Hess's law is stated as follows: the enthalpy change of a chemical reaction is the same whether the reaction takes place in one step or several steps. Hess's law is important because it allows calculation of a desired ΔH value from a sequence of thermochemical equations that are combined algebraically to give the desired equation.
35. ΔH° = (sum of ΔH_f° products) - (sum of ΔH_f° reactants).
36. The equation consists of one mole of substance plus $O_2(g)$ as reactants and $CO_2(g)$ and $H_2O(l)$ as products; (a) $C_6H_6(l) + \frac{15}{2}O_2(g) \rightarrow 6CO_2(g) + 3H_2O(l)$, (b) $C(s) + O_2(g) \rightarrow CO_2(g)$, (c) $CH_3OH(l) + \frac{3}{2}O_2(g) \rightarrow CO_2(g) + 2H_2O(l)$, (d) $CH_3COOH(l) + 2O_2(g) \rightarrow 2CO_2 + 2H_2O(l)$.
37. As energy is added or removed during a phase change, the temperature remains the same.
38. The energy released is the result of the formation of intermolecular bonds; the energy is used to break intermolecular bonds.
39. (a) fusion or melting; (b) sublimation; (c) freezing or crystallization, deposition.
40. $\Delta H_{fus} = -\Delta H_{cond}$.
41. There is a larger change in intermolecular forces during boiling than during melting.
42. If it were possible to melt a substance and then vaporize it at the same temperature, ΔH_{sub} would equal $\Delta H_{vap} + \Delta H_{fus}$.

SKILLS

Skill 7.1 Favorable and unfavorable energy changes

The sign and magnitude of the enthalpy change can be used to decide whether or not a chemical process involves a favorable energy change. Favorable energy changes are exothermic (negative ΔH values) and unfavorable energy changes are endothermic (positive ΔH values). For example, the enthalpy changes for both of the following reactions are favorable.

$$C(s) + \frac{1}{2}O_2(g) \rightarrow CO(g) \qquad \Delta H^\circ_{298} = -110.5 \text{ kJ}$$
$$C(s) + O_2(g) \rightarrow CO_2(g) \qquad \Delta H^\circ_{298} = -393.5 \text{ kJ}$$

Skill Exercise: Indicate whether or not the following enthalpy changes are favorable: (a) ΔH°_{298} = 280 kJ for $AsI_3(s) \rightarrow AsI_3(g)$, (b) ΔH°_{298} = 300.4 kJ for $As_4(g) \rightarrow 2As_2(g)$, (c) ΔH°_{298} = -15 kJ for As(yellow) → As(gray), (d) ΔH°_{298} = -574 kJ for $2P(s) + 3Cl_2(g) \rightarrow 2PCl_3(g)$ and (e) ΔH°_{298} = -749.8 kJ for $2P(s) + 5Cl_2(g) \rightarrow 2PCl_5(g)$. *Answers*: (a) unfavorable, (b) unfavorable, (c) favorable, (d) favorable, (e) favorable.

Skill 7.2 Using Hess's law

Hess's law states that the enthalpy change of a chemical reaction is the same whether the reaction takes place in one step or several steps. In other words, the heat of reaction is the same no matter by what route the reaction takes place. Thus, ΔH for a reaction can be calculated from a sequence of thermochemical equations that can be combined algebraically to give the desired equation.

The five general steps that can be used to apply Hess's law are:

1. Write the desired thermochemical equation.
2. Reverse (if necessary) some of the known thermochemical equations so that the major reactants and products in the desired reaction are on the reactants and products side.
3. Multiply (if necessary) the known equations by appropriate coefficients so that the major reactants and products have the same coefficients as they do in the desired equation.
4. Include (if necessary) additional equations to eliminate unwanted reactants or products that appear after the third step.
5. Add the known equations and the values of ΔH°.

For example, Hess's law can be used to find the heat of formation of acetic acid from the knowledge of the heats of combustion of acetic acid, carbon, and hydrogen:

$$C(s) + O_2(g) \rightarrow CO_2(g) \qquad \Delta H^\circ = -393.5\ kJ$$

$$H_2(g) + \frac{1}{2}O_2(g) \rightarrow H_2O(l) \qquad \Delta H^\circ = -285.8\ kJ$$

$$CH_3COOH(l) + 2O_2(g) \rightarrow 2CO_2(g) + 2H_2O(l) \qquad \Delta H^\circ = -874.5\ kJ$$

1. Write the desired thermochemical equation. The equation for the formation of CH_3COOH is

$$2C(s) + 2H_2(g) + O_2(g) \rightarrow CH_3COOH(l) \qquad \Delta H^\circ = ?$$

2. Reverse (if necessary) some of the known thermochemical equations. The reaction for acetic acid is reversed so that CH_3COOH, a major product in the desired equation, is on the products side. The carbon and hydrogen equations are not reversed because these major reactants are on the reactants side in the desired equation.

$$2CO_2(g) + 2H_2O(l) \rightarrow CH_3COOH(l) + 2CO_2(g) \qquad \Delta H^\circ = 874.5\text{ kJ}$$

3. Multiply (if necessary) the known equations by appropriate coefficients. Both the carbon reaction and the hydrogen reaction need to be multiplied by 2.

$$2C(s) + 2O_2(g) \rightarrow 2CO_2(g) \qquad \Delta H^\circ = -787.0\text{ kJ}$$

$$2H_2(g) + O_2(g) \rightarrow 2H_2O(l) \qquad \Delta H^\circ = -571.6\text{ kJ}$$

4. Include (if necessary) additional equations to eliminate unwanted reactants or products. There are no species that need to be eliminated.
5. Add the known equations and the values of ΔH°.

$$\begin{array}{rcll} 2CO_2(g) + 2H_2O(l) & \rightarrow & CH_3COOH(l) + 2O_2(g) & \Delta H^\circ = 874.5\text{ kJ} \\ 2C(s) + 2O_2(g) & \rightarrow & 2CO_2 & \Delta H^\circ = -787.0\text{ kJ} \\ 2H_2(g) + O_2(g) & \rightarrow & 2H_2O(l) & \Delta H^\circ = -571.6\text{ kJ} \\ \hline 2C(s) + 2H_2(g) + O_2(g) & \rightarrow & CH_3COOH(l) & \Delta H^\circ = -484.1\text{ kJ} \end{array}$$

Skill Exercise: Use the following thermochemical equations to calculate ΔH_f° for AgCl(s):

$$H_2(g) + \tfrac{1}{2}O_2(g) \rightarrow H_2O(l) \qquad \Delta H^\circ = -285.8\text{ kJ}$$

$$2Ag(s) + \tfrac{1}{2}O_2(g) \rightarrow Ag_2O(s) \qquad \Delta H^\circ = -31.1\text{ kJ}$$

$$\tfrac{1}{2}H_2(g) + \tfrac{1}{2}Cl_2(g) \rightarrow HCl(g) \qquad \Delta H^\circ = -92.3\text{ kJ}$$

$$Ag_2O(s) + 2HCl(g) \rightarrow 2AgCl(s) + H_2O(l) \qquad \Delta H^\circ = -324.3\text{ kJ}$$

Answer: $\Delta H_f^\circ = -127.1$ kJ/mol.

A special application of Hess's law can be used to calculate the heat of reaction if all of the known data are heats of formation

$$\Delta H^\circ = (\text{sum of } \Delta H_f^\circ \text{ products}) - (\text{sum of } \Delta H_f^\circ \text{ reactants})$$

For example, given $\Delta H_f^\circ = -411$ kJ/mol for NaCl(s), -814 kJ/mol for $H_2SO_4(l)$, -1387 kJ/mol for $Na_2SO_4(s)$, and -92 kJ/mol for HCl(g), find ΔH° for

$$H_2SO_4(l) + 2NaCl(s) \rightarrow Na_2SO_4(s) + 2HCl(g) \qquad \Delta H^\circ = ?$$

The desired heat of reaction is

$$\Delta H^\circ = [(1\text{ mol})\Delta H^\circ_f(Na_2SO_4) + (2\text{ mol})\Delta H^\circ_f(HCl)]$$
$$- [(1\text{ mol})\Delta H^\circ_f(H_2SO_4) + (2\text{ mol})\Delta H^\circ_f(NaCl)]$$
$$= [(1\text{ mol})(-1387\text{ kJ/mol}) + (2\text{ mol})(-92\text{ kJ/mol})]$$
$$- [(1\text{ mol})(-814\text{ kJ/mol}) + (2\text{ mol})(-411\text{ kJ/mol})]$$
$$= 65\text{ kJ}$$

Skill Exercise: Use the following standard heats of formation: -285.8 kJ/mol for $H_2O(l)$, -393.5 kJ/mol for $CO_2(g)$, and 226.7 kJ/mol for $C_2H_2(g)$ to calculate the standard heat of combustion for $C_2H_2(g)$. *Answer*: $C_2H_2(g) + \frac{5}{2}O_2(g) \rightarrow 2CO_2(g) + H_2O(l)$, $\Delta H^\circ_c = -1299.5$ kJ.

Another special application of Hess's law can be used to calculate the heat of reaction involving gaseous reactants and products if all of the known data are bond energies (see Chapter 9):

$$\Delta H^\circ = (\text{sum of bond energies of all bonds in reactants}) - (\text{sum of bond energies of all bonds in products})$$

Sample calculations using this relationship are presented in *Example 10.3*.

The use of Hess's law is not restricted to ΔH° values. The law can also be used to calculate values of ΔE° for chemical reactions, as well as for calculating values of ΔS° and ΔG° for chemical reactions (see Chapter 23).

NUMERICAL EXERCISES

*Example 7.1 Energy, heat, and work**

A system is to be held at constant internal energy. If the surroundings perform 6.3 kJ of work on the system, what is the value of the heat flow? In which direction does heat flow occur?

This problem is analyzed using the comprehensive problem-solving approach presented in *Skill 2.7 Solving problems*.

1. <u>Study the problem and be sure you understand it</u>.
   (a) What is unknown?
   The heat flow between the system and its surroundings.
   (b) What is known?
   The work done on the system by the surroundings and that $\Delta E = 0$ for the system.

2. Decide how to solve the problem.
   (a) What is the connection between the known and the unknown?
   The first law of thermodynamics relates work, heat, and internal energy.
   (b) What is necessary to make the connection?
   The value and sign of q is found from the known values of w and ΔE by using $\Delta E = q + w$
3. Set up the problem and solve it.
   Because the surroundings are doing the work on the system, w = 6.3 kJ. The heat transferred is therefore

$$q = \Delta E - w = 0 - (6.3\ \text{kJ}) = -6.3\ \text{kJ}$$

   The negative sign indicates that the system must transfer 6.3 kJ of heat to the surroundings.
4. Check the result.
   (a) Are significant figures and the location of the decimal point correct?
   Yes.
   (b) Did the answer come out in the correct units?
   Yes.
   (c) Is the answer reasonable?
   Yes. It makes sense that if there is no internal energy change, all work being done on a system must be dissipated by heat loss.

Exercise 7.1 During a complicated process, a system absorbed 65 J of heat from the surroundings and suffered a 218 J loss in internal energy. What is the amount of work involved in this process? *Answer*: w = -283 J, work is being done by the system on the surroundings.

Example 7.2 Heat flow

Exactly 100 J was added to 50.0 g samples of $SO_2(g)$, $H_2S(g)$, and $SOF_2(g)$. The respective specific heats are 0.623 J/K g, 1.008 J/K g, and 0.661 J/K g. What was the change in temperature for each gas?

The amount of heat required to change the temperature of a substance is given by

$$q = (\text{mass})(\text{sp. heat})(\Delta T)$$

Solving for ΔT and substituting the data give

$$\Delta T = \frac{100.0\ \text{J}}{(50.0\ \text{g})(0.623\ \text{J/K g})} = 3.21\ \text{K for } SO_2$$

$$\Delta T = \frac{100.0 \text{ J}}{(50.0 \text{ g})(1.008 \text{ J/K g})} = 1.98 \text{ K for } H_2S$$

$$\Delta T = \frac{100.0 \text{ J}}{(50.0 \text{ g})(0.661 \text{ J/K g})} = 3.03 \text{ K for } SOF_2$$

The respective temperature increases were 3.21 °C, 1.98 °C, and 3.03 °C.

Exercise 7.2 Exactly 100 J was added to one mole samples of $SO_2(g)$, $H_2S(g)$, and $SOF_2(g)$. The respective heat capacities are 39.87 J/K mol, 34.23 J/K mol, and 56.82 J/K mol. What was the respective change in temperature for each gas? *Answers*: 2.508 °C, 2.921 °C, 1.760 °C.

Example 7.3 Calorimetry*

A 20.0 g sample of aluminum metal at 83.2 °C was mixed with 49.30 g of water at 19.7 °C. Assuming no energy loss, what was the final temperature of the aluminum-water mixture? The specific heat of aluminum is 0.900 J/K g and that of water is 4.184 J/K g.

This problem is analyzed using the comprehensive problem-solving approach presented in *Skill 2.7 Solving problems*.

1. Study the problem and be sure you understand it.
   (a) What is unknown?
   The final temperature of the aluminum-water mixture.
   (b) What is known?
   The masses, specific heats, and initial temperatures of the aluminum and of the water.
2. Decide how to solve the problem.
   (a) What is the connection between the known and the unknown?
   According to the law of conservation of energy, the heat gained by the water will equal the heat lost by the aluminum. Heat transfer will occur until the final temperature of the water and of the aluminum becomes the same.
   (b) What is necessary to make the connection?
   The known masses and specific heats can be used to set up expressions for the value of q for the aluminum and for the water by using q = (mass)(sp. heat)(ΔT). The initial temperatures of the aluminum and water are known, and both expressions will include the same unknown--the final temperature. By utilizing the law of conservation of energy, q_{gained} +

$q_{lost} = 0$. The two expressions for q can be combined and solved for the value of the final temperature.

3. <u>Set up the problem and solve it.</u>

   The heat gained by the water is

$$q_{gained} = (49.30 \text{ g})(4.184 \text{ J/K g})(T - 292.9 \text{ K})$$
$$= (206.3 \text{ J/K})T - (60{,}420 \text{ J})$$

   where T is the final temperature. The heat lost by the metal is

$$q_{lost} = (20.0 \text{ g})(0.900 \text{ J/K g})(T - 356.4 \text{ K})$$
$$= (18.0 \text{ J/K})T - (6{,}420 \text{ J})$$

   Substituting these expressions for q_{gained} and q_{lost} into the mathematical statement of the first law of thermodynamics and solving for T give

$$q_{gained} + q_{lost} = 0$$
$$[(206.3 \text{ J/K})T - (60{,}420 \text{ J})] + [(18.0 \text{ J/K})T - (6{,}420 \text{ J})] = 0$$
$$(224.3 \text{ J/K})T = 66{,}840 \text{ J}$$
$$T = 298.0 \text{ K} = 24.9 \text{ °C}$$

4. <u>Check the result.</u>

   (a) Are significant figures and the location of the decimal point correct?

   Yes.

   (b) Did the answer come out in the correct units?

   Yes.

   (c) Is the answer reasonable?

   Yes. The water gains (49.30)(4.184)(298.0 - 292.9) = 1100 J and the aluminum loses (20.0)(0.900)(298.0 - 365.4) = -1100 J. All energy is accounted for.

Exercise 7.3 Equimolar samples of CO(g) and CO_2(g) were mixed. Find the final temperature of the gaseous mixture if the CO was originally at 275 K and the CO_2 at 375 K. The molar heat capacities are 29.142 J/K mol for CO and 37.11 J/K mol for CO_2(g). *Answer*: 331 K.

Example 7.4 Calorimetry

A frozen waterpipe contains 450 g of ice at -5 °C. If this ice is to be thawed by pouring warm water (65 °C) over it, what is the minimum amount of water needed? The specific heat of ice is 2.1 J/K g, the specific heat of water is 4.2 J/K g, and the heat of fusion of water is 335 J/g.

In order to thaw the ice, the ice must first be heated from -5 °C to 0 °C and then melted. The total heat gained for the two-step process is

$$q_{gained} = (450\ g)(2.1\ J/K\ g)(5\ K) + (450\ g)(335\ J/g)$$
$$= 5000\ J + 150{,}000\ J = 160{,}000\ J$$

The heat lost by the warm water as it cools from 65 °C to 0 °C (ideally) is

$$q_{lost} = (mass)(4.2\ J/K\ g)(-65\ K) = (-270\ J/g)(mass)$$

Substituting the expressions for q_{gained} and q_{lost} into the mathematical statement of the first law gives

$$q_{gained} + q_{lost} = 0$$
$$(160{,}000\ J) + (-270\ J/g)(mass) = 0$$
$$mass = \frac{160{,}000\ J}{270\ J/g} = 590\ g$$

A minimum of 590 g of warm water would be needed.

Exercise 7.4 As water evaporates at room temperature, a slight cooling effect is observed for the remaining water. What would be the temperature change for 99 g of water after 1 g of water evaporates? At 25 °C, 43.991 kJ is needed to evaporate one mole of water and the specific heat of water is 4.184 J/K g. *Answer*: -5.9 °C.

Example 7.5 Calorimetry

A 1.03 g sample of graphite was burned in a calorimeter, giving rise to a 3.98 °C temperature increase. The enthalpy of combustion is -393.51 kJ/mol for graphite. Find the calorimeter constant. Assume that the calorimeter is designed so that any contribution of the heat capacities of the products is negligible.

The amount of heat released by the combustion process is

$$q_{lost} = (1.03\ g\ C)\left[\frac{1\ mol\ C}{12.01\ g\ C}\right]\left[\frac{-393.51\ kJ}{1\ mol\ C}\right] = -33.7\ kJ$$

and the heat gained by the calorimeter is

$$q_{gained} = (calorimeter\ constant)(\Delta T) = (calorimeter\ constant)(3.98\ K)$$

Because of the conservation of heat

$$q_{gained} + q_{lost} = 0$$
$$(calorimeter\ constant)(3.98\ K) + (-33.7\ kJ) = 0$$
$$(calorimeter\ constant) = \frac{-33.7\ kJ}{398\ K} = 8.47\ kJ/K$$

The calorimeter constant is 8.47 kJ/K.

Exercise 7.5 A 1.07 g sample of sulfur was burned in the calorimeter described in *Example 7.5*. The resulting temperature change was 1.17 °C. Find the molar heat of combustion for sulfur. *Answer*: -297 kJ/mol.

*Example 7.6 Heats of reaction**

The thermochemical equation for the neutralization of a solution of a strong base by a solution of a strong acid is

$$H^+ + OH^- \rightarrow H_2O(l) \qquad \Delta H° = -55.836 \text{ kJ}$$

What is the enthalpy change by the reaction of 25.0 mL of 0.0926 M HCl with a stoichiometric amount of NaOH solution?

This problem is analyzed using the comprehensive problem-solving approach presented in *Skill 2.7 Solving problems*.

1. Study the problem and be sure you understand it.
   (a) What is unknown?
   The enthalpy change of the neutralization of a given amount of HCl solution.
   (b) What is known?
   The thermochemical equation for the reaction and the volume and concentration of the HCl solution used.
2. Decide how to solve the problem.
   (a) What is the connection between the known and the unknown?
   The thermochemical equation relates the heat of reaction to the molar quantities of reactants and products.
   (b) What is necessary to make the connection?
   Conversion of solution molarity and volume to number of moles (see *Skill 4.4 Finding the number of moles*) allows for a molar interpretation of the thermochemical equation (see *Skill 5.1 Interpreting chemical equations*).
3. Set up the problem and solve it.

$$(25.0 \text{ mL})\left[\frac{1 \text{ L}}{1000 \text{ mL}}\right]\left[\frac{0.0926 \text{ mol HCl}}{1 \text{ L}}\right]\left[\frac{1 \text{ mol H}^+}{1 \text{ mol HCl}}\right] = 2.32 \times 10^{-3} \text{ mol H}^+$$

$$(2.32 \times 10^{-3} \text{ mol H}^+)\left[\frac{-55.836 \text{ kJ}}{1 \text{ mol H}^+}\right]\left[\frac{1000 \text{ J}}{1 \text{ kJ}}\right] = -129 \text{ J}$$

4. Check the result.
   (a) Are significant figures and the location of the decimal point correct?
   Yes. The number of significant figures in the answer is limited by the three significant figures in the concentration and the volume.

(b) Did the answer come out in the correct units?

Yes.

(c) Is the answer reasonable?

Yes. The reaction involved only 2.32 mmol of reactants and so a large enthaply change would not be expected.

Exercise 7.6 Using the thermochemical equation

$$P(\text{white}) \rightarrow P(\text{red}) \qquad \Delta H° = -17.6\ \text{kJ}$$

determine the energy enthalpy change as 1.07 kg of white phosphorus changes to red phosphorus. *Answer*: -608 kJ.

Example 7.7 Heats of reaction

Find $\Delta H°_{293}$ for the reaction

$$C_2H_6(g) + \frac{1}{2}O_2(g) \rightarrow C_2H_5OH(l)$$

given the following thermochemical data at 20 °C:

$$C_2H_6(g) + \frac{7}{2}O_2(g) \rightarrow 2CO_2(g) + 3H_2O(l) \qquad \Delta H° = -1541.4\ \text{kJ}$$

$$C_2H_5OH(l) + 3O_2(g) \rightarrow 2CO_2(g) + 3H_2O(l) \qquad \Delta H° = -1370.7\ \text{kJ}$$

The desired equation and heat are obtained by using Hess's law (see *Skill 7.2 Using Hess's law*):

$$C_2H_6(g) + \frac{7}{2}O_2(g) \rightarrow 2CO_2(g) + 3H_2O(l) \qquad \Delta H° = -1541.4\ \text{kJ}$$

$$2CO_2(g) + 3H_2O(l) \rightarrow C_2H_5OH(l) + 3O_2(g) \qquad \Delta H° = 1370.7\ \text{kJ}$$

$$C_2H_6(g) + \frac{1}{2}O_2(g) \rightarrow C_2H_5OH(l) \qquad \Delta H° = -170.7\ \text{kJ}$$

The $\Delta H°$ for the reaction at 293 K is -170.7 kJ.

Exercise 7.7 Find $\Delta H°$ for dissolving HBr in water,

$$HBr(g) + nH_2O(l) \rightarrow HBr(aq)$$

from the following information.

$$\frac{1}{2}H_2(g) + \frac{1}{2}Br_2(g) \rightarrow HBr(g) \qquad \Delta H° = -36.40\ \text{kJ}$$

$$\frac{1}{2}H_2(g) + \frac{1}{2}Br_2(g) + nH_2O(l) \rightarrow HBr(aq) \qquad \Delta H° = -121.55\ \text{kJ}$$

Answer: -85.15 kJ.

Example 7.8 Heats of reaction

The respective standard enthalpies of formation at 25 °C for $C_2H_4(g)$, $H_2O(l)$, and $C_2H_5OH(l)$ are 52.26, -285.830, and -277.69 kJ/mol. Find $\Delta H°$ for the reaction

$$C_2H_4(g) + H_2O(l) \rightarrow C_2H_5OH(l)$$

Because all of the data are heats of formation, the heat of reaction can be found by using a special case of Hess's law (see *Skill 7.2 Using Hess's law*):

$$\Delta H° = (\text{sum of } \Delta H_f° \text{ products}) - (\text{sum of } \Delta H_f° \text{ reactants})$$

$$= [(1 \text{ mol})\Delta H_f°(C_2H_5OH)] - [(1 \text{ mol})\Delta H_f°(C_2H_4) + (1 \text{ mol})\Delta H_f°(H_2O)]$$

$$= [(1 \text{ mol})(-277.69 \text{ kJ/mol})] - [(1 \text{ mol})(52.56 \text{ kJ/mol})$$

$$+ (1 \text{ mol})(-285.830 \text{ kJ/mol})]$$

$$= -44.42 \text{ kJ}$$

The heat of reaction is -44.42 kJ.

Exercise 7.8 The standard heats of formation at 25 °C for $NO(g)$, $O_3(g)$, $NO_2(g)$, and $O_2(g)$ are 90.25, 142.7, 33.18, and 0 kJ/mol, respectively. Calculate $\Delta H°$ of reaction for

$$NO(g) + O_3(g) \rightarrow NO_2(g) + O_2(g)$$

Answer: -199.8 kJ.

Example 7.9 Energy, heat, and work

For the thermal decomposition of potassium chlorate at constant external pressure

$$2KClO_3(s) \xrightarrow[MnO_2]{\Delta} 2KCl(s) + 3O_2(g)$$

discuss the relationships among ΔE, ΔH, q_p, and w.

The Δ over the arrow shows that this is an endothermic reaction. Therefore, ΔH and q_p both have positive values. With no gaseous reactants and 3 mol gaseous products, the system does work on the surroundings, $P\,\Delta V$ is negative, and the work done is given by $w = -P\,\Delta V$. The internal energy change for this reaction, which involves pressure-volume work, is $\Delta E = q_p - P\,\Delta V$, or $\Delta E = \Delta H - P\,\Delta V$. The value of ΔE will be positive unless the amount of work done by the system is so great that $P\,\Delta V$ is larger than ΔH (which is unlikely).

Exercise 7.9 The combustion of benzoic acid

$$C_6H_5COOH(s) + \frac{15}{2}O_2(g) \rightarrow 7CO_2(g) + 3H_2O(l)$$

was carried out under constant volume conditions. Discuss the relationships among ΔE, ΔH, q_V and w. *Answer*: $w = 0$, $\Delta E = q_V < 0$, $\Delta H < 0$.

PRACTICE TEST

1 (10 points). Consider the six reactions and values of ΔH°_{298} given below.

(i)	$NH_4I(s) + nH_2O(l) \rightarrow NH_4I(aq)$	$\Delta H^\circ = 13.7$ kJ
(ii)	$BiCl_3(s) \rightarrow BiCl_3(g)$	$\Delta H^\circ = 113.4$ kJ
(iii)	$S(g) + e^- \rightarrow S^-(g)$	$\Delta H^\circ = -205.9$ kJ
(iv)	$B(g) \rightarrow B^+(g) + e^-$	$\Delta H^\circ = 806.7$ kJ
(v)	$\frac{1}{2}N_2(g) + \frac{1}{2}O_2(g) + \frac{1}{2}F_2(g) \rightarrow NOF(g)$	$\Delta H^\circ = -66.5$ kJ
(vi)	$CH_3CHO(l) + \frac{5}{2}O_2(g) \rightarrow 2CO_2(g) + 2H_2O(l)$	$\Delta H^\circ = -1166.4$ kJ

Which of these represent standard enthalpy changes for (a) formation, (b) electron affinity, (c) ionization energy, (d) change of state, (e) combustion, and (f) solution? Which reactions are (g) endothermic and (h) exothermic?

2 (10 points). Consider the following thermochemical equations:

$H_2(l) \rightarrow H_2(g)$	$\Delta H^\circ = 0.904$ kJ (for London forces)
$H_2(g) \rightarrow 2H(g)$	$\Delta H^\circ = 435.931$ kJ
$NaCl(s) \rightarrow NaCl(g)$	$\Delta H^\circ = 215.5$ kJ (for an ionic solid)
$NaCl(g) \rightarrow Na^+(g) + Cl^-(g)$	$\Delta H^\circ = 546.8$ kJ
$H_2O(l) \rightarrow H_2O(g)$	$\Delta H^\circ = 40.656$ kJ (for hydrogen bonding, dipole-dipole interactions, and London forces)
$H_2O(g) \rightarrow 2H(g) + O(g)$	$\Delta H^\circ = 926.919$ kJ (for O-H bonds broken)
$ClF(l) \rightarrow ClF(g)$	$\Delta H^\circ = 22.34$ kJ (for dipole-dipole interactions and London forces)
$ClF(g) \rightarrow Cl(g) + F(g)$	$\Delta H^\circ = 255.149$ kJ

List the (a) intermolecular forces and (b) intramolecular forces in order of increasing strength based on the magnitudes of the ΔH° values given for these equations.

3 (10 points). Choose the response that best completes each statement.

(a) The heat of formation of liquid water refers to the equation

(i) $2H_2(g) + O_2(g) \rightarrow 2H_2O(l)$

(ii) $2H(g) + O(g) \rightarrow H_2O(l)$

(iii) $H^+(aq) + OH^- \rightarrow H_2O(l)$

(iv) none of these answers

(b) The reaction A + B → C + D liberates 100 kJ. The reaction 2C + 2D → 2A + 2B

(i) absorbs 100 kJ (ii) releases 100 kJ

(iii) absorbs 200 kJ (iv) releases 200 kJ

(c) The standard heat of combustion of one mole of naphthalene is -5.154×10^3 kJ. Express this heat of combustion in units of joules.

(i) -5.154 J (ii) -515.4 J

(iii) -5.154×10^6 J (iv) -5.154×10^9 J

(d) What is the internal energy change for a system that absorbs 3.00 kJ from its surroundings and performs 750 J of work on the surroundings?

(i) 3.75 kJ (ii) 2.25 kJ

(iii) -2.25 kJ (iv) 753 J

(e) For an exothermic process

(i) *q* for the system is negative

(ii) energy is released by the system to the surroundings

(iii) the process will usually continue, once started

(iv) all of these answers

4 (25 points). Choose the response that best completes each statement.

(a) Which of the following would melt the most ice at 0 °C?

(i) 1.0 kg of metal at 25 °C with a specific heat of 0.50 J/K g

(ii) 25 kg of metal at 1 °C with a specific heat of 0.50 J/K g

(iii) 1.0 kg of water at 2.99 °C with a specific heat of 4.184 J/K g

(iv) All three samples are equally effective.

(b) The heat of formation of liquid water at 25 °C is -285.830 kJ/mol and of gaseous water is -241.818 kJ/mol. What is the heat of vaporization of water at 25 °C?

(i) -44.012 kJ/mol (ii) -241.818 kJ/mol

(iii) 527.648 kJ/mol (iv) 44.012 kJ/mol

(c) When 10.00 mol of acetylene, C_2H_4, burns in oxygen at 25 °C and 100 kPa pressure, 13.00 MJ is evolved. From this information, what is the $\Delta H°$ for the following thermochemical equation?

$$C_2H_2(g) + \frac{5}{2}O_2 \rightarrow 2CO_2(g) + H_2O(g)$$

(i) -1300. kJ (ii) -1.300×10^7 J
(iii) 1300. kJ (iv) none of these answers

(d) When 169 g pure silver metal is heated from 25 °C to 44 °C, 750 J are required. Calculate the specific heat of silver.

(i) 0.23 J/K g (ii) 2.4×10^6 J g K
(iii) 0.064 J/K g (iv) 25.4 J/K mol

(e) How much energy was released when the hydrogen in the Hindenburg dirigible burned? The volume of hydrogen was 2.00×10^8 L at 25 °C and 1.00 atm. The heat of combustion of hydrogen is -286 kJ/mol.

(i) 1.17×10^9 J (ii) 2.34×10^9 J
(iii) 5.72×10^{10} J (iv) none of these answers

5 (5 points). The respective standard state enthalpies of formation for $H_2O(g)$, $H_2S_2O_7(s)$, and $H_2SO_4(l)$ are -241.818, -1273.6, and -813.989 kJ/mol. Find $\Delta H°$ for the reaction

$$2H_2SO_4(l) \rightarrow H_2S_2O_7(s) + H_2O(g)$$

Will heat be released or absorbed?

6 (10 points). Using the following thermochemical data

$$Bi(s) + \frac{1}{2}Cl_2(g) \rightarrow BiCl(s) \qquad \Delta H° = -130.5 \text{ kJ}$$

$$BiCl(s) + Cl_2(g) \rightarrow BiCl_3(g) \qquad \Delta H° = -135.1 \text{ kJ}$$

$$BiCl_3(g) \rightarrow BiCl_3(s) \qquad \Delta H° = -113.4 \text{ kJ}$$

predict $\Delta H_f°$ for $BiCl_3(s)$.

7 (15 points). Samples of copper, silver, and gold, each weighing 50.0 g, were heated to 98.6 °C and placed in 100.0 g samples of water at 22.3 °C. If the respective heat capacities of the metals are 24.435, 25.351, and 25.418 J/K mol and that of water is 75.291 J/K mol, find the final temperature of each mixture.

8 (15 points). A calorimeter was calibrated by dissolving 10.5 g of $MgCl_2(s)$ in 250.0 g of water at 23.7 °C. The final temperature of the solution was 32.4 °C. Assuming the specific heat of the solution to be 4.31 J/K g and the enthalpy change for dissolving $MgCl_2(s)$ to be -155.06 kJ/mol, find the calorimeter constant.

Answers to practice test

1. (a) v; (b) iii; (c) iv; (d) ii; (e) vi; (f) i; (g) i, ii, iv; (h) iii, v, vi.
2. (a) London forces $<$ dipole-dipole interactions and London forces $<$ hydrogen bonding, dipole-dipole interactions, and London forces $<$ ionic solid; (b) Cl-F bond $<$ H-H bond $<$ H-O bond $<$ Na^+-Cl^- bond.
3. (a) iv, (b) iii, (c) iii, (d) ii, (e) iv.
4. (a) iv, (b) iv, (c) i, (d) i, (e) iv.
5. 112.6 kJ, absorbed.
6. -379.0 kJ.
7. 25.65 °C, 24.37 °C, 23.4 °C.
8. 840 J/K.

ADDITIONAL NOTES

CHAPTER 8

ELECTRONIC STRUCTURE AND THE PERIODIC TABLE

CHAPTER OVERVIEW

The chapter begins with a discussion of the dualistic nature of light and matter. This discussion leads into an introduction to quantum theory and its application to atomic structure--the Bohr model and the quantum-mechanical model of the atom. The electronic structure of the atom is then related to the position of the element in the periodic table, demonstrating the importance of electronic structure in determining chemical and physical properties of the elements.

COMPETENCIES

Definitions for the following terms should be learned:

- absorption spectrum
- angular momentum
- atomic orbital
- band spectra
- chemical stability
- continuous spectra
- diamagnetism
- diffraction
- electron configuration
- electronic structure
- emission spectrum
- excited states
- frequency
- ground state
- group, family (periodic table)
- Heisenberg uncertainty principle
- line spectra
- momentum
- noble gas
- outer electron configurations
- paramagnetism
- period (periodic table)
- periodic table
- photoelectric effect
- photon
- quantized
- quantum
- quantum mechanics
- quantum number
- quantum theory
- spectrometer
- spectroscopy
- wave number
- wavelength

General concepts and skills that should be learned include

- the dualistic nature of light and matter
- the quantization of particles and energy
- the Bohr model of the atom
- the quantum-mechanical model of the atom
- the names, symbols, permitted values, and physical interpretations of the quantum numbers
- writing and interpreting atomic electron configurations

Numerical exercises that should be understood include

- calculating the frequency, wave number, energy or wavelength of light when given one of these variables
- calculating the wavelength corresponding to momentum
- using the Rydberg equation to calculate the wave number for an electronic transition for atomic hydrogen or to find the values of the quantum numbers that correspond to a given spectral line

QUESTIONS

1. Complete the following sentences by filling in the various blanks with these terms: atomic orbitals, diamagnetism, electron configurations, family or group, paramagnetism, period, and periodic law.

The placement of the elements in the periodic table is related to their (a)______. Electrons will be occupying the same type of (b)______ for elements in a (c)______. An atom with an electron configuration such that there are unpaired electrons will show (d)______ and those with no unpaired electrons will show (e)______. The repetitive trends in electron configurations and magnetic properties illustrate the (f)______. In a horizontal row of elements in the table, a (g)______, the electrons fill the *ns*, (*n*-1)*d*, and *np* subshells in a regular fashion.

2. Match the correct term to each definition.

(a) the distance between any two points in the same relative locations on adjacent waves

(b) the number of wavelengths passing a given point in a unit of time

(c) the number of wavelengths per unit of length covered

(i) absorption spectrum
(ii) band spectrum
(iii) continuous spectrum
(iv) emission spectrum
(v) excited states
(vi) frequency

(d) radiation at all wavelengths is emitted	(vii) ground state
(e) radiation at only certain wavelengths is emitted from atoms and ions	(viii) line spectrum
	(ix) spectrometer
(f) spectrum of radiation emitted by a substance	(x) wave number
(g) spectrum of radiation after a continuum of radiation has passed through a substance	(xi) wavelength
(h) radiation occurring in clusters of wavelengths from molecules	
(i) an instrument used to produce a spectrum	
(j) the lowest energy orbit of an electron	
(k) states of energy higher than the ground state	

3. Which of the following statements are true? Rewrite each false statement so that it is correct.
   (a) The properties of elements are periodic functions of their atomic masses.
   (b) The distance between any two points in the same place on adjacent waves is the wavelength.
   (c) The atomic number gives the position of the element in the periodic table and is equal to the positive charge on the nucleus of each atom of that element.
   (d) The energy of a photon is directly proportional to the wavelength.
   (e) Wave mechanics is that part of quantum theory that deals with wave-like properties of particles.
   (f) An electron configuration of an atom is a description of the orbitals used by all the electrons in the atom.
   (g) Paramagnetism is the attraction of a substance toward a magnetic field as a result of unpaired electrons.
   (h) Quantum theory allows the simultaneous prediction of the exact location and momentum of an electron.

4. Distinguish clearly between/among (a) emission and absorption spectra; (b) continuous, band, and line spectra; (c) the particle and wave nature of electrons; (d) ground and excited states; (e) orbits and atomic orbitals; (f) paramagnetism and diamagnetism; (g) a period and a group (or family) in the periodic table; (h) wavelength, wave number, and frequency; and (i) quantized and nonquantized properties.

5. Choose the best answer for each statement.

(a) Which of the following deals with the uncertainty in knowing simultaneously both the exact momentum and the exact position of an electron?

(i) Schrödinger wave equation
(ii) Pauli exclusion principle
(iii) Hund rule of maximum multiplicity
(iv) Heisenberg uncertainty principle

(b) The mass times the velocity is the

(i) frequency (ii) stationary state
(iii) momentum (iv) ground state of a particle

(c) The emission of electrons from certain metals caused by shining light on them is the

(i) emission spectrum (ii) photoelectric effect
(iii) line spectrum (iv) wave number

(d) A single quantum of radiant energy is a

(i) photon (ii) ground state
(iii) stationary state (iv) mass number

(e) The symbols n, ℓ, m_ℓ, and m_s represent

(i) electron configurations
(ii) quantum numbers
(iii) isotopes
(iv) neutron, proton, atomic, and mass numbers

6. Each property of light can be best explained in terms of either (a) wave nature or (b) particle nature. Identify the nature that provides the better explanation for each of the following properties: (i) interference, (ii) diffraction, and (iii) the photoelectric effect.

7. Which property of electrons was demonstrated by the Davisson-Germer experiment? How can the experimental results be best explained?

8. Briefly discuss the contributions of the following physicists to the quantum revolution: (a) Max Planck, (b) Albert Einstein, (c) Niels Bohr, (d) Louis de Broglie, (e) Werner Heisenberg, and (f) Erwin Schrödinger.

9. Are the regions of absorption and emission of radiation restricted to the visible region of the spectrum?

10. The yellow light emitted by sodium vapor actually consists of two closely spaced lines (called the sodium D lines) with wavelengths of 589.59 nm and 588.99 nm. Which of these lines is caused by photons of greater energy?

11. Write the equation that can be used to calculate the wave numbers of the lines in the spectrum of the hydrogen atom. What is the relationship between the two intergers n_1 and n_2?

12. List and briefly discuss some of the important consequences of the quantum-mechanical model of atomic theory.

13. What happens when quantum mechanics is applied to large-scale, familiar phenomena?

14. How does the Bohr theory explain the line spectra that are observed for atomic hydrogen?

15. What are quantum numbers?

16. Interpret the following notation: $n\ell^x$.

17. Choose the configurations that represent excited states of atoms: (a) $1s^22s^1$, (b) $1s^22s^22p^5$, (c) $1s^22s^22p^53s^1$, (d) $1s^22s^22p^63s^23p^63d^24s^2$, (e) $1s^22s^22p^63s^23p^63d^34s^1$.

18. The periodic law states that the physical and chemical properties of elements are periodic functions of atomic number. Why is this true?

Answers to questions

1. (a) electron configurations, (b) atomic orbitals, (c) family or group, (d) paramagnetism, (e) diamagnetism, (f) periodic law, (g) period.
2. (a) xi, (b) vi, (c) x, (d) iii, (e) viii, (f) iv, (g) i, (h) ii, (i) ix, (j) vii, (k) v.
3. (a) F, replace *masses* with *numbers*; (b) T; (c) T; (d) F, replace *wave length* with *frequency*; (e) T; (f) T; (g) T; (h) F, replace *allows* with *restricts to within a certain value.*
4. (a) An *emission spectrum* is the spectrum of energy released by an excited species; an *absorption spectrum* is the spectrum of energy absorbed by a species. (b) A *continuous spectrum* consists of all wavelengths; a *band spectrum* consists of certain bands or clusters of continuous wavelengths; a *line spectrum* consists of specific single wavelengths. (c) *Particle*

nature is characterized by such properties as electron transfer; *wave nature* is characterized by such properties as diffraction. (d) The *ground state* is the lowest allowed energy state; an *excited state* refers to a system having extra energy so that it is in a higher energy state. (e) An *orbit* is the Bohr theory terminology for the major energy level described by the quantum number n; an *atomic orbital* is an energy state described by the quantum number m_ℓ. (f) *Paramagnetism* refers to the attraction to a magnetic field; *diamagnetism* refers to the repulsion from a magnetic field. (g) A *period* is a horizontal row in the periodic table; a *group* is a vertical column in the table. (h) *Wavelength* is the distance between similar points on consecutive waves; *wave number* is the number of wavelengths per unit of length; *frequency* is the number of wavelengths passing a given point per unit time. (i) *Quantized properties* are allowed only certain values; *nonquantized properties* have continuous values.

5. (a) iv, (b) iii, (c) ii, (d) i, (e) ii.
6. (a) i, ii; (b) iii.
7. The experiment proved that electrons could be diffracted. Thus, electrons have a wavelike nature.
8. (a) Planck studied the emission of light by heated solids and stated that the emission and absorption of radiation are quantized processes; (b) Einstein studied the equivalence of mass and energy and stated that radiation is a flow of photons; (c) Bohr developed a theory for a quantized hydrogen atom; (d) de Broglie studied the wavelike nature of particles; (e) Heisenberg stated the uncertainty principle; (f) Schrödinger formulated mathematical equations for quantum theory.
9. No, spectral studies are made in all regions of the spectrum.
10. The photon energy is proportional to the frequency which, in turn, is inversely proportional to wavelength. Thus the line with the shorter wavelength will contain the more energetic photons, i.e., 588.99 nm.
11. $\nu = R(1/n_1^2 - 1/n_2^2)$ where $n_2 > n_1$.
12. Two important consequences are that the energy is quantized and that the exact location of electron and momentum cannot be simultaneously determined exactly; the concept of the probability of the location of an electron with a certain energy is used.

13. When quantum mechanics is applied to large-scale, familiar phenomena, the effects are too small to be significant and we are left with the laws of classical mechanics intact.
14. The Bohr theory explains the line spectra that are observed for atomic hydrogen as follows: (1) when the atom absorbs energy, the atom becomes excited and the electron jumps from the lowest energy level of the ground state to higher energy levels; (2) as the electron falls back to the ground state, the extra energy is emitted as a quantum of electromagnetic radiation.
15. Quantum numbers are whole-number multipliers that specify amounts of energy.
16. The symbol n is the principal quantum number, ℓ is the subshell quantum number (usually given as a letter corresponding to the numerical value), and x is the number of electrons in that subshell.
17. The excited states are a, c, and e.
18. Chemical and physical properties are related to the electronic structure of the atom, which is related to atomic number.

SKILLS

Skill 8.1 Understanding quantum numbers for electrons in atoms

The principal quantum number (n) may take any integer value, i.e., $n = 1, 2, 3, \ldots$. In general, the larger the value of n for an electron, the farther from the nucleus the electron may be found and the easier it is to remove it. The value of n corresponds to the Bohr shell. The angular momentum quantum number (ℓ) may take any integer value (including zero) up to n-1, i.e., $\ell = 0, 1, 2, \ldots, (n-1)$. Each value of ℓ corresponds to a subshell that is usually denoted by a letter (s for $\ell = 0$, p for $\ell = 1$, d for $\ell = 2$, f for $\ell = 3$, etc.). Each subshell has a distinct set of shapes: spheres, dumb bells, four-leafed clovers, etc. The magnetic quantum number, m_ℓ, may take any integer value (including zero) from the value of $+\ell$ to $-\ell$ and represents the orientation of the subshell in space. The combination of n, ℓ, and m_ℓ designates a specific atomic orbital that can hold two electrons, one with a spin quantum number (m_s) of +½ and the other with -½.

To illustrate the determination of the values of the four quantum numbers for a given electron, consider the third electron in the 2p subshell. The value of n is 2 and the value of ℓ for the p subshell is 1 (recall that the value of ℓ

is represented by a letter). The corresponding permitted values of m_ℓ are +1, 0, and -1 (recall m_ℓ is limited by the value of ℓ). To find which value of m_ℓ and m_s correspond to the third electron, boxes are drawn representing the three values of m_ℓ:

	[ ]	[ ]	[ ]
m_ℓ =	+1	0	-1

and the appropriate number of electrons are added according to Hund's rule--first one electron with a positive spin (represented by an arrow pointing upwards) in each box before putting in a second electron in any box with a negative spin (represented by an arrow pointing downwards). For three electrons the boxes look like

	[↑]	[↑]	[↑]
m_ℓ =	+1	0	-1

The box containing the third electron, shows that m_ℓ = -1 and m_s = +½. As another example, consider the sixth electron in the 3d subshell. The value of n is 3 and the value of ℓ is 2. The filling of the *d* subshell by six electrons would be

	[↑↓]	[↑]	[↑]	[↑]	[↑]
m_ℓ =	+2	+1	0	-1	-2

giving m_ℓ = +2 and m_s = -½.

Skill Exercise: Determine the values of the four quantum numbers for the (a) second electron in the 7*s* orbital, (b) third electron in the 3*p* subshell, (c) eighth electron in the 4*d* subshell, and (d) eighth electron in the 6*f* subshell. *Answers*: (a) n = 7, ℓ = 0, m_ℓ = 0, m_s = -½; (b) n = 3, ℓ = 1, m_ℓ = -1, m_s = +½; (c) n = 4, ℓ = 2, m_ℓ = 0, m_s = -½, (d) n = 6, ℓ = 3, m_ℓ = +3, m_s = -½.

Skill 8.2 Writing electron configurations for atoms

The general order of filling of subshells is

1*s* 2*s* 2*p* 3*s* 3*p* 4*s* 3*d* 4*p* 5*s* 4*d* 5*p* 6*s* 4*f* 5*d* 6*p* 7*s* 5*f* 6*d* 7*p*

(although there are some exceptions because of the nearness in energy of some of the orbitals, the extra stability of an atom for a half-filled or completely filled subshell, etc.). Each orbital can hold a maximum of two electrons, so that the *s* subshell can hold a maximum of two electrons; the *p* subshell, six; the *d* subshell, ten; and the *f* subshell, fourteen.

The number of electrons in each subshell is written as a following superscript, e.g., $2p^3$ represents three electrons in the $2p$ subshell. The sum of the superscripts must equal the number of electrons in the atom. Consider the following configurations written for the ground states of the elements:

B	Z = 5	5 electrons	$1s^2 2s^2 2p^1$
K	Z = 19	19 electrons	$1s^2 2s^2 2p^6 3s^2 3p^6 4s^1$
Kr	Z = 36	36 electrons	$1s^2 2s^2 2p^6 3s^2 3p^6 3d^{10} 4s^2 4p^6$
I	Z = 53	53 electrons	$1s^2 2s^2 2p^6 3s^2 3p^6 3d^{10} 4s^2 4p^6 4d^{10} 5s^2 5p^5$
La	Z = 57	57 electrons	$1s^2 2s^2 2p^6 3s^2 3p^6 3d^{10} 4s^2 4p^6 4d^{10} 4f^1 5s^2 5p^6 6s^2$

Skill Exercise: Write electron configurations for (a) Mg, (b) Se, (c) Xe, (d) V, and (e) C. *Answers*: (a) $1s^2 2s^2 2p^6 3s^2$, (b) $1s^2 2s^2 2p^6 3s^2 3p^6\ 3d^{10} 4s^2 4p^4$, (c) $1s^2 2s^2 2p^6 3s^2 3p^6 3d^{10} 4s^2 4p^6 4d^{10} 5s^2 5p^6$, (d) $1s^2 2s^2 2p^6 3s^2 3p^6 3d^3 4s^2$, (e) $1s^2 2s^2 2p^2$.

Writing electron configurations can be simplified by using the symbol for the appropriate noble gas to represent the configuration of inner electrons. For example, the configurations above can be written

B	Z = 5	5 electrons	$[He]2s^2 2p^1$
K	Z = 19	19 electrons	$[Ar]4s^1$
Kr	Z = 36	36 electrons	[Kr]
I	Z = 53	53 electrons	$[Kr]4d^{10} 5s^2 5p^5$
La	Z = 57	57 electrons	$[Kr]4d^{10} 4f^1 5s^2 5p^6 6s^2$

Skill Exercise: Write the simplified electron configurations for (a) Mg, (b) Se, (c) Xe, (d) V, and (e) C. *Answers*: (a) $[Ne]3s^2$, (b) $[Ar]3d^{10} 4s^2 4p^4$, (c) [Xe], (d) $[Ar]3d^3 4s^2$, (e) $[He]2s^2 2p^2$.

Some chemists write out the complete set of orbitals when writing electron configurations instead of using the subshell notation as above. For example, $2p^3$ would be written as $2p_x^1 2p_y^1 2p_z^1$. Often this orbital notation is written in the form of orbital diagrams using lines, circles, boxes, etc. For example, $2p^3$ would be

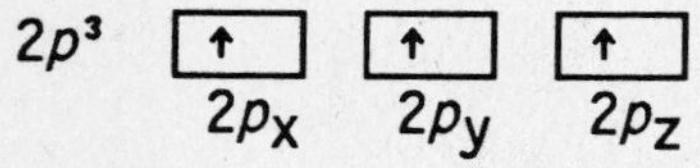

[Note that $2p_x$, $2p_y$, and $2p_z$ do not mean the same thing as m_ℓ = +1, 0, and -1 (see *Skill 8.1 Understanding quantum numbers for electrons in atoms*).]

To write the electron configuration for an atom in an excited state, the ground state configuration is modified so that only the last electron in the atom is written in some subshell other than the one that it occupies in the ground state. For example, the 670.8 nm red line in Li corresponds to an electron making the transition $2p \rightarrow 2s$, which corresponds to the following configurations:

ground state Li	Z = 3	3 electrons	$1s^2 2s^1$
excited state Li	Z = 3	3 electrons	$1s^2 2p^1$

If an atom has unpaired electrons, it will be paramagnetic. If all of the electrons are paired, the atom will show diamagnetism. In the above examples the paramagnetic species are B, K, I, and Li (in both the ground and excited states) and the diamagnetic species is Kr.

Skill Exercise: Predict on the basis of the electron configurations whether the following atoms are paramagnetic or diamagnetic: (a) Mg, (b) Se, (c) Xe, (d) V, and (e) C. *Answers*: (a) diamagnetic, (b) paramagnetic, (c) diamagnetic, (d) paramagnetic, (e) paramagnetic.

NUMERICAL EXERCISES

Example 8.1 Wave nature of light

What is the frequency of light in the red part of the visible spectrum, e.g., λ = 650 nm?

The frequency is related to the wavelength by the speed of light.

$$\nu = \frac{c}{\lambda} = \frac{3.00 \times 10^8 \text{ m/s}}{(650 \text{ nm})(10^{-9} \text{ m/1 nm})} = 4.6 \times 10^{14} \text{ s}^{-1}$$

The frequency is 4.6×10^{14} s^{-1} or 4.6×10^{14} Hz.

Exercise 8.1 What is the frequency of light in the blue part of the visible spectrum, e.g., λ = 470? *Answer*: 6.4×10^{14} s^{-1}.

Example 8.2 Wave nature of light

What is the wave number of the red light described in *Example 8.1*? Express the value in cm^{-1}.

The wave number is the reciprocal of the wavelength:

$$\bar{\nu} = \frac{1}{\lambda} = \frac{1}{(650 \text{ nm})(10^{-9} \text{ m/1 nm})(10^2 \text{ cm/1 m})} = 1.5 \times 10^4 \text{ cm}^{-1}$$

The wave number of the red light is 1.5×10^4 cm^{-1}.

Exercise 8.2 What is the wave number of the blue light described in Exercise 8.1? Express the value in cm^{-1}. *Answer*: $2.1 \times 10^4\ cm^{-1}$.

Example 8.3 Particle nature of light

What is the energy of a photon of the red light described in *Example 8.1*?

The energy of a photon is given by $E = h\nu$ where ν is the frequency.

$$E = h\nu = (6.626 \times 10^{-34}\ J\ s)(4.6 \times 10^{14}\ s^{-1}) = 3.0 \times 10^{-19}\ J$$

The energy associated with a quantum of red light is 3.0×10^{-19} J. Although this may not seem to be much, for a mole of quanta there will be $(3.0 \times 10^{-19}$ J/quantum)$(6.022 \times 10^{23}$ quanta/mol) = 1.8×10^5 J, which is enough energy to heat one liter of water by 45 °C.

Exercise 8.3 What is the energy of a photon of the blue light described in Exercise 8.1? Does a photon in the blue or red region of the spectrum have more energy? *Answers*: 4.2×10^{-19} J, blue.

Example 8.4 Wave nature of particles

An electron, $m = 9.11 \times 10^{-31}$ kg, is traveling at the speed of 0.7 the speed of light. What is the corresponding wavelength of radiation?

The wavelength is given by

$$\lambda = \frac{h}{mu} = \frac{(6.626 \times 10^{-34}\ J\ s)(1\ kg\ m^2\ s^{-2}/1\ J)}{(9.11 \times 10^{-31}\ kg)[(0.7)(3.00 \times 10^8\ m/s)]} = 3 \times 10^{-12}\ m$$

The de Broglie wavelength for the electron is 3×10^{-12} m.

Exercise 8.4 What must be the momentum, *mu*, of a particle that corresponds to microwave radiation, e.g., $\lambda = 10$ cm? *Answer*: 7×10^{-33} kg m/s.

Example 8.5 Rydberg equation

One of the spectral lines of atomic hydrogen in the visible portion of the spectrum (400 nm to 700 nm) is located at $\lambda = 658$ nm. What are the values of the principal quantum numbers involved in this energy transition?

The spectral lines for hydrogen are given by

$$\bar{\nu} = (109{,}677\ cm^{-1})\left[\frac{1}{n_1^2} - \frac{1}{n_2^2}\right]$$

The value of $\bar{\nu}$ for the observed red line is

$$\bar{\nu} = \frac{1}{\lambda} = \frac{1}{(658\ \text{nm})(10^{-9}\ \text{m/1 nm})(10^{2}\ \text{cm/1 m})} = 1.52 \times 10^{4}\ \text{cm}^{-1}$$

which gives

$$\left[\frac{1}{n_1{}^2} - \frac{1}{n_2{}^2}\right] = \frac{1.52 \times 10^{4}\ \text{cm}^{-1}}{109{,}677\ \text{cm}^{-1}} = 0.139$$

To find the values of n_1 and n_2 requires a solution by trial and error. If $n_1 = 1$ and $n_2 = 2, 3, \ldots$, the respective values of $(1/n_1{}^2 - 1/n_2{}^2)$ are 0.750, 0.889, 0.938, ..., and the values approach 1.000. Thus $n_1 \neq 1$. For $n_1 = 2$ and $n_2 = 3$, $(1/n_1{}^2 - 1/n_2{}^2) = 0.139$, which is the desired value.

Exercise 8.5 What is the wavelength of the line in the hydrogen spectrum for a transition between the $n_2 = 4$ and $n_1 = 3$? Will this transition be in the visible spectrum (400 nm to 700 nm)? *Answers*: 1876 nm, no.

PRACTICE TEST

1 (12 points). Match the correct term to each definition.

(a) an array of waves or particles spread out according to the increasing or decreasing magnitude of some physical property
(b) a whole-number multiplier of the basic quantity for the system
(c) something that is restricted to amounts that are multiples of the basic quantity for the system
(d) the lowest energy for a system
(e) the designation of the orbitals occupied by all of the electrons in an atom
(f) a state of higher energy than the ground state

(i) electron configuration
(ii) excited state
(iii) ground state
(iv) quantized
(v) quantum number
(vi) spectrum

2 (32 points). Complete the following table describing the four quantum numbers of an electron:

Symbol	Name	Permitted values	Interpretation
	spin		
m_ℓ			
		0, 1, 2, ..., $(n-1)$	
			size and energy

3 (18 points). Write reasonable electron configurations for (a) Na, $Z = 11$; (b) Ge, $Z = 32$; (c) Zr, $Z = 40$; (d) Pr, $Z = 59$; (e) F, $Z = 9$; and (f) U, $Z = 92$. Which of these are paramagnetic?

4 (10 points). What are the values of the four electronic quantum numbers for the third electron in the $5p$ subshell?

5 (28 points). Choose the reponse that best completes each statement.

(a) As the wavelength increases for light, the energy

(i) increases (ii) remains the same

(iii) decreases (iv) none of these answers

(b) If the momentum of a particle is determined very precisely, the error in its position is

(i) negligible

(ii) given by the Pauli exclusion principle

(iii) independent of the uncertainty in the velocity

(iv) great

(c) How many emission lines for atomic hydrogen would be observed if only the electronic energy levels shown at the right are involved?

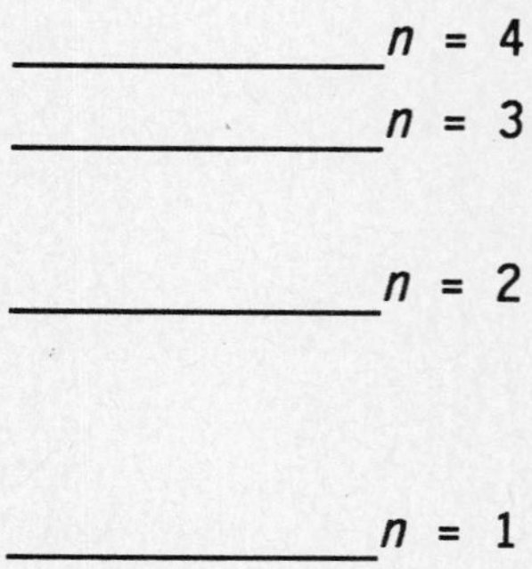

(i) 3 (ii) 4

(iii) 6 (iv) 12

(d) The Bohr theory successfully explained

(i) blackbody radiation

(ii) many-electron molecular spectra

(iii) the magnetic properties of monatomic ions

(iv) atomic hydrogen spectra

(e) Which of the following hydrogen atom transitions *emits* the highest energy light?

(i) $1s \rightarrow 10s$ (ii) $4s \rightarrow 1s$

(iii) $4s \rightarrow 3s$ (iv) $2s \rightarrow 1s$

(f) The Bohr theory for atomic hydrogen included each of the following *except*

(i) electrons in a stable atom do not radiate or absorb energy as long as they remain in the same energy level

(ii) electrons are in stable orbits about the nucleus of an atom

(iii) an electron can radiate quanta of energy when passing from one orbit to another

(iv) it is possible to determine accurately both the momentum and the position of an electron simultaneously

(g) The quantum-mechanical picture of the hydrogen atom

(i) is mainly concerned with the energies associated with electrons

(ii) describes the path traveled by the electron

(iii) proves that the electron moves in an orbit about the nucleus as if it were a satellite

(iv) is easier to visualize than Bohr's theory

(h) The orbital shown in the sketch is a/an

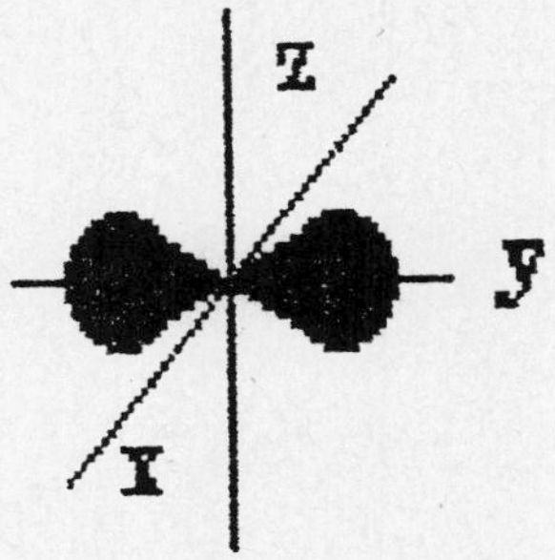

(i) *s* orbital

(ii) d_{y^2} orbital

(iii) p_x orbital

(iv) none of these answers

(i) The orbital quantum number corresponding to a *d* orbital is

(i) $\ell = 2$ (ii) $\ell = 0$

(iii) $\ell = 1$ (iv) $m_\ell = 2$

(j) Which of the following rules must be obeyed in the "lowest energy" process?

(1) Orbitals can hold a maximum of two electrons with parallel spins.

(2) Subshells are used in order of increasing energy.

(3) A subshell is filled before proceeding to fill the next subshell.

(4) Orbitals having the same energy are allocated one electron before pairing occurs.

(5) Single electrons have parallel spins in orbitals of the same energy.

(i) all of the above (ii) only (2), (4), and (5)

(iii) all except (1) (iv) all except (3)

(k) Which is a satisfactory set of quantum numbers (n, ℓ, m_ℓ, m_s) for an electron in a *p* orbital?

(i) 3, 1, 1, +½ (ii) 2, 1, -2, +½

(iii) 3, 0, 1, -½ (iv) none of these answers

(l) The subshell that fills immediately after the 4*s* subshell is

(i) 4*p* (ii) 3*d*

(iii) 5*s* (iv) 3*f*

(m) The maximum number of electrons in a *d* orbital is

(i) 10 (ii) 2

(iii) 6 (iv) 8

(n) An electron configuration of an atom is

(i) a diagram indicating the relative energy states of each orbital

(ii) a picture of the shapes of atomic orbitals

(iii) given by $R(1/n_2^2 - 1/n_1^2)$

(iv) the designation of the subshells and orbitals occupied by the electrons

Answers to practice test

1. (a) vi, (b) v, (c) iv, (d) iii, (e) i, (f) ii.
2. m_s, ±½, angular momentum of electron; orbital (or magnetic) quantum number, 0, ±1, ±2, ..., ±ℓ, orientation of orbital in space; ℓ, subshell (or azimuthal or angular momentum) quantum number, subshell shape; *n*, principal quantum number, 1,2,3,....
3. (a) $1s^22s^22p^63s^1$ or [Ne]$3s^1$, (b) $1s^22s^22p^63s^23p^63d^{10}4s^24p^2$ or [Ar]$3d^{10}4s^24p^2$, (c) $1s^22s^22p^63s^23p^63d^{10}4s^24p^64d^25s^2$ or [Kr]$4d^25s^2$, (d) $1s^22s^22p^63s^23p^63d^{10}$ $4s^24p^64d^{10}4f^35s^25p^66s^2$ or [Xe]$4f^36s^2$, (e) $1s^22s^22p^5$ or [He]$2s^22p^5$, (f) $1s^22s^22p^63s^23p^63d^{10}4s^24p^64d^{10}4f^{14}5s^25p^65d^{10}5f^46s^26p^67s^2$ or [Rn]$5f^47s^2$ (note that actual configuration is [Rn]$5f^36d^17s^2$); Na, Ge, Zr, Pr, F, and U are paramagnetic.
4. $n = 5$, $\ell = 1$, $m_\ell = -1$, $m_s = +½$.
5. (a) iii, (b) iv, (c) iii, (d) iv, (e) ii, (f) iv, (g) i, (h) iv, (i) i, (j) iii, (k) i, (l) ii, (m) ii, (n) iv,

ADDITIONAL NOTES

CHAPTER 9

PERIODIC PERSPECTIVE I: ATOMS AND IONS

CHAPTER OVERVIEW

The various classifications of the elements based on location in the periodic table are presented. The chapter concludes with a discussion of the periodic trends for atomic and ionic radii, ionization energy, and electron affinity.

COMPETENCIES

Definitions for the following terms should be learned:

- actinides
- alkali metals
- alkaline earth metals
- atomic radii
- bond length
- effective nuclear charge
- electron affinity
- halogens
- ionic radii
- ionization energy
- isoelectronic
- lanthanide contraction
- lanthanides
- Lewis symbol
- metals
- noble gas configuration
- nonmetals
- periodic law
- rare earth elements
- representative elements
- screening effect
- semiconducting elements
- transition elements
- transuranium elements
- valence electrons

General concepts and skills that should be learned include

- classifying elements according to their electron configurations
- the periodic law
- writing Lewis symbols for atoms and monatomic ions
- the general trends across a period and down a family of elements of properties such as radii, ionization energies, and electron affinities

- predicting the properties for an element based on known values of the properties of elements surrounding it in the periodic table

QUESTIONS

1. Match the correct term to each definition or representation of the term.

(a) $X(g) + e^- \rightarrow X^-(g)$	(i) atomic radius
(b) $X(g) \rightarrow X^+(g) + e^-$	(ii) electron affinity
(c) $X^+(g) \rightarrow X^{2+}(g) + e^-$	(iii) first ionization energy
(d) F^-, Ne, Na^+	(iv) isoelectronic
(e) the radius of a cation or anion	(v) ionic radius
(f) the radius of an atom in a single covalent bond	(vi) second ionization energy

2. Which of the following statements are true? Rewrite each false statement so that it is correct.
   (a) The ionization energy is the energy for the formation of ions from an ionic material.
   (b) The second ionization energy is usually much larger than the first ionization energy because it is more difficult to remove an electron from a positively charged species than from a neutral species.
   (c) The screening effect is the decrease in nuclear charge acting on an electron due mainly to the effect of other electrons in inner shells.
   (d) The outer-shell electron configuration of elements in Representative Group VI is ns^2nd^4.
   (e) The ions O^{2-}, F^-, Na^+, Mg^{2+}, and Al^{3+} are isoelectronic with Ne.
   (f) Atomic radii are always larger than ionic radii.

3. Complete the following sentences by filling in the various blanks with these terms: actinides, alkali metals, alkaline earth metals, halogens, lanthanides, noble gas, rare earth elements, representative elements, semiconducting elements, transition elements, and transuranium elements.

   There are several ways to classify the elements by location in the periodic table. Each period ends with a (a)_______. These elements filling the *s* or *p* subshells are called (b)_______ and those filling the *d* or *f* subshells are called (c)_______. Several of the groups or families in the periodic table have special names. For example, elements in Representative Group VII (the fluorine family) are called the (d)_______, elements in Representative Group I (the lithium family) are called the

(e) ________, and elements in Representative Group II (the beryllium family) are called the (f) ________. Most of the representative and transition elements are classified as metals, several of the representative elements are classified as nonmetals, and a few of the representative elements are classified as (g) ________. Scandium, yttrium, and all of the sixth period elements from lanthanum to lutetium are sometimes called the (h) ________. The two groups of *f*-transition elements (sometimes called the inner-transition elements) consist of the (i) ________ and (j) ________. The elements having $Z > 92$ are called the (k) ________.

4. Distinguish clearly between (a) ionic and atomic radii and (b) ionization energy and electron affinity.

5. Choose the response that best completes each statement.

(a) Species having identical electron configurations are

(i) catenated (ii) isobaric
(iii) isoelectronic (iv) isothermal

(b) The energy necessary to remove the least tightly bound electron from a gaseous atom is the

(i) effective nuclear charge (ii) first ionization energy
(iii) electron affinity (iv) screening effect

(c) The decrease in the nuclear charge acting on an electron as a result of the effect of other electrons in inner shells is the

(i) screening effect (ii) second ionization energy
(iii) effective nuclear charge (iv) isolation effect

(d) The portion of the nuclear charge that acts on an electron and is always less than the nuclear charge is the

(i) penetration effect (ii) effective nuclear charge
(iii) ionization potential (iv) electronegativity

(e) The element that has the lowest first ionization energy is

(i) Li (ii) K
(iii) Ca (iv) Kr

(f) The species that has the largest radius is

(i) He^{+} (ii) Br
(iii) Br^{-} (iv) Br^{7+}

(g) The element that has the lowest first ionization energy is

(i) He (ii) Ne
(iii) Ar (iv) Kr

(h) The electron that will be removed first in the ionization of a phosphorus atom is

(i) 1*s* (ii) 2*p*

(iii) 3*d* (iv) 3*p*

(i) The electron that will be removed first in the ionization of a chromium atom is

(i) 1*s* (ii) 4*s*

(iii) 3*d* (iv) 3*p*

6. What property of atoms did Mendeleev and Meyer use to organize the elements in their periodic tables? How was this changed by the results obtained by Moseley? How did Mendeleev extend our chemical knowledge using his table?

7. Classify each of the following elements as a (a) metal, (b) nonmetal, or (c) semiconducting element: (i) Br, (ii) Ba, (iii) P, (iv) Si, (v) Fe, (vi) Sb, (vii) Ne, (viii) Mg, (ix) Ge, and (x) W.

8. From the list of elements given in Question 7, identify (a) the transition element(s), (b) the noble gas(es), (c) the halogen(s), (d) the alkali metal(s), (e) the alkaline earth metal(s).

9. What are valence electrons? Compare the energy level of valence electrons to the energy levels of the rest of the electrons in an atom. What is the relationship between the number of outermost electrons and the periodic table group number for representative elements?

10. Compare the outermost electron configurations of the metals, nonmetals, and semiconducting elements.

11. Briefly describe how a Lewis symbol is written for an atom.

12. Write the Lewis symbols for (a) Cs, (b) Ga, (c) Se, (d) He, and (e) N.

13. Interpret the following Lewis symbols
(a) $\cdot\ddot{\underset{\bullet}{S}}:$; (b) $\cdot\dot{Si}:$; (c) Rb·

14. What does the term "atomic radius" mean? How are atomic radii determined?

15. What does the term "ionic radius" mean? How are ionic radii determined?

16. How does the atomic radius compare to the anionic radius for a given element? How does the atomic radius compare to the cationic radius?

17. Name two factors that influence atomic radii.

18. Name the two effects that determine the effective nuclear charge of an atom. How does the effective nuclear charge change as the atomic number increases across a period?

19. Briefly describe the screening effect. Which electrons--those in the *n*, (*n*-1), or less than the (*n*-1) energy level--are most effective in screening the outermost electrons?

20. Which atom would you predict to have the highest effective nuclear charge: Be, N, or Ne? Why?

21. How does the atomic radius depend on the principal quantum number? Explain this effect.

22. Following is a list of atomic radii in nanometers: 0.099, 0.227, 0.143, 0.071, and 0.186. Match these radii with the following atoms: Na, F, Al, K, and Cl. Give reasons for your choices.

23. Match the radii (a) 0.123 nm, (b) 0.118 nm, (c) 0.095 nm, and (d) 0.265 nm to the following atoms: Si, Cs, Ge, and Ar.

24. Define the term "isoelectronic."

25. Place the following species in order of increasing radius; K^+, Ar, Cl^-, and Ca^{2+}.

26. How would the ionic radius of O^{2-} compare to that of O^-?

27. Using the arrangement of elements in the periodic table, match the following ionic radii--(i) 0.068 nm, (ii) 0.136 nm, (iii) 0.196 nm, (iv) 0.112 nm--to the following ions: (a) F^-, (b) Sr^{2+}, (c) Br^-, and (d) Li^+.

28. Define the term "ionization energy". How do the values of the ionization energy change as successive electrons are removed? Why?

29. Why would you expect the second ionization energy of lithium to be very large?

30. Name the two factors that influence the value of the ionization energy for an atom.

31. The first ionization energies of Mn and Zn are abnormally high compared to the values of the elements on either side of them in the fourth period of the

periodic table. How does this observation relate to the electron configurations of the various atoms?

32. Which electrons will be removed first during the ionization of an atom of Ba?

33. Compare the first ionization energies of the oxygen family elements to those of the nitrogen family. Do these values follow the general trends?

34. Define the term "electron affinity." Are the values of the electron affinities of the elements always negative?

35. Do the nonmetals form anions or cations? What determines the charge on the ions that form?

36. Compare the cations formed by the *s*-block representative metals in the fourth through sixth periods to these formed by the Representative Group III elements in these periods.

37. What types of cations do the metals of Group IV and V form?

38. Briefly discuss the formation of cations by the *d*-transition elements. Which transition elements form pseudo-noble gas cations?

39. Compare the electron configurations of K^+ and Cl^-.

40. Write the reasonable electron configurations for (a) Fe^{2+}, (b) Fe^{3+}, (c) Ca^{2+}, (d) Pb^{2+}, (e) Ag^+, (f) Cd^{2+}, and (g) Cr^{3+}.

41. Identify and give a reasonable explanation for the trend in the (a) first ionization energies of the alkaline earth metals--Representative Group II--based on atomic sizes, (b) first ionization energies of the elements in the third period--Na to Ar--based on electron configurations and atomic sizes, and (c) atomic radii of the noble gases based on atomic structure.

42. Identify the general trend across a period as (i) increasing, (ii) decreasing, or (iii) showing no significant change for each of the following properties: (a) atomic radius, (b) ionization energy, (c) electron affinity, (d) positive charge on cation, and (e) negative charge on anion. Repeat this exercise identifying the trend down a family.

Answers to questions

1. (a) ii, (b) iii, (c) vi, (d) iv, (e) v, (f) i.

2. (a) F, replace *ions from ... material* with *cations from neutral gaseous atoms*; (b) T; (c) T; (d) F, change nd^4 to np^4; (e) T; (f) F, replace *ionic radii* with *the radii of the respective cations*.
3. (a) noble gas, (b) representative elements, (c) transition elements, (d) halogens, (e) alkali metals, (f) alkaline earth metals, (g) semiconducting elements, (h) rare earth elements, (i) actinides or lanthanides, (j) lanthanides or actinides, (k) transuranium elements.
4. (a) An *ionic radius* refers to the size of the ion of an element in an ionic bond; an *atomic radius* refers to the size of the atom in a covalent bond. (b) *Ionization energy* is the energy required to remove an electron; *electron affinity* is the energy involved in adding an electron.
5. (a) iii, (b) ii, (c) i, (d) ii, (e) ii, (f) iii, (g) iv, (h) iv, (i) ii.
6. The arrangements by Mendeleev and Meyer were on the basis of increasing atomic mass. Moseley used an arrangement based on increasing atomic number. Mendeleev left gaps in his table and predicted some of the properties of the then-unknown elements.
7. (a) ii, v, viii, x; (b) i, iii, vii; (c) iv, vi, ix.
8. (a) v, x; (b) vii; (c) i; (d) none; (e) ii, viii.
9. Valence electrons are the electrons that are available to take part in chemical bonding. Valence electrons are in the highest occupied energy level, whereas the remainder are buried in the lower levels. The number of outermost electrons in an atom of a representative element equals the group number of that element.
10. For the metals: Li and Be family elements and all elements but one transition metal family have ns^1 or ns^2, eight p block metal elements have ns^2np^X; for the nonmetals: p block nonmetal elements have ns^2np^X; for the semiconducting elements: p block elements with ns^2np^X.
11. The symbol of element represents nucleus plus inner electrons; valence electrons are indicated by dots arranged around the symbol.
12. (a) Cs·, (b) :$\dot{\text{Ga}}$, (c) :$\underset{\cdot}{\dot{\text{Se}}}$:, (d) He:, (e) ·$\underset{\cdot}{\dot{\text{N}}}$:.
13. (a) The sulfur atom has 6 outermost electrons--$3s^23p^4$. The $3s$ electrons and two of the $3p$ electrons are paired; the other two $3p$ electrons are unpaired.
    (b) The silicon atom has 4 outermost electrons--$3s^23p^2$. The $3s$ electrons are paired and the $3p$ electrons are unpaired.
    (c) The rubidium atom has one outermost electron--$5s^1$.

14. The atomic radius is the radius of an atom in a single covalent bond. The distance between bonded atoms is measured and a portion of this distance assigned to each atom.
15. The ionic radius is the radius of an ion in crystalline ionic compounds. The distance in compounds in which similar ions touch is measured and that value is used to find radii of other ions in compounds with other structures.
16. The atomic radius is smaller and the atomic radius is larger.
17. Atomic radii are affected by the effective nuclear change and the quantum level.
18. The two effects are the nuclear charge and the screening effect. The effective nuclear charge increases going across a period.
19. The screening effect is the decrease in nuclear charge acting on an electron due to the effects of other electrons. The less than the (n-1) energy level is the most effective.
20. Neon should have the highest value. All of these atoms have very similar screening constants because they are all in the same period, but Ne has the highest atomic number.
21. The radius increases as n increases because the region of probability of finding an electron is farther from the nucleus as n increases and also the screening effect increases this distance.
22. Fluorine must be the smallest because it is first in a family and last in a period. Three of the elements are in the third period and their atomic radii must decrease in size in the order Na > Al > Cl. Two pairs of elements in the list are in the same families, and for each pair, atoms of the element farther down in the family must be larger, therefore K > Na and Cl > F. Therefore F is 0.071 nm, Cl is 0.099 nm, Al is 0.143 nm, Na is 0.186 nm, and K is 0.227 nm.
23. Cesium is located at the far left and bottom of the periodic table and is the largest. Two of the elements are in the same period and their atomic radii must decrease in the order Si > Ar. Two of the elements are in the same family and their atomic radii must decrease in the order Ge > Si. Therefore (a) Ge is 0.123 nm, (b) Si is 0.118 nm, (c) Ar is 0.095 nm, and (d) Cs is 0.265 nm.
24. Isoelectronic species have the same outer electronic structure.
25. These are isoelectronic species. The size decreases as the nuclear charge increases giving $Ca^{2+} < K^+ < Ar < Cl^-$.

26. The ionic radius of O^{2-} is larger than that of O^- because it has an extra electron present that causes the electron cloud to spread out.
27. The loss of an electron from the small Li atom will give the smallest ion. F^- is smaller than Br^- because size decreases going from botton to top in a periodic table family. Br^- and Sr^{2+} are isoelectronic, with Br^- being the larger ion because size decreases as the charge becomes more positive. The size of F^- relative to that of Sr^{2+} cannot be easily determined by using the general rules. The order of increasing radii is

$$\begin{matrix} Li^+ & < & Sr^{2+} & < & F^- & < & Br^- \\ 0.068\ nm & & 0.112\ nm & & 0.136\ nm & & 0.196\ nm \end{matrix}$$

28. Ionization energy is the energy required for the removal of the least tightly bound electron in a gaseous atom or ion. The values increase because it becomes increasingly more difficult to remove a negatively charged electron from the positively charge ion.
29. The process would involve removing electron from noble gas configuration which is unfavorable.
30. The factors are effective nuclear charge and electron configuration.
31. This indicates that half-filled and completely filled subshells have extra stability.
32. The 6*s* electrons will be removed.
33. The energies are lower. No, the general trend is not followed.
34. The electron affinity is the enthalpy change for the addition of an electron to an atom or ion in the gaseous state. No, values are positive for elements in Representative Groups I and II and in the noble gas family.
35. Nonmetals form anions. The charge is determined by the number of electrons gained by the atom in order to fill the outer level.
36. The cations formed by the *s*-block representative metals in the fourth through sixth periods have noble gas configurations and those formed by the Representative Group III elements in the periods have noble gas configurations plus ten electrons in the same *d* subshell.
37. The metals of Representative Groups IV and V, respectively, form only the +2 and +3 cations that result from loss of the valence electrons from the *np* or outer *p* subshells.
38. Usually +2 cations are formed by the loss of the ns^2 electrons. Elements in the Cu and Zn families from pseudo-noble gas cations.

39. Potassium atoms, $[Ar]4s^1$, lose one electron to give [Ar] and chlorine atoms, $[Ne]3s^2 3p^5$, gain one electron to give [Ar]. These ions are isoelectronic.
40. (a) $[Ar]3d^6$, (b) $[Ar]3d^5$, (c) [Ar], (d) $[Xe]4f^{14}5d^{10}6s^2$, (e) $[Kr]4d^{10}$, (f) $[Kr]4d^{10}$, (g) $[Ar]3d^3$.
41. (a) First ionization energies of the alkaline earth metals decrease as the atomic size increases because the outer electrons are attracted less strongly. (b) First ionization energies increase across the third period as the number of electrons increases because the atomic size is decreasing, so the outer electrons are attracted more strongly. (c) The atomic radii of the noble gases increase down the family because the principal quantum number of the outer electrons increases.
42. The general trends across a period for the following properties are (a) (ii) decreasing atomic radius, (b) (i) increasing ionization energy, (c) (i) increasing electron affinity, (d) (i) increasing positive charge on cation, and (e) (ii) decreasing negative charge on anion.

    The general trends down a family for the following properties are (a) (i) increasing atomic radius, (b) (ii) decreasing ionization energy, (c) (ii) decreasing electron affinity, (d) (iii) showing no change in positive charge on cation, and (e) (iii) showing no change in negative charge on anion.

SKILLS

Skill 9.1 Predicting periodic table position by using electron configuration

The atoms of the representative elements have incomplete filling in the *ns* and *np* subshells, while all inner shells are complete. The atoms of the noble gases have complete filling of all subshells. The transition (or *d*-transition) elements are characterized by electrons filling the $(n-1)d$ subshell and the inner-transition (or *f*-transition) elements by electrons filling the $(n-2)f$ subshell. For example, on the basis of their electron configurations

B	$[He]2s^2 2p^1$
K	$[Ar]4s^1$
Kr	[Kr]
I	$[Kr]4d^{10}5s^2 5p^5$
La	$[Kr]4d^{10}4f^1 5s^2 5p^6 6s^2$

B, K, and I would be classified as representative elements, La would be classified as a transition element, and Kr would be classified as a noble gas.

Skill Exercise: Classify the following elements as noble gases, representative elements, transition elements, or inner transition elements: (a) Mg, (b) Se, (c) Xe, (d) V, and (e) C. *Answers*: (a) representative, (b) representative, (c) noble gas, (d) transition, (e) representative.

Based on their physical and chemical properties, an element can be classified as a metal, a nonmetal, or a semiconducting element. The nonmetals occupy the upper right portion of the periodic table and the metals occupy the lower left portion of the periodic table; see Figure 9.3 of the textbook. The metals and nonmetals are separated by the seven semiconducting elements--B, Si, Ge, As, Sb, Se, and Te. All of the *s* block elements, all of the transition metals, and some of the heavier *p* block elements are metals. The nonmetals consist of the noble gases, the halogens, and five additional *p* block elements--C, N, P, O, and S. For example, the element having the electron configuration

$$[Kr]4d^8 5s^1$$

is classified as a metal because of the partially filled 4d and 5s subshells.

Skill Exercise: Classify each of the following elements from the electron configuration as a nonmetal, a semiconducting metal, or a metal: (a) $[Ar]3d^{10}4s^2 4p^5$, (b) $[Kr]5s^2$, (c) $[Xe]4f^3 6s^2$, and (d) $[Kr]4d^{10}5s^2 5p^4$. *Answers*: (a) nonmetal, (b) metal, (c) metal, (d) semiconducting element.

Skill 9.2 Writing electron configurations and Lewis symbols for ions

The Lewis symbol for a single atom consists of a chemical symbol--representing the nucleus and inner-shell electrons--surrounded by dots representing the valence electrons. For example, the Lewis symbols for atoms of sodium and chlorine would be

Na·   :Cl:

corresponding to the electron configurations of $[Ne]3s^1$ and $[Ne]3s^2 3p^5$ (see *Skill 8.2 Writing electron configurations for atoms*).

Skill Exercise: Write Lewis symbols for (a) Al, (b) Xe, (c) H, (d) C, (e) Ca, and (f) P. *Answers*: (a) :Al·, note the paired dots representing the $3s^2 3p^1$ valence configuration; (b) :Xe:; (c) H·; (d) ·C:; (e) Ca:; (f) ·P:.

To write the electron configuration for an ion, the usual order of filling (see *Skill 8.2 Writing electron configurations for atoms*) is used. For an ion with a charge of $+n$, the number of electrons will be $(Z - n)$; for an ion with a charge of $-n$, the number of electrons will be $(Z + n)$. Consider the following configurations, written for the ground states of the ions:

Li^+	$Z = 3$	3 - 1 = 2 electrons	$1s^2$
O^{2-}	$Z = 8$	8 + 2 = 10 electrons	$1s^2 2s^2 2p^6$

When dealing with positive ions of the transition and inner-transition elements, remember that the electrons that are lost come from the highest principal quantum level. For example, in

Zn^{2+}	$Z = 30$	30 - 2 = 28 electrons	$1s^2 2s^2 2p^6 3s^2 3p^6 3d^{10}$
Fe^{2+}	$Z = 26$	26 - 2 = 24 electrons	$1s^2 2s^2 2p^6 3s^2 3p^6 3d^6$
Fe^{3+}	$Z = 26$	26 - 3 = 23 electrons	$1s^2 2s^2 2p^6 3s^2 3p^6 3d^5$

the electrons lost first are those from the $4s$ subshell.

Skill Exercise: Write the electron configurations for (a) Al^{3+}, (b) H^+, (c) H^-, (d) C^{4-}, (e) Ca^{2+}, (f) P^{3-}, and (g) V^{2+}. *Answers*: (a) $1s^2 2s^2 2p^6$ or [Ne], (b) $1s^0$, (c) $1s^2$, (d) $1s^2 2s^2 2p^6$ or [Ne], (e) $1s^2 2s^2 2p^6 3s^2 3p^6$ or [Ar], (f) $1s^2 2s^2 2p^6 3s^2 3p^6$ or [Ar], (g) $1s^2 2s^2 2p^6 3s^2 3p^6 3d^3$ or [Ar]$3d^3$.

For monatomic ions, the Lewis symbol contains the proper number of electrons and brackets and ionic charges are added. For example, the Lewis formulas for Na^+ and Cl^- would be

$$[Na]^+ \qquad [:\ddot{\underset{..}{Cl}}:]^-$$

Skill Exercise: Write the Lewis symbols for (a) Al^{3+}, (b) H^+, (c) H^-, (d) C^{4-}, (e) Ca^{2+}, and (f) P^{3-}. *Answers*: (a) $[Al]^{3+}$, (b) $[H]^+$, (c) $[H:]^-$, (d) $[:\ddot{\underset{..}{C}}:]^{4-}$, (e) $[Ca]^{2+}$, (f) $[:\ddot{\underset{..}{P}}:]^{3-}$.

NUMERICAL EXERCISES

*Example 9.1 Periodic law**

Early periodic tables did not show the element germanium, Ge. However, several properties for Ge and its compounds were predicted on the basis of the properties of Si and Sn and their compounds. For example, the formula of the chloride was predicted to be "Es"Cl_4 (where "Es" stood for "ekasilicon"), because the formulas of the chlorides of Si and Sn were known to be $SiCl_4$ and

$SnCl_4$, respectively. Likewise, by using the periodic law (which states that chemical and physical properties are periodic functions of atomic numbers) it is possible to estimate the boiling point of the tetrachloride from the known boiling points of $SiCl_4$ (57.57 °C) and $SnCl_4$ (114.1 °C). Predict the boiling point of "Es"Cl_4.

This problem is analyzed by using the comprehensive problem-solving approach presented in *Skill 2.7 Solving problems*.

1. Study the problem and be sure you understand it.
   (a) What is unknown?
   The boiling point of "Es"Cl_4.
   (b) What is known?
   The boiling points of $SiCl_4$ and $SnCl_4$.
2. Decide how to solve the problem.
   (a) What is the connection between the known and the unknown?
   The periodic law states that chemical and physical properties are periodic functions of atomic number.
   (b) What is necessary to make the connection?
   The easiest way to make this estimate is to prepare a graph of boiling point against the atomic number of Si and Sn, draw a line through the data points, and read the value at the atomic number of "Es."
3. Set up the problem and solve it.
   The plot shown in Fig. 9-1 gives the boiling point at Z = 32 as 85.5 °C.
4. Check the result.
   (a) Are significant figures and the location of the decimal point correct?
   Yes.
   (b) Did the answer come out in the correct units?
   Yes.
   (c) Is the answer reasonable?
   The accepted handbook value for $GeCl_4$ is 84 °C, indicated by a "+" in the graph above. These values are in good agreement.

Exercise 9.1 Predict the density of "Es"Cl_4 if the density of $SiCl_4$ is 1.483 g/mL and that of $SnCl_4$ is 2.226 g/mL. *Answer*: 1.85 g/mL, which is in good agreement with the accepted value of 1.8443 g/mL.

Fig. 9-1

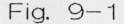

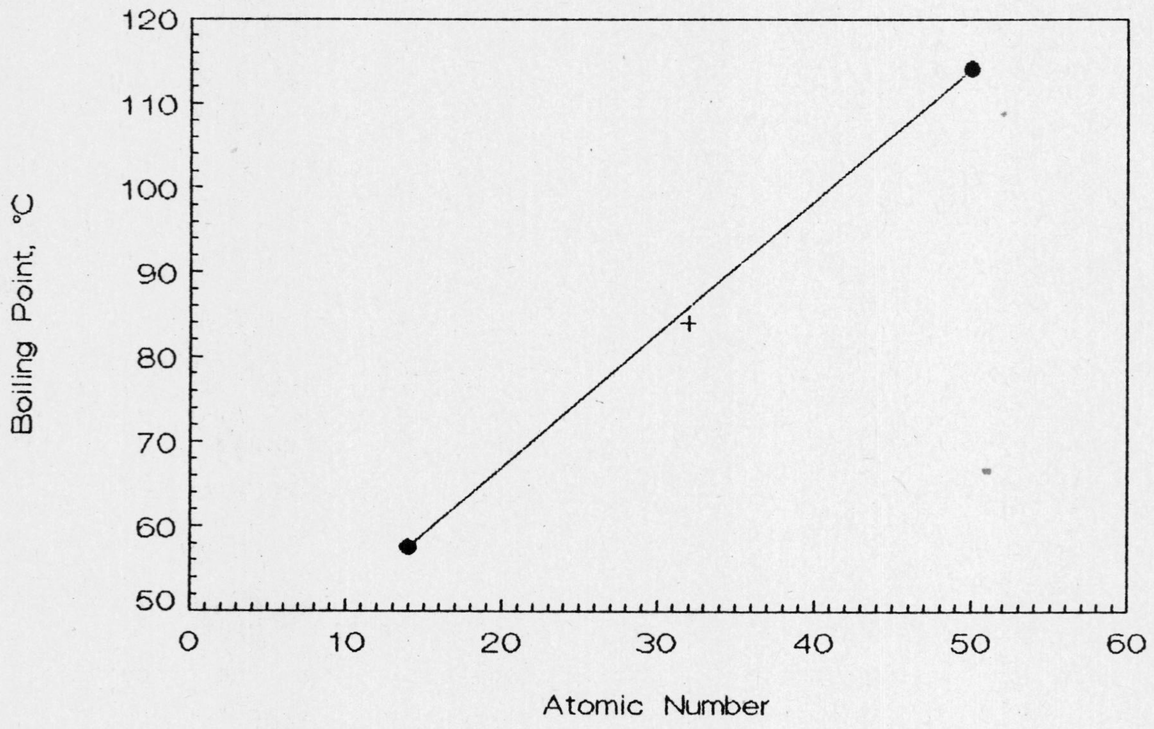

PRACTICE TEST

1 (68 points). Choose the response that best completes each statement.

(a) Na^+ and Ne

(i) have equal ionization energies (ii) are ions

(iii) are isoelectronic (iv) none of these answers

(b) Which of the following is *not* isoelectronic with the others?

(i) Cl (ii) Ca^{2+}

(iii) K^+ (iv) Ar

(c) Arrange the ions F^-, Mg^{2+}, Be^{2+} in order of increasing size.

(i) $Mg^{2+} < Be^{2+} < F^-$ (ii) $Be^{2+} < Mg^{2+} < F^-$

(iii) $Be^{2+} < F^- < Mg^{2+}$ (iv) $F^- < Be^{2+} < Mg^{2+}$

(d) Which ionization energy is the greatest for boron?

(i) first (ii) second

(iii) third (iv) fourth

(e) Across a given period as atomic number increases, the ionization energies of the elements
(i) increase (ii) decrease
(iii) remain the same (iv) increase and then decrease

(f) The elements of Representative Group II
(i) all have two electrons in the valence shell
(ii) have relatively high ionization energies
(iii) tend to form covalent bonds
(iv) are very electronegative

(g) Element 119 will belong to
(i) Representative Group VII (ii) Representative Group VI
(iii) Representative Group II (iv) Representative Group I

(h) Across a given period as atomic number increases, the atomic radii
(i) decrease because the effective nuclear charge increases
(ii) increase because the screening effect increases
(iii) decrease because the effective nuclear charge decreases
(iv) none of these answers

(i) Mendeleev arranged the elements in order of
(i) atomic mass (ii) density
(iii) atomic number (iv) electron configuration

(j) Similar chemical properties are related to
(i) melting points (ii) similar outer shells of electrons
(iii) neutron numbers (iv) none of these answers

(k) Transition elements fill the
(i) *s* subshell (ii) *p* subshell
(iii) *d* and *f* subshells (iv) *g* subshell

(l) An element has an outer electron configuration of ns^2np^4. This element
(i) Ne (ii) F
(iii) O (iv) N

(m) The element with $Z = 11$ has chemical properties similar to the element with
(i) $Z = 19$ (ii) $Z = 12$
(iii) $Z = 10$ (iv) none of these answers

(n) Which is a semiconducting element?
(i) B (ii) Si
(iii) Ge (iv) all of these answers

(o) Which is the electron configuration for Al^{3+}?

(i) $[Ne]3s^2 3p^1$ (ii) [Ne]

(iii) $[Ne]3s^2$ (iv) [Ar]

(p) Which is the electron configuration for O^{2-}?

(i) $1s^2 2s^2 2p^4$ (ii) $1s^2 2s^2 2p^6$

(iii) $1s^2 2s^2$ (iv) $1s^2 2s^2 2p^2$

(q) Which of the following has the largest number of *s* electrons?

(i) Ar (ii) K^+

(iii) Cl^- (iv) all the same

2 (16 points). In each case, decide which has the largest, highest, or greatest

(a) size: H, H^+, H^-

(b) size: H^-, He, Na^+

(c) size: Cl, Br, I

(d) size: Mg, Al, Si

(e) first ionization energy: F, Ne, Na

(f) first ionization energy: Ca, Sr, Ba

(g) electron affinity: N, O, F

(h) second ionization energy: Na, P, S

3 (16 points). Classify each of the elements (a) Na, (b) Ge, (c) Zr, (d) Pr, and (e) F as (i) a noble gas, (ii) a representative element, (iii) a transition element, or (iv) an inner-transition element based on their electronic structures.

Answers to practice test

1. (a) iii, (b) i, (c) ii, (d) iv, (e) i, (f) i, (g) iv, (h) i, (i) i, (j) ii, (k) iii, (l) iii, (m) i, (n) iv, (o) ii, (p) ii, (q) iv.
2. (a) H^-, (b) H^-, (c) I, (d) Mg, (e) Ne, (f) Ca, (g) F, (h) Na.
3. (a) ii, (b) ii, (c) iii, (d) iv, (e) ii.

ADDITIONAL NOTES

CHAPTER 10

PERIODIC PERSPECTIVE II: CHEMICAL BONDS

CHAPTER OVERVIEW

In this chapter the three basic types of chemical bonds--metallic, ionic, and covalent--are defined and the properties associated with each type of bond are discussed. The properties of the bonds themselves--bond energies, bond lengths, and bond polarity (resulting from polarization or differences in electronegativity)--are considered. Oxidation number for atoms in molecules and ions are introduced and inorganic nomenclature is reviewed.

COMPETENCIES

Definitions for the following terms should be learned:

- acceptor atom
- bond dissociation energy
- bond energy
- chemical bond
- coordinate covalent bond
- covalent bonding
- d^{10} configuration, pseudo-noble gas configuration
- dipole
- donor atom
- double covalent bond
- electronegative atom
- electronegativity
- electropositive atom
- ionic bonding
- Lewis structures
- lone electron pairs
- metallic bonding
- multiple covalent bond
- network covalent substance
- nonpolar covalent bond
- octet rule
- oxidation number, oxidation state
- polar covalent bond
- polarization of an ion
- single covalent bond
- triple covalent bond

General concepts and skills that should be learned include

- predicting the type of bonding between two atoms
- the general properties of substances with the various types of bonding

- the use and limitations of the octet rule for determining bonding in a substance
- the general properties of covalent bonds--bond length, bond strength, and polarity
- assigning oxidation numbers to atoms in molecules and polyatomic ions
- naming and writing formulas for inorganic compounds

Numerical exercises that should be understood include

- calculating the bond dissociation energy and average bond energy from given thermochemical data
- using bond energies to calculate the enthalpy change of a gas-phase reaction

QUESTIONS

1. Complete the following sentences by filling in the various blanks with these terms: acceptor atom, chemical bond, coordinate covalent bond, covalent bonding, donor atom, double covalent bonds, ionic bonding, metallic bonding, nonpolar covalent bond, octet rule, polar covalent bond, single covalent bond, triple covalent bonds, and valence electrons.

   A (a)______ is a force that acts strongly enough between two atoms or groups of atoms to hold them together in a stable species having measurable properties. If (b)______ are transferred from one atom to the other, the species will be held together by (c)______. In (d)______, the bonding results from the electrostatic attraction between cations and a "sea of electrons." In (e)______, the atoms share electrons to form the bond. If the sharing is done equally, the bond is called a (f)______ and if unequally, a (g)______. A general guiding principle, although often violated, which helps in writing the Lewis symbols and structures for molecules and ions is the (h)______. Many molecules contain (i)______, which is a way that fewer electrons can satisfy this general principle, and some molecules contain (j)______, in which six electrons are shared between the atoms. If one atom, the (k)______, supplies both electrons in a (l)______ to the (m)______, the bond is called a (n)______.

2. Match the correct term to each definition.

(a) structures in which Lewis symbols are combined so that the bonding and nonbonding outer electrons are indicated
(b) the average enthalpy per mole for breaking one bond of the same type per molecule
(c) three-dimensional arrays of covalently bonded atoms
(d) a symbol in which the outer electrons of an atom are shown by dots arranged around the atomic symbol
(e) the enthalpy per mole for breaking exactly one bond of the same type per molecule
(f) a pair of opposite charges of equal magnitude at a specific distance from each other
(g) a bond in which more than one pair of electrons are shared between the same two atoms
(h) the distance between the nuclei of two atoms in a stable compound

(i) bond dissociation energy
(ii) bond energy
(iii) bond length
(iv) dipole
(v) Lewis structures
(vi) Lewis symbol
(vii) multiple covalent bond
(viii) network covalent substance

3. Which of the following statements are true? Rewrite each false statement so that it is correct.

(a) The chemical bonds in hydrogen chloride, HCl, and deuterium chloride, DCl, are essentially the same.
(b) Once formed, a coordinate covalent bond is no different than an ordinary covalent bond.
(c) Electrons are shared equally in a nonpolar covalent bond, shared unequally in a polar covalent bond, and transferred in an ionic bond.
(d) Because of intramolecular vibrations, we might expect the bond lengths of the four C-Cl bonds in CCl_4 to be slightly different at any given instant.
(e) The covalent bonds in the so-called network covalent substances are considerably different from those in ordinary molecules.
(f) All valence electrons in a covalent compound are involved in bonding.
(g) Oxygen is the only element sufficiently electronegative for the fluorine atom to have a positive oxidation state when combined with it.

4. Distinguish clearly between (a) covalent and ionic bonding, (b) polar and nonpolar covalent bonding, (c) covalent and coordinate covalent bonding, (d) electrical conduction by a metal and by an ionic substance, (e) donor and acceptor atoms, (f) bond dissociation energy and average bond energy, (g) electropositive and electronegative atoms.

5. Choose the best answer for each statement.

(a) Electrons that usually take part in chemical bonding are
  (i) found in the valence shell
  (ii) lone pairs
  (iii) from an electronegative atom
  (iv) from network covalent substance

(b) Pairs of valence electrons not involved in the bonding are
  (i) donor pairs (ii) lone or nonbonding electron pairs
  (iii) single covalent bonds (iv) dipoles

(c) A bond containing more than one pair of electrons being shared between the same two atoms is a
  (i) multiple covalent bond (ii) dipole-dipole interaction
  (iii) ionic bond (iv) coordinate covalent bond

(d) The unequal sharing of electrons in a polar covalent bond results from differences in
  (i) atomic radii
  (ii) bond length
  (iii) electronegativities of the atoms
  (iv) oxidation numbers

(e) The compound that has the most ionic character in its bonding is
  (i) NaCl (ii) $MgCl_2$
  (iii) HCl (iv) CCl_4

(f) The substance that contains the "least polar" covalent bonding is
  (i) H_2O (ii) CO_2
  (iii) NO_2 (iv) O_2

(g) A pair of opposite charges of equal magnitude at a specific distance from each other is
  (i) the bond length (ii) a dipole
  (iii) an ionic bond (iv) the bond angle

(h) The number that represents the positive or negative character of an atom in a compound is known as the

   (i) oxidation number   (ii) atomic number

   (iii) electronegativity   (iv) none of these answers

6. What is meant by the term "continuum" of bond types? Briefly discuss the ionic-to-covalent continuum.

7. Describe the types of bonding in ammonium sulfate

$$\left[\begin{array}{c} H \\ | \\ H-N-H \\ | \\ H \end{array}\right]^{+} \qquad \left[\begin{array}{c} :\!\ddot{O}: \\ | \\ :\!\ddot{O}-S-\ddot{O}: \\ | \\ :\!\ddot{O}: \end{array}\right]^{2-}$$

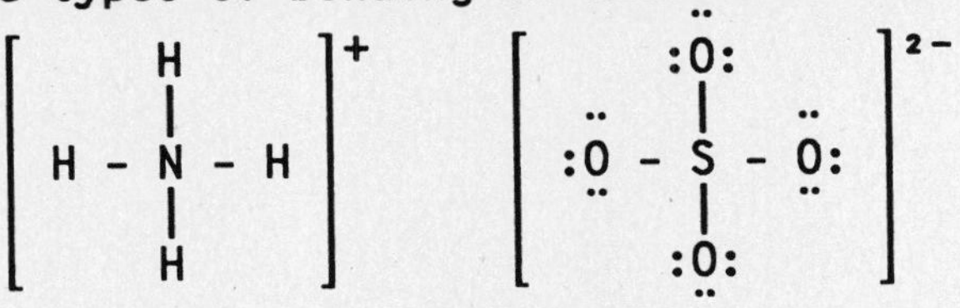

8. Which electrons are involved in metallic bonding for the elements in (a) Representative Groups I and II, (b) Representative Groups III to VI, and (c) the transition metals?

9. Why are metallic and ionic bonds nondirectional?

10. What type of electrical charge (positive or negative) will ions of metallic elements have? What type of electrical charge will ions of nonmetallic elements have? Why is there this difference in chemical behavior between metals and nonmetals?

11. Show with Lewis symbols and electron configurations the changes that occur when (a) Sc and Cl combine to form $ScCl_3$, (b) Ag and O combine to form Ag_2O, (c) Be and C combine to form Be_2C, and (d) Li and N combine to form Li_3N.

12. The outermost electron configuration of the nitrogen family elements is ns^2np^3. How can an atom of each of these elements attain a noble gas electron configuration?

13. What type of bonding is usually formed between atoms of nonmetals? How does this type of bonding help satisfy the octet rule for the atoms?

14. A sketch representing a hydrogen molecule is shown in Fig. 10-1. Identify the four forces of attraction and the two forces of repulsion in the hydrogen molecule.

15. Name the three categories of stable compounds in which the central atom is not surrounded by eight electrons. Give an example of each.

Fig. 10-1

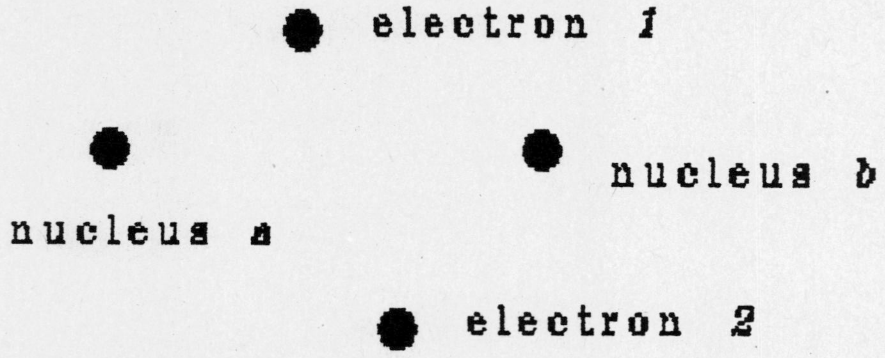

16. What is meant by the term "multiple covalent bond"? Which elements can form these bonds?

17. How many electrons are shared between the two atoms that form a double covalent bond? How many for a triple covalent bond?

18. What properties of cations influence the amount of polarization that will occur in an anion?

19. Choose the cation that would be most effective in polarizing a given anion: K^+, ionic radius = 0.133 nm; Mg^{2+}, ionic radius = 0.066 nm; or Cs^+, ionic radius = 0.167 nm.

20. Arrange the following bonds in order of increasing polarity: (a) Li - O, (b) Li - N, (c) Li - F, and (d) Li - I.

21. Which compound, PbF_2 or SiF_4, would conduct electricity better in the molten state? Explain your reasoning.

22. Compare the values of the electronegativities for the metals with those of the nonmetals. In what way are the chemical properties of metals and nonmetals influenced by the electronegativity?

23. Why do the electronegativities increase across a period from left to right with increasing atomic number, but decrease down a group with increasing atomic number?

24. What is the oxidation number of an element in the free state? Is the oxidation number the same or different for a polyatomic element?

25. What is the oxidation number of a monatomic ion?

26. What is the usual oxidation number of an oxygen atom in a compound? What are the exceptions to this rule?

27. What is the usual oxidation number of fluorine in a compound?

28. What is the usual oxidation number of hydrogen in a compound? Write the electron configuration for the hydrogen ion in this oxidation state. In addition to zero, what is the other oxidation number for hydrogen? Write the electron configuration for this hydrogen ion.

29. For a polyatomic molecule or ion, what does the sum of the oxidation numbers of all of the atoms equal?

30. Identify and give a reasonable explanation for the trend in the (a) electronegativities of the halogens--Representative Group VII--based on atomic sizes, and (b) amount of ionic character in chemical bonds between F and the elements in Representative Group III based on electronegativities.

31. Compare the following general properties of metals and nonmetals: (a) physical state, (b) density, (c) hardness, (d) conduction of heat and electricity, and (e) boiling points.

32. Once discovered, element 113 should be placed below thallium in the periodic table. Predict some of its characteristics such as electronegativity, size, oxidation numbers, and physical state.

Answers to questions

1. (a) chemical bond, (b) valence electrons, (c) ionic bonding, (d) metallic bonding, (e) covalent bonding, (f) nonpolar covalent bond, (g) polar covalent bond, (h) octet rule, (i) double covalent bonds, (j) triple covalent bonds, (k) donor atom, (l) single covalent bond, (m) acceptor atom, (n) coordinate covalent bond.
2. (a) v, (b) ii, (c) viii, (d) vi, (e) i, (f) iv, (g) vii, (h) iii.
3. (a) T; (b) T; (c) T; (d) T; (e) F, change *considerably different from* to *the same as*; (f) F, insert *or are nonbonding* after *bonding*; (g) F, switch *oxygen* with *fluorine*.
4. (a) *Covalent bonding* is the sharing of electrons; *ionic bonding* involves the transfer of electrons. (b) *Polar covalent bonding* involves unequal sharing of electrons; *nonpolar covalent bonding* involves equal sharing of electrons. (c) In *covalent bonding* each atom donates one electron to the bond; in

coordinate covalent bonding one atom donates both electrons. (d) A *metal* conducts electricity by the flow of electrons; an *ionic substance* conducts electricity by the flow of cations and anions. (e) A *donor atom* provides both electrons in a coordinate covalent bond; an *acceptor atom* accepts both electrons. (f) The *bond dissociation energy* is the energy involved for the breaking of one bond per molecule in a mole of substance; the *average bond energy* is the average energy involved for the breaking of several bonds of the same type in a molecule for a mole of substance. (g) An *electropositive atom* is the atom that acquires a partial positive charge in a covalent bond or that forms a cation; an *electronegative atom* is the atom that acquires a partial negative charge in a covalent bond or that forms an anion.

5. (a) i, (b) ii, (c) i, (d) iii, (e) i, (f) iv, (g) ii, (h) i.
6. Most bonds are not 100 % ionic, covalent, nor metallic, but somewhere in between. The bonds range from ionic to polarized ionic to polar covalent to covalent.
7. The bonding in $(NH_4)_2SO_4$ is ionic bonding between NH_4^+ and SO_4^{2-} ions and polar covalent bonding between N and H atoms and between S and O atoms.
8. (a) *ns*, (b) *ns* and *np*, (c) *ns* and (*n*-1)*d*.
9. These bonds are formed by electrostatic attractions that completely surround a given ion.
10. Metallic elements form positive ions and nonmetallic elements form negative ions. The energy needed for metals to lose electrons is not very high and the energy for nonmetals to gain electrons is favorable.
11. Lewis symbols and electron configurations showing the changes that occur as (a) Sc and Cl combine to form $ScCl_3$ are

Sc:	+	3 ·C̈l:	→	$[Sc]^{3+}$	3 $[:\ddot{C}l:]^-$
$[Ar]3d^14s^2$		$[Ne]3s^23p^5$		[Ar]	[Ar]

Note that Sc has only two valence electrons (as shown in the Lewis symbol), but readily forms the +3 ion by the loss of the 3d electron.

(b) Ag and O combine to form Ag_2O are

2 Ag·	+	·Ö:	→	2 $[Ag]^+$	$[:\ddot{O}:]^{2-}$
$[Kr]3d^{10}4s^1$		$[He]2s^22p^4$		$[Kr]3d^{10}$	[Ne]

(c) Be and C combine to form Be_2C are

2 Be:	+	·Ċ:	→	2 $[Be]^{2+}$	$[:\ddot{C}:]^{4-}$
$[He]2s^2$		$[He]2s^22p^2$		[He]	[Ne]

(d) Li and N combine to form Li_3N are

3 Li· + ·N: → 3 $[Li]^+$ $[:N:]^{3-}$

$[He]2s^1$ $[He]2s^2 2p^3$ [He] [Ne]

12. Nonmetals (N, P) and semiconducting elements (As, Sb) gain three electrons forming -3 ions, or share electrons to attain $ns^2 sp^6$ configurations.
13. These atoms form covalent bonds. The sharing of valence electrons satisfies both atoms.
14. The four forces of attraction are between nucleus *a* and electron *1*, nucleus *a* and electron *2*, nucleus *b* and electron *1*, and nucleus *b* and electron *2*. The two forces of repulsion occur between nucleus *a* and nucleus *b* and between electron *1* and electron *2*.
15. The categories are compounds of Be, B, and Al such as $BeCl_2$; compounds of P, S, Cl, and elements of the third period and beyond such as SF_6; and compounds containing unpaired electrons such as ClO_2.
16. Multiple bonding is the sharing of more than one pair of electrons between the same two atoms. The elements that readily form these bonds are C, N, and O (P, S, and Se to a lesser extent).
17. The number of electrons is 4 in a double bond and 6 in a triple bond.
18. The properties are the electron configuration and the combined effect of charge-to-size ratio.
19. The cation that would be most effective in polarizing a given anion is Mg^{2+}, because it has the largest charge-to-radius ratio.
20. The order of increasing bond polarity, with electronegativity differences shown, is

(d) Li-I <	(b) Li-N <	(a) Li-O <	(c) Li-F
1.5	2.0	2.5	3.0

21. The better conductor will be PbF_2. The bonds in Si-F are covalent because Si^{4+} (if it existed) would be a highly polarizing ion.
22. Metals have lower values than nonmetals. Metals lose electrons and nonmetals gain electrons.
23. Electronegativity is inversely related to the radius--the radius decreases across a period and increases down a group.
24. The oxidation number is 0. The value is the same.
25. The oxidation number equals the charge on the ion.
26. The usual oxidation number is -2. The exceptions are for oxygen in the free state, O-F bonds, peroxides, superoxides, and ozonides.

27. The usual oxidation number is -1.
28. The usual oxidation number is +1; $1s^0$; -1, $1s^2$.
29. The sum equals the charge on the species.
30. (a) Electronegativities of the halogens decrease down the family as the size increases, because the outer electrons are attracted less strongly. (b) The amount of ionic character in chemical bonds between F and the elements in Representative Group III increases down the family because the electronegativities decrease, making the difference between F greater.
31. See Table 10.1 of the text.
32. Element 113 would have an electronegativity of about 1.8; an atomic radius greater than 0.17 nm; oxidation numbers of 0, +1 and +3; and be a solid metal.

SKILLS

Skill 10.1 Predicting types of bonds between atoms

If both atoms are those of metallic elements, usually the atoms will be held by metallic bonding in a solid or liquid pure metal or alloy. If both atoms are those of nonmetallic elements, covalent bonding will occur, with nonpolar covalent bonding between atoms of the same element (excluding the noble gases) and with polar covalent bonding between dissimilar nonmetal atoms. If one atom is a metal and one a nonmetal, the bonding is often ionic, but could be anywhere on the ionic-polar covalent continuum. There will be no bonding between atoms of the noble gases.

Skill Exercise: Identify the most likely type of bonding, if any, between the following pairs of atoms: (a) K-Cl, (b) H-O, (c) P-P, (d) Na-Hg, (e) Ar-Ar, (f) Xe-F, (g) Ca-Ca, (h) O-O, (i) Cl-F, and (j) Ca-O. *Answers*: (a) ionic, (b) polar covalent, (c) nonpolar covalent, (d) metallic, (e) no bonding, (f) polar covalent, (g) metallic, (h) nonpolar covalent, (i) polar covalent, (j) ionic.

Skill 10.2 Determining oxidation numbers

The oxidation number is a number that represents the positive or negative character of atoms in compounds. The rules for assigning oxidation numbers are presented in Section 10.10 of the text and are summarized below:

(1) The oxidation number of any element in the free state is 0.

(2) The oxidation number of a monatomic ion is equal to the charge on the ion.

(3) Fluorine in combination with other elements always has the oxidation number -1.

(4) Oxygen usually has an oxidation number of -2.

(5) Hydrogen usually has an oxidation number of +1.

(6) For neutral compounds, the algebraic sum of the oxidation numbers is 0.

(7) For a polyatomic ion, the algebraic sum of the oxidation number is equal to the charge on the ion.

The following examples demonstrate the use of these rules to determine the oxidation number of the underlined element in each species:

$\underline{P}_4$, $\underline{Mg}$ Rule (1) gives the oxidation number as 0 for each element.

$\underline{Mg}^{2+}$, $\underline{P}^{3-}$ Rule (2) gives the oxidation numbers as +2 and -3, respectively.

$\underline{Ca}O$ Rules (4) and (6) give the value for Ca as +2 using the simple equation $x + (-2) = 0$.

$H\underline{Cl}$ Rules (5) and (6) give the value for Cl as -1 using the equation $(+1) + x = 0$.

$\underline{S}_2O_3^{2-}$ Rules (4) and (7) give the average value for S as +2 using the equation $2x + 3(-2) = -2$.

Skill Exercise: Determine the oxidation number for the underlined element in each of the following species: (a) $\underline{Na}$, (b) $\underline{S}_8$, (c) $\underline{Cl}^-$, (d) $\underline{Fe}^{3+}$, (e) $\underline{Fe}_2O_3$, (f) $H\underline{Br}$, (g) $\underline{C}H_4$, (h) $\underline{Cl}_2O_7$, (i) $\underline{S}O_4^{2-}$, (j) $\underline{N}H_4^+$, (k) $\underline{Mn}O_4^-$, and (l) $K_2\underline{Cr}_2O_7$. *Answers*: (a) 0, (b) 0, (c) -1, (d) +3, (e) +3, (f) -1, (g) -4, (h) +7, (i) +6, (j) -3, (k) +7, (l) +6.

NUMERICAL EXERCISES

Example 10.1 Bond dissociation energy

The standard enthalpy of formation at 25 °C is 38.95 kJ/mol for OH(g), 217.965 kJ/mol for H(g), and 249.170 kJ/mol for O(g). Find the bond dissociation energy for the O-H bond.

The bond dissociation energy is the heat of reaction per mole involved in the breaking of exactly one bond of the same type per molecule. For this case the bond dissociation energy is ΔH° for the reaction

$$OH(g) \rightarrow O(g) + H(g)$$

The heat of reaction can be found by using Hess's law (see *Skill 7.2 Using Hess's law*) for the heat of formation data.

$$\begin{aligned}\Delta H^\circ &= (\text{sum of } \Delta H^\circ_f \text{ of products}) - (\text{sum of } \Delta H^\circ_f \text{ of reactants})\\ &= [(1\ \text{mol})\Delta H^\circ_f(\text{O}) + (1\ \text{mol})\Delta H^\circ_f(\text{H})] - [(1\ \text{mol})\Delta H^\circ_f(\text{OH})]\\ &= [(1\ \text{mol})(249.170\ \text{kJ/mol}) + (1\ \text{mol})(217.965\ \text{kJ/mol})]\\ &\quad - [(1\ \text{mol})(38.95\ \text{kJ/mol})]\\ &= 428.19\ \text{kJ}\end{aligned}$$

The bond dissociation energy for the O-H bond is 428.19 kJ. Note that this value is slightly different from the value in Table 10.5 of the text, which is an average bond energy.

Exercise 10.1 Find the bond dissociation energy for the N-H bond. The standard enthalpy of formation is 351.5 kJ/mol for NH(g), 217.965 kJ/mol for H(g), and 472.704 kJ/mol for N(g). *Answer*: 339.2 kJ/mol.

Example 10.2 Bond energy

The standard enthalpy of formation at 25 °C is -241.818 kJ/mol for $H_2O(g)$, 249.170 kJ/mol for O(g), and 217.965 kJ/mol for H(g). Calculate the bond energy for the O-H bond.

The bond energy is the average enthalpy per mole involved in the breaking of one bond of the same type per molecule. For this case, the desired answer is one half the ΔH° for

$$H_2O(g) \rightarrow O(g) + 2H(g)$$

$$\begin{aligned}\Delta H^\circ &= (\text{sum of } \Delta H^\circ_f \text{ of products}) - (\text{sum of } \Delta H^\circ_f \text{ of reactants})\\ &= [(1\ \text{mol})\Delta H^\circ_f(\text{O}) + (2\ \text{mol})\Delta H^\circ_f(\text{H})] - [(1\ \text{mol})\Delta H^\circ_f(\text{H}_2\text{O})]\\ &= [(1\ \text{mol})(249.170\ \text{kJ/mol}) + (2\ \text{mol})(217.965\ \text{kJ/mol})]\\ &\quad - [(1\ \text{mol})(-241.818\ \text{kJ/mol})]\\ &= 926.918\ \text{kJ}\end{aligned}$$

The O-H bond energy is (926.918 kJ/2 mol) = 463.459 kJ/mol.

Exercise 10.2 Find the N-H bond energy. The standard enthalpy of formation is 217.965 kJ/mol for H(g), 472.704 kJ/mol for N(g), and -46.11 kJ/mol for $NH_3(g)$. *Answer*: 390.90 kJ/mol.

*Example 10.3 Heats of reaction**

Using the bond energies in Table 10.5 of the text, calculate ΔH° for

$$Cl_2(g) + F_2(g) \rightarrow 2ClF(g)$$

This problem is analyzed by using the comprehensive problem-solving approach presented in *Skill 2.7 Solving problems*.

1. Study the problem and be sure you understand it.
   (a) What is unknown?
   The heat of reaction.
   (b) What is known?
   The bond energies of Cl-Cl (243 kJ/mol), F-F (159 kJ/mol), and Cl-F (255 kJ/mol).
2. Decide how to solve the problem
   (a) What is the connection between the known and the unknown?
   For gas-phase reactions, energy is required to break the bonds in the reactants and energy is released by forming the bonds in the products. The sum of these energies represents the heat of reaction.
   (b) What is necessary to make the connection?
   For a gas-phase reaction (see *Skill 7.2 Using Hess's law*)
   $$\Delta H° = \text{(sum of bond energies of all bonds in reactants)} - \text{(sum of bond energies of all bonds in products)}$$
3. Set up the problem and solve it.
   In this case one Cl-Cl and one F-F bond are being broken and two Cl-F bonds are being formed. Thus
   $$\begin{aligned}\Delta H° &= [(1\text{ mol})BE_{Cl-Cl} + (1\text{ mol})BE_{F-F}] - [(2\text{ mol})BE_{Cl-F}]\\ &= [1\text{ mol})(243\text{ kJ/mol}) + (1\text{ mol})(159\text{ kJ/mol})]\\ &\quad - [(2\text{ mol})(255\text{ kJ/mol})]\\ &= -108\text{ kJ}\end{aligned}$$
4. Check the result.
   (a) Are significant figures and the location of the decimal point correct?
   Yes.
   (b) Did the answer come out in the correct units?
   Yes.
   (c) Is the answer reasonable?
   Yes. The calculated heat of reaction, -108 kJ, is close to the true value of -108.96 kJ. Note that it is important to use the bond energies that correspond to the actual bonding in the molecules.

Exercise 10.3 Using the bond energies in Table 10.5 of the text, calculate $\Delta H°$ for

$$Cl_2(g) + O_2(g) \rightarrow 2ClO(g)$$

Answer: 331 kJ.

PRACTICE TEST

1 (15 points). Match the correct term to each definition.

(a) a force that acts strongly enough between two atoms or groups of atoms to hold them together in a stable species having measurable properties

(b) the electrostatic attraction between cations and the surrounding sea of electrons

(c) the electrostatic attraction between cations and anions

(d) the sharing of valence electrons

(e) a single covalent bond in which both electrons in the shared pair come from the same atom

(f) equal sharing of the electrons in this covalent bond

(g) unequal sharing of the electrons in this covalent bond

(i) chemical bond
(ii) coordinate covalent bond
(iii) covalent bonding
(iv) ionic bonding
(v) metallic bonding
(vi) nonpolar covalent bond
(vii) polar covalent bond

2 (30 points). Choose the response that best completes each statement.

(a) Which best describes a covalent bond?
 (i) two nuclei sharing a pair of electrons that is located exactly between the nuclei
 (ii) an anion held electrostatically to a cation
 (iii) three atoms held by two electrons
 (iv) two atoms sharing an electron pair--each atom donates one electron to the other atom

(b) What type of bond would be formed between the carbon and nitrogen atoms in the HCN molecule?
 (i) single covalent (ii) double covalent
 (iii) triple covalent (iv) ionic

(c) The sharing of more than two electrons between atoms is a
 (i) multiple covalent bond (ii) lone pair
 (iii) coordinate covalent bond (iv) polar covalent bond

(d) Which species violates the octet rule?
 (i) CO (ii) NH_3
 (iii) IF (iv) BF_3

(e) Which Lewis structure best represents ethene, C_2H_4?

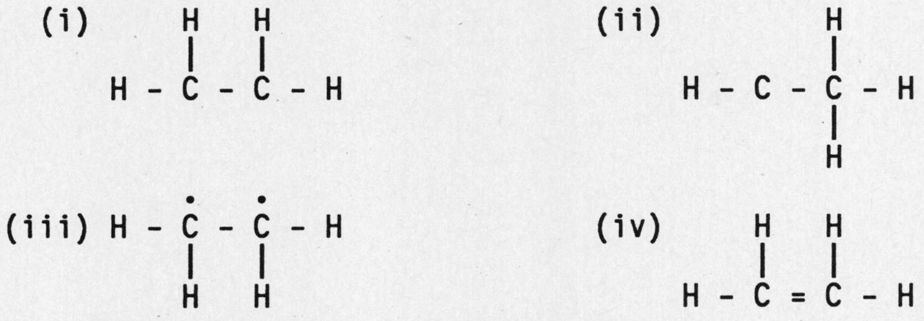

(f) The electronegativities of the elements

    (i) increase across a period as atomic number increases

    (ii) decrease across a period as atomic number increases

    (iii) increase down a family as atomic number increases

    (iv) remain the same in a family

(g) What type of bonding is expected between two nonmetal atoms having similar electronegativities?

    (i) ionic   (ii) polar covalent

    (iii) nonpolar covalent   (iv) metallic

(h) Which substance has the most pronounced covalent bonding?

    (i) $BeCl_2$ (ii) Cl_2

    (iii) NO_2 (iv) $CaCl_2$

(i) The bonds within CCl_4 are

    (i) metallic   (ii) polar covalent

    (iii) nonpolar covalent   (iv) ionic

(j) Which ion would have the greatest polarizing effect on an anion?

    (i) Cs^+ (ii) Ca^{2+}

    (iii) Na^+ (iv) Al^{3+}

(k) The heat of formation of $NH_3(g)$ is -46 kJ/mol, the H-H bond energy is 435 kJ/mol, and the N-H bond energy is 389 kJ/mol. Calculate the $N{\equiv}N$ bond energy.

    (i) 876 kJ/mol   (ii) -46 kJ/mol

    (iii) 937 kJ/mol   (iv) 824 kJ/mol

(l) Calculate $\Delta H°$ for

```
    H   H              H   H
    |   |              |   |
H - C - C - H(g)   →   H - C = C - H(g) + H - H(g)
    |   |
    H   H
```

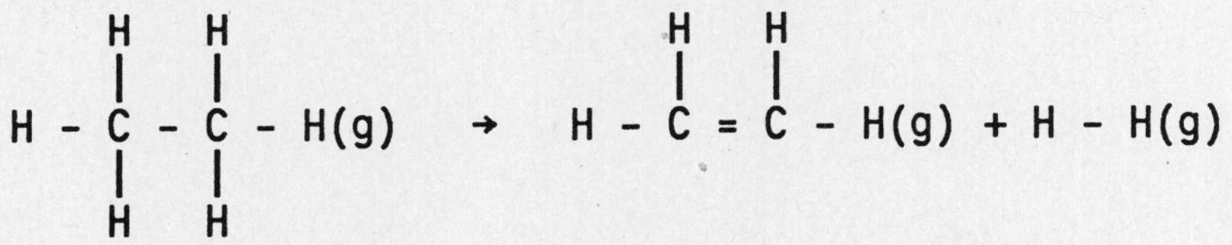

using the following bond energy data:

C-H	414 kJ/mol	H-H	435 kJ/mol
C=C	590 kJ/mol	C-C	331 kJ/mol
C≡C	812 kJ/mol		

(i) 134 kJ (ii) -134 kJ
(iii) 259 kJ (iv) 1791 kJ

(m) The oxidation number of Mn in $KMnO_4$ is
(i) +7 (ii) +2
(iii) +3 (iv) -7

(n) The oxidation number of Cl in $(NH_4)(ClO)$ is
(i) +7 (ii) +5
(iii) +3 (iv) +1

(o) Which element forms two fluorides?
(i) Mg (ii) Al
(iii) P (iv) Na

3 (10 points). Describe the bond type between the pairs of atoms given as mainly (i) metallic, (ii) ionic, (iii) polar covalent, or (iv) nonpolar covalent: (a) Cr-W, (b) S-O, (c) K-F, (d) P-P, and (e) Xe-O.

4 (15 points). The value of ΔH°_f is 249 kJ/mol for O(g), 218 kJ/mol for H(g), -242 kJ/mol for H_2O(g), and -136 kJ/mol for H_2O_2(g). Calculate (a) the O-H average bond energy and (b) the O-O bond energy.

5 (10 points). The average bond energy is 151 kJ/mol for I-I, 159 kJ/mol for F-F, and 280. kJ/mol for I-F. Calculate ΔH° for

$$I_2(g) + 7F_2(g) \rightarrow 2IF_7(g)$$

6 (10 points). Choose the most electronegative element in each case:
(a) Cs, K, Li (b) N, O, F

7 (10 points). Which element has all of the following properties: is a nonmetal that bonds to itself, has a first ionization energy between Al and Ar, has an electronegativity between O and Sn, forms oxides having the empirical formulas E_2O_3 and E_2O_5, and can form binary ionic compounds containing E^{3-}?

Answers to practice test

1. (a) i, (b) v, (c) iv, (d) iii, (e) ii, (f) vi, (g) vii.
2. (a) iv, (b) iii, (c) i, (d) iv, (e) iv, (f) i, (g) ii, (h) ii, (i) ii, (j) iv, (k) iii, (l) i, (m) i, (n) iv, (o) iii.

3. (a) i, (b) iii, (c) ii, (d) iv, (e) iii.
4. (a) 464 kJ/mol, (b) 142 kJ/mol.
5. -2660 kJ.
6. (a) Li, (b) F.
7. P.

ADDITIONAL NOTES

CHAPTER 11

COVALENT BONDING AND PROPERTIES OF MOLECULES

CHAPTER OVERVIEW

The chapter begins with a discussion on writing Lewis structures for molecules and polyatomic ions. Molecular geometry based on the electron-pair repulsion theory is presented. The valence bond theory, which describes covalent bond formation in terms of the overlap of atomic orbitals, is discussed. The concepts of hybridization of atomic orbitals and delocalized bonding are introduced. The chapter concludes with a discussion about intermolecular forces.

COMPETENCIES

Definitions for the following terms should be learned:

- bond angle
- bond axis
- delocalized electrons
- dipole-dipole interaction
- dipole moment
- hybridization
- hydrogen bond
- London forces
- molecular geometry
- π bond
- resonance
- resonance hybrid
- σ bond
- valence bond theory
- valence shell electron-pair repulsion theory
- van der Waals forces

General concepts and skills that should be learned include

- writing Lewis structures for molecules and polyatomic ions
- identifying the shape of molecules and polyatomic ions
- describing covalent bonding in terms of the overlap of atomic orbitals
- identifying the types of hybridization that correspond to various molecular geometries

- interpreting resonance in terms of delocalized electrons
- predicting intermolecular forces between molecules

QUESTIONS

1. Complete the following sentences by filling in the various blanks with these terms: bond axis, delocalization, hybridization, π bonding, σ bonding, and valence bond theory.

   According to the (a)_______, a chemical bond is described by the sharing of electrons in overlapping atomic orbitals along the (b)_______. The atomic orbitals involved in the bonding can be either the regular atomic orbitals--*s*, *p*, *d*, etc.,--or those that have undergone (c)_______--*sp*, sp^2, etc. If a high electron density is between the nuclei along the bond axis, the bonding is known as (d)_______ and if the high electron density between the nuclei does not surround the bond axis but occurs in locations above and below the bond axis, the bonding is known as (e)_______. The concept of resonance is best explained by (f)_______ of electrons in the π orbitals.

2. Match the correct term to each definition.

   (a) the arrangement of valence electrons in molecules or ions for which several Lewis structures can be written
   (b) the actual, single structure of a molecule or ion for which resonance structures can be written
   (c) orbitals not confined to a region near two nuclei but are spread over several nuclei
   (d) collective name for intermolecular forces excluding hydrogen bonding
   (e) the electrostatic attraction of a hydrogen atom covalently bonded to an electronegative atom for a second electronegative atom
   (f) the attractions between fluctuating dipoles in atoms and molecules that are very close together
   (g) the angle between two atoms that are both bonded to the same third atom

   (i) bond angle
   (ii) delocalized orbitals
   (iii) hydrogen bond
   (iv) London forces
   (v) resonance
   (vi) resonance hybrid
   (vii) van der Waals forces

3. Which of the following statements are true? Rewrite each false statement so that it is correct.

(a) The valence bond theory considers a chemical bond to be formed by electrons occupying orbitals that are characteristic of the whole molecule.

(b) As hybridization of atomic orbitals occurs, the newly formed orbitals are identical to each other in shape and energy and are equal in number to the original atomic orbitals that underwent hybridization.

(c) Compared to multiple bonds, the atoms involved in a single bond can rotate rather freely about the internuclear axis.

(d) The Lewis structures shown for a case of resonance really represent the limiting situations, and the true electronic structure falls somewhere between these limits.

(e) Delocalization of electrons in a molecule usually makes a molecule less stable than expected.

(f) If a molecule contains a polar covalent bond, it will be a dipole.

4. Distinguish clearly between (a) intermolecular and intramolecular forces, (b) valence bond theory and molecular orbital theory, (c) σ and π bonds, (d) double and triple covalent bonds, (e) localized and delocalized π bonding, and (g) equatorial and axial positions on an AB_5 molecule.

5. Choose the response that best completes each statement.

(a) The intermediate electronic state of a molecule for which several electronic arrangements are possible is

(i) donor-acceptor atoms (ii) electronegativity-electropositivity
(iii) resonance hybrid (iv) coordinate covalent bonding

(b) The increased thermal stability of a substance in which delocalized bonding in present is

(i) the result of the London forces
(ii) known as the forbidden energy gap
(iii) the valence band
(iv) the delocalization energy

(c) Which type of overlap will *not* produce a σ bond?

(i) two *s* orbitals
(ii) an *s* orbital with a *p* orbital
(iii) two *p* orbitals that are parallel
(iv) two *p* orbitals that lie along the same bond axis

(d) Which form of hybridization gives trigonal planar geometry?
  (i) *sp* (ii) sp^2
  (iii) sp^2d (iv) sp^3

(e) The electrostatic attraction between molecules having a permanent dipole moment is a
  (i) dipole-dipole interaction (ii) London force
  (iii) hydrogen bond (iv) metallic bond

6. List and briefly discuss the five rules that help in writing a Lewis structure for a covalent compound.

7. Write the Lewis structures for (a) OH^-, (b) C_2H_3Cl, (c) SeF_4, (d) ICl_4^-, (e) NO, (f) H_2S, (g) ONCl, (h) $AsCl_5$, (i) F_2O, and (j) HSO_4^-.

8. List the rules that are used to identify whether or not two or more structures are resonance structures.

9. Write the possible resonance forms for the (a) carbonate ion, CO_3^{2-}; (b) nitrate ion, NO_3^-; (c) nitric acid molecule, HNO_3 (the oxygen atom that is bonded to the nitrogen atom and the hydrogen atom is not part of the resonance system); and (d) hydrazoic acid molecule, HN_3.

10. There are five compounds having the formula CH_xCl_y. Write the Lewis structures for these substances.

11. Indicate what is incorrect in each of the following Lewis structures:

(a) B_2H_6

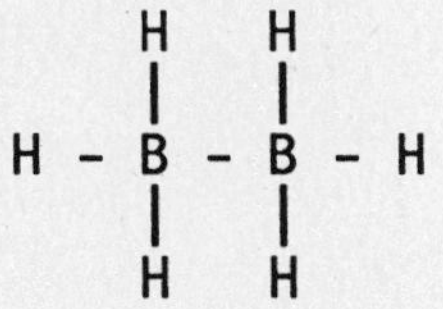

(b) $CaSO_3$

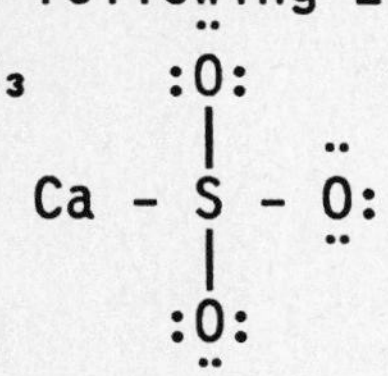

(c) $AlCl_3$

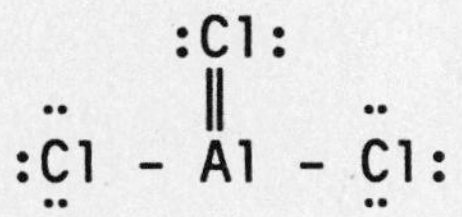

(d) NH_4^+

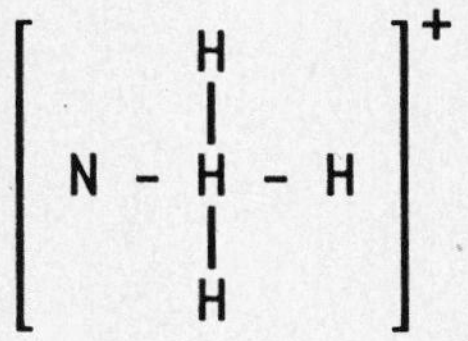

(e) XeF_2

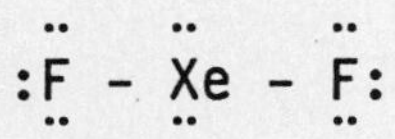

(f) NO_2^-

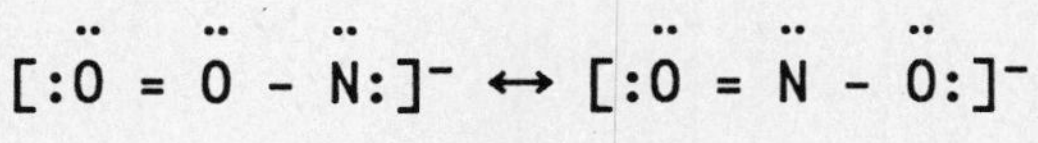

12. Define the term "molecular geometry." What are the two physical parameters that are used to describe the geometry of a molecule?

13. What is the basis for the valence-shell electron-pair repulsion theory (VSEPR) for describing molecular geometry?

14. What are the values of *n* in the formula AB_n that correspond to a molecule having (a) a tetrahedral and (b) an octahedral geometric structure?

15. Sketch the triangular bipyramidal geometrical structure. What are the values of the bond angles?

16. What geometrical structures would be predicted for molecules having the following formulas: (a) AB_3E, (b) AB_2E_3, (c) AB_5E, and (d) $A_2B_2E_2$?

17. Which general formulas, AB_nE_m for $(m+n) \leq 6$, give rise to a linear molecule?

18. Briefly describe the valence bond theory. What are the requirements for successful bond formation?

19. Describe the bonding in the carbon monoxide molecule using the valence bond theory. Assume that both the carbon atom and the oxygen atom are hybridized.

20. Using the valence bond theory, describe the bonding in the hydrogen cyanide molecule, HCN.

21. What types of bonds are usually found in (a) single bonds, (b) double bonds, and (c) triple bonds?

22. What causes a covalent bond to be polar? Are the O-H bonds in a water molecule polar? If the water molecule were linear, would it be a dipole? Why is water a dipole?

23. What is the relationship between bond polarity and the difference in electronegativities of the atoms?

24. Choose the molecular geometries which would give rise to polar molecules: (a) AB_3E, (b) AB_2E_3, (c) AB_6, and (d) AB_3E_2.

25. Is the PF_3 molecule a dipole? Is the PF_5 molecule a dipole? Why?

26. Usually hydrogen bonds are considered to be intermolecular forces. What would the term "intramolecular hydrogen bonding" mean?

Answers to questions

1. (a) valence bond theory, (b) bond axis, (c) hybridization, (d) σ bonding, (e) π bonding, (f) delocalization.
2. (a) v, (b) vi, (c) ii, (d) vii, (e) iii, (f) iv, (g) i.
3. (a) F, replace *valence bond* with *molecular orbital*; (b) T; (c) T; (d) T; (e) F, change *less* to *more*; (f) F, add *if the geometry of the molecule is such*

that the polarity is not offset by an equal polarity in the opposite direction.

4. (a) *Intermolecular forces* are forces between molecules; *intramolecular forces* are forces within a molecule (interatomic forces). (b) *Valence bond theory* describes bond formation in terms of the overlap of atomic orbitals; *molecular orbital theory* describes bonding in terms of orbitals belonging to the entire molecule. (c) σ *Bonds* have a circular cross section of electron density along the bond axis; π *bonds* have two areas of electron density on each side of the bond axis. (d) A *double covalent bond* consists of a σ bond and one π bond; a *triple covalent bond* consists of a σ bond and two π bonds. (e) *Localized* π *bonding* refers to π bonding that involves two atoms; *delocalized* π *bonding* refers to a π bond that is spread out over three or more atoms. (f) The *equatorial positions* in AB₅ geometry refer to the three positions on the equilateral triangle; the *axial positions* refer to the two positions on the normal to the equilateral triangle.

5. (a) iii, (b) iv, (c) iii, (d) ii (e) i.

6. Write correct arrangement of atoms; find total number of valence electrons; assign two electrons to each single covalent bond between atoms; distribute remaining electrons as lone pairs; use multiple bonding if necessary.

7. (a) $[:\overset{..}{\underset{..}{O}} - H]^-$

    (b)
    $$H - \overset{\overset{H}{|}}{C} = \overset{\overset{H}{|}}{C} - \overset{..}{\underset{..}{C}}l:$$

    (c) $:\overset{..}{\underset{..}{F}} - \overset{..}{Se} - \overset{..}{\underset{..}{F}}:$ with $:\overset{..}{\underset{..}{F}}:$ $:\overset{..}{\underset{..}{F}}:$

    (d)
    $$\left[\begin{array}{c} :\overset{..}{\underset{..}{C}}l: \\ | \\ :\overset{..}{\underset{..}{C}}l - \overset{..}{\underset{..}{I}} - \overset{..}{\underset{..}{C}}l: \\ | \\ :\overset{..}{\underset{..}{C}}l: \end{array} \right]^-$$

    (e) $\cdot\overset{..}{N} = \overset{..}{\underset{..}{O}}:$

    (f) $H - \overset{..}{\underset{..}{S}} - H$

    (g) $\overset{..}{\underset{..}{O}} = \overset{..}{N} - \overset{..}{\underset{..}{C}}l:$

    (h)
    $$\begin{array}{c} :\overset{..}{\underset{..}{C}}l: \\ | \\ :\overset{..}{\underset{..}{C}}l - As - \overset{..}{\underset{..}{C}}l: \\ \diagup \quad \diagdown \\ :\overset{..}{\underset{..}{C}}l: \quad :\overset{..}{\underset{..}{C}}l: \end{array}$$

    (i) $:\overset{..}{\underset{..}{F}} - \overset{..}{\underset{..}{O}} - \overset{..}{\underset{..}{F}}:$

    (j)
    $$\left[\begin{array}{c} :\overset{..}{\underset{..}{O}}: \\ | \\ :\overset{..}{\underset{..}{O}} - S - \overset{..}{\underset{..}{O}} - H \\ | \\ :\overset{..}{\underset{..}{O}}: \end{array} \right]^-$$

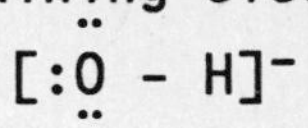

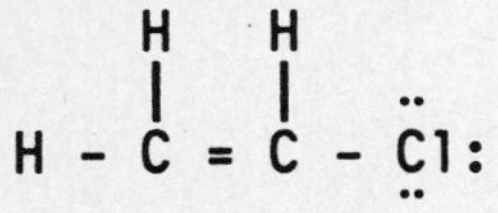

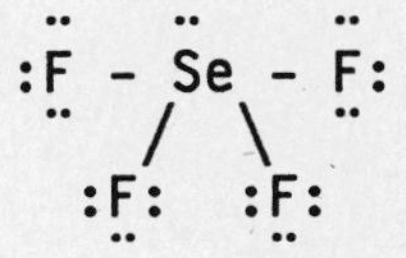

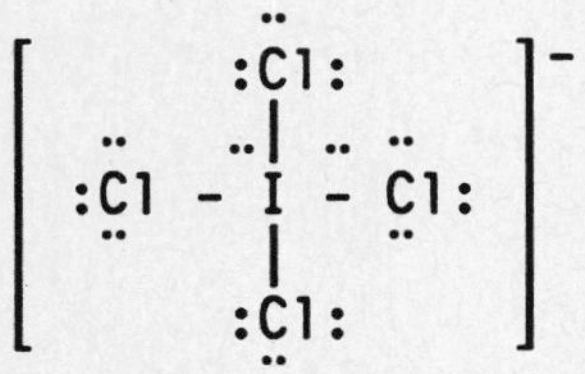

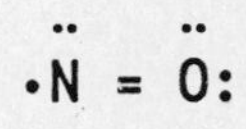

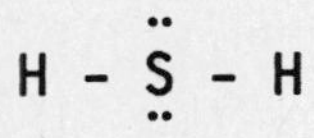

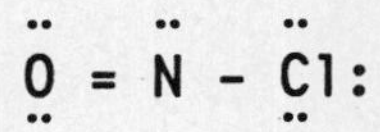

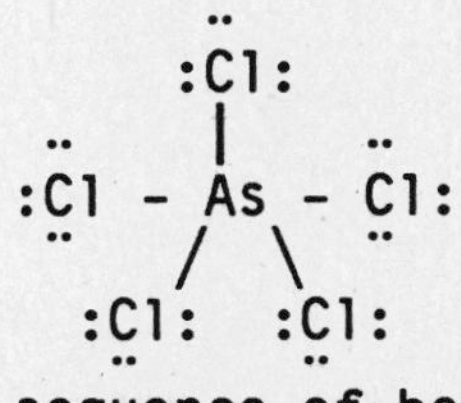

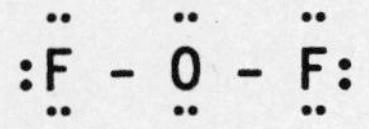

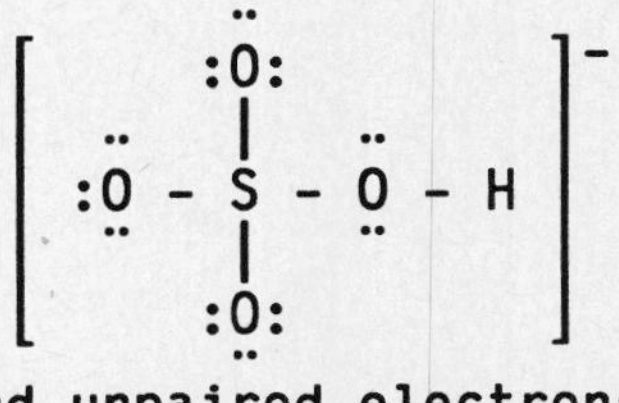

8. The sequence of bonds and the number of paired and unpaired electrons must be the same.

9. (a)
$$\left[\begin{array}{c} :\overset{..}{O}: \\ \| \\ :\overset{..}{\underset{..}{O}} - C - \overset{..}{\underset{..}{O}}: \end{array} \right]^{2-} \leftrightarrow \left[\begin{array}{c} :\overset{..}{\underset{..}{O}}: \\ | \\ :\overset{..}{\underset{..}{O}} - C = \overset{..}{\underset{..}{O}}: \end{array} \right]^{2-} \leftrightarrow \left[\begin{array}{c} :\overset{..}{\underset{..}{O}}: \\ | \\ :\overset{..}{\underset{..}{O}} = C - \overset{..}{\underset{..}{O}}: \end{array} \right]^{2-}$$

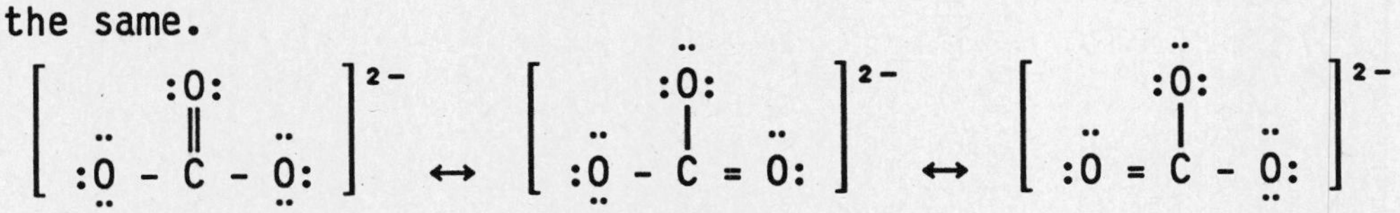

(b)
$$\left[\begin{array}{c} \ddot{:}\overset{..}{O}\text{:} \\ \| \\ \ddot{:}\overset{..}{O} - N - \overset{..}{O}\ddot{:} \end{array} \right]^{-} \leftrightarrow \left[\begin{array}{c} \overset{..}{\underset{..}{O}}\text{:} \\ | \\ \ddot{:}\overset{..}{O} - N = \overset{..}{O}\text{:} \end{array} \right]^{-} \leftrightarrow \left[\begin{array}{c} \overset{..}{\underset{..}{O}}\text{:} \\ | \\ \ddot{:}O = N - \overset{..}{O}\ddot{:} \end{array} \right]^{-}$$

(c)
$$\begin{array}{ccc} & \overset{..}{O}\text{:} & \\ & \| & \\ H - \overset{..}{\underset{..}{O}} - N - \overset{..}{O}\ddot{:} \end{array} \leftrightarrow \begin{array}{ccc} & \overset{..}{\underset{..}{O}}\text{:} & \\ & | & \\ H - \overset{..}{\underset{..}{O}} - N = \overset{..}{O}\text{:} \end{array}$$

(d) $H - \overset{..}{N} = N = \overset{..}{\underset{..}{N}}\text{:} \leftrightarrow H - \overset{..}{\underset{..}{N}} \equiv N - \overset{..}{\underset{..}{N}}\text{:} \leftrightarrow H - \overset{..}{\underset{..}{N}} - N \equiv N\text{:}$

10.
$$\begin{array}{ccccc} H & H & H & \ddot{:}\overset{..}{Cl}\text{:} & \ddot{:}\overset{..}{Cl}\text{:} \\ | & | & | & | & | \\ H - C - H \quad & H - C - \overset{..}{\underset{..}{Cl}}\text{:} \quad & H - C - \overset{..}{\underset{..}{Cl}}\text{:} \quad & H - C - \overset{..}{\underset{..}{Cl}}\text{:} \quad & \ddot{:}\overset{..}{Cl} - C - \overset{..}{\underset{..}{Cl}}\text{:} \\ | & | & | & | & | \\ H & H & \ddot{:}\overset{..}{\underset{..}{Cl}}\text{:} & \ddot{:}\overset{..}{\underset{..}{Cl}}\text{:} & \ddot{:}\overset{..}{\underset{..}{Cl}}\text{:} \end{array}$$

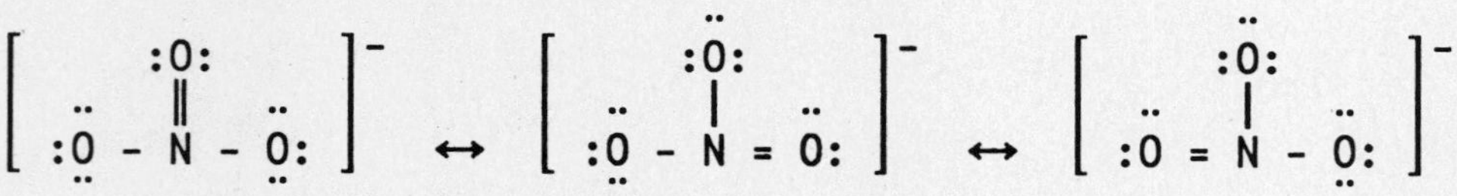

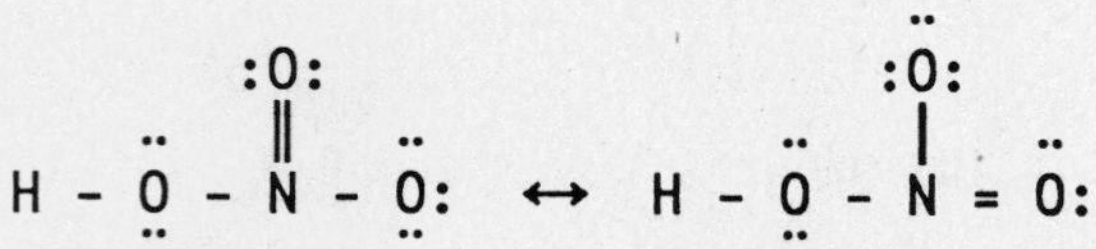

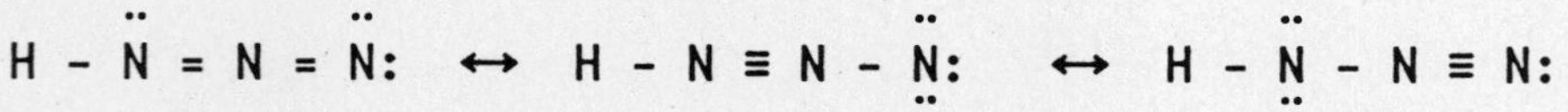

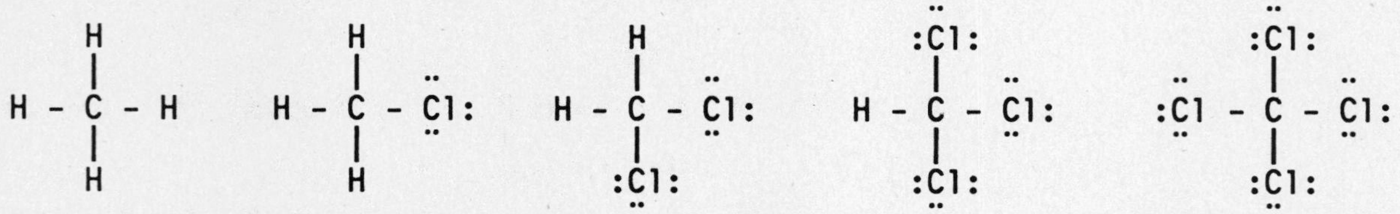

11. (a) too many electrons, (b) Ca should have an ionic bond to SO_3^{2-} ion, (c) Al and Cl do not form double bonds, (d) N should be in center, (e) missing one pair of nonbonding electrons on Xe, (f) atoms cannot be in different positions in resonance forms.

12. Molecular geometry is the three-dimensional arrangement of the atoms in a molecule. The two parameters are bond lengths and bond angles.

13. Electron pairs repel each other so as to assume a geometric arrangement in which they are as far apart as possible.

14. (a) 4, (b) 6.

15. See Table 11.1 of the text for the sketch. The bond angles are 120° and 90°.

16. The following molecules would have the predicted geometric structures indicated:

(a) AB_3E, triangular pyramidal   (b) AB_2E_3, linear

(c) AB_5E, square pyramidal   (d) $A_2B_2E_2$, bent

17. The linear structures will be AB_1E_m, Ab_2E_0, AB_2E_3, and AB_2E_4.

18. Bonds are formed by the overlap of atomic orbitals. For successful bond formation, the orbitals involved in bonding must have similar energies; they must overlap sufficiently; and they must have the same symmetry with respect to the two atoms.

19. The Lewis structure is $:C \equiv O:$ and the sketch is shown in Fig. 11-1. Each atom is sp-hybridized, generating two equivalent sp orbitals, one for the lone pair of electrons and one for the σ bond with the other atom. There are two unhybridized p orbitals on each atom that can overlap with the two unhybridized p orbitals of the other atom, producing two π bonds.

Fig. 11-1

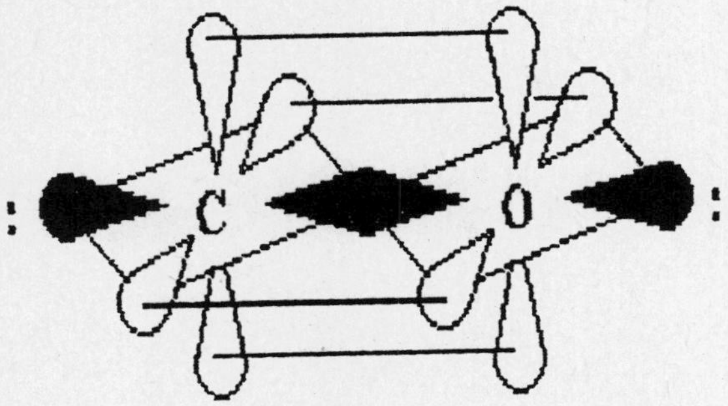

20. The Lewis structure is H - C ≡ N: and the sketch is shown in Fig. 11-2. The carbon atom and the nitrogen atom are sp-hybridized. The two hybridized orbitals on the C are used for the σ bonds between C and H, and C and N. The two unhybridized *p* orbitals form two π bonds with the two unhybridized *p* orbitals on the nitrogen atom. The additonal hybridized orbital on the N has the lone pair of electrons in it.
21. (a) σ, (b) σ and π, (c) σ and two π.
22. A bond is polar because of the difference in the electronegativities of the atoms. Yes, the bonds are polar, but the molecule would not be polar if it were linear. Water is a nonlinear molecule.
23. More polar bonds have greater difference in the electronegativities.
24. The molecular geometries which would give rise to polar molecules are (a) (a) AB_3E and (b) AB_3E_2.
25. Yes, the PF_3 molecule was a triangular pyramidal geometry (type AB_3E). No, the molecule PF_5 has a triangular bipyramidal geometry (type AB_5).
26. The hydrogen bonding would be within the same molecule.

SKILLS

Skill 11.1 Writing Lewis structures

A correctly written Lewis structure is a two-dimensional representation of a molecule or polyatomic ion that shows the bonding and nonbonding outer electrons.

The first step in writing the Lewis structure for polyatomic molecules or ions is to write the "skeleton" of the structure--the arrangement of the atoms with single bonds between them. There are a few guidelines that help if nothing is known about the structure: (1) usually smaller, more electronegative non-metal

Fig. 11-2

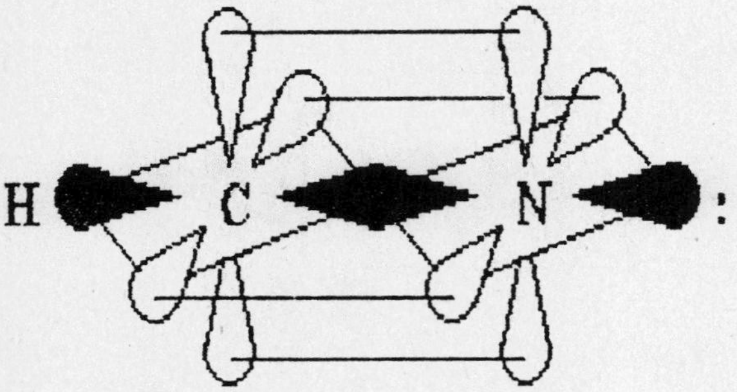

atoms surround larger, less electronegative nonmetal atoms, (2) usually the molecule will show a considerable amount of symmetry, (3) usually carbon atoms will be surrounded by other atoms, and (4) usually hydrogen atoms will not be found at the center of the molecule. For example, the skeletons for HCl, CCl_4, SO_4^{2-}, $S_2O_3^{2-}$, IF_3, XeF_4, BCl_3, HCN, CO_2, CO_3^{2-}, N_2O, and NO_2 are

```
                    Cl                   O              O
                    |                    |              |
  H - Cl       Cl - C - Cl           O - S - O      O - S - S
                    |                    |              |
                    Cl                   O              O

                    F                   Cl
                    |                    |
 F - I - F      F - Xe - F          Cl - B - Cl      H - C - N
     |              |
     F              F

                    O
                    |
 O - C - O      O - C - O            N - N - O       O - N - O
```

All of the above skeletons were written using the guidelines except for $S_2O_3^{2-}$, in which it is known that the general pattern is similar to SO_4^{2-}.

Skill Exercise: Write the skeletons for (a) H_2O, (b) NH_4^+, (c) ClF, (d) IF_4^+, (e) IF_6^-, (f) $AlBr_3$, (g) CO, (h) $COCl_2$, (i) O_3, (j) NO_3^-, and (k) ClO_2 (which has a chlorine atom in the center). *Answers*:

```
(a)                (b)      H          (c)              (d)      F
                            |                                    |
    H - O - H           H - N - H          Cl - F            F - I - F
                            |                                    |
                            H                                    F
```

(e) F   F   (f)   Br   (g)   (h)   O
      \ / | |
   F - I - F   Br - Al - Br   C - O   Cl - C - Cl
      / \
     F   F

(i)   (j)   O   (k)
                                     |
   O - O - O   O - N - O   O - Cl - O

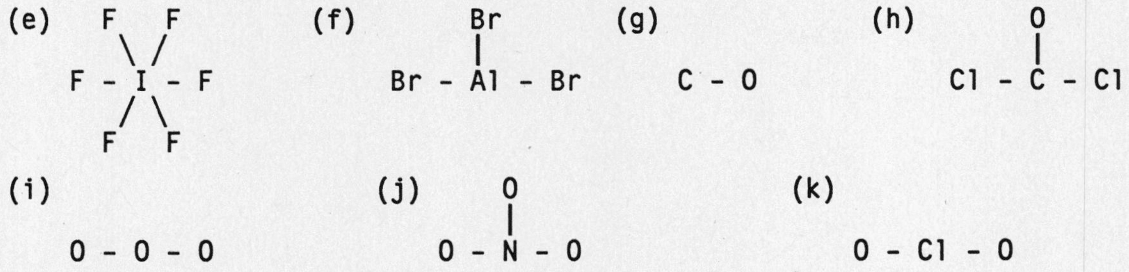

The next step is to determine the number of valence electrons that are available for the species. This is done by adding the number of valence electrons for each atom in the species, and if the species is ionic, adding one electron for each negative charge or subtracting one electron for each positive charge. For example, in HCl there are $1 + 7 = 8$ valence electrons, in CCl_4 there are $4 + 4(7) = 32$, in SO_4^{2-} there are $6 + 4(6) + 2 = 32$, in $S_2O_3^{2-}$ there are $2(6) + 3(6) + 2 = 32$, in IF_3 there are $7 + 3(7) = 28$, in XeF_4 there are $8 + 4(7) = 36$, in BCl_3 there are $3 + 3(7) = 24$, in HCN there are $1 + 4 + 5 = 10$, in CO_2 there are $4 + 2(6) = 16$, in CO_3^{2-} there are $4 + 3(6) + 2 = 24$, in N_2O there are $2(5) + 6 = 16$, and in NO_2 there are $5 + 2(6) = 17$.

Skill Exercise: Determine the number of available valence electrons (hence the number that will appear in the Lewis formula) for the following species: (a) H_2O, (b) NH_4^+, (c) ClF, (d) IF_4^+, (e) IF_6^-, (f) $AlBr_3$, (g) CO, (h) $COCl_2$, (i) O_3, (j) NO_3^-, and (k) ClO_2. *Answers*: (a) 8, (b) 8, (c) 14, (d) 34, (e) 50, (f) 24, (g) 10, (h) 24, (i) 18, (j) 24, (k) 19.

Each single bond in the skeleton accounts for two electrons. After subtracting the number of electrons used for these bonds from the number of valence electrons, the next step in writing Lewis structures for polyatomic molecules or ions is to put in the remainder of the available electrons trying to satisfy the octets of all atoms. The only number of electrons that may appear in the structure is the number of valence electrons (no more or no less-- no matter how desirable). If the species is paramagnetic, some single electrons must appear. In "saturated" species, only single bonds and complete octets will appear in the formula. For example

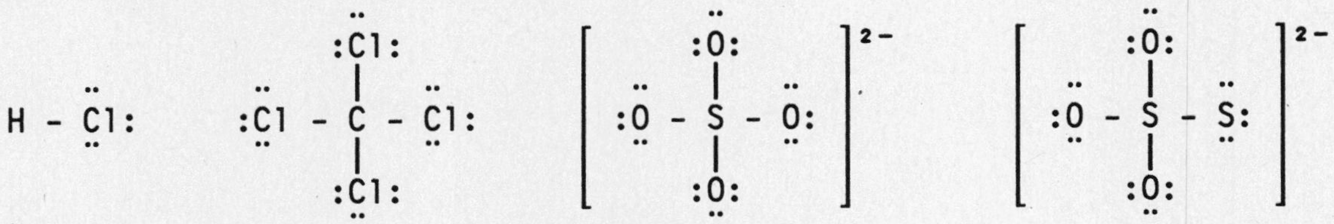

In "unsaturated" species--those that do not have enough electrons to satisfy all octets of atoms and have only single bonds--the octet may be left incomplete on the central atom or multiple bonds may be used between certain elements (particularly C, N, and O) to satisfy the octets.   For example

$$\ddot{:}\overset{\displaystyle \overset{..}{:}\overset{..}{\underset{\displaystyle |}{C}l}\overset{..}{:}}{\underset{\displaystyle :\overset{..}{\underset{..}{C}}l - B - \overset{..}{\underset{..}{C}}l:}{}} \qquad\qquad H - C \equiv N: \qquad\qquad :\overset{..}{\underset{..}{O}} = C = \overset{..}{\underset{..}{O}}:$$

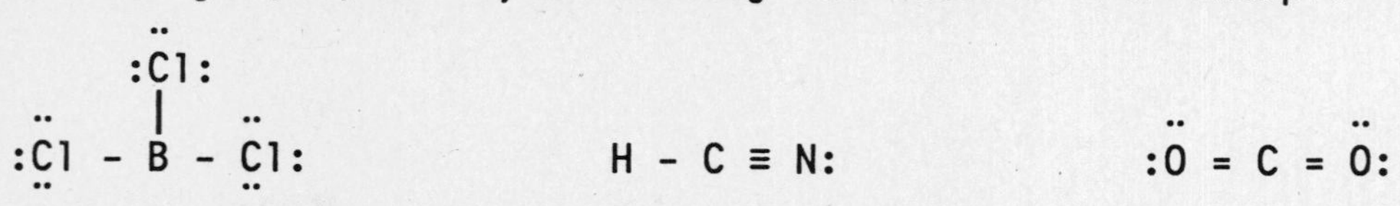

For those species that seem to have too many electrons, the octet is violated for the central atom by putting more than 8 electrons around it. [It is assumed that these extra electrons are in the empty *nd* subshell on the central atom.] For example

$$\overset{..}{:}\overset{\displaystyle :\overset{..}{F} - \overset{..}{\underset{\displaystyle |}{I}} - \overset{..}{F}:}{\underset{\displaystyle :\overset{..}{F}:}{}} \qquad\qquad :\overset{..}{F} - \overset{\displaystyle Xe}{\underset{\displaystyle / \quad \backslash}{}} - \overset{..}{F}:$$
$$\qquad\qquad\qquad\qquad :\overset{..}{F}: \quad :\overset{..}{F}:$$

   If two or more Lewis structures can be written that have the same arrangement of atoms and number of bonds, a resonance situation exists and all formulas are used.

$$\overset{..}{N} = N = \overset{..}{\underset{..}{O}} \quad\leftrightarrow\quad :N \equiv N - \overset{..}{\underset{..}{O}}: \qquad :\overset{..}{\underset{..}{O}} - \overset{.}{N} = \overset{..}{O} \quad\leftrightarrow\quad \overset{.}{O} = \overset{.}{N} - \overset{..}{\underset{..}{O}}:$$

$$\left[\; :\overset{..}{\underset{..}{O}} - \overset{\displaystyle \overset{..}{:O:} }{\underset{\displaystyle |\!|}{C}} - \overset{..}{\underset{..}{O}}: \;\right]^{2-} \leftrightarrow \left[\; :\overset{..}{\underset{..}{O}} - \overset{\displaystyle \overset{..}{:O:}}{\underset{\displaystyle |}{C}} = \overset{..}{\underset{..}{O}} \;\right]^{2-} \leftrightarrow \left[\; \overset{..}{\underset{..}{O}} = \overset{\displaystyle \overset{..}{:O:}}{\underset{\displaystyle |}{C}} - \overset{..}{\underset{..}{O}}: \;\right]^{2-}$$

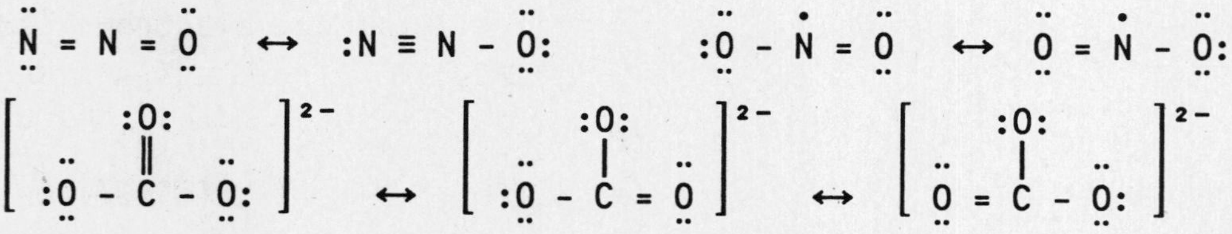

Sometimes these resonance structures are represented by

$$N \overset{\bullet\bullet\bullet}{=\!=\!=} N \overset{\bullet\bullet\bullet}{-\!-\!-} O \qquad\qquad O \overset{\bullet\bullet\bullet}{-\!-\!-} N \overset{\bullet\bullet\bullet}{-\!-\!-} O \qquad\qquad \left[\; O \overset{\bullet\bullet\bullet}{-\!-\!-} \overset{\displaystyle \overset{..}{O:}}{\underset{\displaystyle |}{C}} \overset{\bullet\bullet\bullet}{-\!-\!-} O \;\right]^{2-}$$

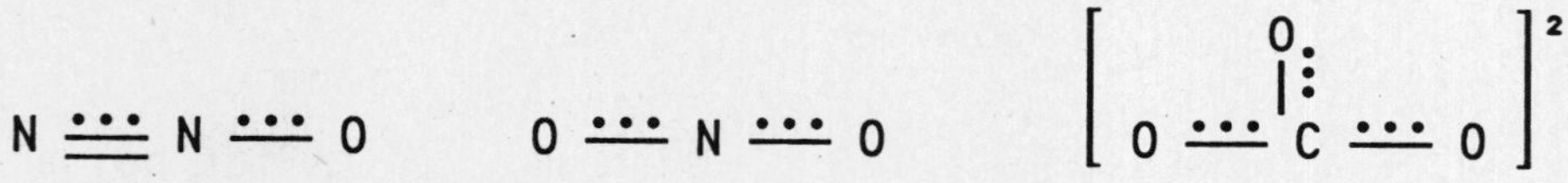

Skill Exercise: Write the complete Lewis electron-dot structures for (a) H_2O, (b) NH_4^+, (c) ClF, (d) IF_4^+, (e) IF_6^-, (f) $AlBr_3$, (g) CO, (h) $COCl_2$, (i) O_3, (j) NO_3^-, and (k) ClO_2. *Answers:*

(a)

$$H - \overset{..}{\underset{..}{O}} - H$$

(b)

$$\left[\; \overset{\displaystyle H}{\underset{\displaystyle H}{H - \overset{\displaystyle |}{\underset{\displaystyle |}{N}} - H}} \;\right]^{+}$$

(c)

$$:\overset{..}{\underset{..}{C}}l - \overset{..}{\underset{..}{F}}:$$

(d)

$$\left[\; \overset{\displaystyle :\overset{..}{F} - \overset{..}{\underset{\displaystyle |}{I}} - \overset{..}{F}:}{\underset{\displaystyle :\overset{..}{F}: \quad :\overset{..}{F}:}{/ \quad \backslash}} \;\right]^{+}$$

(e)

$$\left[\; \overset{\displaystyle :\overset{..}{F}: \;\; :\overset{..}{F}:}{\underset{\displaystyle :\overset{..}{F}: \;\; :\overset{..}{F}:}{:\overset{..}{F} - \overset{\displaystyle \backslash \cdot\cdot /}{\underset{\displaystyle / \quad \backslash}{I}} - \overset{..}{F}:}} \;\right]^{-}$$

(f)

$$\overset{\displaystyle :\overset{..}{B}r:}{\underset{\displaystyle :\overset{..}{B}r - Al - \overset{..}{B}r:}{|}}$$

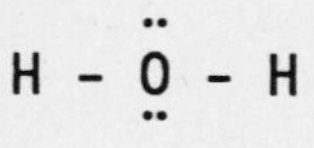

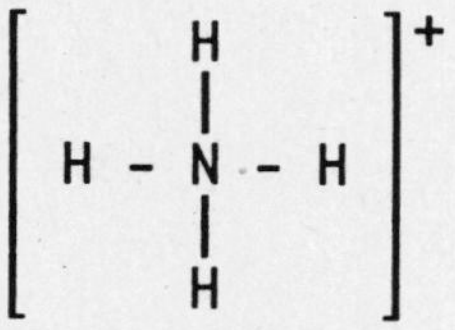

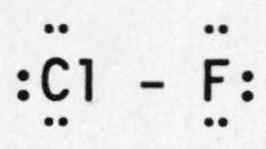

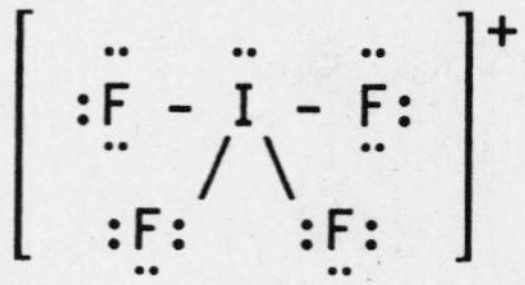

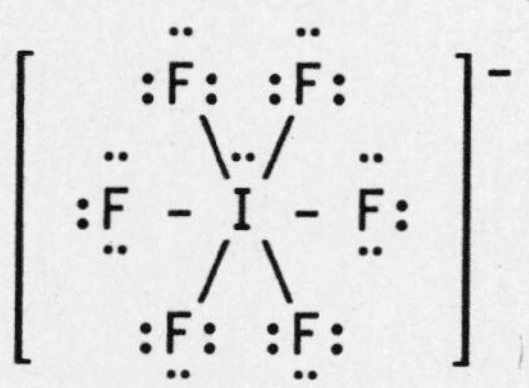

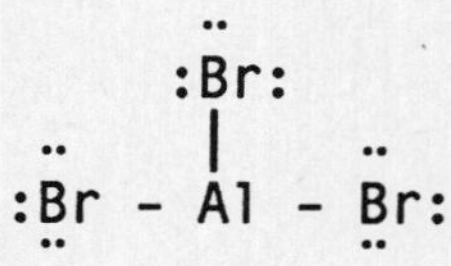

(g)   (h)   :O: (i)
                                   ‖
    :C ≡ O: :Cl - C - Cl: O = O - O: ↔ :O - O = O

(j) ⎡   :O: ⎤⁻   ⎡   :O: ⎤⁻   ⎡   :O: ⎤⁻   (k)
     ⎢   ‖   ⎥   ⎢ | ⎥   ⎢ | ⎥
     ⎢ :O - N - O: ⎥ ↔ ⎢ :O - N = O   ⎥ ↔ ⎢ O = N - O: ⎥   :O - Cl - O:
     ⎣   ⎦   ⎣   ⎦   ⎣   ⎦

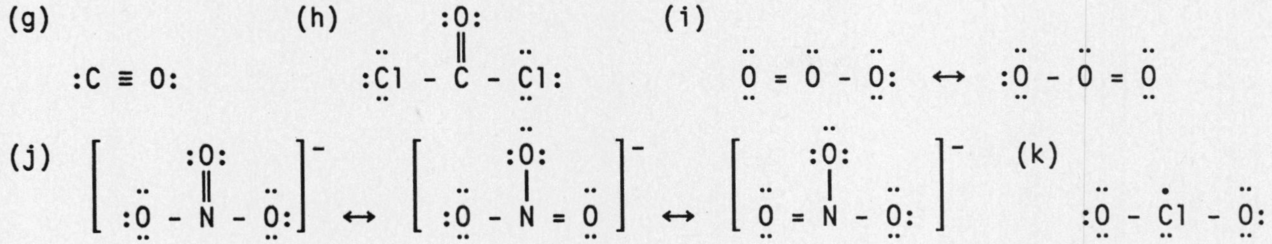

Skill 11.2 Identifying molecular geometry

The geometric shape of a molecule is determined by the location of the
atoms in space. Thus any diatomic molecule will be linear because the two
nuclei define a straight line. When more than one atom is attached to a given
atom, the atoms and lone pairs usually will take the idealized molecular
geometries about the center atom that are shown in Figure 11.3 of the text, and
the molecular shape, which depends on the number of atoms around the center atom
are shown in the figure.

For example, based on the Lewis structures determined in *Skill 11.1 Writing
Lewis structures*, HCl is predicted to be linear because it is a diatomic
molecule (type AB); CCl_4 and SO_4^{2-} are predicted to be tetrahedral (type AB_4)
with four bonded and no nonbonded pairs of electrons; $S_2O_3^{2-}$ is predicted to
resemble a tetrahedron for the same reasons as SO_4^{2-}, but because there are one
S-S bond and three S-O bonds, each with distinctive lengths, the tetrahedron
becomes distorted into a trigonal pyramid; HCN and NO_2 are predicted to be
linear (type AB_2) because they have two bonded atoms and no lone pairs; IF_3 is
predicted to be T-shaped (type AB_3E_2) because it has three bonded pairs and two
lone pairs of electrons; XeF_4 is predicted to be square planar (type AB_4E_2)
because it has four bonded pairs and two lone pairs of electrons; CO_2 is
predicted to be linear (type AB_2) because it has no lone pairs of electrons;
NO_2 is predicted to be angular (type AB_2, bond angle about 120°) because it has
two bonded pairs of electrons and a single unshared electron which acts as a
lone pair when determining structure; and BCl_3 and CO_3^{2-} are predicted to be
triangular planar (type AB_3) because they have no lone pairs of electrons. Note
that an entire multiple bond acts as a bonded pair of electrons when determining
molecular geometry.

Sketches of these molecules are shown in Fig. 11-3. In these sketches,
usually part of the ideal molecular geometry is outlined around the central atom
and the code to the three-dimensional perspective is as follows: A solid line
is in the plane of the paper, a diminishing wedge moves into the plane of the
paper, and an expanding wedge comes out of the plane of the paper.

Fig. 11-3

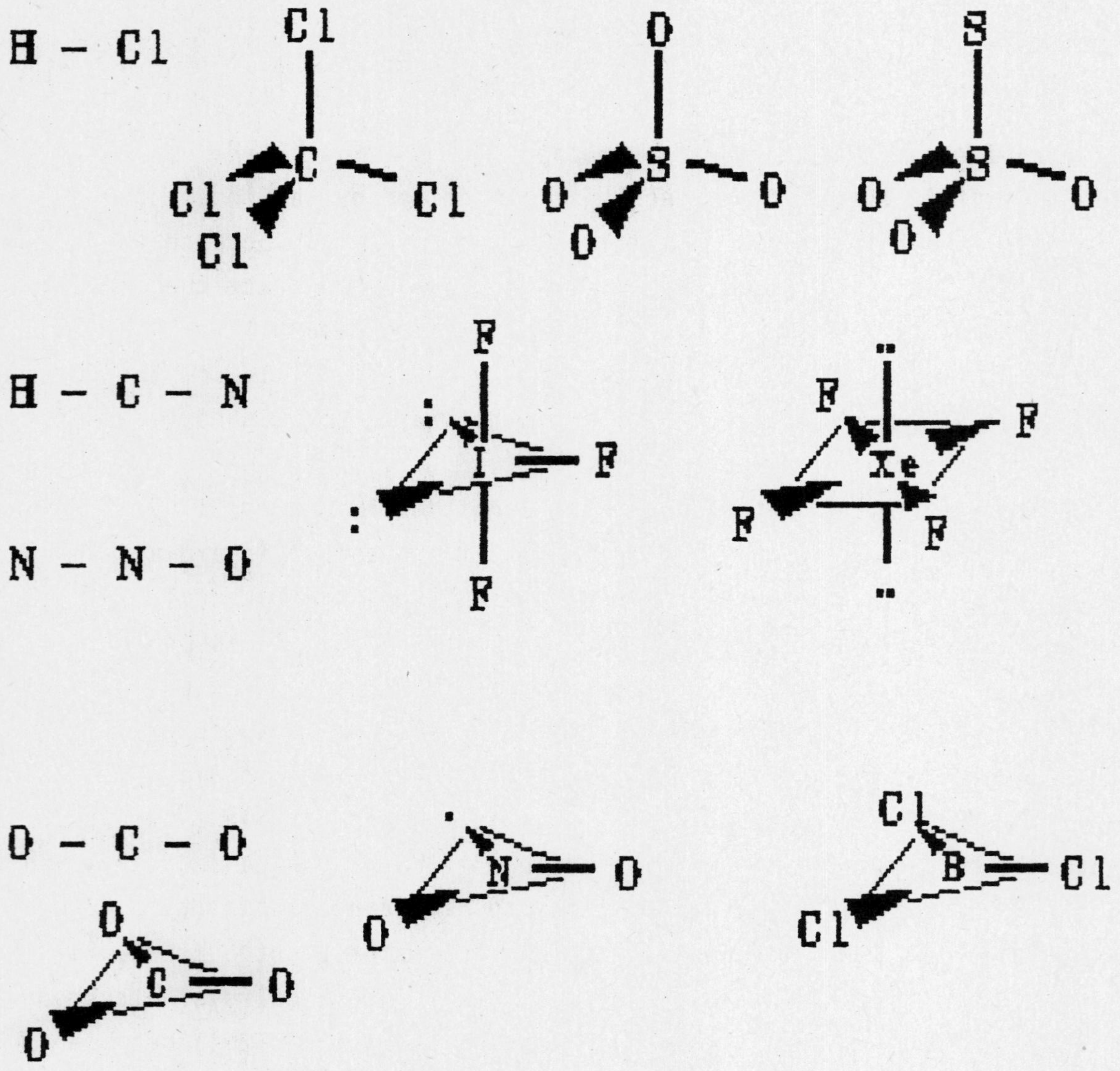

Skill Exercise: Predict the molecular shapes and prepare sketches for (a) H_2O, (b) NH_4^+, (c) ClF, (d) IF_4^+, (e) IF_6^-, (f) $AlBr_3$, (g) CO, (h) $COCl_2$, (i) O_3, (j) NO_3^-, and (k) ClO_2. *Answers*: See Fig. 11-4 for the sketches. The shapes are (a) angular (type AB_2E_2), (b) tetrahedral (type AB_4), (c) linear (type AB), (d) seesaw-shaped (type AB_4E), (e) distorted octahedron (type AB_6E), (f) triangular (type AB_3), (g) linear (type AB), (h) triangular (type AB_3, isosceles), (i) angular or bent or nonlinear (type AB_2E), (j) triangular (type AB_3), (k) angular (type AB_2E_2),

Fig. 11-4

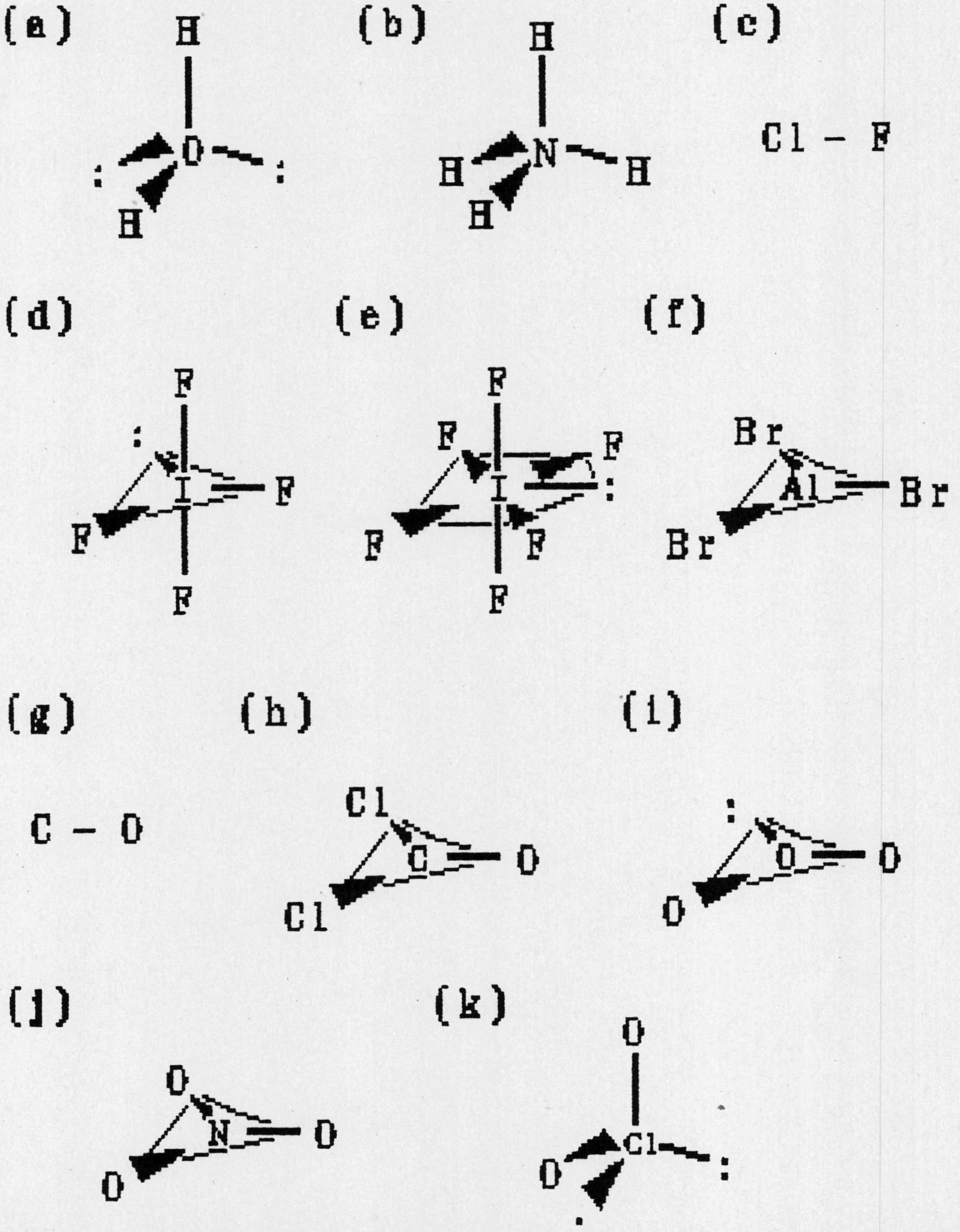
(a)
H
O
H
(b)
H
N
H
H
H
(c)
Cl - F
(d)
F
I
F
F
F
(e)
F
F
F
I
F
F
F
(f)
Br
Al
Br
Br
(g)
C - O
(h)
Cl
C
O
Cl
(i)
O
O
O
(j)
O
N
O
O
(k)
O
Cl
O

Skill 11.3 Identifying types of hybridization

The concept of molecular geometry is discussed in *Skill 11.2 Identifying molecular geometry*. The basis of that discussion is that electron pairs--bonding and nonbonding--occupy locations in space as far apart as possible. Using the concept of hybridization of atomic orbitals, these same geometries can be derived mathematically from quantum mechanics. Thus to each geometric arrangement there corresponds a given type of hybridization (and vice versa), as shown in the following table.

Total number of bonded and lone pairs	Molecular geometry about the center atom	Type of hybridization
2	linear	*sp*
3	triangular planar	sp^2
4	tetrahedral square planar	sp^3 sp^2d or dsp^2
5	triangular bipyramidal square pyramidal	sp^3d or dsp^3 sp^3d or dsp^3
6	octahedral	d^2sp^3 or sp^3d^2

For example, CCl_4 has four bonding pairs and no lone pairs of electrons around the carbon atom, and so the carbon must be sp^3 hybridized. Likewise, H_2O has two bonding pairs and two lone pairs of electrons around the oxygen atom, and so the oxygen must be sp^3 hybridized.

Skill Exercise: Identify the type of hybridization for each underlined atom in the following species: (a) $\underline{C}H_3Cl$, (b) $\underline{CO}_3{}^{2-}$, (c) $H\underline{Cl}$, (d) $H\underline{CN}$, (e) $\underline{Cu}(\underline{N}H_3)_4{}^{2+}$, (f) $[\underline{Cr}(H_2O)_6]^{3+}$, (g) $\underline{S}O_4{}^{2-}$, and (h) $\underline{P}Cl_5$. *Answers*: (a) sp^3, (b) sp^2, sp^2; (c) sp^3, (d) *sp*, *sp*; (e) sp^2d, sp^3; (f) sp^3d^2; (g) sp^3; (h) sp^3d.

Skill 11.4 Predicting types of intermolecular forces

Van der Waals forces (dipole-dipole interactions and London forces) and hydrogen bonding involve the interactions of one molecule with another. London forces are always present even if other intermolecular forces are present as well. Dipole-dipole interactions can take place only if the molecule has a permanent dipole moment--a separation or inbalance of the electron distribution. Hydrogen bonding takes place between a hydrogen atom bonded to a fluorine, oxygen, or nitrogen atom (or, to a much lesser extent, a chlorine atom) and a lone pair of electrons on a similar atom on another molecule. For example, HCl and HCN would have a very small amount of hydrogen bonding, dipole-dipole interactions, and London forces; N_2O, IF_3, and NO_2 would have dipole-dipole interactions and London forces; and CO_2, CCl_4, XeF_4, and BCl_3 would have London forces. [Sometimes the forces involved in ionic and metallic bonding are considered as intermolecular forces, but these were considered in *Skill 10.1 Predicting types of bonds between atoms* and will not be considered further in this skill.]

Skill Exercise: Predict the intermolecular forces in (a) H_2O, (b) ClF, (c) $AlBr_3$, (d) CO, (e) $COCl_2$, (f) O_3, and (g) ClO_2. *Answers*: (a) hydrogen bonding, dipole-dipole interactions, London forces; (b) dipole-dipole interactions, London forces; (c) London forces; (d) dipole-dipole interactions, London forces; (e) dipole-dipole interactions, London forces; (f) dipole-dipole interactions, London forces; (g) dipole-dipole interactions, London forces.

PRACTICE TEST

1 (20 points). Write the Lewis structures, determine the molecular geometries, and draw three-dimensional sketches for (a) PO_4^{3-}, (b) CO, (c) PCl_5, (d) SO_2, and (e) SO_3^{2-}.

2 (10 points). Draw sketches representing (a) a σ bond formed from 1*s* orbitals, (b) a π bond formed from two p_x orbitals, and (c) a set of sp^3d hybridized orbitals.

3 (15 points). Identify the type of hybridization for the underlined atoms in (a) H - <u>C</u> ≡ C - H, (b) <u>N</u>H_3, (c) [<u>Co</u>Cl_6]$^{3-}$, (d) H_2<u>CO</u>, and (e) <u>NO</u>$_3^-$.

4 (20 points). Describe the molecular shape and intermolecular forces for the molecules (a) CH_4, (b) CH_3Cl, (c) CH_2Cl_2, (d) $CHCl_3$, and (e) CCl_4. In each case the H and Cl atoms are "tetrahedrally" placed around the C atom.

5 (35 points). Choose the response that best completes each statement.

(a) Which best represents N_2O, laughing gas (which replaced a shot of whiskey and biting the bullet as an anesthetic in the Old West)?

```
              ..   ..                         ..
   (i) :N ≡ N - O:          (ii) :N ≡ N = O:
                   ..
         ..   ..   ..                         ..
 (iii) :N = N = O:          (iv) :N = N = O:
                                              ..
```

(b) Which of the following Lewis structures is incorrect?

(i)
```
        :O:
         |      ..
    K - Cl - O:
         |
        :O:
         ..
```

(ii)
```
              :O:
      ..       ‖      ..
    :Cl - C - Cl:
      ..              ..
```

(iii)
```
        H   O
        |   ‖    ..
    H - C - C - O - H
        |        ..
        H
```

(iv)
```
    ⎡        :O:       ⎤ 2-
    ⎢  ..     |    ..  ⎥
    ⎢ :O - S - O:     ⎥
    ⎢  ..     |    ..  ⎥
    ⎣        :O:       ⎦
             ..
```

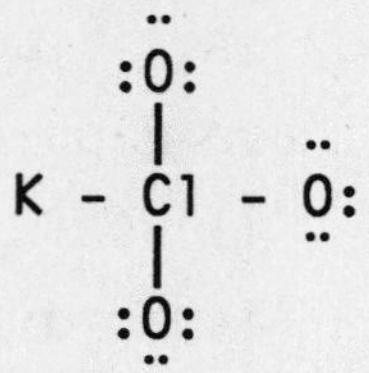

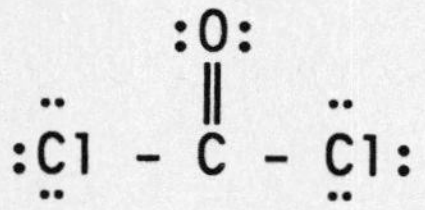

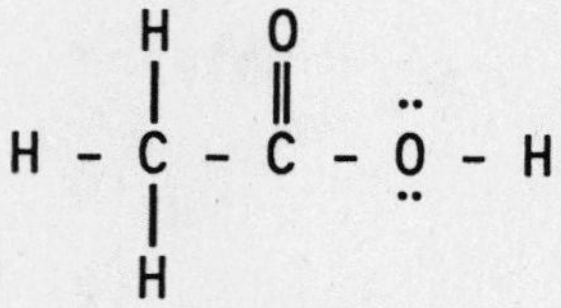

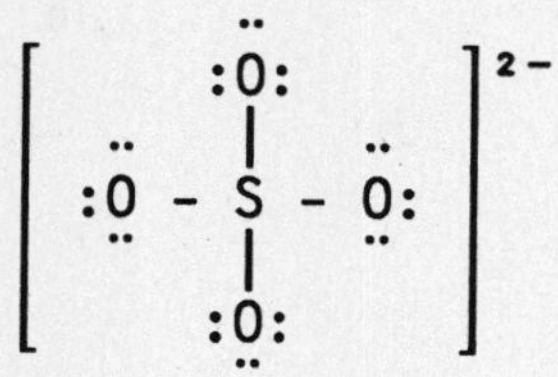

(c) Sulfur reacts to form a compound with an empirical formula of SCl. The Lewis structure is

(i) :S̈ - C̈l·

(ii) :C̈l - S̈ - C̈l - S̈:

(iii) :C̈l - S̈ - S̈ - C̈l:

(iv) :S̈ = C̈l:

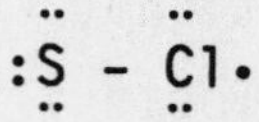

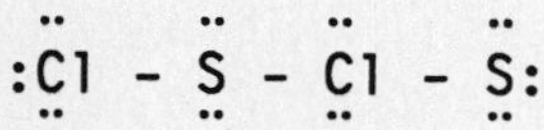

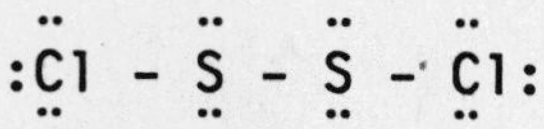

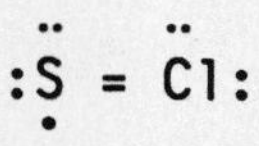

(d) Which of these Lewis structures best represents the azide ion, N_3^-?

(i) [:N = N = N:]⁻

(ii) [:N̈ - N - N̈:]⁻

(iii) [:N̈ - N = N̈:]⁻

(iv) [:N̈ - N̈ - N̈:]⁻

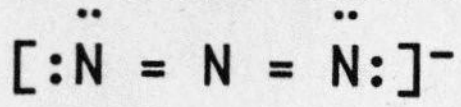

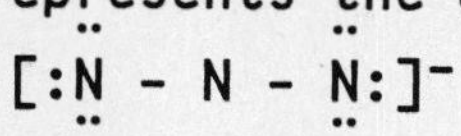

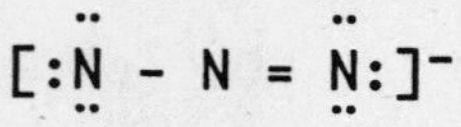

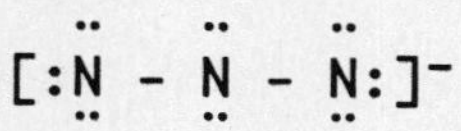

(e) Delocalization of electrons usually
  (i) makes a molecule more stable
  (ii) makes a molecule less stable
  (iii) gives rise to structural isomerism
  (iv) none of these answers

(f) π Bonds are formed by
  (i) two *p* orbitals that lie parallel to each other
  (ii) two *p* orbitals that lie along the internuclear axis
  (iii) an *s* and *p* orbital
  (iv) two *sp²* orbitals

(g) Which molecule has the shortest carbon-carbon bond length?
  (i) CH_3CH_3 (ii) CH_2CH_2
  (iii) CHCH   (iv) $CH_3CH_2CH_3$

(h) Which molecule has the largest carbon-carbon bond energy?
  (i) CH_3CH_3 (ii) CH_2CH_2
  (iii) CHCH   (iv) $CH_3CH_2CH_3$

(i) How does the BAB bond angle in AB_3 compare to the BAB bond angle in AB_2E?
- (i) greater than
- (ii) less than
- (iii) the same as
- (iv) none of these answers

(j) How many hybridized orbitals are formed by sp^3d hybridization?
- (i) 1
- (ii) 7
- (iii) 5
- (iv) 14

(k) What is the hybridization of the nitrogen atom in CH_3NH_2?
- (i) *sp*
- (ii) sp^2
- (iii) sp^3
- (iv) sp^3d

(l) The geometry of the NH_3 molecule is best described as
- (i) angular
- (ii) tetrahedral
- (iii) trigonal planar
- (iv) pyramidal

(m) The bond angle between sp^3 hybrid orbitals is about
- (i) 90°
- (ii) 120°
- (iii) 109°
- (iv) 180°

(n) The shape of a molecule classified as AB_2E_2 is
- (i) tetrahedral
- (ii) linear
- (iii) square planar
- (iv) bent

(o) Which type of intermolecular forces accounts for the increase in boiling points for the noble gases as the atomic number increases?
- (i) hydrogen bonding
- (ii) network covalent bonding
- (iii) dipole-dipole interactions
- (iv) London forces

(p) Which substance has the highest boiling point?
- (i) CH_4
- (ii) CH_3CH_3
- (iii) $CH_3CH_2CH_3$
- (iv) $CH_3CH_2CH_2CH_3$

(q) Which of the following has the lowest boiling point?
- (i) H_2O
- (ii) H_2Se
- (iii) H_2S
- (iv) H_2Te

(r) ICl has a higher boiling point than Br_2 because
- (i) ICl has a dipole moment
- (ii) ICl has more electrons
- (iii) Br_2 has a dipole moment
- (iv) I_2 is a solid

Answers to practice test

1. (a) $\begin{bmatrix} & \overset{\cdot\cdot}{:O:} & \\ & | & \\ :\!\overset{\cdot\cdot}{\underset{\cdot\cdot}{O}} - P - \overset{\cdot\cdot}{\underset{\cdot\cdot}{O}}\!: \\ & | & \\ & \overset{}{\underset{\cdot\cdot}{:O:}} & \end{bmatrix}^{3-}$ (b) $:C \equiv O:$ (c) $\quad\overset{\cdot\cdot}{:}\quad :\overset{\cdot\cdot}{Cl}: \quad\overset{\cdot\cdot}{:}$
       $\qquad :Cl \quad | \quad Cl:$
       $\qquad\qquad\quad P$

       $\qquad\qquad :\underset{\cdot}{Cl}: :\underset{\cdot}{Cl}:$

    (d) $:\underset{\cdot\cdot}{O} = \overset{\cdot\cdot}{S} - \overset{\cdot\cdot}{\underset{\cdot\cdot}{O}}: \leftrightarrow :\overset{\cdot\cdot}{\underset{\cdot\cdot}{O}} - \overset{\cdot\cdot}{S} = \underset{\cdot\cdot}{O}:$ (e) $\begin{bmatrix} & \overset{\cdot\cdot}{:O:} & \\ & | & \\ :\!\overset{\cdot\cdot}{\underset{\cdot\cdot}{O}} - \overset{\cdot\cdot}{S} - \overset{\cdot\cdot}{\underset{\cdot\cdot}{O}}\!: \end{bmatrix}^{2-}$

    (a) tetrahedral (type AB_4), (b) linear (type AB), (c) triangular bipyramidal (type AB_5), (d) angular (type AB_2E), (e) triangular pyramidal (type AB_3E). See Fig. 11-5 for the sketches.

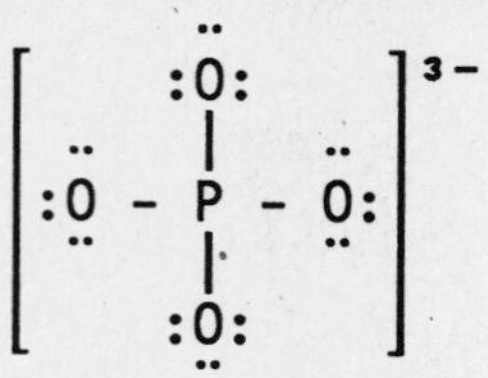

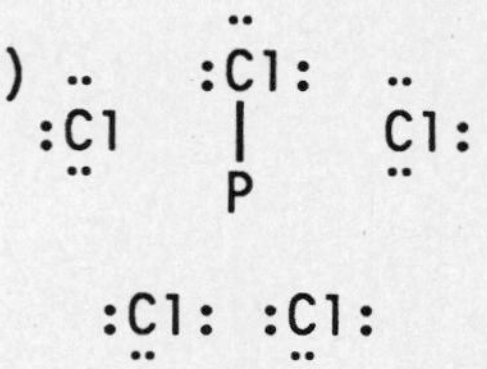

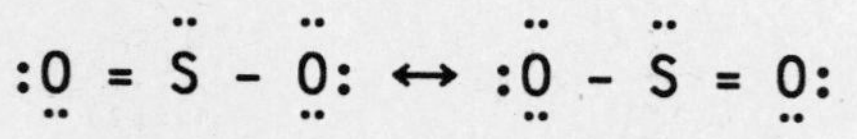

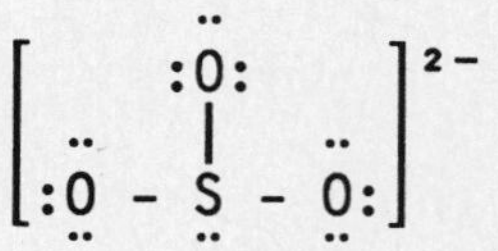

2. See Fig. 11-6.

3. (a) sp; (b) sp^3; (c) sp^3d^2; (d) sp^2 for C, sp^2 for O; (e) sp^2 for N, sp^2 for O.

4. (a) tetrahedral, London forces; (b) triangular pyramidal, dipole-dipole interactions and London forces; (c) distorted tetrahedron (disphenoidal), dipole-dipole interactions and London forces; (d) triangular pyramidal, dipole-dipole interactions and London forces; (e) tetrahedral, London forces.

5. (a) ii, (b) i, (c), iii, (d) ii, (e) i, (f) i, (g) iii, (h) iii, (i) i, (j) iii, (k) iii, (l) iv, (m) iii, (n) iv, (o) iv, (p) iv, (q) iii, (r) i.

ADDITIONAL NOTES

Fig. 11-5

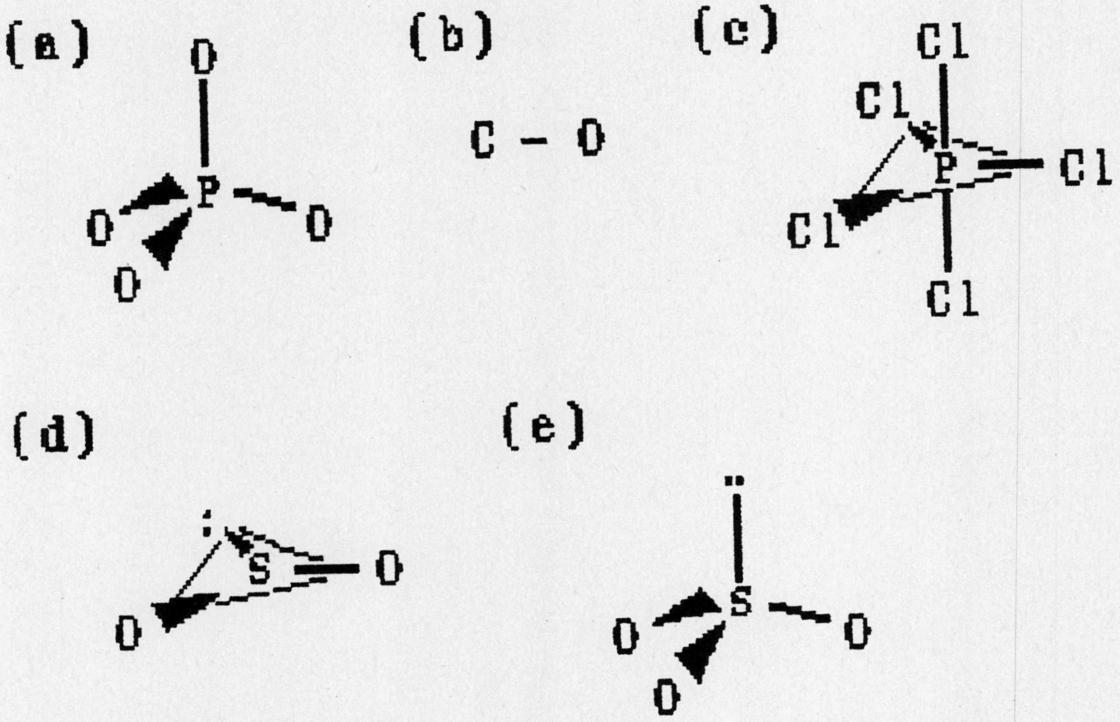

Fig. 11-6

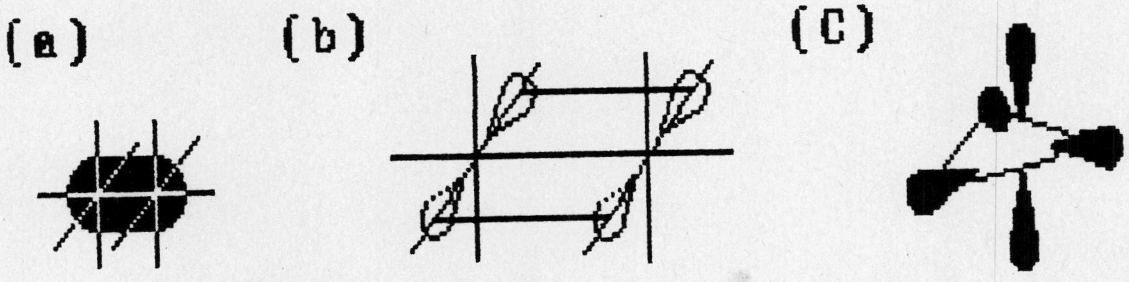

CHAPTER 12

NUCLEAR CHEMISTRY

CHAPTER OVERVIEW

This chapter considers the forces within the nucleus, the stability of nuclides, and the nuclear changes of radioactive isotopes (α, β, and γ decay). Radioisotope dating is discussed and a brief discussion of the interaction of radiation with matter is given. The various types of nuclear reactions (bombardment, nuclear fusion, and nuclear fission) are discussed. The chapter concludes with a discussion of nuclear energy--present and future.

COMPETENCIES

Definitions for the following terms should be learned:

- α decay
- β decay
- γ decay
- activity (radioactive)
- band of stability
- binding energy per nucleon
- bombardment reaction
- breeder reactor
- chain reaction
- critical mass
- electron capture
- half-life
- isomeric transition
- magic numbers
- mass defect
- nuclear binding energy
- nuclear chemistry
- nuclear fission
- nuclear force
- nuclear fusion
- nuclear reaction
- nuclear reactor
- nucleon
- nuclide
- positron
- radioactivity
- radionuclide
- thermonuclear reaction

General concepts and skills that should be learned include

- using binding energy and neutron-proton ratio to discuss the stability of a nuclide

- predicting modes of decay for nuclides based on the neutron-proton ratio
- writing nuclear equations
- determining and interpreting a decay scheme describing a decay process

Numerical exercises that should be understood include

- calculating the binding energy and the binding energy per nucleon for a nucleus
- calculating the energy involved in a nuclear reaction
- calculating the time, half-life, or amount of nuclide given two of the three variables

QUESTIONS

1. Complete the following sentences by filling in the various blanks with these terms: artificial isotopes, artificial radioactivity, natural isotopes, natural radioactivity, nucleons, nuclides, positron, radioactive isotopes, stable isotopes, and transuranium elements.

About 80% of the elements occur naturally as mixtures of isotopes. These are known as (a)_______ and can be either (b)_______ or (c)_______. About 33% of these elements have isotopes that show (d)_______. The (e)_______ are all (f)_______ and show (g)_______. The study of nuclear decay of certain (h)_______ showed that protons and neutrons, or (i)_______, were being transmuted into each other; for example, a proton became a neutron by the emission of a (j)_______ and a neutron became a proton by the emission of an electron.

2. Match the correct term to each definition.

(a) the emission of a helium nucleus by a radioactive nuclide	(i) α decay
(b) the emission of an electron or positron or the capture of an electron by a radioactive nuclide	(ii) β decay
(c) a nuclear reaction in which an inner shell electron is used to convert a proton to a neutron	(iii) chain reaction
(d) a heavy isotope splitting into two atoms of intermediate mass and several neutrons	(iv) half-life
(e) a reaction or series of reaction steps that initiates repetition of itself	(v) electron capture
	(vi) isomeric transition
	(vii) nuclear fission
	(viii) nuclear fusion
	(ix) nuclear reaction
	(x) thermonuclear reaction

(f) nuclear reactions at very high temperatures
(g) the time required for one-half of the nuclei in a radioactive sample to decay
(h) the decay of γ ray emission from an excited nuclide
(i) the combination of two light nuclei to give a heavier nucleus
(j) reactions that result in changes in the atomic number, mass number, or energy of nuclei

3. Which of the following statements are true? Rewrite each false statement so that it is correct.
(a) Radioactivity is the spontaneous emission of particles or electromagnetic radiation by unstable nuclei.
(b) Natural isotopes cannot be radioactive.
(c) At the end of one half-life 50% of a radioactive substance remains, and at the end of the second half-life the remaining 50% has undergone radioactive decay, leaving no original sample.
(d) Certain combinations of atomic numbers and neutron numbers (known as magic numbers) result in extremely stable nuclei.
(e) A chain reaction involves the emission of γ radiation from excited nuclei.
(f) A self-sustaining nuclear fission reaction generates a sufficient number of neutrons, equal to or larger than the number of neutrons absorbed by the fissioning nuclei and lost to the surroundings.
(g) The critical mass is the neutral mixture of ions and electrons at a high temperature undergoing a thermonuclear reaction.

4. Distinguish clearly between/among (a) α, β, and γ radiation; (b) fusion and fission; (c) binding energy and binding energy per nucleon; (d) natural and artificial radioactivity; (e) radioactive and stable isotopes; (f) positrons and electrons; (g) somatic and genetic effects; and (h) γ emission and isomeric transition.

5. Choose the response that best completes each statement.
(a) An excellent way to compare the stability of nuclides is by looking at the
(i) positron-neutron ratio (ii) critical mass
(iii) binding energy per nucleon (iv) nuclear force

(b) The energy that would be released in the combination of nucleons to form a nucleus is the
   (i) nuclear binding energy   (ii) band of stability
   (iii) magic number   (iv) thermonuclear energy

(c) A nuclear reactor that generates more fissionable fuel than it consumes is a
   (i) light water reactor   (ii) fusion reactor
   (iii) breeder reactor   (iv) heavy water reactor

(d) The smallest mass of a fissionable material that will support a self-sustaining chain reaction is
   (i) the magic number   (ii) binding energy per nucleon
   (iii) critical mass   (iv) stoichiometric amount

(e) The process in which a nucleus is struck by a moving particle is known as
   (i) electron capture   (ii) artificial radioactivity
   (iii) bombardment   (iv) isomeric transition

(f) The force of attraction or repulsion between charged particles is a/an
   (i) Coulombic force   (ii) nuclear force
   (iii) isomeric transition   (iv) critical mass

(g) Certain numbers of nucleons that produce stable nuclides are
   (i) neutron-proton ratios   (ii) critical masses
   (iii) magic numbers   (iv) half-lives

6. With what type of reactions does nuclear chemistry deal? What are the two aspects of nuclear chemistry that make it notably different from "ordinary" chemistry?

7. How is the name given and the symbol written for a particular isotope of an element?

8. What is the name of the force that attracts nucleons? What are the two general properties of this force?

9. What does the term "radioactivity" mean? What are radionuclides that are found on the earth called? What are radionuclides that are made only in the laboratory called?

10. Complete the table describing the changes in atomic number, neutron number, and mass number during the nuclear events listed

Event	ΔZ	ΔN	ΔA
α decay			
Electron emission			
Positron emission			
Electron capture			
Gamma decay			

11. What would be the change in the neutron-proton ratio following the decay of a radioactive nuclide that originally had too many protons? What would be the change for a nuclide having too many neutrons?

12. Briefly discuss the contributions to the understanding of nuclear chemistry made by (a) Antoine Henri Becquerel, (b) Marie and Pierre Curie, (c) Ernest Rutherford, and (d) Irène and Frédéric Joliot-Curie.

13. What are the four major types of nuclear reactions? Briefly describe the types of nuclei that undergo each type of reaction and the typical outcomes of these reactions.

14. Write the symbol that would be used in a nuclear equation for (a) an α particle, (b) a β particle, (c) a positron, (d) a proton, and (e) a neutron.

15. What are the two rules that are used to write a nuclear equation correctly?

16. What is a radioactive decay series? What is the end product of the three natural series?

17. What does the term "nuclear reactor" mean?

Answers to questions

1. (a) natural isotopes, (b) radioactive isotopes or stable isotopes, (c) stable isotopes or radioactive isotopes, (d) natural radioactivity, (e) transuranium elements, (f) artificial isotopes, (g) artificial radioactivity, (h) nuclides, (i) nucleons, (j) positron.
2. (a) i, (b) ii, (c) v, (d) vii, (e) iii, (f) x, (g) iv, (h) vi, (i) viii, (j) ix.
3. (a) T; (b) F, replace *cannot* with *can* and add *or stable*; (c) F, replace *the remaining ... no* with *50% of the remaining substance undergoes radioactive*

decay, leaving 25% *of the*; (d) T; (e) F, replace *A chain reaction* by *Isomeric transition*; (f) T; (g) F, replace *The critical mass* with *Plasma*.

4. (a) α *Radiation* is the emission of a helium nucleus by a radioactive nucleus; β *radiation* is the emission of an electron; γ *radiation* is the emission of a photon. (b) *Fusion* is the building up of intermediate nuclei from light nuclei; *fission* is the splitting up of heavy nuclei to form intermediate nuclei. (c) *Binding energy* is the total energy released in forming a nucleus as a result of the mass defect; *binding energy per nucleon* is the average binding energy on the basis of the number of nucleons present. (d) *Natural radioactivity* occurs in nature; *artificial radioactivity* is manmade. (e) *Radioactive isotopes* are unstable nuclides that undergo decay; *stable isotopes* are stable nuclides. (f) *Positrons* have positive electrical charge; *electrons* have negative electrical charge. (g) *Somatic effects* are limited to the individual organism; *genetic effects* are passed on to future generations. (h) γ *Emission* is the immediate release of γ radiation; *isomeric transition* is the delayed release.
5. (a) iii, (b) i, (c) iii, (d) iii, (e) iii, (f) i, (g) iii.
6. Nuclear chemistry deals with reactions in which changes in the nuclei occur. The two aspects are that mass-energy conversions and changes in the composition of the nucleus must be considered.
7. The name consists of the element name followed by the mass number. The symbol is the element symbol preceded by superscript for mass number and by subscript for atomic number.
8. The name of the force is the nuclear force. This force acts at very short range and is very strong.
9. Radioactivity is the spontaneous emission of particles and/or radiation by unstable nuclei. The terms are natural radionuclides and artificial radionuclides, respectively.
10. α decay: -2, -2, -4; electron emission: +1, -1, 0; positron emission: -1, +1, 0; electron capture: -1, +1, 0; gamma decay: 0, 0, 0.
11. The n/p ratio increases because Z decreases. The n/p ratio decreases because N decreases.
12. (a) Becquerel discovered radioactivity. (b) The Curies demonstrated the presence of and separated new radioactive elements. (c) Rutherford studied types of radiation and transmutation (with Soddy). (d) The Joliet-Curies produced artificial radioisotopes and aided in the discovery of the neutron.

13. The types are (1) spontaneous decay of radioactive nuclei to produce more stable nuclides that do not differ greatly in atomic number from the original nuclide; (2) bombardment reactions in which target nuclei are bombarded with others to give heavier nuclides (which, in turn, might undergo decay); (3) fission of heavy nuclides to give two nuclides of intermediate mass and some small fragments; and (4) fusion reactions in which light nuclides are fused into heavier nuclides.
14. The symbol that would be used in anuclear equation for (a) an α particle is ${}^{4}_{2}\alpha$ or ${}^{4}_{2}He$, (b) a β particle is ${}^{0}_{-1}\beta$ or ${}^{0}_{-1}e$, (c) a positron is ${}^{0}_{+1}\beta$ or ${}^{0}_{+1}e$, (d) a proton is ${}^{1}_{1}p$ or ${}^{1}_{1}H$, and (e) a neutron is ${}^{1}_{0}n$.
15. The equation must demonstrate the conservation of mass number and the conservation of atomic number.
16. A radioactive decay series consists of a long-lived parent and the series of nuclides down to the final stable isotope formed through a sequence of reactions, most of which are α or β particle emission. The end products of the three natural series are stable isotopes of lead.
17. A nuclear reactor is the equipment in which nuclear fission is carried out at a controlled rate.

SKILLS

Skill 12.1 Writing nuclear equations

When writing the equation describing a nuclear process, two criteria must be met: conservation of atomic number and conservation of mass number. For example, in the equation for the β^- emission of ${}^{3}_{1}H$,

$${}^{3}_{1}H \rightarrow {}^{0}_{-1}e + {}^{3}_{2}He$$

the sum of the atomic numbers is the same on both sides, 1 = (-1) + 2, and the sum of the mass numbers is the same on both sides, 3 = 0 + 3. Likewise, for the α decay of ${}^{204}_{86}Rn$,

$${}^{204}_{86}Rn \rightarrow {}^{4}_{2}He + {}^{200}_{84}Po$$

the sum of the mass numbers is the same, 204 = 4 + 200, and the sum of the atomic numbers is the same, 86 = 2 + 84. Similarly, for a bombardment reaction such as

$${}^{27}_{13}Al + {}^{1}_{0}n \rightarrow {}^{24}_{11}Na + {}^{4}_{2}He$$

the sums of the mass numbers are the same, 27 + 1 = 28 = 24 + 4, and the sums of the atomic numbers are the same, 13 + 0 = 13 = 11 + 2.

For bombardment processes, a shorthand notation that is often used to represent the reaction is

target nuclide(bombarding particle,product particle)product nuclide

For example, $^{14}_{7}N(\alpha,p)^{17}_{8}O$ corresponds to the nuclear equation

$$^{14}_{7}N + ^{4}_{2}He \rightarrow ^{1}_{1}H + ^{17}_{8}O$$

If one of the entries in the shorthand notation is missing, it can be deduced from the incomplete nuclear equation. For example, the product particle in the reaction $^{27}_{13}Al(\alpha,X)^{30}_{15}P$ has to be a neutron so that in the equation

$$^{27}_{13}Al + ^{4}_{2}He \rightarrow ^{30}_{15}P + ^{1}_{0}n$$

the mass and atomic numbers are conserved. Occasionally two or more product particles can be released, for example, $^{27}_{13}Al(n,2n)^{26}_{13}Al$.

Skill Exercise: Write nuclear equations for the following processes: (a) positron emission of $^{14}_{8}O$, (b) positron emission of $^{60}_{29}Cu$, (c) $^{60}_{29}Cu$ undergoing electron capture, (d) electron emission of $^{75}_{31}Ga$, (e) alpha emission of $^{204}_{87}Fr$, (f) $^{14}_{7}N(n,p)^{14}_{6}C$, (g) $^{6}_{3}Li(X,\alpha)^{3}_{1}H$ and identify X, and (h) $^{189m}_{76}Os$ undergoing γ emission. *Answers*: (a) $^{14}_{8}O \rightarrow ^{0}_{+1}e + ^{14}_{7}N$; (b) $^{60}_{29}Cu \rightarrow ^{0}_{+1}e + ^{60}_{28}Ni$; (c) $^{60}_{29}Cu + ^{0}_{-1}e \rightarrow ^{60}_{28}Ni$; (d) $^{75}_{31}Ga \rightarrow ^{0}_{-1}e + ^{75}_{32}Ge$; (e) $^{204}_{87}Fr \rightarrow ^{4}_{2}He + ^{200}_{85}At$; (f) $^{14}_{7}N + ^{1}_{0}n \rightarrow ^{1}_{1}H + ^{14}_{6}C$; (g) $^{6}_{3}Li + ^{1}_{0}n \rightarrow ^{4}_{2}He + ^{3}_{1}H$, X = n; (h) $^{189m}_{76}Os \rightarrow ^{189}_{76}Os + \gamma$.

Skill 12.2 Predicting modes of decay for radioactive nuclei

Nuclides having a neutron-proton ratio that is higher than that of the stable isotopes will decay by electron or β^- emission (in which a neutron essentially changes to a proton):

$$^{1}_{0}n \rightarrow ^{1}_{1}H^+ + ^{0}_{-1}e$$

Lighter nuclides having a low neutron-proton ratio will usually decay by positron or β^+ emission (in which a proton essentially changes to a neutron):

$$^{1}_{1}H^+ \rightarrow ^{1}_{0}n + ^{0}_{+1}e$$

while heavier nuclides either undergo electron capture to effect the same change:

$$^{1}_{1}H^+ + ^{0}_{-1}e \rightarrow ^{1}_{0}n$$

or undergo α emission to increase the neutron-proton ratio. Nuclei with too much energy will emit the energy in the form of γ rays.

Thus tritium, $^{3}_{1}H$, which has a neutron-proton ratio of 2, would be predicted to undergo β⁻ emission:

$$^{3}_{1}H \rightarrow {}^{0}_{-1}e + {}^{3}_{2}He$$

$^{20}_{12}Mg$ would be predicted to undergo β⁺ emission (0.67 ratio compared to the stable value of 1.08):

$$^{20}_{12}Mg \rightarrow {}^{0}_{+1}e + {}^{20}_{11}Na$$

$^{77m}_{34}Se$ would be predicted to undergo γ emission to the stable $^{77}_{34}Se$ nucleus:

$$^{77m}_{34}Se \rightarrow \gamma + {}^{77}_{34}Se$$

$^{144}_{63}Eu$ would be predicted to undergo either β⁺ emission or electron capture (1.29 ratio compared to the stable 1.41 value):

$$^{144}_{63}Eu \rightarrow {}^{0}_{+1}e + {}^{144}_{62}Sm \quad \text{or} \quad {}^{144}_{63}Eu + {}^{0}_{-1}e \rightarrow {}^{144}_{62}Sm$$

and $^{204}_{86}Rn$ to undergo α emission (1.37 ratio compared to the more stable 1.45 value):

$$^{204}_{86}Rn \rightarrow {}^{4}_{2}He + {}^{200}_{84}Po$$

Skill Exercise: Calculate the neutron-proton ratio, compare the value to the given value for the more stable isotopes, and predict the mode of decay for each of the following nuclides: (a) $^{14}_{8}O$, 1.000; (b) $^{60}_{29}Cu$, 1.207; (c) $^{75}_{31}Ga$, 1.258; (d) $^{204}_{87}Fr$, 1.437; (e) $^{246}_{94}Pu$, 1.574; and (f) $^{189m}_{76}Os$. *Answers*: (a) 0.750, β⁺; (b) 1.069, β⁺ or EC; (c) 1.419, β⁻, (d) 1.345, α; (e) 1.617, β⁻; (f) γ emission.

Skill 12.3 Determining decay schemes for radioactive processes

A decay scheme is a sketch of energy against atomic number showing the nuclides and particles involved in the nuclear process. Usually these are drawn by trial and error from known experimental values of the energies of the particles being emitted, but with the restriction that the energy along any path must be conserved.

For example, as $^{186}_{75}Re$ undergoes β⁻ emission, β⁻ particles having energies of 0.30 MeV, 0.933 MeV, and 1.07 MeV are observed and γ rays having energies of 0.13716 MeV, 0.632 MeV, and 0.768 MeV are also observed. The decay scheme is shown in Fig. 12-1. An interpretation of the diagram shows that $^{186}_{76}Re$ undergoes emission to $^{186}_{76}Os$ by emitting either a 1.071 MeV β⁻ particle, a 0.933 MeV β⁻ particle followed by a 0.13716 MeV γ ray, or a 0.30 MeV β⁻ particle followed by either a 0.768 MeV γ ray or 0.632 and 0.13716 MeV γ rays.

Fig. 12-1

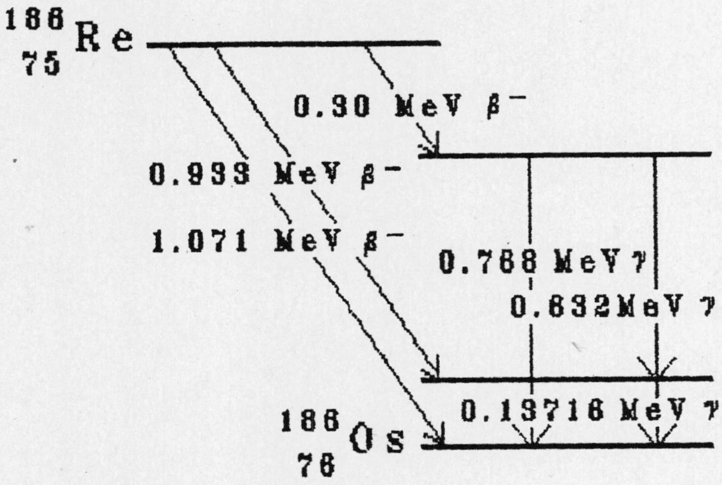

Skill Exercise: Prepare decay schemes for the following processes: (a) β^- decay of $^{39}_{17}Cl$ in which β^- energies of 3.45, 2.18, and 1.91 MeV and energies of 0.246, 1.27, and 1.52 MeV are observed; (b) $^{73m}_{32}Ge$ undergoing γ emission by emitting 0.054 and 0.0135 MeV γ rays; and (c) α decay of $^{224}_{90}Th$ in which 7.17, 6.99, 6.77, and 6.70 MeV α particles are emitted along with 0.177, 0.235, 0.297, and 0.410 MeV γ rays. *Answers*: See Fig. 12-2.

NUMERICAL EXERCISES

*Example 12.1 Nuclear binding energy**

The atomic mass of $^{223}_{90}Th$ is 223.0209 u. Calculate the binding energy for this nuclide.

This problem is solved by using the comprehensive problem-solving approach presented in *Skill 2.7 Solving problems.*

1. Study the problem and be sure you understand it.
   (a) What is unknown?
   The binding energy.
   (b) What is known?
   The mass of the atom and the masses of the neutrons, protons, and electrons that make up the atom.
2. Decide how to solve the problem.
   (a) What is the connection between the known and the unknown?
   The binding energy of a nuclide is the energy equivalent to the mass

Fig. 12-2

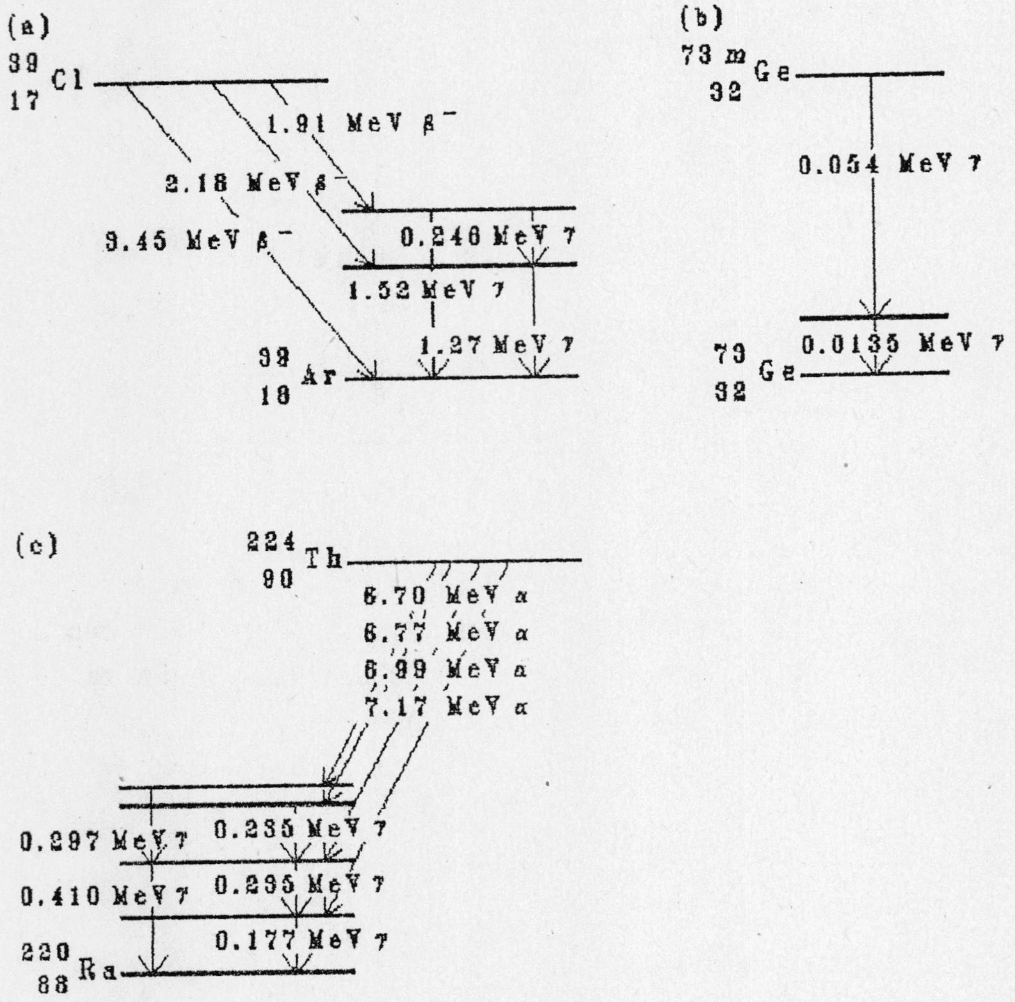

defect--the difference between the mass of the atom and the sum of the masses of the individual constituents of the atom.

(b) What is necessary to make the connection?

Determine the number of protons, neutrons, and electrons present in one atom, then calculate the mass defect and use $E = mc^2$ to find energy.

3. Set up the problem and solve it.

For this nuclide Z = 90, N = 133, and the number of electrons is 90 (see *Examples 3.1 and 3.2*). The mass defect is

mass defect = (nuclide mass) - (Z)(proton mass) + (Z)(electron mass) + (N)(neutron mass)

$$\text{mass defect} = (223.0209\ \text{u}) - [(90\ \text{protons})(1.007276\ \text{u/1 proton}) + (90\ \text{electrons})(0.00054858\ \text{u/1 electron}) + (133\ \text{neutrons})(1.008665\ \text{u/1 neutron})]$$
$$= -1.8358\ \text{u}$$

Converting this value to kilograms yields

$$(-1.8358\ \text{u})\left[\frac{1.6605655 \times 10^{-27}\ \text{kg}}{1\ \text{u}}\right] = -3.0485 \times 10^{-27}\ \text{kg}$$

and to energy gives the binding energy as

$$E = mc^2 = (-3.0485 \times 10^{-27}\ \text{kg})(2.9979 \times 10^{8}\ \text{m/s})^2\left[\frac{1\ \text{J}}{1\ \text{kg m}^2/\text{s}^2}\right]$$
$$= -2.7398 \times 10^{-10}\ \text{J}$$

4. Check the result.
   (a) Are significant figures and the location of the decimal point correct?
   Yes.
   (b) Did the answer come out in the correct units?
   Yes.
   (c) Is the answer reasonable?
   Yes.

Exercise 12.1 Calculate the binding energy for $^{230}_{90}Th$. The atomic mass of this nuclide is 230.0331 u. *Answer*: -2.8120×10^{-10} J.

Example 12.2 Average binding energy

Calculate the binding energy per nucleon in $^{223}_{90}Th$.

The total binding energy for $^{223}_{90}Th$ is -2.7398×10^{-10} J (see *Example 12.1*). The average binding energy is

$$\frac{-2.7398 \times 10^{-10}\ \text{J}}{223\ \text{nucleons}} = -1.2286 \times 10^{-12}\ \text{J/nucleon}$$

Exercise 12.2 Calculate the average binding energy in $^{230}_{90}Th$. See Exercise 12.1 for the total binding energy. *Answer*: -1.2226×10^{-12} J/nucleon.

Example 12.3 Half-life*

The half-life of $^{161}_{65}Tb$ is 6.9 days. What fraction of a radioactive sample of this isotope would remain after 10.0 days?

This problem is solved by using the comprehensive problem-solving approach presented in *Skill 2.7 Solving problems*.

1. Study the problem and be sure you understand it.
   (a) What is unknown?
   The fraction of nuclide remaining, (q/q_0).
   (b) What is known?
   The half-life and the time of decay.
2. Decide how to solve the problem.
   (a) What is the connection between the known and the unknown?
   The fraction of a nuclide remaining after a given period of time is dependent upon the rate constant for the specific reaction. The rate constant is related to the half-life.
   (b) What is necessary to make the connection?
   The rate constant is calculated from the known half-life value by using

$$k = \frac{0.693}{t_{1/2}}$$

   and the fraction remaining is determined from the value of k and the time by using

$$\log\left[\frac{q_0}{q}\right] = \frac{kt}{2.303}$$

3. Set up the problem and solve it.
   The nuclear decay rate constant is

$$k = \frac{0.693}{t_{1/2}} = \frac{0.693}{6.9\ \text{day}} = 0.10\ \text{day}^{-1}$$

   The fraction of isotope left after 10.0 days is

$$\log\left[\frac{q_0}{q}\right] = \frac{kt}{2.303} = \frac{(0.10\ \text{day}^{-1})(10.0\ \text{day})}{2.303} = 0.43$$

$$\frac{q_0}{q} = 2.7$$

$$\frac{q}{q_0} = 0.37$$

   Only 37 % of the isotope remained.
4. Check the result.
   (a) Are significant figures and the location of the decimal point correct?
   Yes.

(b) Did the answer come out in the correct units?

Yes. The quantity (q/q_0) is a dimensionless quantity. Note that units of time with a -1 sign are often used in these types of calculations ($1\ day^{-1} = 1/day$).

(c) Is the answer reasonable?

Yes. At the end of one half-life period, 50% of the original sample remains; at the end of two half-life periods, 25% of the original sample remains; etc. The half-life of this nuclide is 6.9 days and the time period of decay is 10.0 days--between one and two half-life periods. Thus 25% to 50% of the sample should be remaining.

Exercise 12.3 The half-life of $^{162}_{65}Tb$ is 7.5 min. What fraction of a sample remains at the end of 1.0 h? *Answer*: 0.3 %.

Example 12.4 Half-life

A sample of $^{200}_{82}Pb$ was monitored for a period of 1.59 h. During this time, exactly 5 % of the radioactive lead underwent decay. What is the half-life of this nuclide?

At t = 1.59 h, 95% of the sample remained unchanged. Thus $q = (0.95)q_0$. Solving for the nuclear decay rate constant gives

$$k = \frac{(2.303)\ \log\left[\frac{q_0}{q}\right]}{t} = \frac{(2.303)\ \log\left[\frac{q_0}{(0.95)q_0}\right]}{1.59\ \text{h}}$$

$$= 0.0323\ \text{h}^{-1}$$

so the half-life is

$$t_{1/2} = \frac{0.693}{k} = \frac{0.693}{0.0323\ \text{h}^{-1}} = 21.5\ \text{h}$$

Exercise 12.4 A sample of $^{137}_{55}Cs$ was monitored for 160 days. During this time exactly 1 % of the radioactive cesium underwent decay. What is the half-life of this nuclide? *Answer*: 1.1×10^4 day or 30. yr.

Example 12.5 Half-life

Calculate the age of a sample of glacial wood found to have a ^{14}C activity of 2.73 $min^{-1}\ (g\ C)^{-1}$, disintegrations per minute per gram of carbon. The half-life of ^{14}C is 5730 yr and the current ^{14}C activity is 15.3 disintegration per minute per gram of carbon.

The rate constant for the decay of ^{14}C is

$$k = \frac{0.693}{t_{1/2}} = \frac{0.693}{5730\ \text{yr}} = 1.21 \times 10^{-4}\ \text{yr}^{-1}$$

The equation relating the activity of radioactive decay, a, to k and t is of the same form as for q with k and t. Substituting $a = 2.73\ \text{min}^{-1}\ (\text{g C})^{-1}$, $a_0 = 15.3\ \text{min}^{-1}\ (\text{g C})^{-1}$, and $k = 1.21 \times 10^{-4}\ \text{yr}^{-1}$ gives

$$t = \frac{(2.203)\ \log\left[\frac{a_0}{a}\right]}{k} = \frac{(2.303)\ \log\left[\frac{15.3\ \text{min}^{-1}\ (\text{g C})^{-1}}{2.73\ \text{min}^{-1}\ (\text{g C})^{-1}}\right]}{1.21 \times 10^{-4}\ \text{yr}^{-1}}$$
$$= 14{,}200\ \text{yr}$$

The wood is 14,200 yr old.

Exercise 12.5 The β^- activity of carbon from the Dead Sea scrolls was measured as $12.0\ \text{min}^{-1}\ (\text{g C})^{-1}$. If the activity of modern carbon is $15.3\ \text{min}^{-1}\ (\text{g C})^{-1}$, how long ago were these scrolls written? *Answer*: 2010 yr ago.

Example 12.6 Energy of nuclear reaction

Calculate the energy released in the fusion process

$$^{3}_{2}\text{He} + ^{2}_{1}\text{H} \rightarrow ^{4}_{2}\text{He} + ^{1}_{1}\text{H}$$

for each fusion event and per mole of reactants. The respective nuclide masses are 3.01603 u, 2.0140 u, 4.00260 u, and 1.007825 u.

The change in mass for the fusion process is

$$\begin{aligned}\text{mass change} &= (\text{mass of products}) - (\text{mass of reactants}) \\ &= (4.00260\ \text{u} + 1.007825\ \text{u}) - (3.01603\ \text{u} + 2.0140\ \text{u}) \\ &= -0.0196\ \text{u}\end{aligned}$$

which corresponds to

$$(-0.9196\ \text{u})\left[\frac{1.661 \times 10^{-27}\ \text{kg}}{1\ \text{u}}\right] = -3.26 \times 10^{-29}\ \text{kg}$$

According to the Einstein mass-energy relationship, the energy equivalent to this mass is

$$\begin{aligned}E &= mc^2 \\ &= (-3.26 \times 10^{-29}\ \text{kg})(3.00 \times 10^{8}\ \text{m/s})^2\left[\frac{1\ \text{J}}{1\ \text{kg m}^2/\text{s}^2}\right] \\ &= -2.93 \times 10^{-12}\ \text{J}\end{aligned}$$

The energy released by the fusion process is 2.93×10^{-12} J. For one mole of each of the reactants, the energy released is

$$(-2.93 \times 10^{-12}\ \text{J})(6.022 \times 10^{23}\ \text{nuclei/1 mol}) = -1.76 \times 10^{12}\ \text{J}$$

Exercise 12.6 Calculate the energy released for the fusion process

$$^{3}_{2}He + ^{3}_{1}H \rightarrow ^{4}_{2}He + ^{2}_{1}H$$

for each fusion event. The nuclide mass of tritium is 3.01605 u. *Answer*: -2.32×10^{-12} J.

PRACTICE TEST

1 (10 points). Write nuclear equations showing the nuclide $^{A}_{Z}X$ undergoing each of the following processes: (a) electron capture, (b) positron decay, (c) electron decay, (d) α decay, and (e) γ decay.

2 (15 points). Calculate the neutron-proton ratio, compare the value to the value given for more stable isotopes, predict the mode of decay, and write the nuclear equation for the decay process for each of the following nuclides: (a) $^{8}_{2}He$, 1.00; (b) $^{15}_{8}O$, 1.00; and (c) $^{230}_{92}U$, 1.533.

3 (15 points). Write the nuclear equation and identify X for each of the following transmutations: (a) $^{238}_{92}U(\alpha,3n)X$, (b) $^{238}_{92}U(X,5n)^{249}_{100}Fm$, and (c) $^{7}_{3}Li(p,X)^{4}_{2}He$.

4 (10 points). The actual masses of $^{210}_{84}Po$, $^{4}_{2}He$, and $^{206}_{82}Pb$ are 209.9828 u, 4.00260 u, and 205.9745 u, respectively. Calculate the energy given off as $^{210}_{84}Po$ undergoes α decay.

5 (10 points). The actual masses of $^{10}_{4}Be$, $^{11}_{4}Be$, $^{1}_{1}H^{+}$, e^{-}, and $^{1}_{0}n$ are 10.0135 u, 11.0216 u, 1.007276 u, and 0.00054858 u, and 1.008665 u, respectively. Calculate the binding energy per nucleon for each isotope of Be. Is there a significant difference in the stability of these nuclides?

6 (10 points). During the decomposition of $^{15}_{6}C$, the following particles are observed: 9.82 MeV β^{-}, 4.51 MeV β^{-}, and 5.299 MeV γ. Prepare a decay scheme describing the β^{-} decay of $^{15}_{6}C$.

7 (10 points). The activity of Indian artifacts from a dig in Illinois had a ^{14}C activity of 4.1 min^{-1} $(g\ C)^{-1}$. The half-life of ^{14}C is 5730 yr and the current ^{14}C activity is 15.3 min^{-1} $(g\ C)^{-1}$. How old is the artifact?

8 (20 points). Choose the response that best completes each statement.

(a) Which has the greatest rest mass?

(i) proton (ii) electron

(iii) alpha particle (iv) neutron

(b) The force holding nucleons in the nucleus is the
(i) frictional force (ii) Coulombic force
(iii) nuclear force (iv) band of stability

(c) If 1 mg of a radioactive sample is left after a 4 mg sample is allowed to decay for 60 min, what is the half-life?
(i) 15 min (ii) 30 min
(iii) 20 min (iv) 60 min

(d) A mole of uranium has a mass of 238 g. The average mass of an atom of uranium is 238 u. What is the conversion factor between atomic mass units and grams?
(i) 1 u/1 g (ii) 1 u/(1 g/mol)
(iii) (1.661×10^{-24} g/1 u) (iv) none of these answers

(e) Elements with low atomic mass undergo
(i) fusion processes (ii) α decay
(iii) fission processes (iv) none of these answers

(f) Which of the following decay processes change the atomic number?
(i) electron capture (ii) β^- emission
(iii) β^+ emission (iv) all of the above

(g) The most stable nuclide with A = 16 is expected to be
(i) ^{16}N (ii) ^{16}C
(iii) ^{16}O (iv) ^{16}F

(h) The most stable nuclide of mass number 101 is Ru. The mode of decay of ^{101}Mo is
(i) β^+ emission (ii) γ emission
(iii) β^- emission (iv) electron capture

(i) Identify the missing substance, X, in the reaction $^{6}Li(n,\alpha)X$.
(i) ^{7}Be (ii) T
(iii) ^{7}Li (iv) α

(j) Identify the missing substance, X, in the equation

$$^{238}_{92}U \rightarrow\ ^{234}_{90}Th + X$$

(i) $2p$ (ii) $2n$
(iii) α (iv) none of these answers

Answers to practice test

1. (a) $^{A}_{Z}X +\ ^{0}_{-1}e \rightarrow\ ^{A}_{Z-1}X$, (b) $^{A}_{Z}X \rightarrow\ ^{0}_{+1}e +\ ^{A}_{Z-1}X$, (c) $^{A}_{Z}X \rightarrow\ ^{0}_{-1}e +\ ^{A}_{Z+1}X$, (d) $^{A}_{Z}X \rightarrow\ ^{4}_{2}He +\ ^{A-4}_{Z-2}X$, (e) $^{Am}_{Z}X \rightarrow\ ^{A}_{Z}X + \gamma$.

Fig. 12-3

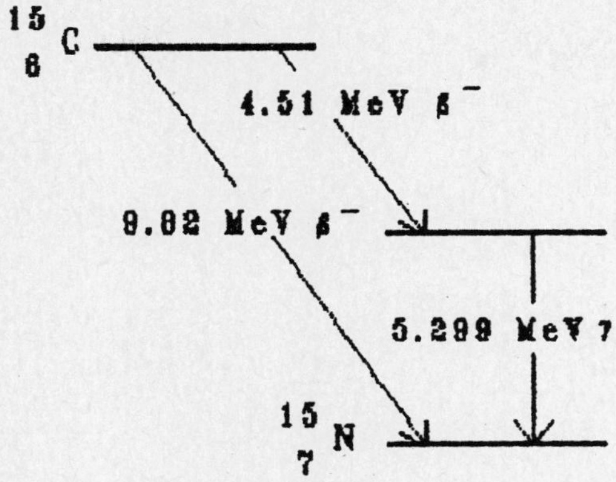

2. (a) 3.00, β^-, ${}^{8}_{2}He \rightarrow {}^{0}_{-1}e + {}^{8}_{3}Li$; (b) 0.875, β^+, ${}^{15}_{8}O \rightarrow {}^{0}_{+1}e + {}^{15}_{7}N$; (c) 1.500, α, ${}^{230}_{92}U \rightarrow {}^{4}_{2}He + {}^{226}_{90}Th$.
3. (a) ${}^{238}_{92}U + {}^{4}_{2}He \rightarrow 3{}^{1}_{0}n + {}^{239}_{94}Pu$, X = ${}^{239}_{94}Pu$; (b) ${}^{238}_{92}U + {}^{16}_{8}O \rightarrow 5{}^{1}_{0}n + {}^{249}_{100}Fm$, X = ${}^{16}_{8}O$; (c) ${}^{7}_{3}Li + {}^{1}_{1}H \rightarrow 2{}^{4}_{2}He$, X = α.
4. -8.5×10^{-13} J.
5. -1.04×10^{-12} J/nucleon, -9.55×10^{-13} J/nucleon, ${}^{10}B$ is more stable.
6. See Fig. 12-3.
7. 11,000 yr.
8. (a) iii, (b) iii, (c) ii, (d) iii, (e) i, (f) iv, (g) iii, (h) iii, (i) ii, (j) iii.

ADDITIONAL NOTES

CHAPTER 13

LIQUID AND SOLID STATES; CHANGES OF STATE

CHAPTER OVERVIEW

The chapter begins with a discussion of the application of the kinetic-molecular theory to solids and liquids. Some general properties of liquids and the solid state are discussed. The relationships among the gaseous, liquid, and solid phases are presented. The topic of crystal structure is explored and the internal arrangements of several types of solids are presented.

COMPETENCIES

Definitions for the following terms should be learned:

- amorphous solid
- boiling point
- condensation, liquefaction
- coordination number (crystal)
- critical point
- critical pressure
- critical temperature
- crystal
- crystal structure
- crystalline solid
- cubic closest packing
- evaporation
- fusion
- hexagonal closest packing
- isomorphous
- lattice energy
- melting point
- multiple unit cell
- nonstoichiometric compound
- normal boiling point
- normal freezing point
- polymorphous
- primitive unit cell
- space lattice
- sublimation
- supercooled
- superheated
- surface tension
- theoretical density
- triple point
- unit cell
- vaporization
- viscosity

General concepts and skills that should be learned include

- interpreting the kinetic-molecular theory for solids and liquids
- the general properties of liquids and solids
- interpreting a phase diagram for a substance
- using the unit cell to describe crystalline compounds
- predicting the type of unit cell that an ionic substance might form

Numerical exercises that should be understood include

- calculating the heat of vaporization, vapor pressure, or temperature when all but one of these variables are given
- calculating the size of an atom or ion from unit cell dimensions
- determining the unit cell content
- calculating the efficiency of packing in a unit cell
- calculating the theoretical density of a crystalline material
- calculating lattice energies by using the Born-Haber cycle

QUESTIONS

1. Complete the following sentences by filling in the various blanks with these terms: amorphous solid, crystal, crystalline solid, isomorphic, polymorphic, space lattice, and unit cell.

In a (a)______ the atoms, ions, or molecules are in a characteristic, regular repetitive three-dimensional arrangement and in an (b)______ they are in a random nonrepetitive three-dimensional arrangement. A solid that has a shape bounded by plane surfaces intersecting at fixed angles is a (c)______. A system of points representing the sites with identical environments occupied by the particles in a crystal is known as a (d)______. A (e)______ is the part of a space lattice that, if repeated in three dimensions, will generate the entire lattice. Substances that have the same crystal structure are said to be (f)______. If a particular substance crystallizes in more than one crystal system, it is said to be (g)______.

2. Match the correct term to each definition.

(a) a homogeneous part of a system in contact with but separate from the other parts of the system

(b) the escape of molecules from the liquid or solid phase to the gas phase

(i) boiling point
(ii) condensation
(iii) critical point
(iv) critical pressure
(v) critical temperature

(c) the escape of molecules from a liquid in an open container to the gaseous phase

(d) the movement of molecules from the gaseous phase to the liquid phase

(e) the vaporization of a solid followed by the condensation of the vapor to form the solid (or the solid-gas transformation only)

(f) the temperature at which the vapor pressure of a liquid equals the pressure of the gases above the liquid

(g) the temperature at which the vapor pressure of a liquid equals 760 Torr

(h) raised to a temperature above which a phase transformation should have occurred

(i) the temperature at which the solid and liquid phases of a substance are at equilibrium

(j) the phase change from a solid to a liquid

(k) the point above which no amount of pressure is great enough to cause liquefaction of a gas

(l) the point at which three phases exist under dynamic equilibrium conditions

(m) cooled to a temperature below which a phase transformation should have occurred

(n) the temperature at which a liquid changes to a solid at 760 Torr pressure

(o) the pressure that will cause liquefaction of a gas at the critical temperature

(p) the temperature at the critical point

(q) the amount of heat needed to convert one mole of a liquid substance to its vapor

(r) the amount of heat needed to convert one mole of a substance from the solid to the liquid state

(vi) evaporation
(vii) fusion
(viii) melting point
(ix) molar heat of fusion
(x) molar heat of vaporization
(xi) normal boiling point
(xii) normal freezing point
(xiii) phase
(xiv) sublimation
(xv) supercooled
(xvi) superheated
(xvii) triple point
(xviii) vaporization

3. Which of the following statements are true? Rewrite each false statement so that it is correct.
   (a) Gases are considered to have very little ordering, liquids to have short-range ordering, and solids to have long-range ordering.
   (b) All of the statements of the kinetic-molecular theory of gases are applicable to the liquid and solid states.
   (c) The constant vaporization and condensation taking place once equilibrium has been established between a liquid and its vapor is an example of static equilibrium.
   (d) A phase change is a constant pressure and constant temperature process.
   (e) Covalent bonds connecting atoms are the major interparticle forces in network covalent materials.
   (f) The Bragg equation describes the reflection of x-rays by a crystal.
   (g) The space lattice, if repeated in three dimensions, will generate the entire crystal.

4. Distinguish clearly between (a) the boiling point and the normal boiling point, (b) vaporization and evaporation, (c) a superheated and a supercooled liquid, (d) the triple and critical points, (e) a crystalline and an amorphous solid, (f) anisotropic and isotropic solids, (g) isomorphism and polymorphism, and (h) evaporation and boiling.

5. Choose the response that best completes each statement.
   (a) The number of nearest neighbors of a particle in a particular crystal structure is the
      (i) crystal coordination number (ii) critical point
      (iii) primitive unit cell (iv) crystal lattice
   (b) The complete geometric arrangement of the particles that occupy the space lattice is the
      (i) unit cell (ii) crystal structure
      (iii) refractive index (iv) network covalency
   (c) The arrangement of closest-packed layers of metal atoms in an ABAB... sequence is
      (i) the crystal coordination number (ii) cubic closest packing
      (iii) hexagonal closest packing (iv) a primitive unit cell
   (d) A unit cell in which only the corners are occupied is
      (i) hexagonal closest packing (ii) isomorphic
      (iii) a primitive cell (iv) multiple unit cell

(e) The property of a surface that imparts membrane-like behavior to the surface of a liquid is
   (i) the critical pressure (ii) the surface tension
   (iii) supercooling (iv) the refractive index

(f) The resistance of a liquid to flow is
   (i) sintering (ii) fusion
   (iii) the viscosity (iv) isomorphism

(g) The energy liberated as gaseous ions combine to give a crystalline ionic substance is the
   (i) first ionization energy (ii) lattice energy
   (iii) heat of formation (iv) electron affinity

6. Briefly discuss how the kinetic-molecular theory applies to solids and liquids.

7. If you pour ether on your hand, your hand will feel cold. Explain this effect in terms of the kinetic-molecular theory.

8. Name five general types of crystalline solids. For each type, identify the particles that are present in the crystal lattice and the interparticle forces among these particles.

9. A brown bag is filled with water and placed directly over the flame of a Bunsen burner. Although some surface blackening occurs, the bag does not burn while the water boils inside. Why?

10. Define the term "vaporization." How does vaporization depend on temperature? Why?

11. Name the changes of state that correspond to (a) a solid vaporizing, (b) a liquid changing to a solid, and (c) a solid changing to a liquid.

12. Choose the variable(s) that influence the vapor pressure over a substance: (a) ratio of liquid or solid to empty space, (b) temperature, (c) pressure of air above liquid, and (d) composition of gas above liquid.

13. Name the changes of state that are the reverse of (a) vaporization of a liquid, (b) fusion, and (c) sublimation.

14. What is "superheating"? How can we avoid superheating a liquid?

15. What would happen to the pressure of an ideal gas as the volume is decreased? What would happen to the pressure of a vapor that is in equilibrium with

liquid or solid as the volume is decreased? Why is there a difference in the answers?

16. How does the vapor pressure depend on the intermolecular forces for various substances?

17. Which liquid has the higher vapor pressure at a given temperature, methyl alcohol (CH_3OH) or ethyl alcohol (CH_3CH_2OH)? Explain your answer.

18. How is the heat of vaporization for a substance determined from a plot of the logarithm of vapor pressure versus the inverse of the absolute temperature?

19. What types of information can be obtained from a phase diagram for a substance? What do each of the curves on a phase diagram represent?

20. What are the names of the two points at which the vapor pressure-temperature curve for a liquid terminates on a phase diagram?

21. Most chemists are careful to report the exact pressure at which a boiling point is measured. Yet, chemists seldom worry about this for melting points. Why?

22. What is a solid with a characteristic, regular, and repetitive three-dimensional arrangement of the atoms, molecules, or ions called? What term is used for a solid in which the atoms, molecules, or ions have a random and nonrepetitive three-dimensional arrangement?

23. What is a space lattice? How does this differ from a crystal structure?

24. Draw several equilateral triangles so that they fit together as tightly as possible. Do the same for squares, hexagons, and octagons. Are all equally close packed?

25. What is the contribution of a corner atom to the number of atoms in a cubic unit cell? What is the contribution of a body-centered atom? What is the contribution of a face-centered atom?

26. How is the mass of a unit cell calculated? How is the theoretical density of a crystalline substance calculated?

27. Briefly discuss how the following factors influence ionic crystal structure: (a) the ratio of the numbers of cations and anions, (b) the location of the

anions with respect to the cations, and (c) the location of the anions with respect to each other (or cations with each other).

28. What types of thermochemical equations usually appear in a Born-Haber cycle? Which of these represent endothermic reactions?

Answers to questions

1. (a) crystalline solid, (b) amorphous solid, (c) crystal, (d) space lattice, (e) unit cell, (f) isomorphic, (g) polymorphic.
2. (a) xiii, (b) xviii, (c) vi, (d) ii, (e) xiv, (f) i, (g) xi, (h) xvi, (i) viii, (j) vii, (k) iii, (l) xvii, (m) xv, (n) xii, (o) iv, (p) v, (q) x, (r) ix.
3. (a) T; (b) F, change *All* to *Several*; (c) F, change *static* to *dynamic*; (d) T; (e) T; (f) T; (g) F, replace *space lattice* with *unit cell*.
4. (a) The *boiling point* of a liquid is the temperature at which the vapor pressure of the liquid equals the atmospheric pressure; the *normal boiling point* is the boiling point at 1 atm pressure. (b) *Vaporization* is the liquid to gas transformation in a closed system; *evaporation* is vaporization in an open container resulting in complete vaporization. (c) A *superheated liquid* is a liquid at a temperature above which boiling should occur; a *supercooled liquid* is a liquid at a temperature below which freezing should occur. (d) The *triple point* is the set of pressure-volume-temperature conditions such that three phases are in equilibrium; the *critical point* is the set of pressure-temperature-volume conditions that last define the liquid state for a substance. (e) A *crystalline solid* has a regular repetitive pattern of atoms, molecules, or ions; an *amorphous solid* has no regular repetitive pattern. (f) *Anisotropic* refers to properties of a crystal that differ in different directions along the solid; *isotropic* refers to properties that are the same in all directions. (g) *Isomorphism* is the property of two or more substances having the same crystal form; *polymorphism* is the property of one substance having two or more crystal forms. (h) *Boiling* is vaporization with bubble formation at a pressure equal to the atmospheric pressure; *evaporation* is vaporization below the boiling point.
5. (a) i, (b) ii, (c) iii, (d) iii, (e) ii, (f) iii, (g) ii.
6. The molecules are in constant motion, although restricted in movement. The energy of the molecules is proportional to temperature. The spaces between molecules are not large compared to the size of the molecules.

7. Your hand will feel cold when you pour ether on your hand because as it evaporates, the highest energy molecules escape, leaving behind a collection of molecules with a lower average kinetic energy and, therefore, a lower temperature.
8. The types are (1) pure metals in which cations are held by metallic bonding; (2) ionic solids in which cations and anions are held by ionic electrostatic attraction; (3) alloys, which are often like pure metals; (4) molecular crystals in which molecules are held by van der Waals forces; and (5) network covalent substances in which atoms held by covalent bonding.
9. The temperature of the bag and water will not rise above the boiling pint until all of the water evaporates. The paper bag will not ignite at such a low temperature.
10. Vaporization is the escape of molecules near the surface of a solid or liquid into the gaseous phase. Vaporization increases as temperature increases because at higher temperatures, more molecules have sufficient energy to overcome intermolecular forces within the liquid.
11. (a) sublimation, (b) freezing or crystallization, (c) melting or fusion.
12. The variable that influences the vapor pressure of a substance is (b) temperature.
13. (a) condensation or liquefaction, (b) freezing or crystallization, (c) deposition.
14. Superheating is heating a liquid to a temperature above the boiling point without the occurrence of boiling. Superheating can be avoided by addition of boiling chips.
15. The pressure would increase. The vapor pressure remains the same because some of the vapor changes to liquid or solid.
16. Substances with stronger intermolecular forces have lower vapor pressures.
17. Methyl alcohol has the higher vapor pressure. Both have similar hydrogen bonding and dipole-dipole interactions, but the London forces in CH_3OH are smaller.
18. $\Delta H^\circ_{vap} = -(2.303)R(\text{slope})$.
19. The data include vapor pressure-temperature-freezing point data. The liquid-solid line represents the pressure dependence of the melting point, the liquid-gas line represents the temperature dependence of the vapor pressure, and the solid-gas line represents the temperature dependence of the sublimation pressure.

20. The two points are the critical point and the triple point.
21. Chemists seldom worry about pressure for melting points because the effect of pressure on melting points is very small.
22. The solids are called a crystalline solid and an amorphous solid, respectively.
23. A space lattice is a system of point representing sites with identical environments in the same orientation in a crystal. A crystal structure is the complete geometric arrangement of the particles that occupy the space lattice.
24. The sketches are shown in Fig. 13-1. All drawings are not equally close packed--the packing of the octagons leaves some empty space as shown by the dark areas.
25. 1/8 atom, 1 atom, 1/2 atom.
26. Multiply the unit cell content by the mass of one atom. Divide the mass of a unit cell by the unit cell volume.
27. (a) The ratio of ions must be the same as in the stoichiometric formula in order for electrical neutrality to be maintained. (b) Maximum stability results in having ions of opposite charge near each other. (c) Maximum stability results in having ions of the same charge relatively far from each other.
28. The thermochemical equations are for sublimation (endothermic), ionization (endothermic), bond dissociation energy (endothermic), electron affinity, and lattice energy.

SKILLS

Skill 13.1 Interpreting phase diagrams

The phase diagram for a substance is a pressure-temperature plot of sublimation pressure, vapor pressure, melting point, and various solid-solid phase transition data. At any point not on a line, only one phase exists; at any point along a line, two phases are in equilibrium; and at a triple point, three phases are in equilibrium. A horizontal line indicates what phases and what equilibrium conditions are encountered as the substance undergoes an isobaric temperature change, and a vertical line reflects conditions during an isothermal pressure change.

For example, in the diagram shown in Fig. 13-2, point *a* is the critical point; points *b*, *c*, and *d* are triple points; point *e* is the normal melting point

Fig. 13-1

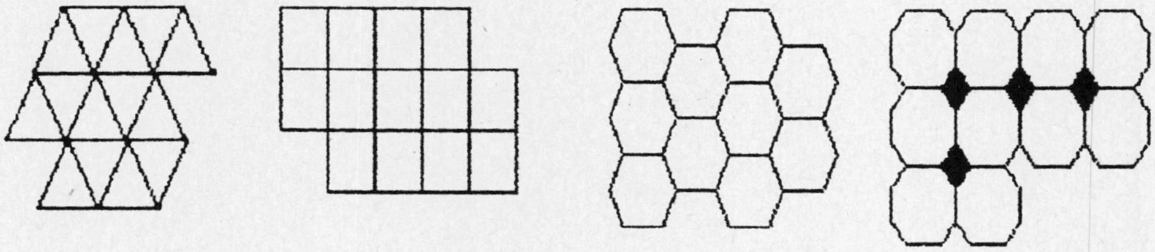

(1 atm pressure); and point *f* is the normal boiling point (1 atm pressure). If a sample of the substance originally at point *g* were heated isobarically, the following would be observed: The β-solid would increase in temperature until it begins to melt; after melting, the liquid would increase in temperature until it reaches the boiling point; and after vaporizing, the gas would continue increasing in temperature. If the pressure on a sample of the substance originally at point *g* were decreased, the following would be observed: The β-solid would expand slightly; if enough time were allowed, the β-solid would change to the α-solid; the α-solid would expand slightly until the pressure reached the sublimation pressure; and after sublimation, the gas would continue to expand.

Skill Exercise: Using the diagram shown in Fig. 13-3 (a) classify points *a*, *b*, *c*, and *d*; (b) describe what would happen as a sample of graphite at 1 atm is heated isobarically; and (c) describe what must be done to make diamonds from graphite quickly. *Answers*: (a) Point *a* is the critical point, points *b* and *c* are triple points, and point *d* is the normal sublimation point; (b) graphite increases in temperature until it begins to sublime, and after sublimation the gas continues to increase in temperature; (c) application of pressure alone takes too long, so heat the graphite so that it undergoes sublimation and then liquefy and freeze at high pressure.

NUMERICAL EXERCISES

Example 13.1 Heat of vaporization

The vapor pressure of CCl_4 is 10.0 Torr at -19.6 °C and 40.0 Torr at 4.3 °C. Calculate the heat of vaporization from these data.

For the vaporization of solute and liquids

$$\log\left[\frac{P_2}{P_1}\right] = \frac{-\Delta H^\circ_{\text{vap}}}{(2.303)R}\left[\frac{1}{T_2} - \frac{1}{T_1}\right]$$

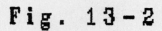

Fig. 13-2

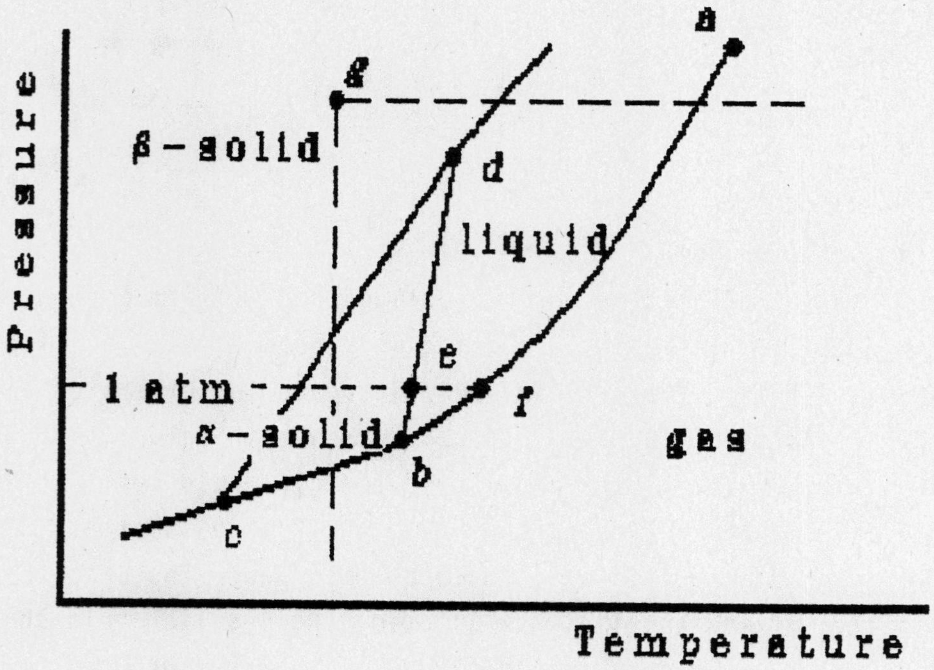

Fig. 13-3

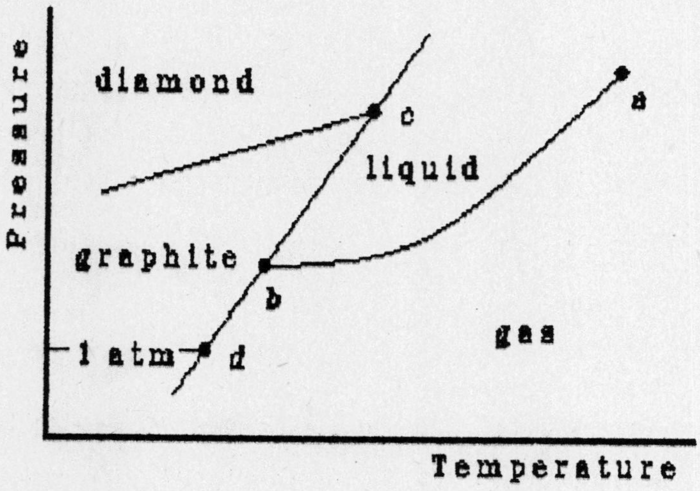

where ΔH_{vap} represents the heat of vaporization. This special equation is known as the Clausius-Clapeyron equation. (As long as both vapor pressures are in the same units, it is not necessary to convert them to atm because the conversion factors would cancel in the (P_2/P_1) term.) Solving the equation for ΔH°_{vap} and substituting the data give

$$\Delta H^\circ_{vap} = \frac{-(2.303)R\ \log(P_2/P_1)}{\left[\frac{1}{T_2} - \frac{1}{T_1}\right]}$$

$$= \frac{-(2.303)(8.314\ \text{J/K mol})\ \log(40.0\ \text{Torr}/10.0\ \text{Torr})}{\left[\frac{1}{277.5\ \text{K}} - \frac{1}{253.6\ \text{K}}\right]}$$

$$= 33{,}900\ \text{J}$$

The heat of vaporization is 33.9 kJ/mol.

Exercise 13.1 Predict the vapor pressure of CCl_4 at 0.0 °C by using the Clausius-Clapeyron equation. See *Example 13.1* for further data. *Answer*: 31.7 Torr.

Example 13.2 Unit cell

Gold crystallizes in a face-centered unit cell with a unit cell length equal to 0.40786 nm. Find the radius of a gold atom.

The relationship between the radius of an atom in a face-centered cubic unit cell and the unit cell length, a, can be seen from the sketch of one face shown in Fig. 13-4. Because the atoms touch, the face diagonal, d, is equal to $4r$. From the Pythagorean theorem, the face diagonal is also given by

$$d^2 = a^2 + a^2 = 2a^2 \qquad \text{so } d = a\sqrt{2}$$

and because $d = 4r$, then $4r = a\sqrt{2}$. Solving for r and substituting values gives

$$r = \frac{a\sqrt{2}}{4} = \frac{(0.40786\ \text{nm})\sqrt{2}}{4} = 0.14420\ \text{nm}$$

The atomic radius is 0.14420 nm.

Exercise 13.2 Iron crystallizes in a body-centered cubic unit cell with a unit cell length, a, equal to 0.28664 nm. Find (a) the relationship between a and the body diagonal, d; (b) the body diagonal and the atomic radius, r; and (c) the atomic radius of an iron atom.
Answers: (a) $d = a\sqrt{3}$, (b) $d = 4r$, (c) 0.12412 nm.

Fig. 13-4

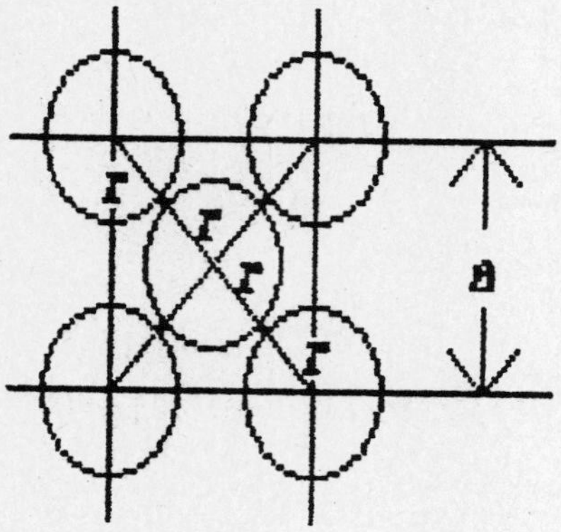

Example 13.3 Unit cell

The unit cell for CsCl consists of eight Cs^+ ions located at the corners of a cube with a Cl^- ion located at the center of the cube. How many Cs^+ ions, Cl^- ions, and formula units are contained in this unit cell?

The contribution of mass and volume that each corner ion makes to the unit cell under consideration is 1/8 atom. The unit cell content for Cs^+, Z_{Cs^+}, is

$$Z_{Cs^+} = (8 \text{ corners})\left[\frac{\frac{1}{8}\text{ atom}}{1 \text{ corner}}\right] = 1 \text{ Cs}^+ \text{ ion}$$

The contribution of mass and volume that each body-centered ion makes to the unit cell is 1 atom. The unit cell content for Cl^-, Z_{Cl^-}, is

$$Z_{Cl^-} = (1 \text{ interior})\left[\frac{1 \text{ atom}}{1 \text{ interior}}\right] = 1 \text{ Cl}^- \text{ ion}$$

Thus in the unit cell there are one Cs^+ ion and one Cl^- ion present, which is equivalent to one CsCl, or $Z_{CsCl} = 1$.

Exercise 13.3 The unit cell for AgBr is like that of NaCl (see Figure 13.25 of the text), in which the Ag^+ ions are located in a face-centered cubic structure with the Br^- ions located midway along each edge and in the center of the cube. Find (a) the contribution from each face-centered atom and each edge-centered atom, (b) the unit cell content for Ag^+, (c) the unit cell content for Br^-, and (d) the number of formula units contained in the unit cell. *Answers*: (a) 1/2, 1/4; (b) 4; (c) 4; (d) $Z_{AgBr} = 4$.

Example 13.4 Unit cell

Silicon crystallizes in a unit cell structure like that of diamond (see Figure 13.20 of the text), in which a face-centered cubic pattern of Si atoms is expanded slightly to allow four additional atoms in the shape of a tetrahedron to be put inside the cube. Calculate the packing efficiency of the Si unit cell.

$$\text{packing efficiency} = \frac{V_{\text{occupied}}}{V_{\text{unit cell}}}$$

The volume actually occupied is the volume of one atom, given by the volume of a sphere of radius r, $(4/3)\pi r^3$, multiplied by the number of atoms in the unit cell, Z. Thus for the Si unit cell having 8 corner atoms, 6 face-centered atoms, and 4 body-centered atoms,

$$Z_{\text{Si}} = (8\ \text{corners})\left[\frac{\frac{1}{8}\ \text{atom}}{1\ \text{corner}}\right] + (6\ \text{faces})\left[\frac{\frac{1}{2}\text{atom}}{1\ \text{face}}\right] + (4\ \text{interior})\left[\frac{1\ \text{atom}}{1\ \text{interior}}\right] = 8\ \text{atoms}$$

$$V_{\text{occupied}} = 8(4/3)\pi r^3$$

The volume of the cubic unit cell is given by a^3, where a is the unit cell length. The relationship between a and r for this unit cell is $2r = a\sqrt{3}/4$. Thus the volume of the unit cell is

$$V_{\text{unit cell}} = a^3 = (8r/\sqrt{3})^3 = (512/3\sqrt{3})r^3$$

Substituting these results for volume into the equation for packing efficiency gives

$$\text{packing efficiency} = \frac{V_{\text{occupied}}}{V_{\text{unit cell}}} = \frac{8(4/3)\pi r^3}{(512/3\sqrt{3})r^3} = 0.340$$

The Si unit cell is 34.0 % occupied and 66.0 % empty space.

Exercise 13.4 What is the packing efficiency for Fe in a body-centered cubic structure? See Exercise 13.2 for details. *Answer*: 0.680.

Example 13.5 Unit cell*

The unit cell length for crystalline NH_4I is 0.7244 nm. The unit cell is isomorphic with that of NaCl. Calculate the theoretical density of NH_4I.

This problem is solved by using the comprehensive problem-solving approach presented in *Skill 2.7 Solving problems*.

1. <u>Study the problem and be sure you understand it</u>.
   (a) What is unknown?
   The theoretical density.

(b) What is known?

The unit cell is like that of the NaCl structure and has a = 0.7244 nm.

2. Decide how to solve the problem.

(a) What is the connection between the known and the unknown?

The density is mass per unit volume. The density is a property that is independent of the amount of a substance, so finding the mass and volume of the unit cell will allow the calculation of density.

(b) What is necessary to make the connection?

The mass is found by determining the number of formula units in the unit cell and multiplying this value by the mass of one formula unit. The volume of the unit cell is simply the cube of the unit cell length.

3. Set up the problem and solve it.

The unit cell content is

$$Z_{NH_4^+} = (8\text{ corners})\left[\frac{\frac{1}{8}\text{ ion}}{1\text{ corner}}\right] + (6\text{ faces})\left[\frac{\frac{1}{2}\text{ ion}}{1\text{ face}}\right] = 4\ NH_4^+\text{ ions}$$

$$Z_{I^-} = (12\text{ edges})\left[\frac{\frac{1}{4}\text{ ion}}{1\text{ edge}}\right] + (1\text{ interior})\left[\frac{1\text{ ion}}{1\text{ interior}}\right] = 4\ I^-\text{ ions}$$

which is equivalent to 4 formula units in one unit cell. The mass of the unit cell is

$$(4\text{ formula units})\left[\frac{144.94\text{ g}}{1\text{ mol}}\right]\left[\frac{1\text{ mol}}{6.022 \times 10^{23}\text{ formula units}}\right] = 9.627 \times 10^{-22}\text{ g}$$

The unit cell volume is

$$(0.7244\text{ nm})^3\left[\frac{1\text{ m}}{10^9\text{ nm}}\right]^3\left[\frac{10^2\text{ cm}}{1\text{ m}}\right]^3 = 3.801 \times 10^{-22}\text{ cm}^3$$

The density is

$$\frac{9.627 \times 10^{-22}\text{ g}}{3.801 \times 10^{-22}\text{ cm}^3} = 2.533\text{ g/cm}^3$$

4. Check the result.

(a) Are significant figures and the location of the decimal point correct?

Yes.

(b) Did the answer come out in the correct units?

Yes.

(c) Is the answer reasonable?

Yes.

Exercise 13.5 Calculate the theoretical density of AgBr given that a = 0.57745 nm. See Exercise 13.3 for details. *Answer*: 6.4774 g/cm³.

Example 13.6 Unit cell

Zinc crystallizes in the hexagonal closest-packed system with a = 0.2665 nm and c = 0.4947 nm. Calculate the theoretical density of this metal.

For the hexagonal closest-packed unit cell, which contains 8 corner atoms (some of which contribute 1/6 of their mass and volume to the unit cell and some of which contribute only 1/12) and 1 body-centered atom,

$$Z = 4(1/12) + 4(1/6) + 1(1) = 2 \text{ atoms}$$

The mass of the unit cell is

$$(2 \text{ atoms})\left[\frac{65.38 \text{ g}}{1 \text{ mol}}\right]\left[\frac{1 \text{ mol}}{6.022 \times 10^{23} \text{ atoms}}\right] = 2.17 \times 10^{-22} \text{ g}$$

The volume of the hexagonal unit cell is

$$\begin{aligned} V &= a^2 c \sin 60° \\ &= (0.2665 \text{ nm})^2(0.4947 \text{ nm})\left[\frac{1 \text{ m}}{10^9 \text{ nm}}\right]^3\left[\frac{10^2 \text{ cm}}{1 \text{ m}}\right]^3 \sin 60° \\ &= 3.043 \times 10^{-23} \text{ cm}^3 \end{aligned}$$

The density is

$$\frac{2.171 \times 10^{-22} \text{ g}}{3.043 \times 10^{-23} \text{ cm}^3} = 7.134 \text{ g/cm}^3$$

The density of pure Zn would be slightly less than this value because of lattice defects, and that of impure Zn would be greater or smaller than this value, depending on the impurities present.

Exercise 13.6 The unit cell length for crystalline MgO is 0.42117 nm. The density of the cubic substance is 3.58 g/cm³. Use these data to determine the unit cell content. Do your results confirm that the unit cell structure of MgO is like that of NaCl? *Answer*: 4 formula units, yes.

*Example 13.7 Born-Haber cycle**

Develop a Born-Haber cycle for the formation of $AlCl_3$(s) from Al(s) and Cl_2(g). Use this cycle to calculate ΔH°_f for $AlCl_3$(s) from the following ΔH° data: 326.4 kJ/mol for the sublimation of Al, -367.8 kJ/mol for the electron affinity of Cl, 243.4 kJ/mol for the Cl_2 bond energy, -5450.1 kJ/mol for the lattice energy of $AlCl_3(s)$, and 5157.6 kJ/mol for the ionization of Al to Al^{3+}. Does the value of the heat of formation indicate that $AlCl_3(s)$ will be a stable compound?

This problem is solved by using the comprehensive problem-solving approach presented in *Skill 2.7 Solving problems.*

1. Study the problem and be sure you understand it.
   (a) What is unknown?
   The heat of formation of $AlCl_3(s)$.
   (b) What is known?
   Various thermodynamic data to use in the Born-Haber cycle.
2. Decide how to solve the problem.
   (a) What is the connection between the known and the unknown?
   According to Hess's law the heat of reaction is independent of the number of steps involved in a reaction.
   (b) What is necessary to make the connection?
   Writing a series of steps in a Born-Haber cycle will give the relationship between the desired heat of reaction and the given data.
3. Set up the problem and solve it.
   The series of steps are shown in Fig. 13-5.

$$\Delta H^\circ_f = \Delta H^\circ_1 + \Delta H^\circ_2 + \Delta H^\circ_3 + \Delta H^\circ_4 + \Delta H^\circ_5$$
$$= \Delta H^\circ(\text{sub,Al}) + \Delta H^\circ(\text{ionization,Al}) + \left(\tfrac{3}{2}\right)\Delta H^\circ(\text{bond,}Cl_2)$$
$$+ 3\Delta H^\circ(\text{electron affinity,Cl}) + \Delta H^\circ(\text{lattice,}AlCl_3)$$
$$= (326.4\ \text{kJ}) + (5157.6\ \text{kJ}) + \left(\tfrac{3}{2}\right)(243.4\ \text{kJ})$$
$$+ (3)(-367.8\ \text{kJ}) + (-5450.1\ \text{kJ})$$
$$= -704.4\ \text{kJ}$$

   This is a favorable heat of reaction. Most likely, $AlCl_3$ is a stable compound.
4. Check the result.
   (a) Are significant figures and the location of the decimal point correct?
   Yes.
   (b) Did the answer come out in the correct units?
   Yes.
   (c) Is the answer reasonable?
   Yes.

Fig. 13-5

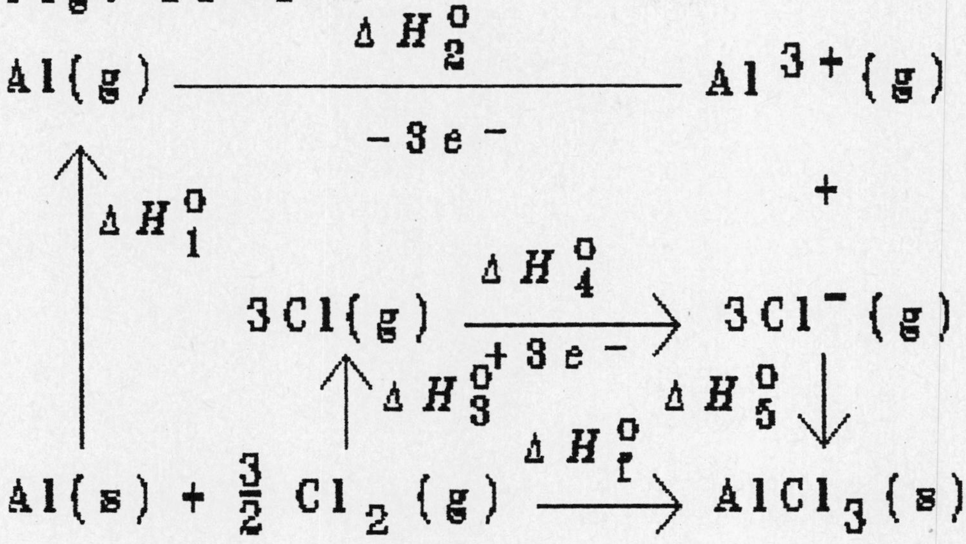

Exercise 13.7 Use a Born-Haber cycle to calculate the heat of formation of $AlI_3(s)$. The electron affinity of I is -303.4 kJ/mol, the bond dissociation energy of I_2 is 151.2 kJ/mol, the heat of sublimation of I_2 is 62.4 kJ/mol, and the lattice energy is -4600 kJ/mol. Is this a favorable enthalpy change?

Answer: $\Delta H^\circ_f = \Delta H^\circ(\text{sub,Al}) + \Delta H^\circ(\text{ionization,Al}) + (\frac{3}{2})\Delta H^\circ(\text{sub},I_2) + (\frac{3}{2})\Delta H(\text{bond},I_2) + (3)\Delta H^\circ(\text{electron affinity,I}) + \Delta H^\circ(\text{lattice})$ = 300 kJ, no.

PRACTICE TEST

1 (10 points). Match the correct term to each definition.

(a) a solid changing directly to a gas
(b) a liquid changing to a gas
(c) temperature at which the vapor pressure of a liquid equals 1 atm
(d) conditions under which three phases are at equilibrium
(e) temperature at which the vapor pressure of a liquid equals atmospheric pressure
(f) a gas changing to a liquid
(g) temperature at which a liquid changes to a solid at 1 atm pressure
(h) a solid changing to a liquid

(i) boiling point
(ii) condensation
(iii) fusion
(iv) normal boiling point
(v) normal freezing point
(vi) sublimation
(vii) triple point
(viii) vaporization

2 (10 points). The vapor pressure of chloroform is 10.0 Torr at -29.7 °C and 40.0 Torr at -7.1 °C. Calculate (a) the heat of vaporization of $CHCl_3$ and (b) the temperature to which the chloroform must be heated in order for the vapor pressure to be 100.0 Torr.

3 (15 points). Lead crystallizes in a face-centered cubic unit cell with the unit cell length, *a*, equal to 0.49505 nm. (a) Sketch the unit cell. (b) What is the unit cell content? (c) What is the value of the crystalline coordination number of an atom? (d) What is the radius of a lead atom? (e) Calculate the theoretical density of lead.

4 (15 points). Using the hypothetical phase diagram for Pb shown in Fig. 13-6, (a) identify points *a*, *b*, *c*, and *d*. Describe what will happen to a sample of lead originally at point *e* if it is (b) isobarically cooled and (c) isothermally compressed.

5 (10 points). For a mole of lead, calculate (a) the volume actually occupied by lead atoms using your answer to question 3d, (b) the volume occupied by the lead in the crystalline form using your answer to question 3e, and (c) the packing efficiency of the face-centered cubic system.

6 (30 points). Choose the response that best completes each statement.

(a) The movement of molecules from the gaseous phase to the liquid phase is

(i) long-range ordering (ii) evaporation
(iii) condensation (iv) the critical pressure

(b) A liquid at a temperature greater than the boiling point is

(i) supercooled (ii) superheated
(iii) critical (iv) amorphous

(c) Which is not a phase transition?

(i) vaporization (ii) liquefaction
(iii) freezing (iv) crystal defect

(d) The energy that must be exceeded to break down a crystal to gaseous ions is the

(i) lattice energy (ii) molar heat of vaporization
(iii) molar heat of fusion (iv) critical point energy

(e) Which is not a type of point defect?

(i) screw dislocation (ii) impurity defect
(iii) vacancy defect (iv) interstitial defect

Fig. 13-6

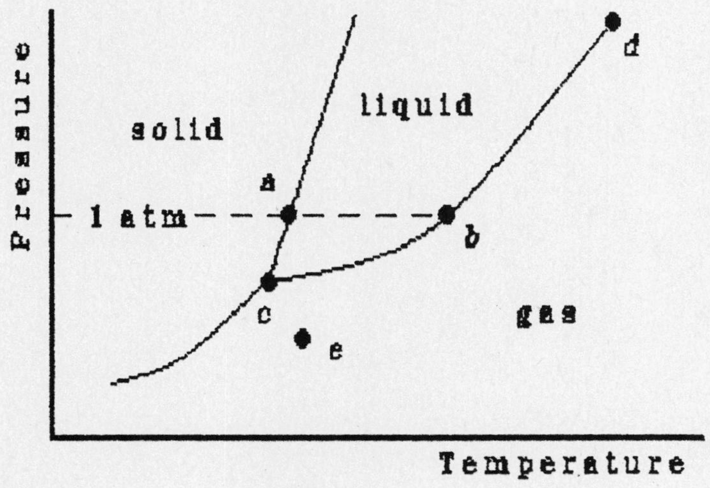

(f) The particles present in a crystal lattice of a pure metal are

(i) molecules (ii) atoms

(iii) polyatomic ions (iv) cations and electrons

(g) The interparticle forces responsible for the bonding in alloys are usually

(i) ionic (ii) metallic

(iii) intermolecular (iv) covalent in nature

(h) Which chemical equation would be written to represent the lattice energy?

(i) $NaCl(s) \xrightarrow{H_2O} NaCl(aq)$ (ii) $Na^+(g) + Cl^- \rightarrow NaCl(s)$

(iii) $Na(g) \rightarrow Na^+(g) + e^-$ (iv) $Cl(g) + e^- \rightarrow Cl^-(g)$

(i) Which of the following liquids is most viscous?

(i) corn syrup (ii) gasoline

(iii) water (iv) glass

(j) The fact that carbon exists as both graphite and diamond makes it an example of what type of solid?

(i) isomorphic (ii) amorphous

(iii) polymorphic (iv) isotropic

(k) Which is true for a liquid?

(i) The vapor pressure increases with increasing temperature and the normal boiling point is constant.

(ii) The vapor pressure is constant and the normal boiling point increases with increasing temperature.

(iii) Both the vapor pressure and the normal boiling point increase with increasing temperature.

(iv) The vapor pressure increases but the normal boiling point decreases with increasing temperature.

(l) What will happen as heat is added to a sample of a pure substance just beginning to melt?

(i) A temperature increase is observed.

(ii) The temperature will fall.

(iii) The temperature will remain constant.

(iv) The critical pressure will increase.

(m) What is true about a substance below the critical temperature?

(i) All gases are liquids.

(ii) A gas can be liquefied under certain pressure conditions.

(iii) Diatomic molecules are unreactive.

(iv) The distinction between a liquid and gas ceases to exist.

(n) The bonding in an ionic crystalline solid can be described as

(i) the atoms being held together by covalent bonds

(ii) a sea of electrons

(iii) ions arranged so that like-charged particles are as far as possible from each other and unlike-charged particles are packed as closely as possible

(iv) ions arranged so that like-charged particles are packed as closely as possible and unlike-charged ions are as far as possible from each other

(o) Which of the following is *not* a crystallographic system?

(i) orthorhombic (ii) rhombohedral

(iii) hexagonal (iv) octahedral

7 (10 points). Use the following thermochemical equations to calculate the lattice energy for $AlCl_3$.

$$Al(s) \rightarrow Al(g) \qquad \Delta H^\circ = 326 \text{ kJ}$$

$$Al(g) \rightarrow Al^{3+}(g) + 3e^- \qquad \Delta H^\circ = 5158 \text{ kJ}$$

$$Cl_2(g) \rightarrow 2Cl(g) \qquad \Delta H^\circ = 243 \text{ kJ}$$

$$Cl(g) + e^- \rightarrow Cl^-(g) \qquad \Delta H^\circ = -368 \text{ kJ}$$

$$Al(s) + \frac{3}{2}Cl_2(g) \rightarrow AlCl_3(s) \qquad \Delta H^\circ = -705 \text{ kJ}$$

Answers to practice test

1. (a) vi, (b) viii, (c) iv, (d) vii, (e) i, (f) ii, (g) v, (h) iii.
2. 33.1 kJ/mol, 10.3 °C.
3. (a) See Figure 13.21 of the text, (b) 4 atoms, (c) 12, (d) 0.17503 nm, (e) 11.34 g/cm³.
4. (a) normal melting (or freezing) point, normal boiling point, triple point, critical point; (b) gas cools, deposits to form solid, solid cools; (c) gas volume decreases, condenses, liquid volume decreases, crystallizes, solid volume decreases.
5. (a) 13.526 cm³, (b) 18.265 cm³, (c) 74.0 %.
6. (a) iii, (b) ii, (c) iv, (d) i, (e) i, (f) iv, (g) ii, (h) ii, (i) iv, (j) iii, (k) i, (l) iii, (m) ii, (n) iii, (o) iv.
7. -5450. kJ/mol

ADDITIONAL NOTES

CHAPTER 14

WATER AND SOLUTIONS IN WATER

CHAPTER OVERVIEW

The first part of this chapter presents the descriptive chemistry of water--its properties and reactions--and the chemistry of the hydrates. An introduction to ions in aqueous solution is presented. The water-ion definitions of acids, bases, and neutralization are given and an introduction to complexes is given. The chapter concludes with a discussion of natural waters, water pollution, water softening, and sewage treatment.

COMPETENCIES

Definitions for the following terms should be learned:

- acid (water-ion)
- acidic aqueous solution
- alkaline aqueous solution
- base (water-ion)
- carbonate hardness, temporary hardness
- chemical equilibrium
- complex ion
- coordination compounds
- deliquescent
- efflorescence
- electrolytes
- hard water
- heat of solution
- hydrated
- hydrates
- hydration
- hydrolysis
- hygroscopic
- ion exchange
- ligands
- neutralization
- noncarbonate hardness, permanent hardness
- nonelectrolytes
- pollutant
- polyprotic acids
- solvation
- strong electrolytes
- water softening
- weak electrolytes

General concepts and skills that should be learned include

- relating molecular structure of water to some of its properties (e.g., intermolecular forces and solution formation)
- the general physical and chemical properties of water
- the properties of hydrates
- interpreting chemical equilibrium
- the general properites of strong and weak electrolytes and of nonelectrolytes
- the general properties and definitions of acids, bases, and neutralization
- the chemistry involved in water pollution and water and sewage treatment
- the properties and causes of hard water and the methods of softening water

QUESTIONS

1. Complete the following sentences by filling in the various blanks with these terms: deionized water, desalination, hard water, ion exchange, permanently hard, temporarily hard, and water softening.

   Water containing divalent ions that cause scaling and scum when mixed with soap is known as (a)______. Water that can be purified by boiling to remove the impurities is called (b)______ water, and water that cannot be purified by boiling is called (c)______ water. The removal of the impurities, called (d)______, can be done by adding certain chemicals, by reverse osmosis, or by (e)______ techniques. Water that has passed through both cation- and anion-exchange systems so that all of the impurities have been removed is called (f)______. The process of removing ions (particularly Na^+ and Cl^- from seawater) is called (g)______ and can be done by some of the water-softening techniques, by crystallization, by evaporation, or by electrodialysis.

2. Match the correct term to each definition.

(a) substances that have combined with a definite amount of water and have a definite composition	(i) acid
	(ii) base
(b) the loss of water of hydration	(iii) deliquescence
(c) the taking up of water from the air	(iv) efflorescence
(d) the taking up of water from the air so as to dissolve the substance	(v) hydrates
	(vi) hydration

(e) the interaction of water with other substances without splitting of the water molecule
(f) the reaction in which the water molecule is split and the hydrogen and oxygen or OH group from the water are added to the product
(g) a substance that in aqueous solution increases the H^+ concentration
(h) a substance that in aqueous solution increases the OH^- concentration
(i) the reaction of an acid with a base

(vii) hydrolysis
(viii) hygroscopic action
(ix) neutralization

3. Which of the following statements are true? Rewrite each false statement so that it is correct.

(a) The chemical formula for a hydrate represents the average number of water molecules present.
(b) A hydrate that takes up enough water from the air to dissolve is undergoing efflorescence.
(c) Molecules of water that are chemically bonded to a hydrated salt can be held by different types of bonding.
(d) Compounds that conduct electricity by the movement of ions are called metallic conductors.
(e) Compounds that are 100% dissociated or ionized in aqueous solution are called strong electrolytes.
(f) The molecules or ions bonded to the central atom or cation in a ligand are the complex ion species.
(g) Hard water contains dipositive ions that form precipitates with soap or upon boiling.
(h) All ions can be removed from water if it is doubly distilled and then passed through a reverse osmosis chamber.
(i) The decomposition of organic matter by bacteria in the absence of oxygen is known as aerobic decomposition.
(j) Primary sewage treatment is essentially a process to eliminate specific pollutants by a series of chemical reactions.

4. Distinguish clearly between/among (a) anhydrous and hydrated salts, (b) hygroscopic and deliquescent substances, (c) hard and soft water, (d) temporary and permanent hard water, (e) hydration and hydrolysis, (f) strong and weak electrolytes, (g) electrolytes and nonelectrolytes, (h) acids and bases according to the water-ion theory, (i) efflorescent and hygroscopic substances, (j) aerobic and anaerobic decomposition, and (k) primary, secondary, and tertiary sewage treatment.

5. Choose the response that best completes each statement.

(a) The H-O-H bond angle in water is
   (i) a tetrahedral angle of 109.67°
   (ii) about 120°
   (iii) 180°
   (iv) slightly smaller than a tetrahedral angle because of the lone pair-bonded pair interactions

(b) The major intermolecular force in water is
   (i) a London induced dipole
   (ii) a dipole-dipole interaction
   (iii) extended covalent bonding
   (iv) hydrogen bonding

(c) As polar and ionic compounds dissolve in water
   (i) the density increases to maximum at 3.98 °C and then decreases
   (ii) negative and positive centers in the dissolving substance are attracted, respectively, to the positive and negative centers in the water molecules
   (iii) the water undergoes decomposition to hydrogen and oxygen
   (iv) efflorescence occurs

(d) Which equation does not demonstrate hydrolysis?
   (i) $CaSO_4(s) + H_2O(l) \rightarrow CaSO_4 \cdot H_2O(s)$
   (ii) $Al_2S_3(s) + 6H_2O(l) \rightarrow 2Al(OH)_3(s) + 3H_2S(aq)$
   (iii) $PBr_3(l) + 3H_2O(l) \rightarrow H_3PO_3(aq) + 3HBr(aq)$
   (iv) $CO_2(g) + H_2O(l) \rightleftarrows HCO_3^- + H^+$

(e) An aqueous solution containing an equal number of H^+ and OH^- ions is
   (i) an acidic aqueous solution
   (ii) an alkaline aqueous solution
   (iii) a neutral aqueous solution
   (iv) a hydrated solution

(f) Which equation does not illustrate a water-softening technique?
   (i) $M^{2+} + 2HCO_3^- \xrightarrow{\Delta} MCO_3(s) + CO_2(g) + H_2O(g)$
   (ii) $M^{2+} + 2NaZ(s) \rightarrow 2Na^+ + MZ_2(s)$
   (iii) $CaCO_3(s) + H^+ \rightarrow Ca^{2+} + HCO_3^-$
   (iv) $2ROH(s) + X^{2-} \rightarrow R_2X(s) + 2OH^-$
   followed by $2(R')H(s) + M^{2+} \rightarrow (R')_2M(s) + 2H^+$

(g) Which impurities would you predict could be added to water by the atmosphere?
- (i) dust particles, inert gases, and gases that react with the water
- (ii) suspended particles, ions from minerals, and microorganisms
- (iii) dust particles, water-soluble gases, and water-soluble organic material
- (iv) catalysts for hydrogenation reactions

(h) Which impurities would you predict could be added to water by land runoff?
- (i) microorganisms, water-soluble organic material, water-soluble ions and suspended particles
- (ii) soluble gases and dust particles
- (iii) inert gases, dust particles, and gases that react with the water
- (iv) fluorocarbons to react with dissolved ozone

(i) Which is *not* a type of water pollutant?
- (i) substances that harm humans or animals by causing disease or physical damage
- (ii) substances or situations that decrease the oxygen content of water leading to anaerobic decay
- (iii) substances that are indirectly harmful
- (iv) ion exchange

(j) Which is *not* an oxygen-demanding pollutant?
- (i) organic wastes
- (ii) overabundance of microorganisms
- (iii) thermal pollution
- (iv) a high concentration of Ca^{2+}

(k) Which is *not* a toxic or harmful pollutant?
- (i) bacteria and viruses
- (ii) sediments
- (iii) insecticides and pesticides
- (iv) heavy metal ions

(l) The removal from watewater of special impurities such as ions, pesticides, and radioactive materials is
- (i) primary treatment
- (ii) secondary treatment
- (iii) tertiary treatment
- (iv) hydrogenation

6. List the following in the order of increasing volume: (a) one mole of steam at 1 atm and 150 °C, (b) one mole of ice at -5 °C and 1 atm, (c) one mole of liquid water at 30 °C and 1 atm.

7. Why are there many more chuckholes in roads and cracks in sidewalks after a severe winter in which there has been much snow and repeated freezing and thawing than after a mild, relatively dry winter?

8. Water and hydrogen selenide, H_2Se, are comparable compounds--both oxygen and selenium are members of the same family of elements in the periodic table. Even though the molar mass of water is slightly less than one fourth that of H_2Se, the melting and boiling points for water are much higher than for H_2Se. Why?

9. What are chemical compounds that include water molecules called? How are the formulas for these compounds usually written?

10. What determines whether a substance is described as efflorescent or hygroscopic?

11. Write the chemical equation for the self-ionization of water. Is a large percentage of water molecules ionized at a given time?

12. Define the term "hydrolysis". Write two chemical equations illustrating hydrolysis reactions.

13. What three types of interactions between solvent and solute particles determine whether dissolution will occur? Which of these are endothermic and which are exothermic? Describe the interactions involved in producing a solution of NaCl in H_2O.

14. An ideal solution can be defined as having a heat of solution equal to zero. Interpret this statement in terms of solvent-solvent, solute-solute, and solvent-solute interactions.

15. Define the term "electrolyte." How does an electrolyte conduct electricity? Why do some substances act as nonelectrolytes in aqueous solution?

16. What is wrong with the following statement: "Only aqueous solutions of ionic substances can be strong electrolytes"?

17. What are strong electrolytes? Write chemical equations illustrating how NaBr(s), NaOH(s), and HBr(g) act as strong electrolytes.

18. What types of substances are weak electrolytes in aqueous solution? Write a chemical equation showing that hydrogen fluoride is a weak electrolyte when dissolved in water.

19. Define an "acidic aqueous solution." How does this differ from the definition of an "alkaline aqueous solution"?

20. Write the formula for a common strong binary acid. Write the chemical equation showing that it is a strong electrolyte.

21. Write the formula for a common strong base. Write the chemical equation showing that it is a strong electrolyte.

22. Write the formula for a weak acid. Write the chemical equation showing that it is a weak electrolyte.

23. Identify which of the following can act as water-ion acids: (a) CH_3CH_2OH, (b) CH_3COOH, (c) HCN, (d) $Mg(OH)_2$, (e) NH_3, (f) HF, (g) H_2SO_4, and (h) HClO.

24. How does the net ionic equation for a "neutralization" reaction involving a weak acid and a strong base differ from that for the reaction involving a strong acid and a strong base?

25. Write chemical equations showing the various stages of ionization of phosphoric acid, H_3PO_4, a triprotic acid. Write chemical equations for all possible reactions of H_3PO_4 with calcium hydroxide, $Ca(OH)_2$.

26. How is the concept of hydration of ionic solutes connected with complex ion formation?

27. What is the general name for the process in which the undesirable ions in hard water are removed? Name several ways in which these ions can be removed.

28. Write the chemical equations showing how hydrated (or slaked) lime and washing soda can be used to soften hard water containing Ca^{2+} and Mg^{2+}.

29. List the four steps that make up "nature's water treatment system".

30. Identify which sewage treatment stage would eliminate (a) pesticides, (b) debris and grit, (c) up to 90 % of the oxygen-demanding wastes, (d) organic compounds, and (e) organic and inorganic solids.

Answers to questions

1. (a) hard water, (b) temporarily hard, (c) permanently hard, (d) water softening, (e) ion exchange, (f) deionized water, (g) desalination.
2. (a) v, (b) iv, (c) viii, (d) iii, (e) vi, (f) vii, (g) i, (h) ii, (i) ix.

3. (a) F, delete *average*; (b) F, replace *efflorescence* with *deliquescence*; (c) T; (d) F, replace *metallic conductors* with *electrolytes*; (e) T; (f) F, interchange *ligand* and *complex ion*; (g) T; (h) F, insert *impurity* between *All ions*; (i) F, change *aerobic* to *anaerobic*; (j) F, replace *Primary* with *Tertiary*.
4. (a) *Anhydrous salts* do not contain hydrated water molecules; *hydrated salts* contain hydrated water molecules. (b) *Hygroscopic substances* absorb water from the air; *deliquescent substances* absorb sufficient water to form a solution. (c) *Hard water* contains ions that make the water undesirable; *soft water* does not contain these ions. (d) *Temporary hard water* contains the HCO_3^- ion, which can be removed by boiling; *permanent hard water* contains other anions that cannot be easily removed by boiling. (e) In *hydration* processes, water molecules are added to the substance intact; in *hydrolysis* processes, the water molecules are split. (f) *Strong electrolytes* are good conductors of electricity; *weak electrolytes* are poor conductors. (g) *Electrolytes* conduct electricity by movement of ions; *nonelectrolytes* are nonconductors. (h) *Acids* are substances that contain hydrogen and form H^+ in water; *bases* contain hydroxide ion and form OH^- in water. (i) *Efflorescent substances* lose water to the air; *hygroscopic substances* absorb water from the air. (j) *Aerobic decomposition* is the decomposition of organic matter in the presence of air or oxygen; *anaerobic decomposition* takes place in the absence of air or oxygen. (k) *Primary sewage treatment* steps involve the removal of solids and purification of sludge; *secondary sewage treatment* removes oxygen-demanding wastes; *tertiary sewage treatment* removes specific pollutants.
5. (a) iv, (b) iv, (c) ii, (d) i, (e) iii, (f) iii, (g) i, (h) i, (i) iv, (j) iv, (k) ii, (l) iii.
6. $c < b < a$.
7. The snow melts and the water flows into the crack or small hole in the pavement. The water freezes and expands causing damage to the road or sidewalk and then the process repeats itself.
8. Water has strong hydrogen bonding that is not present in H_2Se.
9. The compounds are called hydrates. The formulas are written by separating the formula for the anhydrous compound from the combined symbol of a number representing the number of water molecules and the formula for water by a centered dot.

10. The vapor pressure of the hydrate will be greater than the partial pressure of water vapor in the air for an efflorescent substance and will be less for a hygroscopic substance.
11. $H_2O(l) \rightleftarrows H^+ + OH^-$, no.
12. Hydrolysis is a term for reactions in which the water molecule is split. Chemical equations illustrating hydrolysis reactions include

    $CN^- + H_2O(l) \rightleftarrows HCN(aq) + OH^-$ $2Na(s) + 2H_2O(l) \rightarrow 2NaOH(aq) + H_2(g)$

    $SO_3(g) + H_2O(l) \rightleftarrows H_2SO_4(aq)$ $SiI_4(s) + 2H_2O(l) \rightarrow SiO_2(s) + 4HI(aq)$
13. The types are solvent-solvent, solute-solute, and solvent-solute interactions. Overcoming solvent-solvent and solute-solute interactions are endothermic processes and overcoming solute-solvent interactions is an exothermic process. The interactions include ionic and London forces in NaCl; hydrogen bonding, dipole-dipole interactions, and London forces in H_2O; and dipole-ion interactions and London forces in the solution.
14. $\Delta H°$(solvent-solvent) + $\Delta H°$(solute-solute) = $\Delta H°$(solvent-solute).
15. Electrolytes are substances that conduct electricity by movement of ions. Nonelectrolytes do not form ions in solution.
16. Many molecular compounds undergo 100 % ionization in water to become strong electrolytes.
17. Strong electrolytes are compounds that are 100 % dissociated or ionized in aqueous solution.

    $NaBr(s) \xrightarrow{H_2O} Na^+ + Br^-$ $NaOH(s) \xrightarrow{H_2O} Na^+ + OH^-$

    $HBr(g) \xrightarrow{H_2O} H^+ + Br^-$ or $HBr(g) + H_2O(l) \rightarrow H_3O^+ + Br^-$
18. Substances that are weak electrolytes in aqueous solution are molecular substances that contain polar covalent bonds that are not completely ionized.

    $HF(aq) \rightleftarrows H^+ + F^-$ or $HF(aq) + H_2O(l) \rightleftarrows H_3O^+ + F^-$
19. An acidic aqueous solution contains a greater concentration of H^+ than OH^-. This definition is just opposite that of an alkaline aqueous solution.
20. HCl, $HCl(aq) \xrightarrow{H_2O} H^+ + Cl^-$ or $HCl(aq) + H_2O(l) \rightarrow H_3O^+ + Cl^-$.
21. KOH, $KOH(s) \xrightarrow{H_2O} K^+ + OH^-$.
22. CH_3COOH, $CH_3COOH(aq) \overset{H_2O}{\rightleftarrows} CH_3COO^- + H^+$ or $CH_3COOH(aq) + H_2O(l) \rightleftarrows CH_3COO^- + H_3O^+$.
23. b, c, f, g, h.

24. The net ionic equation for a "neutralization" reaction involving a weak acid differs from that involving a strong acid because the anion of the acid is not a spectator ion. Thus the anion is not cancelled from both sides of the equation and appears in the net ionic equation.
25. The chemical equations showing the various stages or ionization of H_3PO_4 are

$$H_3PO_4(aq) \overset{H_2O}{\rightleftharpoons} H^+ + H_2PO_4^- \quad \text{or} \quad H_3PO_4(aq) + H_2O(l) \rightleftharpoons H_3O^+ + H_2PO_4^-$$
$$H_2PO_4^- \overset{H_2O}{\rightleftharpoons} H^+ + HPO_4^{2-} \quad \text{or} \quad H_2PO_4^- + H_2O(l) \rightleftharpoons H_3O^+ + HPO_4^{2-}$$
$$HPO_4^{2-} \overset{H_2O}{\rightleftharpoons} H^+ + PO_4^{3-} \quad \text{or} \quad HPO_4^{2-} + H_2O(l) \rightleftharpoons H_3O^+ + PO_4^{3-}$$

The chemical equations for all possible reactions of H_3PO_4 with $Ca(OH)_2$ are

$$2H_3PO_4(aq) + Ca(OH)_2(aq) \rightarrow Ca(H_2PO_4)_2(aq) + 2H_2O(l)$$
$$H_3PO_4(aq) + Ca(OH)_2(aq) \rightarrow Ca(HPO_4)(aq) + 2H_2O(l)$$
$$2H_3PO_4(aq) + 3Ca(OH)_2(aq) \rightarrow Ca_3(PO_4)_2(aq) + 6H_2O(l)$$
$$H_3PO_4(aq) + 2Ca(OH)_2(aq) \rightarrow Ca_2(OH)(PO_4)(aq) + 3H_2O(l)$$

26. A hydrated metal cation is a complex ion having water molecules as ligands.
27. The general name for the process in which the undesirable ions in hard water are removed is water softening. It can be accomplished by precipitation, complexation, or ion exchange.
28. $Ca^{2+} + 2HCO_3^- + Ca(OH)_2(s) \rightarrow 2CaCO_3(s) + 2H_2O(l)$, $Mg^{2+} + Ca(OH)_2(s) + Na_2CO_3(s) \rightarrow Mg(OH)_2(s) + CaCO_3(s) + 2Na^+$.
29. The four steps include: (i) flushing of gases by air, (ii) filtering or settling of solids, (iii) dilution, and (iv) aerobic and anaerobic decomposition.
30. (a) tertiary, (b) primary, (c) secondary, (d) tertiary, (e) primary.

SKILLS

Skill 14.1 Electrolytes and nonelectrolyes

Electrolytes are pure substances or solutes that conduct electricity by the movement of ions. Any solute that ionizes or dissociates into ions in aqueous solution is an electrolyte. Strong electrolytes are substances that exist completely as ions in aqueous solution. Most salts (see Table 14.5 of the text) and bases are strong electrolytes, as are the strong acids. The interactions of strong electrolytes such as NaCl, LiOH, and HCl with water can be represented by the following chemical equations

$$NaCl(s) \overset{H_2O}{\rightarrow} Na^+ + Cl^-$$

$$LiOH(s) \xrightarrow{H_2O} Li^+ + OH^-$$

$$HCl(g) + H_2O(l) \rightarrow H_3O^+ + Cl^-$$

If there is more solute present than can be dissolved in a given amount of water, a chemical equilibrium is established between the undissolved solute and the ions in aqueous solution. The interaction of a strong electrolyte for which the solubility has been exceeded such as AgCl with water can be represented by the following chemical equation

$$AgCl(s) \overset{H_2O}{\rightleftarrows} Ag^+ + Cl^-$$

Weak electrolytes are substances that are only partially ionized in aqueous solution. An equilibrium is established between the aqueous molecules and the aqueous ions. The interaction of a weak electrolyte such as acetic acid with water can be represented by the following chemical equation

$$CH_3COOH(aq) + H_2O(l) \rightleftarrows H_3O^+ + CH_3COO^-$$

Substances that dissolve in water to give solutions that do not conduct electricity are called nonelectrolytes. Nonelectrolytes remain as molecules in solution and their solutions do not conduct electricity. The interaction of a nonelectrolyte such as ethyl alcohol with water can be represented by the following chemical equation

$$CH_3CH_2OH(l) \xrightarrow{H_2O} CH_3CH_2OH(aq)$$

Skill Exercise: Write chemical equations representing the interactions of the following solutes with water: (a) $CBr_4(l)$, nonionic, 0.024 g/100 mL H_2O; (b) $PbSO_4(s)$, 0.004 g/100 mL H_2O; (c) $CH_3CH_2COOH(s)$, a weak acid, infinitely soluble; (d) $LaSO_4(s)$, 3 g/100 mL H_2O; and (e) $CHCl_2CH_2OH(l)$, nonionic, slightly soluble. Assume 1 g of each solute is mixed with 100 mL H_2O (the solubility of each substance is given above). *Answers*: (a) $CBr_4(l) \rightleftarrows CBr_4(aq)$, (b) $PbSO_4(s) \overset{H_2O}{\rightleftarrows} Pb^{2+} + SO_4{}^{2-}$, (c) $CH_3CH_2COOH(aq) + H_2O(l) \rightleftarrows H_3O^+ + CH_3CH_2COO^-$, (d) $LaSO_4(s) \xrightarrow{H_2O} La^{2+} + SO_4{}^{2-}$, (e) $CHCl_2CH_2OH(l) \overset{H_2O}{\rightleftarrows} CHCl_2OH(aq)$.

PRACTICE TEST

1 (25 points). Write the Lewis structure for water. Describe the bonding in the molecule. Prepare a three-dimensional sketch for the molecule. Identify the molecular geometry. What intermolecular forces are present in water?

2 (15 points). The mineral melanterite contains 20.1 mass % Fe, 11.5 mass % S, 63.3 mass % O, and 5.1 mass % H. Find the empirical formula and write a reasonable formula for the hydrate.

3 (15 points). Write the chemical equation for the partner-exchange reaction between washing soda ($Na_2CO_3 \cdot 10H_2O$) and Mg^{2+} during the water-softening process of a solution containing Mg^{2+}. What mass of washing soda is needed to react with 1.00 L of hard water containing 1.5 ppm Mg^{2+}? Note: 1 ppm ≃ 1 mg/L.

4 (15 points). Calculate the mass of ice that could be melted at 0 °C by the heat liberated as 100.0 g of steam is condensed to liquid at 100 °C. The heat of fusion of water is 6.0095 kJ/mol and the heat of vaporization of water is 40.6563 kJ/mol.

5 (30 points). Choose the response that best completes each statement.

(a) Which will have the highest vapor pressure?

(i) $H_2O(l)$ (ii) $D_2O(l)$

(iii) $H_2O(s)$ (iv) $D_2O(s)$

(b) The solid-to-gas phase transformation for water at 10 °F is

(i) efflorescence (ii) sublimation

(iii) hydration (iv) complexation

(c) Water is a remarkably versatile solvent because

(i) electrons repel each other

(ii) oxygen is highly electronegative

(iii) strong dipole-dipole bonding and hydrogen bonding between solvent and solute are possible

(iv) none of these answers

(d) The heat of hydration of an ion

(i) increases with decreasing size of the ion

(ii) increases with increasing charge of the ion

(iii) is exothermic

(iv) all of these answers

(e) Once chemical equilibrium has been established in a system,

(i) equal amount of reactants and products are present

(ii) no change occurs

(iii) the rate of the forward and reverse reactions are the same
(iv) all of these answers

(f) A reaction in which the water molecule is split is
(i) hydration (ii) hydrolysis
(iii) neutralization (iv) complexation

(g) A compound in which a water molecule is incorporated into a crystalline lattice is
(i) a deliquescent compound (ii) an efflorescent compound
(iii) an anhydrous compound (iv) a hydrate

(h) What is the mass percent of water in $CuSO_4 \cdot 5H_2O$?
(i) 36.1 % (ii) 7.21 %
(iii) 71.4 % (iv) 10.1 %

(i) The major type of solvent-solute intermolecular force as methyl alcohol dissolve in water is
(i) London forces (ii) dipole-dipole interaction
(iii) ion-dipole forces (iv) hydrogen bonding

(j) The term used to designate the process by which ions are formed as a molecular substance dissolves is
(i) ionization (ii) dissociation
(iii) efflorescence (iv) deliquescence

(k) Which is *not* true in a solution of a nonelectrolyte?
(i) Ions move freely in solution.
(ii) Electrodes attract oppositely-charged ions.
(iii) Solvation of the solute occurs.
(iv) answers (i) and (ii)

(l) Which is *not* true concerning the properties of acids, bases, and salts?
(i) Acids have a sour taste.
(ii) Acids are neutralized by bases.
(iii) Bases turn blue litmus red.
(iv) Aqueous solutions of bases are electrolytes.

(m) A neutral aqueous solution is one in which
(i) red litmus turns blue
(ii) blue litmus turns red
(iii) the molar concentration of hydroxide ions equals the molar concentration of hydrogen ions
(iv) none of these answers

Fig. 14-1

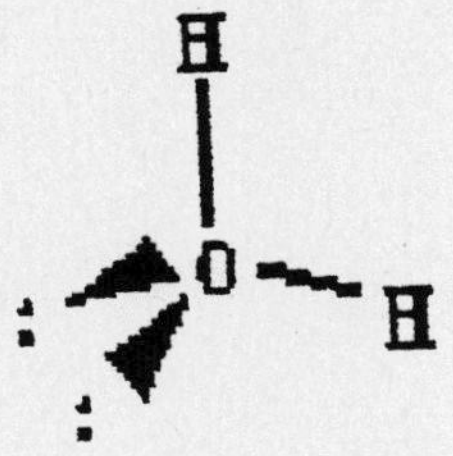

(n) Acetic acid, CH_3COOH, is a weak acid because

(i) it does not ionize

(ii) it ionizes to only a small extent

(iii) it forms OH^- ions in solution

(iv) it does not dissolve readily in water

(o) Which of the following is *not* a base according to the water-ion theory?

(i) $NaNO_3$ (ii) $Mg(OH)_2$

(iii) KOH (iv) all are bases

Answers to practice test

1. $H-\overset{\cdot\cdot}{\underset{\cdot\cdot}{O}}-H$; polar covalent; see Fig. 14-1; angular or bent (type AB_2E_2); London forces, dipole-dipole interactions, and hydrogen bonding.
2. $FeSO_{11}H_{14}$ or $FeSO_4 \cdot 7H_2O$.
3. $Mg^{2+} + Na_2CO_3.10H_2O(s) \rightarrow 2Na^+ + MgCO_3(s) + 10H_2O(l)$, 0.018 g.
4. 676.6 g.
5. (a) i, (b) ii, (c) iii, (d) iv, (e) iii, (f) ii, (g) iv, (h) i, (i) iv, (j) i, (k) iv, (l) iii, (m) iii, (n) ii, (o) i.

ADDITIONAL NOTES

CHAPTER 15

SOLUTIONS AND COLLOIDS

CHAPTER OVERVIEW

This chapter presents a qualitative description of the types of solutions (gases in gases, gases in liquids, etc.) and of the properties of ideal and real solutions of electrolytes and nonelectrolytes. Next, ways of expressing solution concentrations--mass percent, molarity, mole fraction, and molality--are considered, and problems involving dilution are explained. The next sections treat the colligative properties of liquid solutions. Finally, the properties and types of colloids are discussed.

COMPETENCIES

Definitions for the following terms should be learned:

- adsorption
- aerosol
- azeotropes
- colligative property
- colloid
- colloidal dispersion
- distillate
- distillation
- emulsion
- foam
- fractional distillation
- gel
- Henry's law
- ideal solution of a molecular solute
- ideal solution of an ionic solute
- immiscible
- infinitely miscible
- mass percent, weight percent
- miscibility
- molality
- osmosis
- osmotic pressure
- Raoult's law
- refluxing
- saturated solution
- semipermeable membranes
- sol
- solubility
- supersaturated solution
- unsaturated solution
- vapor pressure lowering

General concepts and skills that should be learned include

- the types of solutions and colloids
- the mechanisms by which solutions and colloids form
- the dependence of solubility on pressure and temperature
- the differences in properties of solutions compared to the pure solvent
- how nonideal (or real) solutions differ from ideal solutions
- the differences in properties between solutions of electrolytes and nonelectrolytes

Numerical exercises that should be understood include

- calculating the concentration of a gas in solution by using Henry's law
- expressing the concentration of a solution in units of mass percent, molarity, mole fraction, and molality
- calculating the vapor pressure above a solution by using Raoult's law
- calculating the colligative properties--vapor pressure lowering, boiling point elevation, freezing point depression, and osmotic pressure--of a solution from a known concentration
- determining the molar mass of a compound from measured colligative properties
- calculating and interpreting colligative properties for electrolytic solutions

QUESTIONS

1. Complete the following sentences by filling in the various blanks with these terms: colligative property, Henry's law, ideal solution, immiscible, infinitely miscible, osmosis, osmotic pressure, partially miscible, Raoult's law, saturated solution, semipermeable membrane, solubility, supersaturated solution, and unsaturated solution.

Depending on the amount of solute, a solution can be classified as an (a)______ if more solute can be dissolved, as a (b)______ if more than the equilibrium amount is dissolved, and as a (c)______ if the amount of solute dissolved is equal to that amount which would be in dynamic equilibrium with excess undissolved solute. If the solute and solvent are both liquids, the terminology used to describe the (d)______ is (e)______ if the liquids are mutually soluble in all proportions, (f)______ if mutally insoluble in all proportions, and (g)________ if there is a limited range of mutual solubility. The solubility of a gas in a liquid is quantitatively described by (h)________.

An (i)________ can be defined in several ways, but one criterion it must obey is (j)________, which describes the vapor pressure of the solution. The vapor pressure lowering of a solution is an example of a (k)________ (a property of a solution that depends on the number, not the type, of solute particles dissolved). Another such property is (l)________, in which solvent particles move through a (m)________ from a solution having a higher concentration of solvent to one having a lesser concentration. This solvent flow can be stopped by adding pressure to the side of the system containing the more concentrated solution. This pressure is called the (n)________.

2. Match the correct term to each definition.

(a) a mixture in which particles remain suspended and cannot be removed by filtration
(b) a brief name for a colloidal dispersion
(c) gas bubbles suspended in a liquid
(d) liquid particles suspended in a liquid
(e) a special type of sol in which the solid particles unite in a random and intertwined structure that gives rigidity to the mixture
(f) a colloidal dispersion in which a gas is the suspending medium
(g) solid particles suspended in a liquid

(i) aerosol
(ii) colloid
(iii) colloidal dispersion
(iv) emulsion
(v) foam
(vi) gel
(vii) sol

3. Match the correct term to each equation.

(a) $\frac{\text{mass of solute}}{\text{mass of solution}} \times 100$

(b) $\frac{\text{no. of moles of solute}}{\text{soln vol. in L}}$

(c) $\frac{\text{no. of moles of solute}}{\text{solvent mass in kg}}$

(d) $\frac{\text{no. of moles of solute}}{\text{no. of moles of solute + no. of moles of solvent}}$

(e) $P = kC$

(f) $P = XP°$

(i) Henry's law
(ii) mass percent
(iii) molality
(iv) molarity
(v) mole fraction
(vi) Raoult's law

4. Which of the following statements are true? Rewrite each false statement so that it is correct.

(a) The solubility of naphthalene would be predicted to be greater in water than in benzene.

(b) As a solution freezes, (i) the solvent crystallizes, (ii) the solution becomes more concentrated, (iii) the freezing point is depressed even more, and (iv) the freezing takes place over a range of temperatures.

(c) An emulsifying agent stabilizes a colloidal dispersion of solid in a liquid.

(d) Solubility is the maximum amount of solute that can be dissolved in a solvent.

(e) Nonideal liquid-liquid solutions show positive and/or negative deviations from ideality.

5. Distinguish clearly among/between (a) a suspension, a solution, and a colloidal dispersion; (b) solvent and solute; (c) infinitely miscible, partially miscible, and immiscible substances; (d) saturated, unsaturated, and supersaturated solutions; (e) a sol and a gel; and (f) smoke and fog.

6. Choose the response that best completes each statement.

(a) The process of vapor moving up a distillation column, condensing, and trickling down the column is called

(i) fractional distillation (ii) refluxing
(iii) osmosis (iv) colloidalal dispersion

(b) The unit for expressing concentration in terms of moles of solute per kilogram of solvent is

(i) mole fraction (ii) molarity
(iii) molality (iv) mass percentage

(c) A constant-boiling mixture that distills without a change in composition is

(i) an azeotrope (ii) fractional distillation
(iii) a nonelectrolyte (iv) a gel

(d) The product of distillation is called the

(i) reflux (ii) aerosol
(iii) distillate (iv) foam

(e) The separating of the components of a solution by the heating to boiling and the collecting and condensing of vapors is

(i) distillation (ii) fractional distillation
(iii) supersaturation (iv) refluxing

7. Identify the solvent and the solute(s) in each of the following solutions: (a) household ammonia, (b) a carbonated soft drink, and (c) the coolant in an automobile radiator.

8. A solution may be defined as "a homogeneous mixture of two or more substances." What does the word "homogeneous" mean? Does a mixture have variable or fixed composition? What is the name given to the substance in a solution that is usually present in the larger amount?

9. By what terms are solutions referred to in which the concentration of solute is (a) greater than, (b) equal to, or (c) less than the solubility?

10. State Henry's law in words and in equation form. Several textbooks write Henry's law in the form

$$C = k'P_i$$

What is the relationship between k' and k, the two different Henry's law constants?

11. Why are very few solutions considered to be ideal solutions? As the solute in a real solution is diluted, would you expect the solution to become more or less ideal?

12. Write an equation that defines the mole fraction of solute. What is the relationship between the mole fraction of solute and the mole fraction of solvent in a binary solution?

13. During a dilution process, which component of the solution remains constant in terms of mass (or number of moles)? Many times the equation

$$C_1V_1 = C_2V_2$$

is used to relate the initial concentration (C_1) and volume (V_1) to the final, dilute concentration (C_2) and volume (V_2). With which concentration unit(s) will this equation be valid?

14. Define the term "colligative property." What colligative properties do we consider in this chapter? How do colligative properties depend on the nature of the solute particles?

15. Is the vapor pressure of a solution containing a nonvolatile solute greater or less than that of the pure solvent? How does this change in vapor pressure affect the boiling point of the solution?

16. State Raoult's law in words and in equation form. How is the total vapor pressure of a solution calculated?

17. Describe what happens as a solution begins to freeze. How does the freezing point of a solution compare to that of the pure solvent?

18. Write the equations used for calculating the boiling point elevation and the freezing point depression of a solution. Define all symbols.

19. Explain how measurements of colligative properties can be used to determine the molar mass of a nonionic substance.

20. What is a semipermeable membrane? What is the role of these membranes during osmosis?

21. What is meant by the term "osmotic pressure of a solution?" How is it related to the concentration of the solution?

22. If NaCl is completely dissociated in aqueous solution, it gives less than twice the molal freezing point lowering than does an equivalent number of moles of a nonelectrolyte. Why?

Answers to questions

1. (a) unsaturated solution, (b) supersaturated solution, (c) saturated solution, (d) solubility, (e) infinitely miscible, (f) immiscible, (g) partially miscible, (h) Henry's law, (i) ideal solution, (j) Raoult's law, (k) colligative property, (l) osmosis, (m) semipermeable membrane, (n) osmotic pressure.
2. (a) iii, (b) ii, (c) v, (d) iv, (e) vi, (f) i, (g) vii.
3. (a) ii, (b) iv, (c) iii, (d) v, (e) i, (f) vi.
4. (a) F, replace *greater* by *less*; (b) T; (c) F, replace *solid* by *liquid*; (d) F, add *under equilibrium conditions* after *solvent*; (e) T.
5. (a) A *suspension* settles quickly; a *solution* does not settle because it contains dispersed atoms, molecules, and ions; a *colloid* does not settle because it is stabilized by electrical charges. (b) The *solvent* is the major substance in solution; the *solute* dissolves in the solvent. (c) *Infinitely miscible substances* mix in all proportions; *partially miscible substances* form solutions within limited ranges; *immiscible substances* are insoluble. (d) *Saturated solutions* are equilibrium systems (or potentially so) between the solution and undissolved solute; *unsaturated solutions* can dissolve more solute; *supersaturated solutions* contain more solute than the equilibrium

amount. (e) A *sol* is a colloidal dispersion of a solid in a liquid; a *gel* is a special type of sol in which the solid particles unite to form a rigid structure. (f) *Smoke* is a suspension of a solid in a gas; *fog* is a suspension of a liquid in a gas.

6. (a) ii, (b) iii, (c) i, (d) iii, (e) i.
7. (a) NH_3 (solute) in H_2O (solvent); (b) CO_2, flavorings, and coloring agents (solutes) in H_2O (solvent); (c) ethylene glycol and coloring (solutes) in H_2O (solvent).
8. Homogeneous means having a uniform composition throughout a phase. A mixture has a variable composition. The substance in a solution that is usually present in the larger amount is the solvent.
9. Solutions in which the concentration of solute is (a) greater than the solubility are supersaturated, (b) equal to the solubility are saturated, and (c) less than the solubility are unsaturated.
10. The partial pressure of a gas above a solution is proportional to the concentration of the gas in the solution; $P_i = kC$; $k' = (1/k)$.
11. Interactions between solute particles cause nonideal behavior. The solution will become more ideal.
12. $$X_B = \frac{n_B}{n_A + n_B} \qquad X_B = 1 - X_A$$
13. The amount of solute remains constant. The equation is valid for any volume-based concentration unit such as molarity.
14. A colligative property is any property of a solution that depends on the relative numbers of solute and solvent particles. The colligative properties considered in Chapter 15 are vapor pressure lowering, boiling point elevation, freezing point depression, and osmotic pressure. Colligative properties depend only on the quantity of the solute particles, and not on their nature.
15. The vapor pressure of a solution containing a nonvolatile solute is less than that of the pure solvent. The lowering of the vapor pressure makes the boiling point of the solution higher than that of the pure solvent.
16. The partial pressure of a gas above a solution is given by the product of the mole fraction of that component in the solution and the vapor pressure of that component if it were pure; $P_i = X_i P_i^\circ$. Dalton's law, $P_t = \Sigma P_i$, is used to calculate the total pressure.

17. As a solution begins to freeze, crystals of pure solvent form. The freezing point of a solution is lower than that of the pure solvent.
18. $\Delta T_b = K_b\, m$ where ΔT_b is the increase in the boiling point, K_b is the boiling point constant of the solvent and m is the molality of the solution; $\Delta T_f = K_f\, m$ where ΔT_f is the decrease in the freezing point, K_f is the freezing point constant of the solvent and m is the molality.
19. Calculate the molality from the colligative property, calculate the number of moles of solute from the molality, and calculate the molar mass from the number of moles.
20. A semipermeable membrane allows the passage of some types of molecules, but not others. In this case the membrane allows solvent molecules to pass from the more dilute solution or pure solvent to the more concentrated solution while solute molecules do not pass through the membrane.
21. The osmotic pressure of a solution is the external pressure exactly sufficient to oppose osmosis and stop it. Its relationship to the concentration of a solution is expressed by $\Pi = MRT$.
22. The ions interact with each other in solution to form clusters and the effective ionic concentration is somewhat less than twice the concentration of the salt.

SKILLS

Skill 15.1 Concentration of solutions

Because a solution is a mixture that can have a variable composition, it is necessary to specify the concentration when discussing the solution. Many concentration units are used for this purpose. Of primary interest to the chemist are three units that are related to mass:

$$\text{mass percent} = \frac{\text{mass of solute}}{\text{mass of solute + mass of solvent}} \times 100$$

$$\text{molality} = \frac{\text{number of moles of solute}}{\text{mass of solvent in kilograms}}$$

$$\text{mole fraction} = \frac{\text{number of moles of solute}}{\text{number of moles of solute + number of moles of solvent}}$$

and one unit related to volume of solution:

$$\text{molarity} = \frac{\text{number of moles of solute}}{\text{solution volume in liters}}$$

The definitions above are all given for binary solutions.

The conversion from the mass-based units to molarity is accomplished by using the density of the solution (see *Example 2.13*). As an example of the conversion, consider the so-called dilute nitric acid solution commonly found in general chemistry laboratories. An analysis would show that it contains 32 mass % HNO_3 and has a density of 1.19 g/mL. A "basis" is chosen to do the calculations. Any choice is possible, but usually 1000 g of solution or 1000 mL of solution is the most convenient. In this case, choosing 1000 g of solution immediately gives the following values, which can be used in the definitions of the concentration units:

$$(1000 \text{ g solution})\left[\frac{32 \text{ g } HNO_3}{100 \text{ g solution}}\right] = 320 \text{ g } HNO_3$$

$$1000 \text{ g solution} - 320 \text{ g } HNO_3 = 680 \text{ g } H_2O = 0.68 \text{ kg } H_2O$$

$$(1000 \text{ g solution})\left[\frac{1 \text{ mL}}{1.19 \text{ g solution}}\right]\left[\frac{1 \text{ L}}{1000 \text{ mL}}\right] = 0.840 \text{ L solution}$$

$$(320 \text{ g } HNO_3)\left[\frac{1 \text{ mol } HNO_3}{63.0 \text{ g } HNO_3}\right] = 5.1 \text{ mol } HNO_3$$

$$(680 \text{ g } H_2O)\left[\frac{1 \text{ mol } H_2O}{18.0 \text{ g } H_2O}\right] = 38 \text{ mol } H_2O$$

The concentration of dilute nitric acid in the various units is

$$\text{molality} = \frac{\text{no. of moles of } HNO_3}{\text{mass of solvent in kg}} = \frac{5.1 \text{ mol}}{0.68 \text{ kg}} = 7.5 \text{ mol/kg} = 7.5 \text{ m}$$

$$\text{mole fraction} = X = \frac{\text{no. of moles of } HNO_3}{\text{no. of moles of } HNO_3 + \text{no. of moles of } H_2O}$$

$$= \frac{5.1 \text{ mol}}{5.1 \text{ mol} + 38 \text{ mol}} = 0.12$$

$$\text{molarity} = \frac{\text{no. of moles of } HNO_3}{\text{soln vol. in L}} = \frac{5.1}{0.84 \text{ L}} = 6.1 \text{ mol/L} = 6.1 \text{ M}$$

Skill Exercise: Express the concentration in mass percent, molarity, molality, and mole fraction of $Pb(NO_3)_2$ for a solution containing 10.00 g $Pb(NO_3)_2$ in 100.00 g solution. The density of the solution is 1.093 g/mL. *Answers*: 10.00 mass %, 0.3300 M, 0.3354 m, $X_{Pb(NO_3)_2} = 0.006009$.

NUMERICAL EXERCISES

Example 15.1 Solubility: Henry's law

The value of the constant in Henry's law is 5.99×10^4 L atm/mol for solutions of O_2 in water at 25 °C. Assuming $P_{O_2} = 0.20$ atm for air, calculate the concentration of O_2 in water that is in equilibrium with air.

Substituting the values into Henry's law gives

$$C = \frac{P}{k} = \frac{0.20 \text{ atm}}{5.99 \times 10^4 \text{ L atm/mol}} = 3.3 \times 10^{-6} \text{ mol/L}$$

The concentration of O_2 in water that is in equilibrium with air is 3.3 μM.

Exercise 15.1 The solubility at 25 °C of HCl in toluene is 0.117 mol/L at $P_{HCl} = 0.282$ atm and 0.0655 mol/L at $P_{HCl} = 0.158$ atm. Using these data, find the value of the constant in Henry's law for this temperature. *Answer*: 2.41 L atm/mol.

Example 15.2 Solution concentration

A solution is prepared by mixing 10.00 g of ethanol, C_2H_5OH, in 90.00 g of H_2O. Express the concentration of this solution using mole fractions of both components.

The number of moles of each component is

$$(10.00 \text{ g } C_2H_5OH)\left[\frac{1 \text{ mol } C_2H_5OH}{46.07 \text{ g } C_2H_5OH}\right] = 0.2171 \text{ mol } C_2H_5OH$$

$$(90.00 \text{ g } H_2O)\left[\frac{1 \text{ mol } H_2O}{18.02 \text{ g } H_2O}\right] = 4.994 \text{ mol } H_2O$$

The respective mole fractions are

$$X_{C_2H_5OH} = \frac{0.2171 \text{ mol}}{0.2171 \text{ mol} + 4.994 \text{ mol}} = 0.04166$$

$$X_{H_2O} = \frac{4.994 \text{ mol}}{0.2171 \text{ mol} + 4.994 \text{ mol}} = 0.9583$$

Note that $X_{C_2H_5OH} + X_{H_2O} = 1$.

Exercise 15.2 Express the concentration of a solution containing 5.50 g of methanol, CH_3OH, in 94.50 g of H_2O by using mole fractions of both components. *Answer*: $X_{CH_3OH} = 0.0318$, $X_{H_2O} = 0.9682$.

Example 15.3 Solution concentration

Express the concentration of the solution described in *Example 15.2* in terms of the molality of the solute.

For this solution, $n_{C_2H_5OH}$ = 0.2171 and the mass of the solvent in kilograms is 0.09000 kg. Substituting these data into the expression for molality gives

$$\text{molality} = \frac{0.2171\ \text{mol}}{0.09000\ \text{kg}} = 2.412\ \text{mol/kg} = 2.412\ \text{m}$$

The solution is 2.412 m in ethanol.

Exercise 15.3 What is the concentration of the solution described in Exercise 15.2 expressed in terms of molality of the solute? *Answer*: 1.82 m.

Example 15.4 Solution concentration

Express the concentration of the solution described in *Example 15.2* in terms of the molarity of the solute. The density of the solution is 0.984 g/mL.

The volume of solution is

$$(10.00\ \text{g}\ C_2H_5OH + 90.00\ \text{g}\ H_2O)\left[\frac{1\ \text{mL}}{0.984\ \text{g solution}}\right]\left[\frac{1\ \text{L}}{1000\ \text{mL}}\right] = 0.102\ \text{L}$$

and the molarity is

$$\text{molarity} = \frac{0.2171\ \text{mol}\ C_2H_5OH}{0.102\ \text{L}} = 2.13\ \text{mol/L} = 2.13\ \text{M}$$

Exercise 15.4 Express the concentration of the solution described in Exercise 15.2 in terms of molarity. The density of the solution is 0.991 g/mL. *Answer*: 1.70 mol/L.

Example 15.5 Solution concentration*

A stockroom attendant wants to prepare 3.00 L of a 5.0 mass % solution of hydrogen peroxide from the commercially available solution, which has a concentration of 30.0 mass % and a density of 1.1122 g/mL. The density of the dilute solution will be 1.0131 g/mL. What must be done?

This problem is solved by using the comprehensive problem-solving approach presented in *Skill 2.7 Solving problems*.

1. Study the problem and be sure you understand it.
   (a) What is unknown?
   The volume of the concentrated solution to be diluted with water to make 3.00 L of dilute solution.

(b) What is known?

The concentrations and densities of the two solutions.

2. Decide how to solve the problem.

(a) What is the connection between the known and the unknown?

The mass of H_2O_2 in the volume of concentrated solution to be diluted must be the same as it will be in the dilute solution. To describe the dilution process in terms of volume of solutions, the densities and mass percent concentrations of the solutions must be used to connect masses and volumes.

(b) What is necessary to make the connection?

Use the density and mass percent of the dilute solution to find the mass of H_2O_2 that it must contain and use the mass percent and density of the concentrated solution to find the volume of concentrated solution to be diluted.

3. Set up the problem and solve it.

The mass of H_2O_2 in 3.00 L of the dilute solution is

$$(3.00\ \text{L})\left[\frac{1000\ \text{mL}}{1\ \text{L}}\right]\left[\frac{1.0131\ \text{g}}{1\ \text{mL}}\right] = 3040\ \text{g solution}$$

$$(3040\ \text{g solution})\left[\frac{5.0\ \text{g}\ H_2O_2}{100.0\ \text{g solution}}\right] = 150\ \text{g}\ H_2O_2$$

The volume of the concentrated solution that contains 150 g H_2O_2 is

$$(150\ \text{g}\ H_2O_2)\left[\frac{100.0\ \text{g solution}}{30.0\ \text{g}\ H_2O_2}\right] = 5.0 \times 10^2\ \text{g solution}$$

$$(5.0 \times 10^2\ \text{g solution})\left[\frac{1\ \text{mL}}{1.1122\ \text{g solution}}\right] = 450\ \text{mL solution}$$

The stockroom attendant must take 450 mL of the concentrated solution and add enough water to make 3.00 L of solution.

4. Check the result.

(a) Are significant figures and the location of the decimal point correct?

Yes.

(b) Did the answer come out in the correct units?

Yes.

(c) Is the answer reasonable?

Yes. The dilute solution is roughly one-sixth the concentration of the concentrated solution, so using 450 mL and diluting to 3.00 L seems reasonable.

Exercise 15.5 Exactly 10 mL of a NaOH solution of unknown concentration was diluted with water to 100.00 mL. The NaOH concentration of the latter solution was determined to be 0.1026 M. What was the concentration of NaOH in the original solution? *Answer*: 1.026 M.

Example 15.6 Raoult's law

The respective vapor pressures of pure ethanol and water at 30 °C are 72 Torr and 18 Torr. Calculate the partial pressure of each substance over the solution described in *Example 15.2* and find the total vapor pressure above the solution.

The partial pressure for each component of the solution, according to Raoult's law, is

$$P_{C_2H_5OH} = X_{C_2H_5OH}\, P^\circ_{C_2H_5OH} = (0.04165)(72\text{ Torr}) = 3.0\text{ Torr}$$

$$P_{H_2O} = X_{H_2O}\, P^\circ_{H_2O} = (0.9584)(18\text{ Torr}) = 17\text{ Torr}$$

and, according to Dalton's law,

$$P_t = P_{C_2H_5OH} + P_{H_2O} = 3.0\text{ Torr} + 17\text{ Torr} = 20.\text{ Torr}$$

The respective partial pressures are 3.0 Torr and 17 Torr and the total pressure is 20. Torr.

Exercise 15.6 The respective vapor pressures of pure methanol and water at 30 °C are 147 Torr and 18 Torr. Calculate the partial pressure of each substance over the solution described in Exercise 15.2 and find the total vapor pressure above the solution. *Answer*: P_{CH_3OH} = 4.67 Torr, P_{H_2O} = 17 Torr, P_t = 22 Torr.

Example 15.7 Colligative properties

Calculate the freezing point, boiling point, and osmotic pressure at 0 °C of a solution containing 4.36 g of D-fructose, $C_6H_{12}O_6$, in 100.00 g of solution. The density of the solution is 1.015 g/mL and the freezing point and boiling point constants for water are 1.86 °C kg/mol and 0.512 °C kg/mol, respectively.

The molality of the solution is

$$(4.36\text{ g } C_6H_{12}O_6)\left[\frac{1\text{ mol } C_6H_{12}O_6}{180.16\text{ g } C_6H_{12}O_6}\right] = 0.0242\text{ mol } C_6H_{12}O_6$$

$$\frac{0.0242\text{ mol } C_2H_{12}O_6}{(100.00\text{ g solution} - 4.36\text{ g } C_6H_{12}O_6)}\left[\frac{1000\text{ g}}{1\text{ kg}}\right] = 0.253\text{ mol/kg}$$

The change in the boiling and freezing points are

$$\Delta T_b = K_b\, m = (0.512\ °C\ kg/mol)(0.253\ mol/kg) = 0.130\ °C$$

$$\Delta T_f = K_f\, m = (1.86\ °C\ kg/mol)(0.253\ mol/kg) = 0.471\ °C$$

Thus the normal boiling and freezing points of the solution will be

$$T_{b,soln} = T_{b,solv} + \Delta T_b = 100.000\ °C + 0.130\ °C = 100.130\ °C$$

$$T_{f,soln} = T_{f,solv} - \Delta T_f = 0.000\ °C - 0.471\ °C = -0.471\ °C$$

The molarity of the solution is found by using the density of the solution

$$(100.00\ g\ solution)\left[\frac{1\ mL}{1.015\ g}\right]\left[\frac{1\ L}{1000\ mL}\right] = 0.09852\ L$$

$$\frac{0.0242\ mol\ C_6H_{12}O_6}{0.09852\ L} = 0.246\ mol/L$$

and the osmotic pressure is

$$\Pi = MRT = \left[\frac{0.246\ mol}{1\ L}\right](0.0821\ L\ atm/K\ mol)(273\ K) = 5.51\ atm$$

For the solution, the normal boiling point is 100.130 °C, the freezing point is -0.471 °C, and the osmotic pressure is 5.51 atm.

Exercise 15.7 The density of a 10.00 mass % solution of sucrose, $C_{12}H_{22}O_{11}$, in water is 1.038 g/mL. Calculate the (a) molality and (b) molarity and determine the (c) freezing point, (d) normal boiling point, and (e) osmotic pressure at 35 °C for the solution. The freezing point constant is 1.86 °C kg/mol and the boiling point constant is 0.512 °C kg/mol for water. *Answers*: (a) 0.3246 m, (b) 0.3032 M, (c) -0.604 °C, (d) 100.166 °C, (e) 7.67 atm.

Example 15.8 Colligative properties

What is the molar mass of an unknown that changes the freezing point of benzene from 5.72 °C to 3.54 °C when 0.415 g of the sample is dissolved in 25.0 g of benzene? K_f = 4.90 °C kg/mol for C_6H_6.

The freezing point depression is

$$\Delta T_f = 5.72\ °C - 3.54\ °C = 2.18\ °C$$

which corresponds to a concentration of

$$m = \frac{\Delta T_f}{K_f} = \frac{2.18\ °C}{4.90\ °C\ kg/mol} = 0.445\ mol/kg$$

The number of moles of solute in the solution is

$$\left[\frac{0.455 \text{ mol solute}}{1 \text{ kg solvent}}\right](0.0250 \text{ kg solvent}) = 0.0111 \text{ mol}$$

which corresponds to

$$\frac{0.415 \text{ g}}{0.0111 \text{ mol}} = 37.4 \text{ g/mol}$$

The molar mass of the unknown is 37.4 g/mol.

Exercise 15.8 The same amount of compound described in *Example 15.8* was dissolved in 25.0 g of water (K_f = 1.86 °C kg/mol) and a freezing point change of 1.57 °C was observed. Calculate the molar mass of the substance from these data and explain the difference in answers. *Answer*: 19.7 g/mol, ionized in water into two particles.

Example 15.9 Colligative properties*

The freezing point depression of a 0.100 m solution of the weak base $M(OH)_2$ is 0.203 °C. What fraction of the base is ionized? What is the percentage ionization of the base? K_f = 1.86 °C kg/mol for H_2O.

This problem is solved by using the comprehensive problem-solving approach presented in *Skill 2.7 Solving problems*.

1. Study the problem and be sure you understand it.
   (a) What is unknown?
   The fraction of $M(OH)_2$ that ionizes.
   (b) What is known?
   The freezing point depression for the solution and the fact that one M^{2+} and two OH^- ions are produced for each $M(OH)_2$ that ionizes. Also, the concentration of the solution and the freezing point constant for water.
2. Decide how to solve the problem.
   (a) What is the connection between the known and the unknown?
   Colligative properties are dependent on the total number of solute particles, not on their nature. The molality of a solution is directly proportional to the freezing point depression.
   (b) What is necessary to make the connection?
   Determine the molality of the solute particles in terms of the fraction of molecules that are ionized and use this concentration in the equation for freezing point depression to solve for the value of the fraction that are ionized.

3. Set up the problem and solve it.

If X represents the fraction of $M(OH)_2$ that ionizes, then the concentrations of the solute particles are

$$\left[\frac{0.100 \text{ mol } M(OH)_2}{1 \text{ kg}}\right]\left[\frac{X \text{ mol } M^{2+}}{1 \text{ mol } M(OH)_2}\right] = (0.100)X \text{ mol } M^{2+}/\text{kg}$$

$$\left[\frac{0.100 \text{ mol } M(OH)_2}{1 \text{ kg}}\right]\left[\frac{2X \text{ mol } OH^-}{1 \text{ mol } M(OH)_2}\right] = (0.200)X \text{ mol } OH^-/\text{kg}$$

$$\left[\frac{0.100 \text{ mol } M(OH)_2}{1 \text{ kg}}\right](1 - X) = [0.100 - (0.100)X] \text{ mol } M(OH)_2/\text{kg}$$

and the total concentration of particles is

$$(0.100)X + (0.200)X + [0.100 - (0.100)X] = [0.100 + (0.200)X] \text{ mol/kg}$$

The freezing point depression is a function of the total concentration.

$$\Delta T_f = K_f \, m$$

$$0.203 \text{ °C} = (1.86 \text{ °C kg/mol})\{[0.100 + (0.200)X] \text{ mol/kg}\}$$

$$X = 0.046$$

The fraction of molecules of the base ionized is 0.046. The base is 4.6% ionized in this solution.

4. Check the result.

(a) Are significant figures and the location of the decimal point correct?

Yes.

(b) Did the answer come out in the correct units?

Yes.

(c) Is the answer reasonable?

Yes. If the $M(OH)_2$ were fully ionized, the freezing point depression would have been (1.86)(0.300) = 0.558 °C, and if no ionization occurred, the depression would have been (1.86)(0.100) = 0.186 °C. Because the observed depression is 0.203 °C, the fraction should be quite low.

Exercise 15.9 Acetic acid (a monoprotic acid) is 4.2 % ionized in a 0.0100 m aqueous solution. Predict the freezing point of the solution given K_f = 1.86 °C kg/mol. *Answer*: -0.0194 °C.

PRACTICE TEST

1 (10 points). Match the correct term to each definition.

(a) the attraction of a substance to the surface of a solid

(b) the separation of the components of a solution by the heating to boiling and the collecting and condensing of the vapors

(c) a constant-boiling mixture that distills without a change in composition

(d) the product of distillation

(e) process that separates liquid mixture into fractions that differ in boiling points

(f) a property of the solvent that is affected in proportion to the concentration of solute particles but is independent of the nature of those particles

(g) a measure of the amount of solute that can dissolve in a given amount of solvent

(h) a solution in which the concentration of dissolved solute is equal to that which would be in equilibrium with undissolved solute under the conditions

(i) a solution that can still dissolve more solute than would be in equilibrium with undissolved solute

(j) a solution that holds more dissolved solute than would be in equilibrium with undissolved solute

(i) adsorption
(ii) azeotrope
(iii) colligative property
(iv) distillate
(v) distillation
(vi) fractional distillation
(vii) saturated solution
(viii) solubility
(ix) supersaturated solution
(x) unsaturated solution

2 (15 points). Complete the following table describing colloids.

Dispersing medium	Dispersed substance	Type of colloid	Example
liquid	gas		
liquid	liquid		
liquid	solid		
gas	liquid		
gas	solid		
solid	gas		
solid	liquid		
solid	solid		

3 (20 points). Choose the response that best completes each statement.

(a) The substance with the greatest solubility in water is

(i) $CH_3CH_2CH_2CH_2CH_3$ (ii) $CH_3CH_2CH_2CH_2OH$

(iii) CH_3OCH_3 (iv) KNO_3

(b) Sodium chloride is less soluble in ethyl alcohol than in water at the same temperature because

(i) alcohol has a lower boiling point than water

(ii) alcohol is less effective than water in solvating the ions

(iii) alcohol is soluble in water

(iv) none of these answers

(c) The solubility of a certain salt in water is 25 g/L at 25 °C and 75 g/L at 50 °C. A 500 mL saturated solution of this salt at 50 °C is cooled to 25 °C. What mass of salt will crystallize out of solution?

(i) 50 g (ii) 75 g

(iii) 25 g (iv) 100 g

(d) A solution was prepared by mixing 25 mL of 0.10 M NaCl and 35 mL of 0.25 M $NaNO_3$. The concentration of chloride ion in the final solution is

(i) 0.042 mol/L (ii) 0.050 mol/L

(iii) 0.10 mol/L (iv) 0.021 mol/L

(e) A solution contains 5.0 g KNO_3 dissolved in 95.0 g H_2O. The concentration is

(i) 5.0 mass % (ii) 5.0 m

(iii) 5.0 M (iv) 0.050 M

(f) A 1 M solution of KNO_3 was mixed with an equal amount of water. The number of moles of solute in the final solution is

(i) one-half the original value

(ii) the same as in the original solution

(iii) twice the original value

(iv) none of these answers

(g) A 0.10 m solution of a substance has a boiling point elevation of 3.0 °C. Another solution of the same compound has a freezing point depression of 6.0 °C. What is the concentration of the second solution?

(i) 0.20 m (ii) 0.10 m

(iii) 0.050 m (iv) indeterminate from the data

(h) Raoult's law states that the vapor pressure of a solute

(i) depends on the solvent

(ii) is equal to the vapor pressure of the pure solute times the mole fraction of solute

(iii) equals the vapor pressure of the solvent

(iv) none of these answers

(i) Which aqueous solution would have the greatest osmotic pressure?

(i) 0.01 M CH_3OH (ii) 0.01 M K_2SO_4

(iii) 0.01 M NaCl (iv) 0.01 M K_3PO_4

(j) Which of these is *not* a property of colloids?

(i) Brownian movement

(ii) the Tyndall effect

(iii) precipitation by addition of electrolytes

(iv) small surface area

4 (25 points). The vapor pressure of pure ethanol, C_2H_5OH, is 72 Torr and that of pure methanol, CH_3OH, is 147 Torr at 30 °C. (a) Calculate the mole fraction of methanol in a solution containing 10.0 g of each alcohol. (b) Calculate the partial pressures of each alcohol above the solution. (c) Find the total vapor pressure above the solution. (d) Find the mole fraction of methanol in the vapor. (e) Comment on your answers for parts (a) and (d).

5 (10 points). The lattice energy for $FeCl_2(s)$ is -2598 kJ/mol and the hydration energy for $Fe^{2+}(g)$ is -2841 kJ/mol and for $Cl^-(g)$ is 79 kJ/mol. Calculate the heat of solution for $FeCl_2(s)$. Will the solubility of this salt increase or decrease with increasing temperature?

6 (20 points). The density of a 1.00 mass % solution of acetic acid, CH_3COOH, in water is 1.0014 g/mL. Express the concentration in terms of (a) molality and (b) molarity. The freezing point of this solution is -0.317 °C. (c) Calculate the fraction of molecules ionized. K_f = 1.86 °C kg/mol for water.

Answers to practice test

1. (a) i, (b) v, (c) ii, (d) iv, (e) vi, (f) iii, (g) viii, (h) vii, (i) x, (j) ix.
2. See Table 15.4 of the text.
3. (a) iv, (b) ii, (c) iii, (d) i, (e) i, (f) ii, (g) iv, (h) ii, (i) iv, (j) iv.

4. (a) 0.590; (b) P_{CH_3OH} = 86.7 Torr, $P_{C_2H_5OH}$ = 30. Torr; (c) P_t = 117 Torr; (d) X_{CH_3OH} = 0.741; (e) vapor above a liquid is richer in the more volatile component.
5. -85 kJ/mol, decrease.
6. (a) 0.169 mol/kg, (b) 0.167 mol/L, (c) 0.01.

ADDITIONAL NOTES

CHAPTER 16

HYDROGEN AND OXYGEN; OXIDATION-REDUCTION REACTIONS

CHAPTER OVERVIEW

In this chapter the terms oxidation, reduction, and redox are defined, and an introduction to oxidation-reduction is given. Balancing redox equations using the oxidation number method is discussed. The properties, reactions, preparations, and uses of hydrogen and oxygen are presented. Compounds of each element--the hydrides, hydroxides, and oxides--are examined, as are ozone, hydrogen peroxide, and various peroxides, superoxides, and ozonides.

COMPETENCIES

Definitions for the following terms should be learned:

- absorption
- acidic anhydride, acidic oxide
- allotropes
- amphoteric
- basic anhydride, basic oxide
- fossil fuels
- hydrocarbons
- hydrogenation
- oxidation
- oxidation-reduction reactions, redox reactions
- oxidizing agent
- reducing agent
- reduction

General concepts and skills that should be learned include

- identifying and interpreting oxidation-reduction reactions
- balancing redox equations by the oxidation number method
- the general physical and chemical properties; industrial and laboratory preparations; uses; and reactions with metals, nonmetals, and compounds for H_2 and O_2
- the general physical and chemical properties of ozone and hydrogen peroxide

QUESTIONS

1. Complete the following sentences by filling in the various blanks with these terms: acidic anhydride, basic anhydride, oxidation-reduction, oxide, oxidizing agent, and reducing agent.

   A reaction in which a change in oxidation numbers occurs is called an (a)_______ reaction. If the reactants are molecular oxygen reacting with another element, an (b)_______ will usually result. In this case oxygen is the (c)_______ and the other element is the (d)_______. If a metal oxide is added to water, usually an alkaline solution will result and the oxide is known as a (e)_______. If a nonmetal oxide is added to water, an acidic solution will result and the oxide is known as an (f)_______.

2. Match the correct term to each definition.

(a) capable of acting as either an acid or a base	(i) absorption
(b) compounds of carbon and hydrogen	(ii) acidic anhydride
(c) the addition of the two atoms of hydrogen to a compound	(iii) allotropes
(d) different forms of the same element in the same state	(iv) amphoteric
(e) an atom, molecule, or ion that causes another substance to undergo a decrease in oxidation state (reduction) and is itself oxidized	(v) basic anhydride
	(vi) hydrocarbons
	(vii) hydrogenation
	(viii) oxidizing agent
	(ix) reducing agent

   (f) an atom, molecule, or ion that causes another substance to undergo an increase in oxidation state (oxidation) and is itself reduced
   (g) a nonmetal oxide that combines with water to give an acid
   (h) the incorporation of one substance into another at the molecular level
   (i) a metal oxide that combines with water to give a hydroxide base

3. Which of the following statements are true? Rewrite each false statement so that it is correct.
   (a) Even though aluminum has many properties of a metal, $Al(OH)_3$ is not a particularly strong base.
   (b) If gaseous atomic oxygen is included, there are three allotropic forms of oxygen.

(c) The covalent hydrides are characterized by having nonstoichiometric formulas.

(d) Oxygen is such a good oxidizing agent that it reacts directly with all metals to form oxides.

(e) An atom of hydrogen can complete its "octet" of electrons by gaining an electron (forming H^-), sharing a pair of electrons in a covalent bond, or losing its electron (forming H^+).

4. Distinguish clearly between/among (a) acidic and basic anhydrides; (b) oxidation and combustion; (c) hydrogenation and reduction; (d) oxidation and reduction; (e) oxidizing and reducing agents; and (f) adsorption and absorption.

5. Choose the response that best completes each statement.

(a) Any process in which an oxidation number increases is

(i) oxidation (ii) reduction
(iii) equilibrium (iv) α decay

(b) An oxidizing agent undergoes

(i) oxidation (ii) reduction
(iii) static equilibrium (iv) isomerization

(c) Which is a *poor* natural source of hydrogen?

(i) the atmosphere
(ii) water
(iii) fossil fuels
(iv) proteins, fats, and carbohydrates

(d) Which is *not* a rich source of oxygen?

(i) water (ii) the atmosphere
(iii) the earth's crust (iv) the cosmos

(e) Which properties do *not* describe both hydrogen and oxygen?

(i) colorless, odorless, tasteless gas
(ii) weak intermolecular forces
(iii) low density compared to air
(iv) slightly soluble in H_2O

(f) What is usually formed as H_2 reacts with nonmetals at elevated temperatures?

(i) covalent compounds (ii) ionic hydrides
(iii) reduced metals (iv) hydrogenated products

(g) Many oxides of metals react with H_2 to give

(i) covalent compounds (ii) acid anhydrides

(iii) water and the metal (iv) liquid fuels

(h) Which is *not* a commercial method for preparing hydrogen?

(i) $C(s) + H_2O \xrightarrow{1000\ °C} H_2(g) + CO(g)$

(ii) $2H_2O(l) \xrightarrow{electrolysis} 2H_2(g) + O_2(g)$

(iii) steam reforming of hydrocarbons

(iv) fractional distillation of liquid air

(i) Which would *not* be a practical laboratory method for preparing hydrogen?

(i) steam reforming

(ii) displacement of H from acids (or water) by active metals

(iii) electrolysis of water

(iv) the reaction of saltlike hydrides with water

(j) Which is a *false* statement concerning binary compounds of hydrogen?

(i) Certain very reactive metals form saltlike hydrides containing H^-.

(ii) All hydrides act as acidic anhydrides.

(iii) Most covalent hydrides are gases at room conditions.

(iv) The thermal stability of covalent hydrides decreases going down a family.

(k) Which is *not* true about direct combination reactions of O_2 with other elements?

(i) The usual oxidation state of O in the products is -2.

(ii) The oxides that are formed range from ionic to covalent compounds.

(iii) $O_2(g) + 2F_2(g) \rightarrow 2OF_2(g)$

(iv) Heat is often needed to initiate the reaction.

(l) Which is a laboratory preparation for oxygen?

(i) fractional distillation of liquid air

(ii) heating iron oxide

(iii) the combustion of methane

(iv) heating $KClO_3$ in the presence of MnO_2

(m) Which is *not* a property of ozone?

(i) characteristic odor

(ii) pale blue gas under room conditions

(iii) dissolves in water and acts as a weak acid

(iv) less stable and more reactive than oxygen

(n) Oxygen reacts with
   (i) nonmetals to form ionic oxides
   (ii) water to form a basic anhydride
   (iii) silicon and boron to form polymeric structures
   (iv) very active metals to form normal oxides

(o) Hydrogen peroxide
   (i) is an allotrope of water
   (ii) acts mainly as an oxidizing agent but can act as a reducing agent under certain conditions as well
   (iii) is prepared by passing O_2 through a high voltage
   (iv) is prepared commercially by reacting BaO_2 with cold concentrated H_2SO_4

(p) Which is *not* a use common to both H_2O_2 and O_3?
   (i) reducing agents
   (ii) bleaching agents
   (iii) disinfectants
   (iv) chemical reagents in chemical processes

6. What are the historical definitions of oxidation and reduction in terms of the gain or loss of oxygen? How do these definitions compare to the definitions based on changes in oxidation number?

7. What is an oxidizing agent? What is a reducing agent? What happens to each during a chemical reaction?

8. How can the formula of an atom, molecule, or ion indicate whether the substance could serve as a potential oxidizing or reducing agent?

9. What must occur together with any oxidation process? What is the relationship between the total increase and decrease in oxidation number for all atoms in an oxidation-reduction reaction?

10. What is wrong with the following chemical equation?

$$Mn(s) + 10OH^- \rightarrow Mn(OH)_2(s) + 2O_2(g) + 4H_2O(l)$$

11. In some reactions more than two elements change oxidation state

(a) $4Fe(CrO_2)_2 + 8Na_2CO_3 + 7O_2 \rightarrow 2Fe_2O_3 + 8Na_2CrO_4 + 8CO_2$

(b) $3Hg_2I_2 + 16HNO_3 \rightarrow 6Hg(NO_3)_2 + 3I_2 + 4NO + 8H_2O$

For each of the reactions above identify the element(s) oxidized and reduced.

12. Hydrogen can conveniently be produced in the laboratory by the reaction of CaH_2 with water to form $Ca(OH)_2$. What is the oxidation number of Ca in (a) CaH_2 and (b) $Ca(OH)_2$; of H in (c) CaH_2, (d) $Ca(OH)_2$, (e) H_2O, and (f) H_2; and of O in (g) $Ca(OH)_2$ and (h) H_2O? (i) Write the balanced overall equation for the reaction. Identify the (j) oxidizing and (k) reducing agents and the substances (l) oxidized and (m) reduced.

13. Which of the following are redox reactions? Identify the oxidizing and reducing agent in each of the redox reactions.

(a) $H_2(g) + Cl_2(g) \rightarrow 2HCl(g)$

(b) $3O_2(g) \rightarrow 2O_3(g)$

(c) $CO_2(g) + 2NaOH(aq) \rightarrow Na_2CO_3(aq) + H_2O(l)$

(d) $2CO(g) + O_2(g) \rightarrow 2CO_2(g)$

(e) $2Fe(OH)_3(s) + 10KOH(aq) + 3Cl_2(g) \rightarrow 2K_2FeO_4(aq) + 6KCl(aq) + 8H_2O(l)$

(f) $Cr_2O_7^{2-} + 3S^{2-} + 14\ H^+ \rightarrow 2Cr^{3+} + 3S(s) + 7H_2O(l)$

14. For each of the following chemical equations, choose the oxidizing agent and the reducing agent and show the change in oxidation state that occurs for each substance:

(a) $Pb(s) + PbO_2(s) + 2H_2SO_4(aq) \rightarrow 2PbSO_4(s) + 2H_2O(l)$

(b) $2Br^- + Cl_2(g) \rightarrow Br_2(aq) + 2Cl^-$

(c) $ClO_4^- + 4H_2(g) \rightarrow Cl^- + 4H_2O(l)$

(d) $3ClO^- \rightarrow ClO_3^- + 2Cl^-$

15. List the five steps used to balance redox reactions by the oxidation number method. What chemical species can be used to balance charges?

16. Complete and balance the following redox reactions by the oxidation number method.

(a) $I^- + HSO_4^- \rightarrow S(s) + I_2(aq)$

(b) $Bi(s) + HSO_4^- \rightarrow Bi_2(SO_4)_3(s) + SO_2(g)$

(c) $SO_2(g) + MnO_4^- \rightarrow Mn^{2+} + SO_4^{2-}$

(d) $Fe^{2+} + MnO_4^- \rightarrow Fe^{3+} + MnO_2(s)$

(e) $CrO_4^{2-} + HSnO_2^- \rightarrow HSnO_3^- + CrO_2^-$

17. Why is the word "hydrogen" appropriate for the element of that name? Write the chemical symbol for this element. Write the chemical formula for molecular hydrogen.

18. The boiling and melting points of molecular hydrogen are very low. Explain this observation in terms of intermolecular forces.

19. Write chemical equations illustrating the reduction of a metal oxide such as FeO and a nonmetal oxide such as SO_2 by molecular hydrogen.

20. Write chemical equations illustrating the direct combination of molecular hydrogen with a nonmetal such as F_2 and an active metal such as Ba. In which process is hydrogen reduced?

21. What does the term "hydrogenation" mean? What are some of the industrial applications of hydrogenation?

22. Write the formula of a binary hydride for each element in the third period (Na-Cl) of the periodic table. Specify the type of bonding in each compound. Draw the Lewis structure for each.

23. List the important intermolecular forces present in the liquid state for each of the following substances: (a) HI, (b) NH_3, (c) CH_4, (d) H_2S, and (e) HF.

24. What type of chemical reaction is common for the hydride ion? How does the reaction between NaH(s) and water illustrate this?

25. Hydrogen is the third most abundant element in the crust of the earth (≈ 15%), but very little molecular hydrogen is found in the atmosphere. Why?

26. What are some of the major uses of hydrogen?

27. Explain why reactions between carbon monoxide and molecular hydrogen are of considerable importance at the present time. Write a chemical equation illustrating your answer.

28. A student suggested to his instructor that hydrogen could economically be stored in the form of a saline hydride until needed and then released by the thermal decomposition of the hydride, for example

$$2Na(s) + H_2(g) \underset{\text{use}}{\overset{\text{storage}}{\rightleftharpoons}} 2NaH(s)$$

What reservations might the instructor have about the proposal?

29. What are the important intermolecular forces present in liquid and solid molecular oxygen? Would you expect O_2 to have high melting and boiling points? Why?

30. In what ways does an oxygen atom complete its outer shell of electrons? Write a chemical formula illustrating each of these ways.

31. Compare the properties of oxygen to those of the other elements in the oxygen family.

32. What is the usual oxidation number change for molecular oxygen when it participates in a combination reaction?

33. What type of bonding will occur between oxygen and nonmetals? What types between oxygen and metals?

34. Draw Lewis structures showing the three types of covalent bonds between carbon and oxygen in (a) CO, (b) CO_2, and (c) CH_3OH.

35. Water reacts with OF_2 to form O_2 and HF. Write the chemical equation for this reaction. What is the oxidizing agent in this reaction?

36. Write the formula of an oxide for each element in the third period (Na-Cl) of the periodic table. What type of bonding would you expect in each compound?

37. What types of intermolecular forces are present in liquid and solid (a) SO_2, (b) P_4O_6, (c) CO_2, and (d) SiO_2?

38. What are oxides that combine with water to give (a) acidic solutions and (b) alkaline solutions called? How do these oxides differ in the nature of their bonding?

39. Write a chemical equation that illustrates the direct reaction between an acidic oxide and an alkaline oxide in the absence of water.

40. Write the formula for the oxide of iron(II) as a hydrate and as a hydroxide.

41. What is the source of practically all of the molecular oxygen in the atmosphere?

42. Write the chemical equation for a laboratory method of preparation of oxygen.

43. What are the four gases that make up over 99.99% by volume of clean, dry air? In what order would these distill from liquefied air?

44. How is molecular oxygen prepared commercially? Briefly describe the process.

45. List some of the important uses of oxygen. What is the major chemical property of oxygen on which most of these depend?

46. What does the term "allotrope" mean? What are the two allotropes of oxygen?

47. Why is it not surprising that ozone is much more reactive than molecular oxygen?

48. Write the chemical equation for the formation of ozone from oxygen. Is this reaction endothermic or exothermic?

49. Write the chemical equations showing the reactions of ozone with (a) PbS(s), (b) Ag(s), and (c) KOH(s).

50. The reaction of ozone with PbS to form $PbSO_4$ is exothermic and the reaction of molecular oxygen with PbS to form $PbSO_4$ is endothermic. Show that the conversion of O_2 to O_3 is endothermic.

Answers to questions

1. (a) oxidation-reduction, (b) oxide, (c) oxidizing agent, (d) reducing agent, (e) basic anhydride, (f) acidic anhydride.
2. (a) iv, (b) vi, (c) vii, (d) iii, (e) ix, (f) viii, (g) ii, (h) i, (i) v.
3. (a) T; (b) T; (c) F, replace *nonstoichiometric* by *stoichiometric*; (d) F, insert *except some of the less active metals* between *metals to*; (e) F, delete *or losing its electron* (*forming* H^+).
4. (a) *Acidic anhydrides* react with water to form acidic solutions (H^+); *basic anhydrides* react with water to form alkaline solutions (OH^-). (b) *Oxidation* is a positive change in oxidation number for an element in an atom, ion, or molecule; *combustion* is a rapid oxidationreduction process in which heat and flame are observed. (c) *Hydrogenation* is the addition of molecular hydrogen to a molecule; *reduction* is a negative change in oxidation number. (d) *Oxidation* is a positive change in oxidation number; *reduction* is a negative change. (e) *Oxidizing agents* bring about oxidation and undergo reduction themselves; *reducing agents* bring about reduction and undergo oxidation themselves. (f) *Adsorption* is a surface phenomenon; *absorption* is a bulk or complete phenomenon.
5. (a) i, (b) ii, (c) i, (d) iv, (e) iii, (f) i, (g) iii, (h) iv, (i) i, (j) ii, (k) iii, (l) iv, (m) iii, (n) iii, (o) ii, (p) i.

6. Oxidation was the addition of oxygen to a substance, which means a positive change in oxidation number (satisfies current definition). Reduction was the production of the free metal from an ore, which means a negative change in oxidation number (satisfies current definition).
7. An oxidation agent is an atom, molecule, or ion that causes an increase in the oxidation state of another substance and is itself reduced. A reducing agent is an atom, molecule, or ion that causes a decrease in the oxidation state of another substance and is itself oxidized. During a chemical reaction the oxidizing agent is reduced and the reducing agent is oxidized.
8. The substance must contain an atom in an oxidation state that will spontaneously change to a different oxidation state--higher for a reducing agent and lower for an oxidizing agent.
9. Reduction must always occur together with any oxidation process. In an oxidation-reduction reaction the total increase in oxidation number for all atoms or ions must equal the total decrease in oxidation number for all atoms or ions.
10. There is no reduction--Mn is oxidized from 0 to +2 and O is oxidized from -2 to 0.
11. (a) The elements oxidized are Fe and Cr and the element reduced is O.
    (b) The elements oxidized are Hg and I and the element reduced is N.
12. (a) +2, (b) +2, (c) -1, (d) +1, (e) +1, (f) 0, (g) -2, (h) -2, (i) $CaH_2(s) + 2H_2O(l) \rightarrow 2H_2(g) + Ca(OH)_2(aq)$, (j) H_2O, (k) CaH_2, (l) CaH_2, (m) H_2O.
13. (a) H_2 is reducing agent, Cl_2 is oxidizing agent; (b) nonredox; (c) nonredox; (d) CO is reducing agent, O_2 is oxidizing agent; (e) Cl_2 is oxidizing agent, $Fe(OH)_3$ is reducing agent; (f) $Cr_2O_7^{2-}$ is oxidizing agent, S^{2-} is reducing agent.
14. (a) PbO_2 is oxidizing agent (+4 to +2 for Pb), Pb is reducing agent (0 to +2 for Pb); (b) Cl_2 is oxidizing agent (0 to -1 for Cl), Br^- is reducing agent (-1 to 0 for Br); (c) ClO_4^- is oxidizing agent (+7 to -1 for Cl), H_2 is reducing agent (0 to +1 for H); (d) OCl^- is both oxidizing agent and reducing agent (+1 to -1 and +5 for Cl).
15. The five steps are write the unbalanced equation for major reactants and products showing total oxidation number changes, add coefficients to balance change in oxidation number, balance the charges, complete balance by inspection, and check for atom and charge balance. The species that can be used are H^+ in acidic solution or OH^- in alkaline solution and H_2O.

16. (a) $6I^- + 7H^+ + HSO_4^- \rightarrow S + 3I_2 + 4H_2O$, (b) $2Bi + 6HSO_4^- + 6H^+ \rightarrow$ $Bi_2(SO_4)_3 + 3SO_2 + 6H_2O$, (c) $5SO_2 + 6MnO_4^- + 28H^+ \rightarrow 6Mn^{2+} + 5SO_4^{2-} +$ $14H_2O$, (d) $3Fe^{2+} + MnO_4^- + 2H_2O \rightarrow 3Fe^{3+} + MnO_2 + 4OH^-$, (e) $2CrO_4^{2-} + 3HSnO_2^-$ $+ H_2O \rightarrow 2CrO_2^- + 3HSnO_3^- + 2OH^-$.

17. Hydrogen means water-former; H, H_2.

18. Molecular hydrogen has low boiling and melting points because London forces are the only type of intermolecular force.

19. $FeO_2 + H_2(g) \overset{\Delta}{\rightarrow} Fe(s) + H_2O(g)$; $SO_2(g) + 2H_2(g) \rightarrow S(s) + 2H_2O(g)$.

20. $H_2(g) + F_2(g) \rightarrow 2HF(g)$, $Ba(s) + H_2(g) \rightarrow BaH_2(s)$; hydrogen is reduced during the reaction with an active metal.

21. Hydrogenation is the introduction of hydrogen into a molecular compound. Applications include making synthetic fuels and hydrogenation of oils to make fats.

22. The following are the binary hydrides for each element in the third period and their types of bonding: NaH, ionic; MgH_2, ionic; AlH_3, polar covalent; SiH_4, polar covalent; PH_3, polar covalent; H_2S, polar covalent; and HCl, polar covalent.

$$Na^+[H:]^- \qquad Mg^{2+}\ 2[H:]^- \qquad H-Al-H \qquad H-\overset{H}{\underset{H}{\underset{|}{\overset{|}{Si}}}}-H$$

$$\overset{H}{\underset{H}{\underset{|}{\overset{|}{:P}}}}-H \qquad H-\ddot{S}-H \qquad H-\ddot{\ddot{C}}l:$$

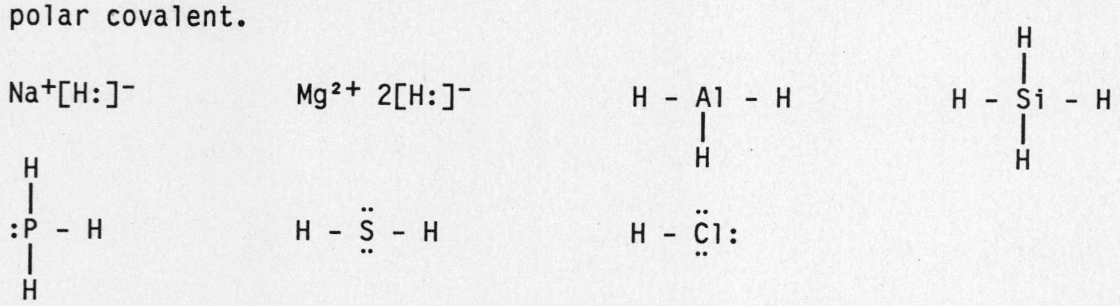

23. (a) London forces, dipole-dipole interactions; (b) London forces, dipole-dipole interactions, hydrogen bonding; (c) London forces; (d) London forces, dipole-dipole interactions; (e) London forces, dipole-dipole interactions, hydrogen bonding.

24. Oxidation-reduction is the type of chemical reaction common for the hydride ion. In the reaction between NaH(s) and H_2O, hydrogen is oxidized from -1 in the hydride to 0 in molecular hydrogen.

25. Molecules of low molecular mass escape earth's gravitational attraction. Hydrogen is retained in the crust in the form of stable compounds because of its high reactivity.

26. See Table 16.5 of the text.

27. Synthetic fuels are produced in this way; $CO(g) + 3H_2(g) \overset{\Delta}{\rightarrow} CH_4(g) + H_2O(g)$.

28. Some problems to overcome include that Na is expensive and needs to be isolated from the atmosphere, the formation of the hydride takes place at high temperatures, and the decomposition of the hydride requires energy.
29. London forces are the only important intermolecular forces present in liquid and solid molecular oxygen. Molecular oxygen is not expected to have high melting or boiling points because there are no strong intermolecular forces.
30. An oxygen atom can gain two electrons, $[:\ddot{\underset{..}{O}}:]^{2-}$, or share electrons, $-\ddot{\underset{..}{O}}-$ or $=\ddot{O}:$ or $\equiv O:$.
31. See Table 16.9 of the text.
32. 0 to -2.
33. The bonding will be polar covalent with nonmetals and mainly ionic (sometimes polar covalent) with metals.
34. (a) $:C \equiv O:$ (b) $:\ddot{O} = C = \ddot{O}:$ (c) $H - \underset{\underset{H}{|}}{\overset{\overset{H}{|}}{C}} - \ddot{\underset{..}{O}} - H$
35. $OF_2 + H_2O \rightarrow O_2 + 2HF$; OF_2 is the oxidizing agent.
36. The following are the oxides for each element in the third period and their types of bonding: Na_2O, ionic; MgO, ionic; Al_2O_3, ionic; SiO_2, polar covalent; P_4O_{10}, polar covalent; SO_3, polar covalent; and Cl_2O_7, polar covalent.
37. (a) London forces, dipole-dipole interactions; (b) London forces; (c) London forces; (d) network covalent bonding.
38. (a) acidic anhydride, polar covalent bonding; (b) basic anhydride, ionic bonding.
39. $CaO(s) + SiO_2(s) \xrightarrow{\Delta} CaSiO_3(s)$.
40. $Fe_2O_3 \cdot xH_2O$, $Fe(OH)_3$.
41. The source is photosynthesis.
42. $2KClO_3(l) \xrightarrow[\Delta]{MnO_2} 2KCl(s) + 3O_2(g)$
43. N_2, O_2, Ar, CO_2; N_2 followed by Ar followed by O_2 followed by CO_2.
44. The process involves the distillation of liquefied air. The air is cooled, compressed, cooled, expanded to cool, recycled until liquefied, and distilled to get O_2.
45. See Table 16.11 of the text. The major property is the ability of O_2 to act as an oxidizing agent.

46. Allotropes are different forms of the same element in the same state. The two allotropes are oxygen and ozone.
47. Most endothermically formed substances are reactive.
48. $3O_2(g) \rightarrow 2O_3(g)$; endothermic.
49. (a) $PbS(s) + 4O_3(g) \rightarrow PbSO_4(s) + 4O_2(g)$
    (b) $2Ag(g) + O_3(g) \rightarrow Ag_2O(s) + O_2(g)$
    (c) $6KOH(s) + 4O_3(g) \rightarrow 4KO_3(s) + 2KOH{\cdot}H_2O(s) + O_2(g)$
50.

$$
\begin{array}{lll}
 & PbS + 2O_2 \rightarrow PbSO_4 & \Delta H^\circ > 0 \\
(-1)[& \underline{PbS + 4O_3 \rightarrow PbSO_4 + 4O_2} & \underline{\Delta H^\circ < 0]} \\
 & 6O_2 \rightarrow 4O_3 & \Delta H^\circ > 0
\end{array}
$$

SKILLS

Skill 16.1 Recognizing oxidation-reduction reactions

Not all chemical reactions involve oxidation-reduction (redox). For example, in the reaction of NH_3 with H_2O,

$$NH_3(g) + H_2O(l) \rightarrow NH_4^+ + OH^-$$

there are no changes in oxidation numbers (N is -3, H is +1, and O is -2 on both sides of the equation), and so the reaction does not involve oxidation-reduction. On the other hand, in the reaction of I_2 with H_2S,

$$I_2(s) + H_2S(aq) \rightarrow 2I^- + S(s) + 2H^+$$

the iodine is changing from 0 to -1 (being reduced) and the sulfur is changing from -2 to 0 (being oxidized), and so the reaction involves oxidation-reduction. According to the definitions, I_2 is the oxidizing agent because it oxidizes the H_2S and is itself reduced and H_2S is the reducing agent because it reduces I_2 and is itself oxidized. Note that the oxidizing and reducing agents are not the elements undergoing the oxidation or reduction, but rather the species (atoms, molecules, or ions) that contain these elements.

Skill Exercise: Identify which of the following equations represent oxidation-reduction processes and, for those that are, identify the oxidizing and reducing agents: (a) $6Ag^+ + AsH_3(g) + 3H_2O(l) \rightarrow 6Ag(s) + H_3AsO_3(aq) + 6H^+$, (b) $KCl(aq) + AgNO_3(aq) \rightarrow AgCl(s) + KNO_3(aq)$, (c) $2MnO_4^- + 16H^+ + 10Cl^- \rightarrow 2Mn^{2+} + 5Cl_2(g) + 8H_2O(l)$, (d) $2H_2O(l) \rightarrow 2H_2(g) + O_2(g)$, (e) $F_2(g) + 2Cl^- \rightarrow Cl_2(g) + 2F^-$, (f) $2Cr(OH)_3(aq) + 3Cl_2(g) + 10OH^- \rightarrow 2CrO_4^{2-} + 8H_2O(l) + 6Cl^-$, and (g)

$HNO_3(aq) + NaOH(aq) \rightarrow H_2O(l) + NaNO_3(aq)$. *Answers*: (a) Ag^+, AsH_3; (b) nonredox; (c) MnO_4^-, Cl^-; (d) H_2O, H_2O; (e) F_2, Cl^-; (f) Cl_2, $Cr(OH)_3$, (g) nonredox.

Skill 16.2 Balancing redox equations: oxidation number method

This method of balancing redox equations is based on the principle that the total increase in oxidation number always equals the total decrease in oxidation number. For example, in the reaction

$$Zn(s) + Ag^+ \rightarrow Zn^{2+} + Ag(s)$$

the zinc is oxidized from 0 to +2 and the silver is reduced from +1 to 0. If the coefficients of 2 are used for Ag^+ and Ag

$$Zn + 2Ag^+ \rightarrow Zn^{2+} + 2Ag$$

so that the decrease in oxidation number (2)(-1) = -2 equals the increase in oxidation number (1)(+2) = +2, the equation is balanced with respect to the numbers of atoms and electrical charge (see *Skill 4.3 Writing and balancing chemical equations* and *Skill 5.3 Writing net ionic equations*).

Most redox reactions cannot be balanced as simply as the one above. The following steps are used for more complicated equations:

A. Write the unbalanced equation for major reactants and products. Determine total oxidation number changes.
B. Add coefficients to balance change in oxidation number.
C. If necessary, balance the charges by adding H^+ for reactions in acidic solution or OH^- for reactions in alkaline solution.
D. Complete balancing by inspection, using water to balance oxygen.
E. Check to be sure that both atoms and charge are balanced.

Three examples are shown below to illustrate the method.

Example (1) $I_2(s) + H_2S(aq) \rightarrow S(s) + I^-$

(The presence of H_2S indicates that the reaction is taking place in an acidic solution.)

Step A. Unbalanced equation showing changes in oxidation number.

$$\overset{0}{I_2} + \overset{+1\ -2}{H_2S} \rightarrow \overset{0}{S} + \overset{-1}{I^-}$$

$$\overset{0}{I_2} \rightarrow 2\overset{-1}{I^-} \quad (2)(-1) = -2 \qquad\qquad \overset{-2}{H_2S} \rightarrow \overset{0}{S} \quad (1)(+2) = +2$$

Step B. Coefficients needed to balance oxidation number changes.

(2)(−1) = −2

$I_2 + H_2S \rightarrow S + 2I^-$

(1)(+2) = +2

Step C. Balance the charges by adding H^+.

$I_2(s) + H_2S(aq) \rightarrow S(s) + 2I^- + 2H^+$

Step D. Complete balancing (balance O by adding H_2O).

This step is not needed.

Step E. Check. On each side of the equation there are 2 I atoms, 2 H atoms, 1 S atom, and no charge.

Example (2) $KNO_3(aq) + I_2(s) \xrightarrow{H^+} KIO_3(aq) + NO_2(g)$

Step A. Unbalanced equation showing changes in oxidation numbers.

$\overset{+5\ -2}{NO_3^-} + \overset{0}{I_2} \rightarrow \overset{+5\ -2}{IO_3^-} + \overset{+4\ -2}{NO_2}$

$\overset{+5}{NO_3^-} \rightarrow \overset{+4}{NO_2}$ (1)(−1) = −1 $\overset{0}{I_2} \rightarrow \overset{+5}{2IO_3^-}$ (2)(+5) = +10

Step B. Coefficients needed to balance oxidation number changes.

(10)(−1) = −10

$10NO_3^- + I_2 \rightarrow 2IO_3^- + 10NO_2$

(2)(+5) +10

Step C. Balance the charges by adding H^+.

$10NO_3^- + I_2 + 8H^+ \rightarrow 2IO_3^- + 10NO_2$

Step D. Complete balancing (balance O by adding H_2O).

$10NO_3^- + I_2(s) + 8H^+ \rightarrow 2IO_3^- + 10NO_2(g) + 4H_2O(l)$

$10KNO_3(aq) + I_2(s) + 8H^+ \rightarrow 2KIO_3(aq) + 10NO_2(g) + 4H_2O(l) + 8K^+$

Step E. Check. On each side of the equation there are 10 K atoms, 10 N atoms, 30 O atoms, 2 I atoms, 8 H atoms, and +8 charge.

Example (3) $I^- + ClO^- \rightarrow I_2(s) + Cl^-$

(The presence of ClO^- indicates an alkaline solution, because HClO would be present instead of ClO^- in an acidic solution.)

Step A. Unbalanced equation showing changes in oxidation numbers.

$$\overset{-1}{I^-} + \overset{+2\ -2}{ClO^-} \rightarrow \overset{0}{I_2} + \overset{-1}{Cl^-}$$

$$2\overset{-1}{I^-} \rightarrow \overset{0}{I_2} \quad (2)(+1) = +2 \qquad \overset{+1}{ClO^-} \rightarrow \overset{-1}{Cl^-} \quad (-1)(-2) = -2$$

Step B. Coefficients needed to balance oxidation number changes.

(2)(+1) + +2

$$2I^- + ClO^- \rightarrow I_2 + Cl^-$$

(1)(-2) = -2

Step C. Balance the charge by adding OH^-.

$$2I^- + ClO^- \rightarrow I_2 + Cl^- + 2OH^-$$

Step D. Complete balancing (balance O by adding H_2O).

$$2I^- + ClO^- + H_2O(l) \rightarrow I_2(s) + Cl^- + 2OH^-$$

Step E. Check. On each side of the equation there are 2 I atoms, 1 Cl atom, 2 O atoms, 2 H atoms, and -3 charge.

Skill Exercise: Balance the following redox equations by the oxidation number method: (a) $Ag^+ + AsH_3(g) \rightarrow Ag(s) + H_3AsO_3(aq)$, (b) $MnO_4^- + HCl(aq) \rightarrow Mn^{2+} + Cl_2(g)$, (c) $H_2O(l) \rightarrow H_2(g) + O_2(g)$, (d) $F_2(g) + Cl^- \rightarrow Cl_2(g) + F^-$, (e) $Cr(OH)_3(aq) + Cl_2(g) \rightarrow CrO_4^{2-} + Cl^-$, and (f) $N_2H_4(g) + ClO^- \rightarrow N_2(g) + Cl^-$. *Answers*: (a) $6Ag^+ + 3H_2O(l) + AsH_3(g) \rightarrow H_3AsO_3(aq) + 6H^+ + 6Ag(s)$, (b) $2MnO_4^- + 16H^+ + 10Cl^- \rightarrow 5Cl_2(g) + 2Mn^{2+} + 8H_2O(l)$, (c) $2H_2O(l) \rightarrow 2H_2(g) + O_2(g)$, (d) $F_2(g) + 2Cl^- \rightarrow 2F^- + Cl_2(g)$, (e) $2Cr(OH)_3(aq) + 10OH^- + 3Cl_2(g) \rightarrow 6Cl^- + 2CrO_4^{2-} + 8H_2O(l)$, (f) $N_2H_2(g) + ClO^- \rightarrow N_2(g) + Cl^- + H_2O(l)$.

PRACTICE TEST

1 (5 points). Match the correct term to each definition.

(a) a chemical reaction in which oxidation numbers change
(b) two different forms of the same element in the same state
(c) the addition of molecular hydrogen to a molecular substance
(d) fuels that are coal, petroleum, or natural gas
(e) capable of acting as either an acid or a base

(i) allotropes
(ii) amphoteric
(iii) fossil fuels
(iv) hydrogenation
(v) redox reactions

2 (15 points). For each of the following molecules, (i) prepare Lewis structures, (ii) determine the molecular geometry, (iii) describe the intermolecular forces, and (iv) give oxidation numbers for each atom: (a) H_2, (b) O_2, (c) H_2O, (d) H_2O_2, (e) O_3, and (f) D_2.

3 (10 points). Write chemical equations describing the reaction between (a) CuO(s) and H_2(g), (b) S(s) and O_2(g), (c) Ag(s) and O_3(g), (d) NaH(s) and H_2O(l), (e) N_2(g) and H_2(g).

4 (10 points). Two reactions that are important in the metallurgy of copper are

$$\text{(i)}\ 2Cu_2S(s) + 3O_2(g) \rightarrow 2Cu_2O(s) + 2SO_2(g)$$

$$\text{(ii)}\ 2Cu_2O(s) + Cu_2S(s) \rightarrow 6Cu(s) + SO_2(g)$$

(a) Identify the substance oxidized in (i). (b) Identify the reducing agent in (ii). What is the oxidation state of Cu in (c) Cu_2S, (d) Cu_2O, and (e) Cu; of O in (f) O_2, (g) Cu_2O, and (h) SO_2; and of S in (i) Cu_2S and (j) SO_2?

5 (10 points). Barium peroxide reacts with water to generate oxygen. What mass of BaO_2 is needed to produce 1.0 L of O_2 measured at 21 °C and 732 Torr?

6 (10 points). Lead(IV) oxide, a brown solid having the formula PbO_2, loses oxygen at 500 °C to form a red solid. A 50.0 g sample of PbO_2 was heated to constant mass and 47.7 g of the red material remained. Determine the formula of the red material.

7 (20 points). Choose the response that best completes each statement.

(a) The process in which iron changes from +2 to +3 is

(i) catalysis (ii) redox
(iii) reduction (iv) oxidation

(b) Which is *not* a redox reaction?

(i) displacement (ii) decomposition
(iii) combination (iv) partner exchange

(c) Which is a physical property of hydrogen?

(i) colored (ii) very slightly soluble in water
(iii) burns with a pale-blue flame (iv) heavy gas

(d) Which is *not* a suitable means for preparing hydrogen from water?

(i) $Cu + 2H_2O \rightarrow Cu(OH)_2 + H_2$ (ii) $Ca + 2H_2O \rightarrow Ca(OH)_2 + H_2$
(iii) $2H_2O \rightarrow 2H_2 + O_2$ (iv) $H_2O + C \rightarrow CO + H_2$

(e) Which is *not* a chemical property of hydrogen?

(i) combination with chlorine (ii) combination with nitrogen

(iii) reduction of oxides (iv) supports combustion

(f) Which is *not* a physical property of oxygen?

(i) colorless and odorless gas under ordinary conditions

(ii) highly soluble in water

(iii) can be liquefied

(iv) slightly heavier than air

(g) Which method of preparing oxygen is used commercially?

(i) distillation of liquid air (ii) decomposition of HgO

(iii) decomposition of $KClO_3$ (iv) action of water on Na_2O_2

(h) Which is *not* a suitable use for oxygen?

(i) an inert atmosphere in light bulbs

(ii) high-temperature torches

(iii) aeration of water

(iv) breathing

(i) Oxides of nonmetals are known as

(i) basic anhydrides (ii) hydrates

(iii) acidic anhydrides (iv) interstitial oxides

(j) Which is *not* a property of ozone?

(i) bleaching agent

(ii) more active than oxygen

(iii) unstable

(iv) less soluble in water than oxygen

8 (20 points). Complete and balance the following equations by the oxidation number method: (a) $MnO_4^{2-} \rightarrow MnO_4^- + MnO_2(s)$, (b) $Ag(s) + HNO_3(aq) \rightarrow Ag^+ + NO(g)$, and (c) $PbO_2(s) + KCl(aq) \rightarrow ClO^- + [Pb(OH)_3]^-$.

Answers to practice test

1. (a) v, (b) i, (c) iv, (d) iii, (e) ii.

2.

H – H	:Ö = Ö:	H – Ö – H
linear (type A_2)	linear (type A_2)	bent (type AB_2E_2)
London forces	London forces	hydrogen bonding
0 for H	0 for O	dipole-dipole interactions
		London forces

		+1 for H, -2 for O
H - Ö - Ö - H	:Ö = Ö - Ö: ↔ :Ö - Ö = Ö:	D - D
nonplanar	bent (type AB_2E_2)	linear (type A_2)
hydrogen bonding	dipole-dipole interactions	London forces
dipole-dipole interactions	London forces	0 for D
London forces	0 for O	
+1 for H, -1 for O		

3. (a) $CuO(s) + H_2(g) \xrightarrow{\Delta} Cu(s) + H_2O(l)$, (b) $S(s) + O_2(g) \rightarrow SO_2(g)$, (c) $2Ag(s) + O_3(g) \rightarrow Ag_2O(s) + O_2(g)$, (d) $NaH(s) + H_2O(l) \rightarrow NaOH(aq) + H_2(g)$, (e) $N_2(g) + 3H_2(g) \xrightarrow{\Delta} 2NH_3(g)$.
4. (a) Cu_2S, (b) Cu_2S, (c) +1, (d) +1, (e) 0, (f) 0, (g) -2, (h) -2, (i) -2, (j) +4.
5. 14 g BaO_2.
6. Pb_3O_4.
7. (a) iv, (b) iv, (c) ii, (d) i, (e) iv, (f) ii, (g) i, (h) i, (i) iii, (j) iv.
8. (a) $3MnO_4^{2-} + 2H_2O(l) \rightarrow 2MnO_4^- + MnO_2(s) + 4OH^-$, (b) $3Ag(s) + HNO_3(aq) + 3H^+ \rightarrow 3Ag^+ + NO(g) + 2H_2O(l)$, (c) $PbO_2(s) + Cl^- + OH^- + H_2O(l) \rightarrow [Pb(OH)_3]^- + ClO^-$.

ADDITIONAL NOTES

CHAPTER 17

CHEMICAL REACTIONS IN PERSPECTIVE

CHAPTER OVERVIEW

This chapter is devoted to discussing types of chemical reactions--redox and nonredox. The equilibria that are established in all chemical reactions and the criteria for a reaction to "go to completion" are discussed. The principle of Le Chatelier is introduced. The classifications of various redox and nonredox reactions are presented in detail. Common oxidizing and reducing agents are considered and some guidelines on predicting possible reaction products are given. A guide is presented for deciding what type of reaction might occur between given reactants and what products might form.

COMPETENCIES

Definitions for the following terms should be learned:

- acid salt
- disproportionation reaction
- internal redox reaction
- Le Chatelier's principle
- oxidizing acid
- oxidizing anion

General concepts and skills that should be learned include

- classifying a chemical reaction as redox or nonredox and classifying the reaction further as displacement, disproportionation, etc.
- predicting whether or not a chemical reaction will take place
- predicting the possible products of a chemical reaction

QUESTIONS

1. Complete the following sentences by filling in the various blanks with these terms: acid salt, disproportionation reaction, internal redox reaction, Le Chatelier's principle, oxidizing acid, and oxidizing anion.

As chlorine gas dissolves in water, the following reaction takes place

$$Cl_2(g) + H_2O(l) \rightleftharpoons HClO(aq) + H^+ + Cl^-$$

This is an example of a (a)________. As the pressure of Cl_2 is increased, the amount of HClO produced is increased. This is an example of (b)________. The decomposition of gold(III) chloride upon heating,

$$2AuCl_3(s) \xrightarrow{\Delta} 2Au(s) + 3Cl_2(g)$$

is an example of an (c)________. At high temperatures, concentrated sulfuric acid undergoes reaction with sulfur:

$$S(s) + 2H_2SO_4(conc) \xrightarrow{\Delta} 3SO_2(g) + 2H_2O(l)$$

In this reaction H_2SO_4 is acting as an (d)________. Dilute sulfuric acid will undergo reaction with sodium hydroxide,

$$H_2SO_4(aq) + NaOH(aq) \rightarrow NaHSO_4(aq) + H_2O(l)$$

to form an (e)________. In the following reaction:

$$IO_3^- + 5I^- + 6H^+ \rightarrow 3I_2(s) + 3H_2O(l)$$

the iodate ion is acting as an (f)________.

2. Match the correct classification of reaction to each equation.

(a) $SnS(s) + S(s) \rightarrow SnS_2(s)$
(b) $CaCO_3(s) \xrightarrow{\Delta} CaO(s) + CO_2(g)$
(c) $NiO(s) + H_2(g) \rightarrow Ni(s) + H_2O(l)$
(d) $Na_2CO_3(aq) + 2HNO_3(aq) \rightarrow 2NaNO_3(aq) + CO_2(g) + H_2O(l)$
(e) $3HNO_2(aq) \rightarrow 2NO(g) + NO_3^- + H_3O^+$
(f) $Na_2O(s) + SO_2(g) \rightarrow Na_2SO_3(s)$
(g) $2FeCl_3(aq) + SnCl_2(aq) \rightarrow 2FeCl_2(aq) + SnCl_4(aq)$
(h) $2Mg(NO_3)_2(s) \xrightarrow{\Delta} 2Mg(NO_2)_2(s) + O_2(g)$
(i) $Na_2CO_3(s) + SiO_2(s) \xrightarrow{\Delta} Na_2SiO_3(l) + CO_2(g)$

(i) nonredox combination
(ii) nonredox decomposition
(iii) nonredox displacement
(iv) nonredox partner exchange
(v) redox combination
(vi) redox decomposition
(vii) redox displacement
(viii) redox disproportionation
(ix) redox electron transfer

3. Which of the following statements are true? Rewrite each false statement so that it is correct.

(a) All chemical reactions fall into two major categories--redox and nonredox.

(b) Any chemical reaction that can proceed in either direction is thought to be reversible and is potentially capable of establishing equilibrium.

(c) Reactions that appear to go to completion continue until equilibrium is established in which practically all of the reactants have been converted to products.

(d) Free elements are never reactants or products in nonredox reactions.

(e) A reaction involving oxidation-reduction cannot be classified as a partner-exchange reaction.

(f) An oxidizing agent contains an element that can undergo a decrease in oxidation number and a reducing agent contains an element that can undergo an increase in oxidation number.

(g) A description of something as "stable" usually means that it is likely to remain unchanged under a certain set of conditions.

4. Distinguish clearly between (a) redox and nonredox combination reactions, (b) internal redox and disproportionation reactions, (c) oxidizing and nonoxidizing acids, and (d) thermodynamic and kinetic stabilities.

5. Choose the response that best completes each statement.

(a) The oxidation number of chromium in $K_2Cr_2O_7$ is

(i) +6 (ii) +4
(iii) +2 (iv) -2

(b) The process in which iron changes from +3 to +2 is

(i) catalysis (ii) redox
(iii) oxidation (iv) reduction

(c) The more reactive a metal, the

(i) less easily it is oxidized
(ii) more easily it is reduced
(iii) more easily it forms positive ions
(iv) none of these answers

(d) What substance is the oxidizing agent in the following reaction?

$MnO_2(s) + 4H^+(aq) + 2Cl^-(aq) \rightarrow Cl_2(g) + Mn^{2+}(aq) + 2H_2O(l)$

(i) H^+ (ii) MnO_2
(iii) Cl^- (iv) Mn^{4+}

(e) Experimentally it is observed that

$$Zn(s) + Cu^{2+} \rightarrow Cu(s) + Zn^{2+}$$
$$Ag(s) + Cu^{2+} \rightarrow \text{no reaction}$$

The correct order of the three metals in the activity series is

(i) Zn > Cu > Ag (ii) Cu > Ag > Zn
(iii) Ag > Cu > Zn (iv) Zn > Ag > Cu

(f) Which substance is the strongest oxidizing agent?

(i) H_2 (ii) F_2

(iii) O_2 (iv) S

(g) Which substance is the strongest oxidizing agent?

(i) CO_2 (ii) CaO

(iii) SiO_2 (iv) PbO_2

(h) The reaction between zinc and sulfur is a

(i) combination reaction (ii) decomposition reaction

(iii) displacement reaction (iv) partner-exchange reaction

(i) The reaction between sodium and water is a

(i) combination reaction (ii) decomposition reaction

(iii) displacement reaction (iv) partner-exchange reaction

(j) Which reaction *cannot* involve oxidation-reduction?

(i) combination reaction (ii) decomposition reaction

(iii) displacement reaction (iv) partner-exchange reaction

(k) Which of the following substances can act as *either* an oxidizing or a reducing agent?

(i) Cl^- (ii) NO_3^-

(iii) Na^+ (iv) NO_2^-

6. Explain what the expression "the reaction goes to completion" means. Do reactions that go to completion reach equilibrium? Why are these reactions driven to completion?

7. Which of the following reactions are likely to go to completion? Explain your reasoning.

(a) $CaCO_3(s) \xrightarrow{\Delta} CaO(s) + CO_2(g)$

(b) $PbCrO_4(s) + 2NaCl(aq) \rightarrow PbCl_2(aq) + Na_2CrO_4(aq)$

(c) $Zn(s) + S(s) \rightarrow ZnS(s)$

(d) $2HgO(s) \xrightarrow{\Delta} 2Hg(l) + O_2(g)$

(e) $2H_2O_2(l) \xrightarrow{\Delta} 2H_2O(l) + O_2(g)$

(f) $LiCl(aq) + NaBr(aq) \rightarrow NaCl(aq) + LiBr(aq)$

8. Briefly define "thermodynamic stability" and "kinetic stability." If a substance is thermodynamically unstable, does that mean that the substance will react readily? Why? A substance can be described as "stable" with respect to a number of conditions. Name some of them.

9. List four classifications of chemical reactions that do not involve oxidation-reduction.

10. Briefly describe what types of reactions can be classified as nonredox combination of compounds. Write a chemical equation to illustrate one of these types.

11. What type of substance is often produced during a nonredox decomposition reaction? Write chemical equations for the nonredox decomposition of $Al(OH)_3(s)$ (to produce Al_2O_3, an abrasive) and $ZnCO_3(s)$ during heating.

12. What are the usual products produced by the nonredox thermal decomposition of (a) ammonium salts; (b) carbonates, except those of the most active metals; (c) many hydroxides; and (d) acid salts?

13. What are two types of nonredox displacement reactions? Write a chemical equation illustrating each type.

14. The boiling point of boron oxide, B_2O_3, lies above 1500 °C, and phosphorus(V) oxide, P_4O_{10}, sublimes at about 250 °C. Will the following displacement reaction occur?

$$6Na_2B_4O_7(s) + P_4O_{10}(s) \xrightarrow{\Delta} 4Na_3PO_4(s) + 12B_2O_3(s)$$

15. Write a general chemical equation representing a partner-exchange reaction between compounds WX and YZ. Name some of the types of products formed when this type of reaction takes place.

16. Choose the nonredox partner-exchange reactions that will take place:
(a) $Pb(NO_3)_2(aq) + H_2SO_4(aq) \rightarrow PbSO_4(s) + 2HNO_3(aq)$
(b) $Hg_2Cl_2(s) + 2KNO_3(aq) \rightarrow Hg_2(NO_3)_2(aq) + 2KCl(aq)$
(c) $BaCrO_4(s) + 2KNO_3(aq) \rightarrow Ba(NO_3)_2(aq) + K_2CrO_4(aq)$
(d) $HCl(aq) + KHSO_3(aq) \rightarrow SO_2(g) + H_2O(l) + KCl(aq)$
(e) $Ni(OH)_2(s) + 2HCl(aq) \rightarrow NiCl_2(aq) + 2H_2O(l)$
Give reasons for your answers.

17. What will be the products of a partner-exchange reaction between water and compounds consisting of two nonmetals (other than oxygen)? With which element will the OH group combine?

18. Complete and balance the following partner-exchange reactions: (a) $PCl_5(s) + H_2O(l) \rightarrow$, (b) $BCl_3(g) + H_2O(l) \rightarrow$, (c) $AsCl_3(s) + H_2O(l) \rightarrow$, and (d) $ICl(l) + H_2O(l) \rightarrow$.

19. Predict the products of each of the following nonredox reactions. If no apparent reaction occurs, so indicate.
(a) $ZnS(s) + HCl(aq) \rightarrow$
(b) $Ca(NO_3)_2(aq) + KCl(aq) \rightarrow$
(c) $NaOH(aq) + CuBr_2(aq) \rightarrow$
(d) $KHSO_3(s) \xrightarrow{\Delta}$
(e) $(NH_4)_2S(aq) + NaI(aq) \rightarrow$

20. What are some of the classifications of redox reactions?

21. Describe what happens to the free element(s) during a combination reaction.

22. What determines which oxidation state is formed by metals that can form more than one oxidation state upon oxidation?

23. Complete and balance the following redox combination reactions:
(a) $Sn(s) + Cl_2(g,excess) \rightarrow$
(b) $Cu(s,excess) + I_2(g) \rightarrow$
(c) $S_8(s,excess) + O_2(g) \xrightarrow{\Delta}$
(d) $Al(s) + S_8(s) \xrightarrow{\Delta}$

24. Identify the products of the redox decomposition reactions of (a) $HgO(s)$ when heated, (b) $MgCl_2(l)$ upon electrolysis, and (c) $KClO_4(s)$ when heated.

25. What are the usual products formed by the thermal decomposition of (a) nitrates of Representative Group I and II metals, and (b) nitrates of less active metals?

26. Describe what happens to the free metal that displaces another metal in a compound. How can the use of an activity series help to predict whether a reaction will occur or not? What restrictions must be kept in mind when using an activity series to predict redox reaction spontaneity?

27. Describe what happens to a halide ion that has been displaced by another halogen in a compound.

28. Name some categories of common oxidizing agents and give an example of each. Repeat the question for common reducing agents.

29. Complete and balance the equations for the following redox displacement reactions: (a) the preparation of chlorine by heating hydrogen chloride with oxygen in the presence of copper chloride as catalyst (the Deacon

process) and (b) the preparation of finely divided elemental iron by passing hydrogen over heated solid iron(III) oxide.

30. What is a disproportionation reaction? How can we recognize the possibility of a given compound undergoing such reaction?

31. Predict the products of each of the following reactions: (a) $Br_2(l) + Fe(s) \rightarrow$, (b) $HClO(aq) + H_2S(aq) \rightarrow$, (c) $MnO_4^- + Mg(s) \xrightarrow{OH^-}$, and (d) $H_2O_2(aq) + SO_3^{2-} \rightarrow$.

Answers to questions

1. (a) disproportionation reaction, (b) Le Chatelier's principle, (c) internal redox reaction, (d) oxidizing acid, (e) acid salt, (f) oxidizing anion.
2. (a) v, (b) ii, (c) vii, (d) iv, (e) viii, (f) i, (g) ix, (h) vi, (i) iii.
3. (a) T, (b) T, (c) T, (d) T, (e) T, (f) T, (g) T.
4. (a) *Redox combination reactions* are combination reactions between chemical species in which oxidation-reduction occurs; *nonredox combination reactions* involve compounds combining to form new compounds without oxidation-reduction occurring. (b) In an *internal redox reaction* the oxidized and reduced forms originate in the same compound; a *disproportionation reaction* is a special case of an internal redox reaction in which the same element in the reactant is both oxidized and reduced. (c) A *nonoxidizing acid* contains H^+ and a nonoxidizing anion; an *oxidizing acid* contains an oxidizing anion in addition to H^+ (which can also act as an oxidizing agent). (d) *Thermodynamic stability* refers to stability against undergoing chemical changes because the substance is already in a state of reasonably low energy; *kinetic stability* refers to stability against reaching a lower energy state by chemical change because the change would occur too slowly.
5. (a) i, (b) iv, (c) iii, (d) ii, (e) i, (f) ii, (g) iv, (h) i, (i) iii, (j) iv, (k) iv.
6. The term completion means that practically all of the reactants are converted to products. All reactions procede until reaching equilibrium. The formation of a product that is "removed" from the system such as a gas, a precipitate, a molecular substance, or a poorer oxidizing and reducing agent than the reactants will drive a reaction to completion.
7. (a) yes, formation of a gas; (b) no, no formation of gas, precipitate, etc.; (c) yes, zinc is good reducing agent and sulfur is an oxidizing agent; (d)

yes, HgO is listed in the activity series as an oxide that is unstable to heat; (e) yes, H_2O_2 can undergo disproportionation; (f) no, all species are strong electrolytes.

8. A thermodynamically stable substance is not likely to undergo chemical changes because it is already in a state of reasonable low energy. A kinetically stable substance might have the potential of reaching a lower energy state by chemical change, but the change would take place too slowly to be useful or interesting. If a substance is thermodynamically unstable, the substance may react slowly because the substance may be kinetically stable. Some of the physical conditions that might determine the stability of a substance include temperature, pressure, acidity or alkalinity, concentration, and the presence of other substances.
9. The four classifications are combination of compounds, decomposition to give compounds, nonredox displacement, and partner-exchange reactions.
10. Acid-base reactions are nonredox reactions; $NaOH(aq) + HCl(aq) \rightarrow NaCl(aq) + H_2O(l)$.
11. A gas is often produced; $2Al(OH)_3(s) \xrightarrow{\Delta} Al_2O_3(s) + 3H_2O(g)$, $ZnCO_3(s) \xrightarrow{\Delta} ZnO(s) + CO_2(g)$.
12. The usual product(s) produced by the nonredox thermal decomposition of (a) ammonium salts is NH_3, (b) carbonates--except those of the most active metals--is CO_2, (c) many hydroxides are oxides and H_2O, and (d) acid salts are gaseous oxides of the nonmetals, the normal salts, and H_2O.
13. The two types of reactions are the formation of volatile oxides and ligand exchange; $CaCO_3(s) + SiO_2(s) \xrightarrow{\Delta} CaSiO_3(l) + CO_2(g)$, $[HgCl_4]^{2-} + 4CN^- \rightarrow [Hg(CN)_4]^{2-} + 4Cl^-$.
14. This reaction is the displacement of one oxide by another. It will not occur because phosphorus(V) oxide is much more volatile than boron oxide, and a more volatile oxide will not displace a less volatile one to form a compound. [Instead, the more volatile oxide will vaporize out of the mixture.]
15. $WX + YZ \rightarrow WZ + YX$; the formation of precipitates, gases, and molecular substances occur when this type of reaction occurs.
16. The nonredox partner-exchange reactions that will take place include (a) because a precipitate forms, (d) because a gas and water form, and (e) because water forms. Neither (b) nor (c) will take place because they contain insoluble reactants.

17. The products will be two acids. The OH group will combine with the lesser electronegative element.
18. (a) $PCl_5(s) + 4H_2O(l) \rightarrow H_3PO_4(aq) + 5HCl(aq)$, (b) $BCl_3(g) + 3H_2O(l) \rightarrow H_3BO_3(aq) + 3HCl(aq)$, (c) $AsCl_3(s) + 3H_2O(l) \rightarrow H_3AsO_3(aq) + 3HCl(aq)$, (d) $ICl(l) + H_2O(l) \rightarrow HCl(aq) + HIO(aq)$.
19. (a) $ZnCl_2(aq) + H_2S(aq)$, (b) no reaction, (c) $Cu(OH)_2(s) + NaBr(aq)$, (d) $SO_2(g) + K_2SO_3(s) + H_2O(g)$, (e) no reaction.
20. Some classifications include displacement, combination, decomposition, electron transfer between monatomic ions in aqueous solution, and disproportionation reactions.
21. The more electronegative element is reduced and the more electropositive element is oxidized.
22. The state is determined by strength of oxidizing agent, concentration, and temperature.
23. (a) $Sn(s) + 2Cl_2(g) \rightarrow SnCl_4(s)$, (b) $2Cu(s) + I_2(g) \rightarrow 2CuI(s)$, (c) $S_8(s) + 8O_2(g) \xrightarrow{\Delta} 8SO_2(g)$, (d) $16Al(s) + 3S_8(s) \xrightarrow{\Delta} 8Al_2S_3(s)$.
24. (a) $Hg(l) + O_2(g)$, (b) $Mg(l) + Cl_2(g)$, (c) $KCl(s) + O_2(g)$.
25. The usual products formed by the thermal decomposition of (a) nitrates of Representative Group I and II metals are metal nitrites plus oxygen and of (b) nitrates of less active metals are oxides plus nitrogen dioxide and oxygen.
26. The free metal is oxidized. If the free metal is listed as a stronger reducing agent than the metal that it will replace, the reaction will occur. The restrictions include physical state, temperature, concentrations, and various oxidation states.
27. The halide ion is oxidized.
28. The common oxidizing agents include elemental nonmetals (F_2), metal and nonmetal anions (MnO_4^-, ClO_4^-), and oxidizing acids (HNO_3). The common reducing agents include free metals (Na) and nonmetals in lowest oxidation states (S^{2-}).
29. (a) $4HCl(g) + O_2(g) \xrightarrow{\Delta} 2Cl_2(g) + 2H_2O(g)$
    (b) $Fe_2O_3(s) + 3H_2(g) \xrightarrow{\Delta} 2Fe(s) + 3H_2O(g)$
30. In a disproportionation reaction an element in one oxidation state is both oxidized and reduced. The possibility of a given substance's undergoing such a reaction is recognized by its containing an element that is capable

of having at least three oxidation states--that in the reactant plus one higher and one lower.

31. (a) $Fe^{2+} + Br^-$, (b) $Cl_2(g) + S(s)$, (c) $MnO_2(s) + Mg^{2+}$, (d) $H_2O(l) + SO_4^{2-}$.

SKILLS

Skill 17.1 Predicting possible reaction products

The various classifications of redox and nonredox reactions are organized into a flow chart given in Table 17.8 of the text. By properly answering the numbered questions, it is usually possible to identify the type of reaction that might occur and to predict what the possible products of the reaction might be. The following examples illustrate the use of the flow chart.

Example (1) $HBrO_4(aq) + KOH(aq) \rightarrow ?$

The first "yes" answer is for question 3--both reactants are ionic in solution. The response to question 4 is "yes"--the formation of water can cause a partner-exchange reaction to occur.

$$HBrO_4(aq) + KOH(aq) \rightarrow KBrO_4(aq) + H_2O(l)$$

Example (2) $KClO(s) \xrightarrow{\Delta} ?$

The answer to question 1 is "yes"--there is only one reactant. There are three types of reactions that could occur: nonredox decomposition, redox decomposition, and disproportionation. The oxidation state of Cl is +1 in ClO^- and could conceivably change to both lower and higher values (disproportionation), forming
conceivably change to both lower and higher values (disproportionation), forming
Cl^- and ClO_3^-.

$$3KClO(s) \xrightarrow{\Delta} KClO_3(s) + 2KCl(s)$$

Example (3) $Cl_2(g) + I^- \rightarrow ?$

The answer to question 2 is "yes"--one of the reactants is an element. In this case a redox displacement can occur because Cl_2 is a strong oxidizing agent:

$$Cl_2(g) + 2I^- \rightarrow I_2(s) + 2Cl^-$$

Example (4) $Al(s) + 3ZnSO_4(aq) \rightarrow ?$

The answer to question 2 is "yes"--one of the reactants is an element. In this case a redox displacement can occur because Al is more active than Zn:

$$2Al(s) + 3ZnSO_4(aq) \rightarrow Al_2(SO_4)_3(aq) + 3Zn(s)$$

Example (5) $KClO_3(aq) + FeSO_4(s) \xrightarrow{H^+}$?

The answer to question 3 is "yes"--both reactants are ionic in solution. A partner-exchange reaction is unlikely (question 4) because both K_2SO_4 and $Fe(ClO_3)_2$ would be soluble compounds. However, ClO_3^- in acidic solution is an oxidizing agent (see Table 17.6 of the text) and Fe^{2+} is a reducing agent (see Table 17.7 of the text). The answer to question 5 is "yes"--a redox reaction of the more complex type can occur. The Fe^{2+} will be oxidized to Fe^{3+} and the ClO_3^- will be reduced to either Cl^- or Cl_2. The K^+ and SO_4^{2-} ions are spectator ions. Choosing Cl^- as the product of the reduction, the equation is

$$6Fe^{2+} + ClO_3^- + 6H^+ \rightarrow 6Fe^{3+} + Cl^- + 3H_2O(l)$$

Example (6) $KNO_3(aq) + PbO_2(s) \rightarrow$?

The answer to question 3 is "yes"--KNO_3 is a strong electrolyte. A partner-exchange reaction is not possible (question 4) because, even if PbO_2 dissolved, the possible products--K_2O and $Pb(NO_3)_2$--are both soluble compounds. Even though PbO_2 can act as an oxidizing agent (see Table 17.6 of the text), neither K^+ nor NO_3^- can be oxidized. The answer to question 5 is "no." Most likely, no reaction will occur.

$$KNO_3(aq) + PbO_2(s) \rightarrow \text{no reaction}$$

Example (7) $HCl(aq) + FeS(s) \rightarrow$?

The answer to question 3 is "yes"--HCl is ionic in aqueous solution. The possible products of a partner-exchange reaction (question 4) would be $FeCl_2$, a soluble compound, and H_2S, a weak electrolyte that could possibly escape from the solution as a gas if a sufficient amount were generated. Thus a partner-exchange reaction will probably occur.

$$2HCl(aq) + FeS(s) \rightarrow FeCl_2(aq) + H_2S(aq)$$

Skill Exercise: Use Table 17.8 of the text to predict possible products for the following reactions: (a) $P_4O_{10}(s) + H_2O(l) \rightarrow$, (b) $NaNO_3(s) \xrightarrow{\Delta}$, (c) $Cu^{2+} + H_2S(aq) \rightarrow$, (d) $C(s) + HNO_3(aq, conc) \xrightarrow{\Delta}$, (e) $Fe_2O_3(s) + CO(g) \xrightarrow{\Delta}$, (f) $CaO(s) + H_2O(l) \rightarrow$, (g) $Cl_2(g) + H_2O(l) \rightarrow$, (h) $Hg_2^{2+} + OH^- \rightarrow$, (i) $CuCl_2(s) + F_2(g) \rightarrow$, (j) $Ca(s) + O_2(g) \rightarrow$, (k) $PbF_2(s) + F_2(g) \rightarrow$, (l) $NaCl(aq) + KI(s) \rightarrow$, (m) $NaOH(aq) + CH_3COOH(aq) \rightarrow$, and (n) $KClO_3(aq) + Na_2SO_3(s) \xrightarrow{H^+}$. Write balanced equations for each reaction. *Answers*: (a) $P_4O_{10}(s) + 6H_2O(l) \rightarrow 4H_3PO_4(aq)$,

(b) $2NaNO_3(s) \xrightarrow{\Delta} 2NaNO_2(s) + O_2(g)$, (c) $Cu^{2+} + H_2S(aq) \rightarrow CuS(s) + 2H^+$, (d) $C(s) + 4HNO_3(aq, conc) \xrightarrow{\Delta} CO_2(g) + 4NO_2(g) + 2H_2O(l)$, (e) $Fe_2O_3(s) + 3CO(g) \xrightarrow{\Delta} 2Fe(s) + 3CO_2(g)$, (f) $CaO(s) + H_2O(l) \rightarrow Ca(OH)_2(s)$, (g) $Cl_2(g) + H_2O(l) \rightarrow HClO(aq) + HCl(aq)$, (h) $Hg^{2+} + 2OH^- \rightarrow HgO(s) + Hg(l) + H_2O(l)$, (i) $CuCl_2(s) + F_2(g) \rightarrow CuF_2(s) + Cl_2(g)$, (j) $2Ca(s) + O_2(g) \rightarrow 2CaO(s)$, (k) $PbF_2(s) + F_2(g) \rightarrow PbF_4(s)$, (l) $NaCl(aq) + KI(s) \rightarrow$ no reaction, (m) $NaOH(aq) + CH_3COOH(aq) \rightarrow Na(CH_3COO)(aq) + H_2O(l)$, (n) $KClO_3(aq) + 3Na_2SO_3(s) \xrightarrow{H^+} K^+ + 6Na^+ + Cl^- + 3SO_4^{2-}$ assuming the formation of Cl^- instead of Cl_2 from the reduction of ClO_3^-.

PRACTICE TEST

1 (20 points). Choose the response that best completes each statement.

(a) The oxidation number of nitrogen in nitric acid is

(i) +7 (ii) +5
(iii) +3 (iv) -3

(b) Use the following activity series (order of increasing oxidizing ability)

$$Zn < H_2 < Ag < Br < Cl$$

to choose the true statement:

(i) H_2 will reduce Zn^{2+} to Zn.
(ii) Ag^+ reacts with Zn, but not with H_2.
(iii) Br_2 reacts with Zn, but not with Cl^-.
(iv) Zn^{2+} and Br^- react to form Zn and Br_2.

(c) The more reactive a nonmetal, the

(i) less easily it is reduced
(ii) more easily it is reduced
(iii) more easily it forms positive ions
(iv) none of these answers

(d) Which is *not* a strong oxidizing agent?

(i) $H_3PO_4(aq)$ (ii) $HNO_3(aq)$
(iii) hot $H_2SO_4(aq, conc)$ (iv) $Cl_2(g)$

(e) The reaction between mercury and oxygen is a

(i) combination reaction (ii) decomposition reaction
(iii) displacement reaction (iv) partner-exchange reaction

(f) The heating of potassium chlorate is a

(i) combination reaction (ii) decomposition reaction
(iii) displacement reaction (iv) partner-exchange reaction

(g) Which type of reaction *cannot* involve oxidation-reduction?

(i) combination reaction (ii) decomposition reaction

(iii) displacement reaction (iv) partner-exchange reaction

(h) Which substance is acting as the reducing agent in the reaction?

$$H_2O_2(aq) + 2HI(aq) \rightarrow I_2(s) + 2H_2O(l)$$

(i) H_2O_2 (ii) I_2

(iii) HI (iv) O^-

(i) Choose the equation that does *not* involve redox.

(i) $Zn(s) + 2HCl(aq) \rightarrow ZnCl_2(aq) + H_2(g)$

(ii) $Al_2(SO_4)_3(aq) + 3Na_2CO_3(aq) + 3H_2O(l) \rightarrow 2Al(OH)_3(s) + 3NaSO_4(aq) + 3CO_2(g)$

(iii) $2K_2 + O_2(g) \rightarrow K_2O_2(s)$

(iv) $ICl(l) \rightarrow H_2O(l) \rightarrow HCl(aq) + HIO(aq)$

(j) As Fe^{2+} is oxidized to Fe^{3+}, $Cr_2O_7^{2-}$ is reduced to Cr^{3+}. The number of moles of Fe^{2+} that are oxidized by one mole of $Cr_2O_7^{2-}$ is

(i) 1 mol Fe^{2+} (ii) 3 mol Fe^{2+}

(iii) 2 mol Fe^{2+} (iv) 6 mol Fe^{2+}

2 (20 points). Classify each of the following reactions as completely as possible.

(a) $2Eu^{2+} + 2H^+ \rightarrow 2Eu^{3+} + H_2(g)$

(b) $2La(s) + N_2(g) \xrightarrow{\Delta} 2LaN(s)$

(c) $[Ag(NH_3)_2]^+ + 2CN^- \rightarrow [Ag(CN)_2]^- + 2NH_3(aq)$

(d) $GeO_2(s) + 4HCl(aq) \xrightarrow{\Delta} GeCl_4(g) + 2H_2O(l)$

(e) $BaGeF_6(s) \xrightarrow{\Delta} GeF_4(g) + BaF_2(g)$

3 (20 points). Predict the products of the following reactions: (a) $Na(g) + As(s) \xrightarrow{\Delta}$, (b) $SiH_4(g) + O_2(g) \rightarrow$, (c) $K(s) + H_2O(l) \rightarrow$, (d) $NaHCO_3(s) \xrightarrow{\Delta}$, and (e) $Fe^{3+} + Cr^{2+} \rightarrow$.

4 (10 points). A solution of $Na_2S_2O_3$ was standardized by treating a known amount of pure $KIO_3(s)$ with excess KI(s) to form $I_2(aq)$ and then reducing the $I_2(aq)$ with the $S_2O_3^{2-}$ in the presence of starch as an indicator. Write the chemical equations for the reactions. (The oxidation of $S_2O_3^{2-}$ gives $S_4O_6^{2-}$.) A 0.111 g sample of KIO_3 required 22.67 mL of $Na_2S_2O_3$ for the reaction. What is the concentration of the $Na_2S_2O_3$ solution?

5 (10 points). A 10.00 mL sample of a commercial bleach containing NaClO was diluted to 100.00 mL with water. Excess KI was added to a 25.00 mL sample of the diluted bleach and the free iodine formed was reduced by using 23.52 mL of a 0.0156 M $Na_2S_2O_3$ solution with starch as an indicator. The density of the original commercial bleach was 1.078 g/mL. What is the mass percent of NaClO in the commercial bleach?

6 (20 points). Nitric acid is a good oxidizing agent. Predict the possible products of the redox reactions between HNO_3(conc) and (a) Cu(s), (b) HCl(conc), and (c) I_2(s). Write the chemical equations for these reactions. How would the products of the reaction with Cu(s) differ if the nitric acid were dilute?

Answers to practice test

1. (a) ii, (b) iii, (c) ii, (d) i, (e) i, (f) ii, (g) iv, (h) iii, (i) ii, (j) iv.
2. (a) redox--electron transfer between monatomic ions in aqueous solution, (b) redox--combination of two elements to give a compound, (c) nonredox--displacement, (d) nonredox--partner exchange, (e) nonredox--decomposition to give compounds.
3. (a) $3Na(g) + As(s) \xrightarrow{\Delta} Na_3As(s)$, (b) $SiH_4(g) + 2O_2(g) \rightarrow SiO_2(s) + 2H_2O(l)$, (c) $2K(s) + 2H_2O(l) \rightarrow 2KOH(aq) + H_2(g)$, (d) $2NaHCO_3(s) \xrightarrow{\Delta} Na_2CO_3(s) + CO_2(g) + H_2O(g)$, (e) $Fe^{3+} + Cr^{2+} \rightarrow Fe^{2+} + Cr^{3+}$.
4. 0.137 mol/L.
5. 5.07 mass %.
6. (a) $4HNO_3(conc) + Cu(s) \rightarrow 2NO_3^- + Cu^{2+} + 2NO_2(g) + 2H_2O(l)$; (b) $HNO_3(conc) + 3HCl(conc) \rightarrow Cl_2(g) + NOCl(g) + 2H_2O(l)$; (c) $10HNO_3(conc) + 3I_2(s) \rightarrow 6HIO_3(aq) + 10NO(g) + 2H_2O$; NO(g) is produced instead of NO_2(g).

ADDITIONAL NOTES

CHAPTER 18

CHEMICAL KINETICS

CHAPTER OVERVIEW

This chapter considers atoms, molecules, and ions as they collide, the conditions under which they might react with each other, and the energy changes involved. The effect of the concentrations of the reactants on the rate of reaction is discussed. The terminology and mathematics of chemical kinetics is introduced. A brief description of reaction mechanisms and the interpretation of rate equations is given. The influence of temperature, of the amount of contact between reacting substances, and of catalysis is discussed.

COMPETENCIES

Definitions for the following terms should be learned:

- activation energy
- bimolecular reaction
- chemical adsorption
- chemical kinetics
- elementary reaction
- free radical
- heterogeneous catalyst
- heterogeneous reaction
- homogeneous catalayst
- homogeneous reaction
- inhibitors
- initial reaction rate
- instantaneous reaction rate
- intermediate
- molecularity
- overall reaction order
- physical adsorption
- promoters
- rate constant
- rate-determining step
- rate equation
- reactant reaction order
- reaction mechanism
- reaction rate
- termolecular reaction
- transition state, activated complex

General concepts and skills that should be learned include

- understanding the role of kinetics in describing the rate and mechanism of a reaction

- interpreting a chemical reaction at the molecular level in terms of collisions between reactants
- understanding the difference between elementary reactions and reactions that occur by more complex mechanisms
- writing and interpreting rate equations and integrated rate equations for zero-, first-, and second-order reactions
- interpreting the role of the activation energy in determining reaction rate
- explaining how a catalyst works
- identifying those experimental conditions that influence reaction rate and explaining their influence

Numerical exercises that should be understood include

- calculating the rate of reaction by using a rate equation
- calculating the concentration, time, or rate constant when two of these variables are given
- determining the order and rate constant for a reaction from initial rate data or concentration-time data
- calculating activation energy, temperature, or rate constant when two of these variables are given

QUESTIONS

1. Complete the following sentences by filling in the various blanks with these terms: activation energy, chemical kinetics, elementary reaction, rate constant, rate equation, reaction order, reaction mechanisms, reaction rate, and transition state.

The study of reaction rates and (a)______ is known as (b)______. An (c)______ occurs only when reacting species collide. Only a very small portion of the collisions result in chemical reaction. The energy that the reacting species must have in order for a reaction to occur is the (d)______. The (e)______ is a combination of reacting molecules intermediate between reactants and products. The (f)______ can be expressed only for a particular moment in time and generally decreases as the reactants are being used up because fewer collisions are possible. A (g)______ is the mathematical relationship between the reaction rate and the concentrations of the reactants. The proportionality constant between the rate and the concentrations is the (h)______ and the sum of the exponents of the concentration terms is the overall (i)______.

2. Match the correct term to each definition.

(a) substances that make a catalyst more effective
(b) a catalyst present in a different phase than the reactants
(c) substances that slow down a catalyzed reaction
(d) a simple reaction that occurs by a single step
(e) a catalyst present in the same phase as the reactants

(i) elementary reaction
(ii) heterogeneous catalyst
(iii) homogeneous catalyst
(iv) inhibitors
(v) promoters

3. Which statements are true? Rewrite each false statement so that it is correct.

(a) An elementary reaction is one that is unimolecular.
(b) The activation energy is released by the products after the reaction has taken place.
(c) A transition state may include the catalyst as part of its structure.
(d) Even though a catalyst is recovered in its original form after the completion of the reaction, it may have actively participated in the reaction and even undergone temporary chemical changes during the reaction.
(e) A promoter, because of its presence, speeds up the rate of a chemical reaction.
(f) Molecularity and reaction order could be the same for a given reaction, but in general are not the same.
(g) The rate of a reaction that occurs in a series of steps will depend on the slowest step in the reaction mechanism.
(h) The rate constant is directly proportional to the absolute temperature.

4. Distinguish clearly between/among (a) activation energy and heat of reaction; (b) homogeneous and heterogeneous reactions; (c) physical and chemical adsorption; (d) an inhibitor, a promoter, and a catalyst; and (e) reaction rate, initial reaction rate, and instantaneous reaction rate.

5. Choose the response that best completes each statement.

(a) The study of kinetics does *not* involve
(i) reaction rates (ii) catalysis
(iii) reaction mechanisms (iv) predicting reaction spontaneity

(b) The number of individual atoms, molecules, or ions that must simultaneously react in an elementary reaction is the

(i) overall reaction (ii) rate constant

(iii) molecularity (iv) reactant reaction order

(c) Rate equations

(i) express the relationship between the rate of reaction and the concentration of reactants

(ii) take different forms for different reactions

(iii) contain the rate constant, a proportionality constant between reactant concentrations and the rate of reaction

(iv) all of the above

(d) The sum of the exponents on the concentration terms in the rate equation is the

(i) molecularity (ii) rate constant

(iii) overall reaction order (iv) transition state

(e) If a reaction is found to be second order in a reactant, doubling the concentration of that reactant would cause the reaction rate to

(i) stay the same (ii) quadruple

(iii) double (iv) none of these answers

(f) To confirm that the reaction rate is first order with respect to [A], a linear plot results if a graph is made of

(i) [A] against t (ii) log [A] against t

(iii) 1/[A] against t (iv) log k against $1/T$

(g) In the linear plot for a first-order reaction, the value of k is

(i) k = (slope)/(2.303) (ii) k = -(2.303)(slope)

(iii) k = (2.303)(slope) (iv) k = intercept of line

(h) For the elementary reaction A + B $\rightarrow$ C, where [A] = [B], which plot would give a straight line?

(i) log [A] against t (ii) $\frac{1}{[A]}$ against t

(iii) log ([A][B]) against t (iv) log(1/[A]) against t

(i) Which is not a factor that influences reaction rate?

(i) nature of reactants

(ii) concentration

(iii) degree of contact between the reacting substances

(iv) reaction order

(j) The rates of most chemical reactions increase as the temperature increases primarily because at higher temperatures
  (i) the activation energy decreases
  (ii) the pressure is higher
  (iii) there are increases in the average distances between atoms within molecules
  (iv) there are more effective collisions involving molecules

(k) If the forces between the adsorbing surface and adsorbate are of the magnitude of chemical bonds, the adsorption is known as
  (i) activation energy (ii) chemical adsorption
  (iii) physical adsorption (iv) homogeneous catalysis

(l) Which is *not* a property of a catalyst?
  (i) increases the reaction rate
  (ii) can lower the activation energy of both the forward and reverse reactions
  (iii) shifts the equilibrium of the reaction toward the products
  (iv) is generally very specific for a particular reaction

6. What information can be derived from a balanced chemical equation about the stoichiometry, thermodynamics, and kinetics of a reaction?

7. What is a reaction mechanism?

8. What is necessary for two molecules to react chemically? How do we describe the initial species which these molecules might form? What can subsequently happen to this species?

9. Explain how elementary reactions are categorized in terms of molecularity. What are the most common molecularities of reactions? Why are termolecular reactions rare?

10. Are there any truly unimolecular reactions (excluding nuclear decay processes)? What is the usual explanation given for the mechanisms of gas-phase decomposition reactions that one might believe to be unimolecular?

11. How do molecular collisions in aqueous solution differ from those in the gas phase?

12. For the reaction of iodide ion with hypochlorite ion,

$$I^- + ClO^- \rightarrow IO^- + Cl^-$$

there are two proposed mechanisms. The first is

$$ClO^- + H_2O \rightleftharpoons HClO(aq) + OH^-$$
$$I^- + HClO(aq) \rightarrow HIO(aq) + Cl^-$$
$$OH^- + HIO(aq) \rightarrow H_2O(l) + IO^-$$

and the second is

$$ClO^- + H_2O \rightleftharpoons HClO(aq) + OH^-$$
$$I^- + HClO(aq) \rightarrow ICl(aq) + OH^-$$
$$ICl(aq) + 2OH^- \rightarrow IO^- + Cl^- + H_2O(l)$$

What is the molecularity of each of these elementary reactions?

13. Define the term "reaction rate." What are typical units for this physical quantity?

14. How does the rate of reaction vary with time for most reactions? Why?

15. What is the relationship between the reaction rate and the concentration of the reactants called? Can this relationship be written down from the stoichiometric chemical equation?

16. What is the name of the proportionality constant that relates reaction rate to concentration and what is its usual symbol? Does this constant always have the same set of units? It is independent of temperature?

17. The rate of a reaction depends on the number of collisions between molecules, which, in turn, depends on the velocity at which the molecules travel. Recalling that at a given temperature, average molecular velocities of gases are inversely related to mass, predict which of the following reactions has the higher value of k, the rate constant:

$$2HI(g) \rightarrow H_2(g) + I_2(g) \qquad 2DI(g) \rightarrow D_2(g) + I_2(g)$$

Assume that the internal energy changes for both reactions are the same. Explain your answer.

18. What is the order of reaction for a reactant? How does this differ from the overall reaction order?

19. Write the rate equation for the following chemical reaction

$$2A + B + C \rightarrow \text{products}$$

Use lowercase letters (m, n, o, ...) to represent the order of reaction for any component that you may not know.

20. The rate equation for the synthesis of urea from ammonium cyanate,

$$NH_4^+ + CNO^- \rightarrow (NH_2)_2CO$$

is

$$\text{Rate} = k[NH_4^+][CNO^-]$$

What is the order of reaction with respect to each reactant and what is the overall order? What happens to the reaction rate as $[NH_4^+]$ increases?

21. Define the term "half-life." What is wrong with the statement, "After one half-life, there is one-half of the original amount of reactant remaining and after the second half-life period, all of the reactant has reacted"?

22. What are the methods that are described in the text for determining an experimental rate equation? Briefly describe each method.

23. How can the initial reaction rate be determined from concentration-time data?

24. What does the term "rate-determining step" mean?

25. What can be said about a proposed mechanism for which the derived rate equation does not agree with the experimental rate equation? What if they agree?

26. Name five factors that influence the rate of reaction.

27. Write the mathematical equation that describes the temperature dependency of the rate constant. What is the name of this equation?

28. Describe how you would determine the activation energy for a chemical reaction.

29. Briefly describe how physical contact between reactants might affect reaction rate.

30. What is a catalyst? Explain how a catalyst increases the reaction rate. Will a catalyst change the relative amounts of reactants and products once equilibrium has been attained? Can a catalyst change a thermodynamically unfavorable reaction into a favorable reaction?

31. Identify what each of the letters in the diagram given in Fig. 18-1 represents.

32. Consider the following reaction mechanism:

$$A + B \xrightarrow{\text{fast}} C$$

Fig. 18-1

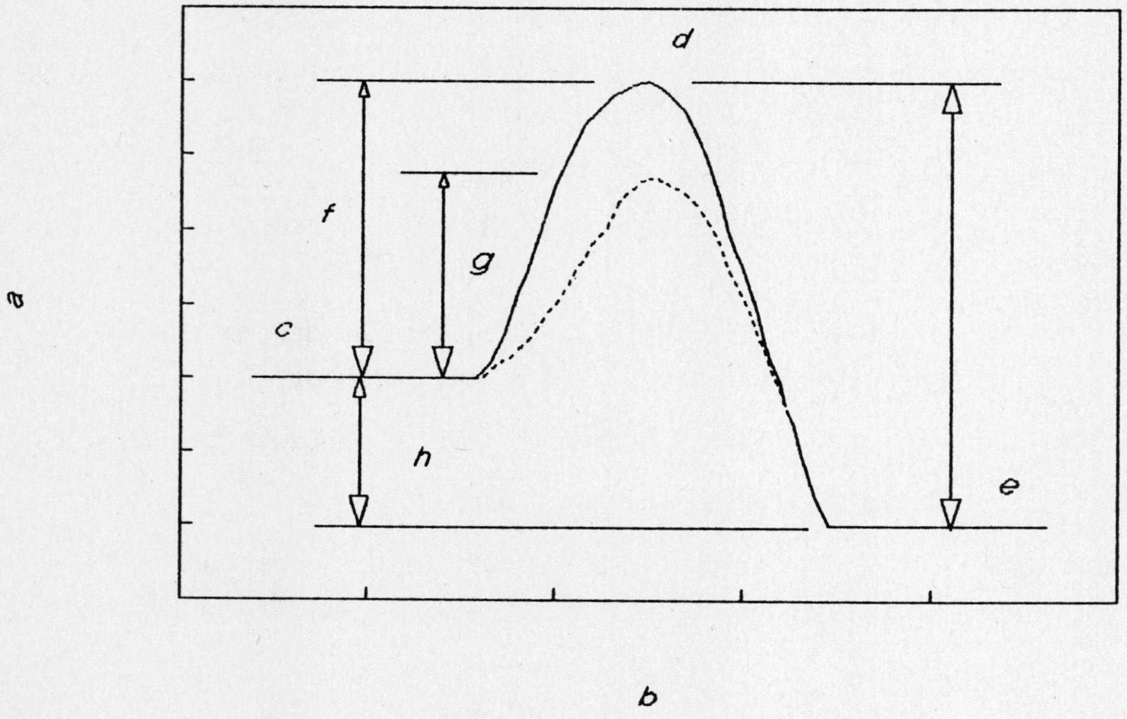

$$B + C \xrightarrow{\text{fast}} D + A$$

$$D + B \xrightarrow{\text{slow}} E$$

Identify the (a) catalyst(s), (b) intermediate(s), (c) reactant(s), (d) product(s). (e) Write the rate equation for the rate-determining step. (f) What is the molecularity of the reaction?

Answers to questions

1. (a) reaction mechanisms, (b) chemical kinetics, (c) elementary reaction, (d) activation energy, (e) transition state, (f) reaction rate, (g) rate equation, (h) rate constant, (i) reaction order.
2. (a) v, (b) ii, (c) iv, (d) i, (e) iii.
3. (a) F, replace *is unimolecular* with *occurs in a single step as written*; (b) F, replace *released by ... taken place*, with *the minimum energy that reactants must have for the reaction to occur*; (c) T; (d) T; (e) F, replace *presence* with *interaction with the catalyst*; (f) T; (g) T; (h) F, replace *directly* with *exponentially*.

4. (a) *Activation energy* is the energy that the reactants must have in order for the reaction to occur; *heat of reaction* is the difference in the energy of the products compared to that of the reactants. (b) *Homogeneous reactions* have reactants in the same phase; *heterogeneous reactions* have reactants in different phases. (c) *Physical adsorption* is adsorption occurring by physical processes that involve forces on the order of van der Waals forces; *chemical adsorption* occurs by chemical forces that are stronger than those of physical adsorption. (d) An *inhibitor* is a substance that slows down a catalyzed reaction; a *promoter* speeds up a reaction by making the catalyst more effective; a *catalyst* is a substance that speeds up a reaction. (e) The *reaction rate* is the change in concentration of a reactant or product for a change in time; the *initial reaction rate* is the reaction rate at the beginning of the reaction; the *instantaneous reaction rate* is the reaction rate at any single instant of time.
5. (a) iv, (b) iii, (c) iv, (d) iii, (e) ii, (f) ii, (g) ii, (h) ii, (i) iv, (j) iv, (k) ii, (l) iii.
6. A balanced chemical equation gives energy and mass relationships but not how fast or by what mechanism the reaction takes place.
7. A reaction mechanism is a single or a series of elementary reactions describing the pathway by which a reaction takes place.
8. Molecular collision is physically necessary for one molecule to react chemcally with another molecule. The formation of a transition state or activated complex results from such a collision between reactants. The transition state or activated complex breaks up to give either the products of the reaction or the reactants.
9. Elementary reactions are categorized in terms of their molecularity--the number of reactant particles involved in the elementary reaction. The most common molecularities of reactions are unimolecular, bimolecular, and termolecular. Termolecular reactions are rare, for the probability of three reacting particles coming together at the same time to form a transition state is low.
10. There are probably not any truly unimolecular reactions (excluding nuclear decay processes). In gas-phase decomposition reactions that one might believe to be unimolecular, the reacting molecules obtain sufficient activation energy by collision with inert molecules or container walls or from radiation.

11. There are fewer collisions in solution than in the gas phase because reactants are slowed down by surrounding solvent molecules. However, once the reactants get close together, they are trapped in the same solvent cage and collide with each other many times in a short space of time.
12. bimolecular, bimolecular, bimolecular; bimolecular, bimolecular, termolecular.
13. Reaction rate is the change in concentration of a reactant or a product in a unit of time. Typical units for reaction rate are moles per liter per second.
14. The rate of reaction decreases with increasing time for most reactions because the reactants are used up and fewer collisions are possible.
15. The relationship is the rate equation; no, it must be determined experimentally.
16. The proportionality constant that relates reaction rate to concentration is the rate constant. Its usual symbol is *k*. The rate constant does not always have the same set of units--it depends upon the overall order of the reaction. The rate constant is not independent of temperature.
17. The HI reaction has the higher value of *k*. The rate of reaction is directly related to the number of collisions, which is directly related to velocity, which is inversely related to mass--thus the faster reaction rate will be for the reaction having the reactants with lower mass.
18. The reactant reaction order is the exponent on the term for that reactant in the rate equation. The overall reaction order is the sum of all of the exponents of the terms that appear in the rate equation.
19. $\text{Rate} = k[A]^m[B]^n[C]^o$.
20. For the synthesis of urea from ammonium cyanate, the reaction is first order with respect to both NH_4^+ and CNO^-. The reaction is second order overall. As $[NH_4^+]$ increases, the reaction rate increases.
21. Half-life is the time it takes for one half of a reactant to undergo a reaction. The second part of the statement is false because during the second half-life period, only one half of the remaining half of the reactant would react, leaving one quarter of the reactant.
22. The methods are the initial rate method (determine order by a ratio of initial rates, varying one reactant at a time) and graphical (determine order by preparing appropriate plots of concentration-time data to obtain a straight-line plot).
23. Prepare a plot of concentration against time. The slope of the curve at time equal to zero is the initial reaction rate.

24. The rate-determining step is the slowest step of a series of elementary reactions in a mechanism.
25. If the derived rate equation does not agree with the experimental rate equation, then the proposed mechanism is wrong. If they agree, the proposed mechanism may be correct.
26. The factors are nature of reactants, temperature, concentration, degree of contact between reactants, catalysis.
27. The mathematical equation that describes the temperature dependency of the rate constant is

$$\log k = \frac{-E_a}{(2.303)RT} + \log A$$

This relationship is known as the Arrhenius equation.
28. The activation energy for a chemical reaction can be determined from the Arrhenius equation by measuring k at several temperatures, plotting log k versus $1/T$, and calculating E_a from the slope of the line [E_a = $-(2.303)R(\text{slope})$].
29. In a homogeneous reaction collisions are frequent, so the question of sufficient contact is not important. In a heterogeneous reaction, however, bringing the reacting molecules or ions together may be difficult, so collisions may be limited in number.
30. A catalyst increases the rate of a chemical reaction but can be recovered in its original form when the reaction is finished. A catalyst increases the reaction rate by providing an alternate and easier pathway from reactants to products. A catalyst will not change the relative amounts of reactants and products once equilibrium has been attained. A catalyst cannot change a thermodynamically nonspontaneous reaction into a spontaneous one.
31. (a) energy, (b) extent of reaction, (c) energy content of reactants, (d) transition state, (e) energy content of products, (f) activation energy for forward reaction, (g) activation energy for catalyzed forward reaction, (h) thermodynamic heat of reaction, (i) activation energy for reverse reaction.
32. (a) A; (b) C, D; (c) B; (d) E; (e) Rate = k[D][B]; (f) bimolecular.

SKILLS

Skill 18.1 Writing and determining rate equations

The rate of reaction is directly proportional to the concentration of the reactants (and occasionally, the concentration of catalyts and products) raised to appropriate powers, that is, for the reaction

$$aA = bB + \ldots \rightarrow \text{products}$$

the rate equation is given by

$$\text{Rate} = k[A]^x[B]^y \ldots$$

where x, y, ... are not necessarily equal to a, b, ... and k is the rate constant. The values of the exponents and the value of k at a given temperature must be determined experimentally. The rate equation for gaseous reactions is often written in terms of partial pressures instead of concentrations.

Skill Exercise: Write rate equations for the following reactions:

(a) $H_2(g) + Cl_2(g) \rightarrow 2HCl(g)$

(b) $Zn(s) + 2H^+ \rightarrow H_2(g) + Zn^{2+}$

(c) $IO_3^- + 8I^- + 6H^+ \rightarrow 3I_3^- + 3H_2O(1)$

(d) $CH_3CH(OC_2H_5)_2(1) + H_2O(1) \xrightarrow{H^+} CH_3CHO(1) + 2C_2H_5OH(1)$

(e)
```
   H₂C——CH        HC——CH
    |    ‖          ‖    ‖
   H₂C  CH(g)  →   HC   CH(g) + H₂(g)
    \  /            \  /
     C               C
     |               |
     H₂              H₂
```

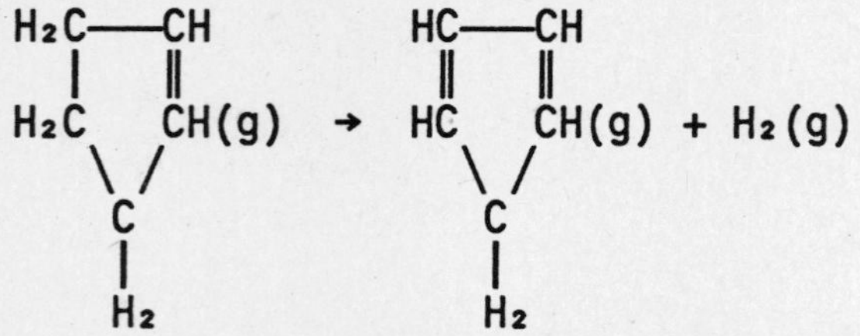

Answers: (a) Rate = $k[H_2]^x[Cl_2]^y$ or Rate = $k' P_{H_2}^x P_{Cl_2}^y$, (b) Rate = $k[H^+]^x$, (c) Rate = $k[IO_3^-]^x[I^-]^y[H^+]^z$, (d) Rate = $k[CH_3CH(OC_2H_5)_2]^x[H_2O]^y$, (e) Rate = $k[C_5H_8]^x$.

A first-order reaction gives a linear plot if log [A] is plotted against time and $k = (-2.303)(\text{slope})$, a second-order reaction gives a linear plot if $1/[A]$ is plotted against time and the slope is equal to k, and a zero-order reaction gives a linear plot if [A] is plotted against time and the slope is equal to $-k$.

An alternative method for determining the reaction order for a reactant is to measure the initial reaction rates for several reactions by varying the initial concentrations of the one reactant, but holding the others constant. This method is illustrated in *Example 18.7 Reaction order*.

NUMERICAL EXERCISES

Example 18.1 Rate equations

Ordinary table sugar, sucrose ($C_{12}H_{22}O_{11}$), rotates polarized light in a given direction. Upon hydrolysis, sucrose forms a mixture of two simpler sugars, glucose and fructose (both having the formula $C_6H_{12}O_6$), which rotates polarized light in the opposite direction. The chemical equation for this "inversion" process is

$$C_{12}H_{22}O_{11}(aq) + H_2O(l) \rightarrow 2C_6H_{12}O_6(aq)$$

The first-order rate equation describing this process at 30 °C is

$$\text{Rate} = (0.0342\ h^{-1})[C_{12}H_{22}O_{11}]$$

Calculate the initial rate of reaction given $[C_{12}H_{22}O_{11}]_0 = 0.010$ mol/L.

Substituting the concentration into the rate equation gives

$$\text{Rate} = (0.0342\ h^{-1})(0.010\ \text{mol/L}) = 3.4 \times 10^{-4}\ \text{mol/L h}$$

The initial rate of inversion of sucrose is 3.4×10^{-4} mol/L h.

Exercise 18.1 The rate equation describing the equation

$$CH_3I + C_2H_5ONa \xrightarrow{\text{ethanol}} CH_3OC_2H_5 + NaI$$

at 30 °C is

$$\text{Rate} = (2.08 \times 10^{-3}\ \text{L/mol s})[CH_3I][C_2H_5ONa]$$

What is the initial rate of reaction given $[CH_3I]_0 = 0.020$ mol/L and $[C_2H_5ONa]_0 = 0.020$ mol/L? *Answer*: 8.3×10^{-7} mol/L s.

Example 18.2 Rate equations

What is the rate of reaction at the end of 1.00 h for the inversion process described in *Example 18.1*?

The amount of sugar left at the end of 1.00 h is found by using the integrated rate equation for a first-order reaction:

$$\log \frac{[C_{12}H_{22}O_{11}]_0}{[C_{12}H_{22}O_{11}]} = \frac{kt}{2.303}$$

$$\log \frac{(0.010\ \text{mol/L})}{[C_{12}H_{22}O_{11}]} = \frac{(0.0342\ h^{-1})(1.00\ h)}{2.303} = 0.0149$$

$$\frac{(0.010\ \text{mol/L})}{[C_{12}H_{22}O_{11}]} = 1.035$$

$$[C_{12}H_{22}O_{11}] = \frac{0.010\ \text{mol/L}}{1.035} = 0.0097\ \text{mol/L}$$

The rate reaction is

$$\text{Rate} = (0.0342\ h^{-1})(0.0097\ \text{mol/L}) = 3.3 \times 10^{-4}\ \text{mol/L h}$$

Exercise 18.2 What is the concentration of CH_3I left after the reaction described in Exercise 18.1 takes place for 1.00 h? *Answer*: 0.018 mol/L

Example 18.3 Rate equations

How long will it take for the concentration of sugar to decrease to 0.0010 mol/L in the process described in *Example 18.1*?

The time required for the concentration to change from 0.010 mol/L to 0.0010 mol/L is

$$t = \frac{(2.303)\ \log\left[\frac{[C_{12}H_{22}O_{11}]_0}{[C_{12}H_{22}O_{11}]}\right]}{k} = \frac{(2.303)\ \log\left[\frac{0.010\ mol/L}{0.0010\ mol/L}\right]}{(0.0342\ h^{-1})}$$

$$= 67.3\ h$$

It would take 67.3 h for the concentration of sugar to decrease to 0.0010 mol/L.

Exercise 18.3 How long will it take for the concentration of CH_3I to decrease to 0.015 mol/L in the reaction described in Exercise 18.1? *Answer*: 8.0×10^3 s or 2.2 h.

Example 18.4 Rate equations: half-life

What is the half-life of the reaction described in *Example 18.1*?

For a first-order reaction, the half-life is given by

$$t_{1/2} = \frac{0.693}{k} = \frac{0.693}{(0.0342\ h^{-1})} = 20.3\ h$$

The half-life of the sugar inversion reaction is 20.3 h.

Exercise 18.4 What is the half-life of the reaction described in Exercise 18.1? *Answer*: 2.4×10^4 s or 6.7 h.

Example 18.5 Reaction order

The reaction

$$2NO(g) + H_2(g) \rightarrow N_2O(g) + H_2O(g)$$

was studied at 820 °C. In the presence of excess NO, the following data were obtained

P_{H_2}, Torr	10.0	8.4	7.0	5.8	4.9	4.1	3.4
t, s	0	5	10	15	20	25	30

Show that this reaction is first order in P_{H_2} under these conditions and determine the rate constant.

The plot of log P_{H_2} against time shown in Fig. 18-2 is linear, proving that the reaction is first order with respect to H_2. The slope of the straight line is $-0.0156\ s^{-1}$ and the rate constant is

$$k = -(2.303)(\text{slope}) = -(2.303)(-0.016\ s^{-1}) = 0.0359\ s^{-1}$$

The first-order rate constant under the conditions of the reaction is $0.037\ s^{-1}$.

Fig. 18-2

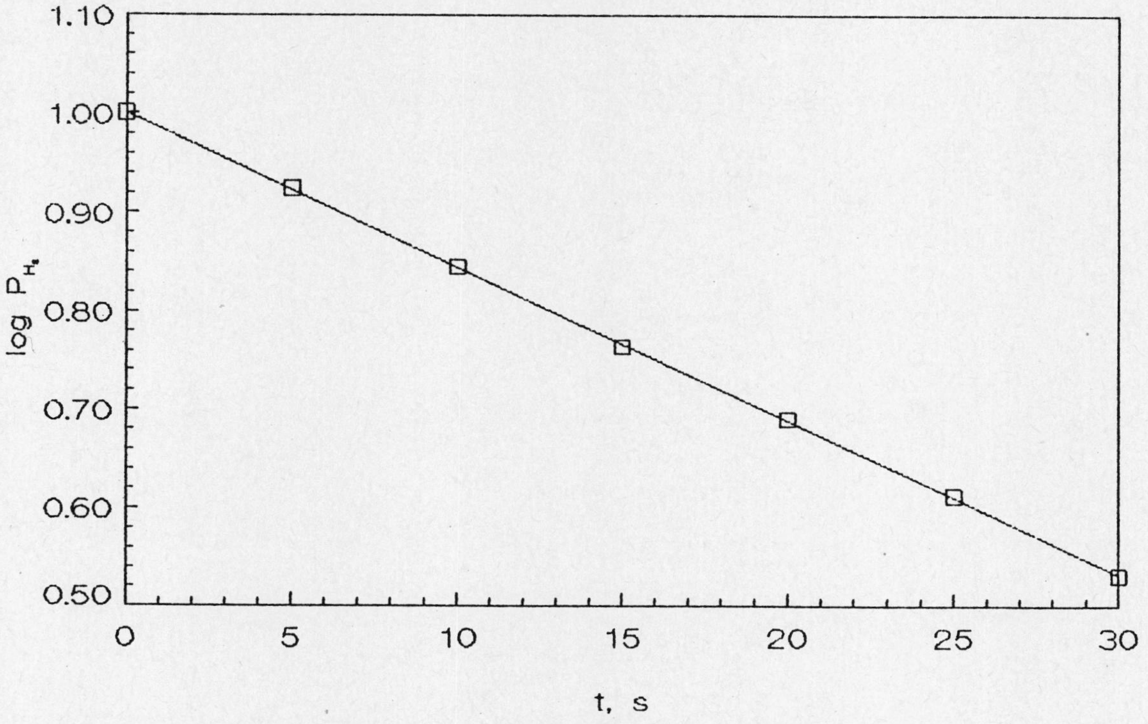

Exercise 18.5 Show that the reaction

$$(CH_3CHO)_3(g) \xrightarrow{260\ °C} 3CH_3CHO(g)$$

is first order from the following data for $P_{(CH_3CHO)_3}$.

$P_{(CH_3CHO)_3}$, Torr	100	64	41	27	17
t, h	0.00	1.00	2.00	3.00	4.00

Determine the rate constant. *Answer*: plot of log $P_{(CH_3CHO)_3}$ against time is linear, $k = 0.440\ h^{-1}$.

Example 18.6 Reaction order

The reaction described in *Example 18.5* was studied in the presence of excess H_2 and the following data were obtained.

P_{NO}, Torr	10.0	8.9	8.1	7.4	6.8
t, s	0	100	200	300	400

Show that this reaction is second order in P_{NO} under these conditions and determine the rate constant.

The plot of $1/P_{NO}$ against time shown in Fig. 18-3, is linear, proving that the reaction is second order under these conditions. The slope of the straight line which is equal to k is 1.16×10^{-4} $Torr^{-1}$ s^{-1}.

Exercise 18.6 The oxidation of CS_2 by O_3 was studied at 29.3 °C in excess CS_2.

$$CS_2(g) + 2O_3(g) \rightarrow CO_2(g) + 2SO_2(g)$$

Using the following data, show that the reaction is second order under these conditions and determine the rate constant.

P_{O_3}, Torr	1.75	0.50	0.29	0.20
t, s	0	2	4	6

Answer: The plot of $1/P_{O_3}$ against time is linear; k = 0.74 $Torr^{-1}$ min^{-1}.

Example 18.7 Reaction order*

The kinetics of the chemical reaction

$$C_2H_4(g) + H_2(g) \xrightarrow[25\,°C]{Cu} C_2H_6(g)$$

was studied by observing initial rates of reaction for various combinations of $P_{C_2H_4}$ and P_{H_2}. Write the general expression for the rate equation and evaluate the exponents for each reactant by using the following data.

Experiment	1	2	3	4
$P_{C_2H_4}$, arbitrary units	1	1	2	2
P_{H_2}, arbitrary units	1	2	1	2
initial rate, arbitrary units	1	2	0.5	1

This problem is solved by using the comprehensive problem-solving approach presented in *Skill 2.7 Solving problems*.

1. Study the problem and be sure you understand it.
   (a) What is unknown?
   The exponents for the pressures in the general rate equation, that is, the reaction order for C_2H_4 and H_2.

Fig. 18-3

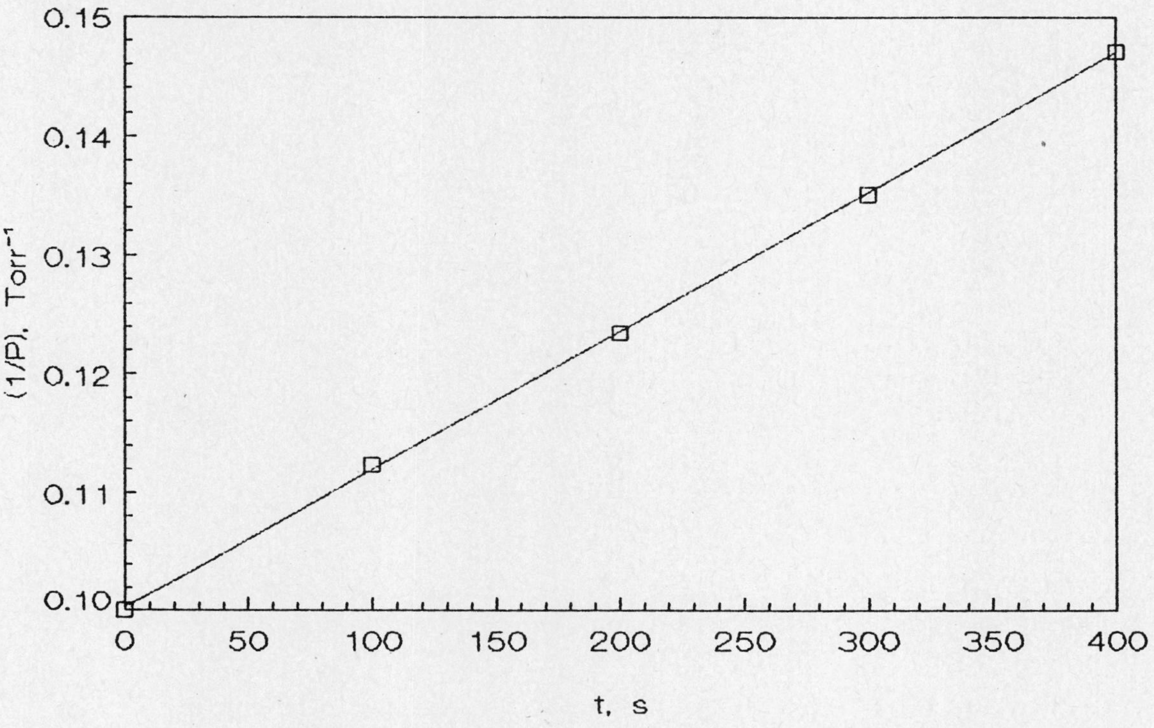

(b) What is known?

Initial rate data for various experiments in which the initial pressure of only one reactant at a time was changed. Also, that many rate equations take the general form

$$\text{Rate} = k\ P_A^x\ P_B^y\ldots$$

2. Decide how to solve the problem.

(a) What is the connection between the known and the unknown?

The rate equation

$$\text{Rate} = k\ P_{H_2}^x\ P_{C_2H_4}^y$$

relates the initial rate data to the order of reaction.

(b) What is necessary to make the connection?

By comparing experiments (1) and (2), the order with respect to H_2 can be determined, and by comparing experiments (1) and (3), the order with respect to C_2H_4 can be determined.

3. Set up the problem and solve it.

Experiments (1) and (2) show that as P_{H_2} is doubled, the rate of reaction is doubled, thus $x = 1$. Experiments (1) and (3) show that as $P_{C_2H_4}$ is doubled, the rate of reaction is halved, thus $y = -1$. The rate equation is

$$\text{Rate} = k\, P_{C_2H_4}^{-1}\, P_{H_2}^{1}$$

4. Check the result.

(a) Is the answer reasonable?

Yes. Comparison of the data given by experiments (1) and (4) verifies that the answer is reasonable. The reaction rate should be doubled by doubling the pressure of H_2 and halved by doubling the pressure of C_2H_4, giving no overall change in rate, which is the result listed in the table.

Exercise 18.7 Using the data shown in the table for the reaction

$$2H_2O_2(aq) + I^- \rightarrow 2H_2O(l) + O_2(g) + I^-$$

Experiment	1	2	3	4
$[H_2O_2]$, mol/L	1	2	1	2
$[I^-]$, mol/L	1	1	2	2
initial rate, arbitrary units	1	2	2	4

determine the rate equation. *Answer*: Rate = $k[H_2O_2][I^-]$.

*Example 18.8 Activation energy and temperature**

A laboratory instructor told a student that raising the temperature of a reaction mixture by 10 °C will double the rate of reaction. This is true only for reactions having a certain value of the activation energy at a given temperature. Find the value of E_a for which this statement holds true if the temperature change is between 25 °C and 35 °C.

This problem is solved by using the comprehensive problem-solving approach presented in *Skill 2.7 Solving problems*.

1. Study the problem and be sure you understand it.

(a) What is unknown?

The value of the activation energy.

(b) What is known?

The reaction rate doubles as the temperature increases from 25 °C to 35 °C.

2. Decide how to solve the problem.
    (a) What is the connection between the known and the unknown?
        The reaction rate is directly proportional to the rate constant. The
        temperature dependence of the rate constant, which is related to the
        value of E_a, is given by the Arrhenius equation,

$$\log\left[\frac{k_2}{k_1}\right] = \frac{E_a}{(2.303)R}\left[\frac{T_2 - T_1}{T_1 T_2}\right]$$

    (b) What is necessary to make the connection?
        Doubling the reaction rate means $(k_2/k_1) = 2$ for $T_2 = 308$ K and $T_1 =$
        298 K. The activation energy can be calculated by using these values in
        the Arrhenius equation.

3. Set up the problem and solve it.
    The activation energy is

$$E_a = \frac{(2.303)RT_1 T_2}{(T_2 - T_1)} = \frac{(2.303)(8.314 \text{ J/K mol})(308 \text{ K})(298 \text{ K}) \log(2)}{(308 \text{ K} - 208 \text{ K})}$$

$$= 53,000 \text{ J/mol} = 53 \text{ kJ/mol}$$

4. Check the result.
    (a) Are significant figures and the location of the decimal point correct?
        Yes. The number of significant figures in the answer is limited to two
        digits because of the number of significant figures from the subtraction
        operation in the denominator.
    (b) Did the answer come out in the correct units?
        Yes.
    (c) Is the answer reasonable?
        Yes. Typical values for activation energies are 10–100 kJ/mol.

Exercise 18.8 Values of the rate constant have been measured for the
decomposition of acetone dicarboxylic acid,

$$\underset{\overset{|}{H}\overset{|}{H}}{HO-\overset{\overset{O}{\|}}{C}-\overset{\overset{H}{|}}{C}-\overset{\overset{O}{\|}}{C}-\overset{\overset{H}{|}}{C}-\overset{\overset{O}{\|}}{C}-OH(aq)} \rightarrow H_3C-\overset{\overset{O}{\|}}{C}-CH_3(aq) + 2CO_2(g)$$

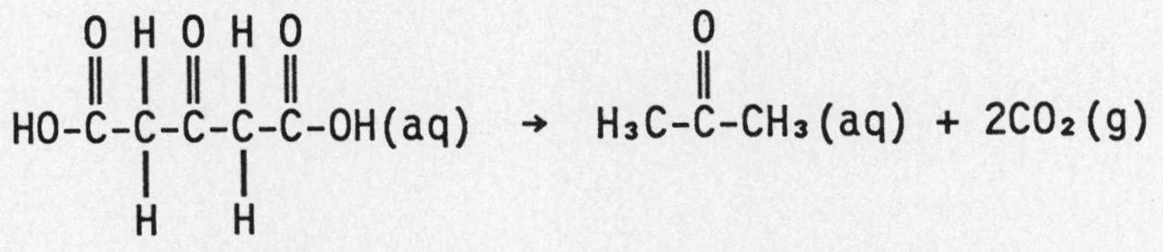

as 1.08×10^{-4} s^{-1} at 10.0 °C and 5.76×10^{-3} s^{-1} at 40.0 °C. Calculate the
activation energy for this reaction. *Answer:* 97.8 kJ/mol.

Example 18.9 Activation energy and temperature

The energy of activation for the reaction

$$H_2(g) + I_2(g) \rightarrow 2HI(g)$$

is 167.4 kJ/mol. If the temperature of the reaction was increased from 500. °C to 600. °C, by what factor would the rate of reaction change?

The desired factor is given by k_2/k_1, which can be found from the Arrhenius equation by

$$\log\left[\frac{k_2}{k_1}\right] = \frac{E_a}{(2.303)R}\left[\frac{T_2 - T_1}{T_1T_2}\right]$$

$$= \frac{(167.4\ \text{kJ/mol})(1000\ \text{J/kJ})}{(2.303)(8.314\ \text{J/K mol})}\left[\frac{873\text{K} - 773\ \text{K}}{(873\ \text{K})(773\ \text{K})}\right]$$

$$= 1.30$$

$$k_2/k_1 = 20.$$

The rate of reaction will increase 20-fold.

Exercise 18.9 The activation energy is 21.5 kJ/mol for the reaction

$$OH(g) + H_2(g) \rightarrow H_2O(g) + H(g)$$

The rate constant at 300. K is 4.1 x 10^6 L/mol s. Calculate the rate constant at 800. K. *Answer*: 9.0 x 10^8 L/mol s.

Example 18.10 Activation energy and temperature

The activation energy is 3.0 kJ/mol for the reaction

$$H(g) + HI(g) \rightarrow H_2(g) + I(g)$$

and the rate constant at 700. K is 3.0 x 10^{10} L/mol s. What is the temperature at which this reaction has a rate constant equal to 3.3 x 10^{10} L/mol s, a 10% increase?

The temperature dependence of the rate constant is given by the Arrhenius equation,

$$\log\left[\frac{k_2}{k_1}\right] = \frac{E_a}{(2.303)R}\left[\frac{1}{T_1} - \frac{1}{T_2}\right]$$

Solving for ($1/T_2$) and substituting the data gives

$$\frac{1}{T_2} = \frac{1}{T_1} - \frac{(2.303)R\ \log(k_2/k_1)}{E_a}$$

$$= \frac{1}{700.\ \text{K}} - \frac{(2.303)(8.314\ \text{J/K mol})\ \log\left[\frac{3.3 \times 10^{10}\ \text{L/mol s}}{3.0 \times 10^{10}\ \text{L/mol s}}\right]}{(3.0\ \text{kJ/mol})(1000\ \text{J/1 kJ})}$$

$$= 0.0012\ \text{K}^{-1}$$

$$T_2 = 860\ \text{K}$$

The new temperature must be 860 K, 160 K higher, for the rate of reaction to increase by 10%.

Exercise 18.10 The activation energy is 15.5 kJ/mol for the reaction

$$H(g) + HBr(g) \rightarrow H_2(g) + Br(g)$$

and the rate constant at 1100. K is 2.0×10^{10} L/mol s. What is the temperature at which this reaction has a rate constant equal to 2.2×10^{10} L/mol s, a 10% increase? Compare the temperature change needed to increase the reaction rate by 10% of this reaction with a large value of E_a to that needed for the reaction in *Example 18.10*, which has a small value of E_a. *Answer*: 1170 K, 70 K higher; the reaction rate of reactions with large activition energies are more temperature sensitive.

PRACTICE TEST

1 (45 points). Choose the response that best completes each statement.

(a) The study of kinetics is *not* involved with

(i) rates of reaction (ii) reaction mechanisms

(iii) catalysis (iv) predicting reaction spontaneity

(b) The rate for a chemical reaction is determined by

(i) performing theoretical calculations of $\Delta H°$

(ii) measuring concentration-time data

(iii) determining the equilibrium constant

(iv) measuring reaction rates at different temperatures

(c) For the reaction A + 2B $\rightarrow$ C + D, the rate equation is

$$\text{Rate} = k[A][B]$$

What is the overall order of reaction?

(i) zero order (ii) first order

(iii) second order (iv) third order

(d) The number of individual atoms, molecules, or ions that must simultaneously react in an elementary reaction is called the

(i) overall reaction order (ii) rate

(iii) molecularity (iv) rate constant

(e) The reaction 2A + 2B $\rightarrow$ C + D proceeds by the following mechanism

$$2A \xrightarrow{fast} A_2$$

$$A_2 + B \xrightarrow{slow} E + C$$

$$E + B \xrightarrow{fast} D$$

The rate equation for the reaction is

(i) Rate = $k[E][B]$ (ii) Rate = $k[A]^2[B]^2$

(iii) Rate = $k[A_2][B]$ (iv) Rate = $k[A]^2$

(f) To confirm that a reaction rate is first order in reactant [A], a straight line results by plotting

(i) [A] against t (ii) $\frac{1}{[A]}$ against t

(iii) log [A] against t (iv) none of these answers

(g) Use the following initial rate data

Experiment	1	2	3
$[NO]_0$, mol/L	1×10^{-2}	3×10^{-2}	3×10^{-2}
$[H_2]_0$, mol/L	1×10^{-3}	1×10^{-3}	2×10^{-3}
initial rate, mol/L s	4.80×10^{-5}	4.32×10^{-4}	8.64×10^{-4}

to determine the rate equation for the reaction

$$2NO(g) + 2H_2(g) \rightarrow N_2(g) + 2H_2O(g)$$

(i) Rate = $k[NO]^2[H_2]$ (ii) Rate = $k[NO][H_2]$

(iii) Rate = $k[NO]^3[H_2]$ (iv) Rate = $k[NO][H_2]^2$

(h) Which of these does *not* influence the rate of a chemical reaction?

(i) the nature of the reactants

(ii) the reaction order

(iii) the concentration of the reactants

(iv) the degree of contact between the reacting substances

(i) Warming bread dough in an oven while it rises is an example of the dependence of the reaction rate on the

(i) concentration (ii) temperature

(iii) surface contact (iv) none of these answers

(j) Which is true about a catalyst?

(i) can initiate and maintain an unfavorable reaction

(ii) cannot change the equilibrium conditions once reached

(iii) increases the value of E_a for the reaction

(iv) generally useful in many reactions

(k) The energy that the reactants must have in order for the reaction to occur is called the

(i) internal energy (ii) hydration energy

(iii) reaction energy (iv) none of these answers

(l) Which one of the following statements is true?

(i) All collisions between atoms, molecules, or ions will result in a reaction if the reactant particles have energies equal to or greater than the activation energy.

(ii) The equation expressing the rate of a chemical reaction is the same as the stoichiometric equation for the reaction.

(iii) Molecularity is the number of individual atoms, molecules, or ions that must simultaneously react in an elementary reaction.

(iv) For a reaction that occurs in a series of successive steps, the rate of reaction will depend on the fastest step in the reaction mechanism.

For questions (m), (n), and (o), refer to the diagram shown in Fig. 18-4 for the reaction

$$2CH_3(g) \rightarrow C_2H_6(g)$$

(m) What is the activation energy for the forward reaction?

(i) 8 kJ/mol (ii) 360. kJ/mol

(iii) 368 kJ/mol (iv) none of these answers

(n) At which point on the diagram would the formula $2CH_3$ appear?

(i) *A* (ii) *B*

(iii) *C* (iv) none of these answers

(o) What is the activation energy for the reverse reaction?

(i) 8 kJ/mol (ii) 360. kJ/mol

(iii) -360 kJ/mol (iv) 368 kJ/mol

2 (25 points). Match the correct term to each definition.

(a) the slowest step in a reaction mechanism

(b) the exact pathway from reactants to products

(c) the proportionality constant between the reaction rate and the reactant concentrations

(d) the speed with which products are produced and reactants are consumed in a specific reaction

(e) a simple reaction that occurs in a single step

(f) the energy required for the reaction to occur

(g) the sum of exponents of concentrations in the rate law

(i) activation energy

(ii) elementary reaction

(iii) molecularity

(iv) rate constant

(v) rate-determining step

(vi) reaction mechanism

(vii) reaction order, overall

(viii) reaction rate

Fig. 18-4

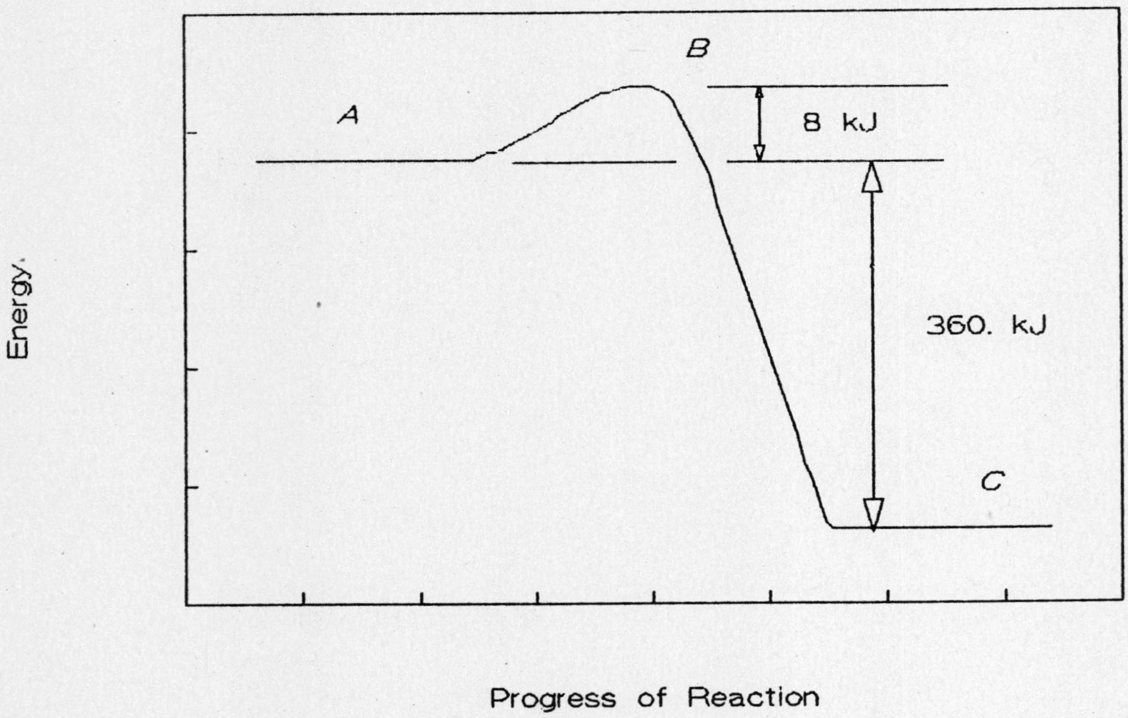

(h) the number of individual atoms, molecules, or ions needed to simultaneously take part in an elementary reaction

3 (15 points). Determine the reaction order and rate constant for the reaction

$$NOCl(g) \xrightarrow{2000\ °C} NO(g) + \frac{1}{2}Cl_2(g)$$

from the following data:

[NOCl], mol/L	0.0200	0.0177	0.0159	0.0144	0.0131	0.0121
t, s	0	100	200	300	400	500

4 (15 points). For the reaction

$$N_2O_5(g) \rightarrow 2NO_2(g) + \frac{1}{2}O_2(g) \qquad \Delta H° = 55.2\ kJ$$

$k = 8.0 \times 10^{-7}\ s^{-1}$ at 0.0 °C and $8.9 \times 10^{-4}\ s^{-1}$ at 50.0 °C. Calculate the energy of activation for this reaction. Prepare a reaction coordinate diagram for this reaction.

Fig. 18-5

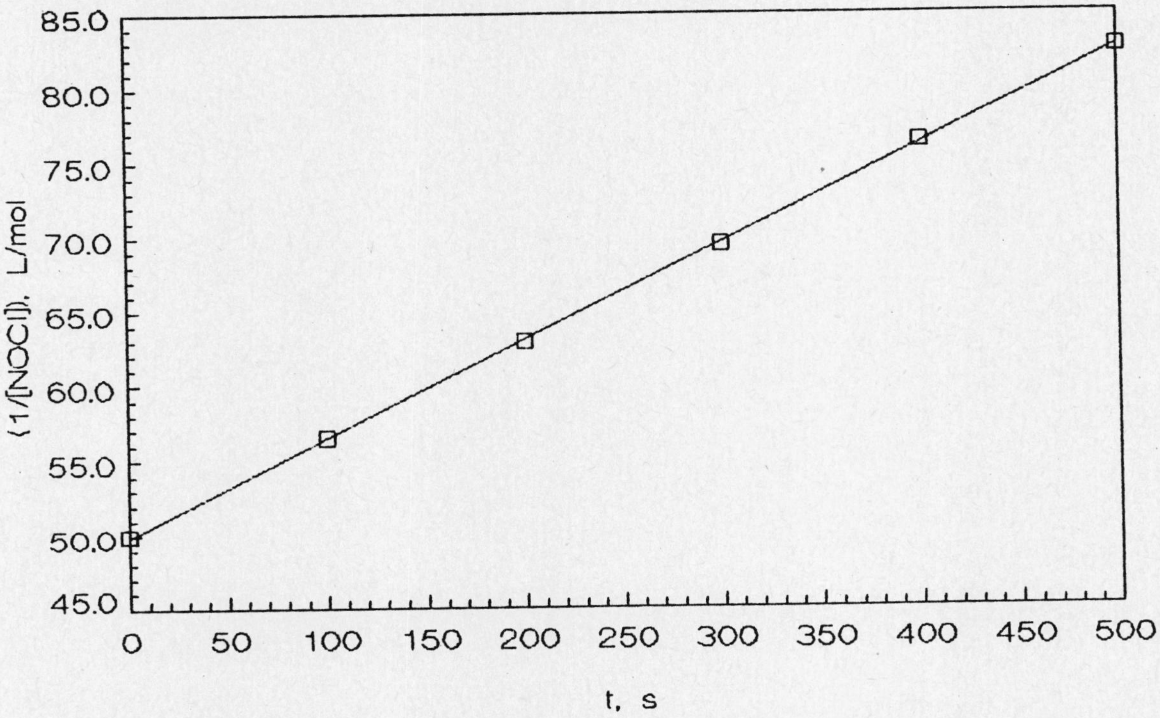

Answers to practice test

1. (a) iv, (b) ii, (c) iii, (d) iii, (e) iii, (f) iii, (g) i, (h) ii, (i) ii, (j) ii, (k) iv, (l) iii, (m) i, (n) i, (o) iv.
2. (a) v, (b) vi, (c) iv, (d) viii, (e) ii, (f) i, (g) vii, (h) iii.
3. see Fig. 18-5, second order, k = 0.0655 L/mol s.
4. 103 kJ/mol, see Fig. 18-6.

ADDITIONAL NOTES

Fig. 18-6

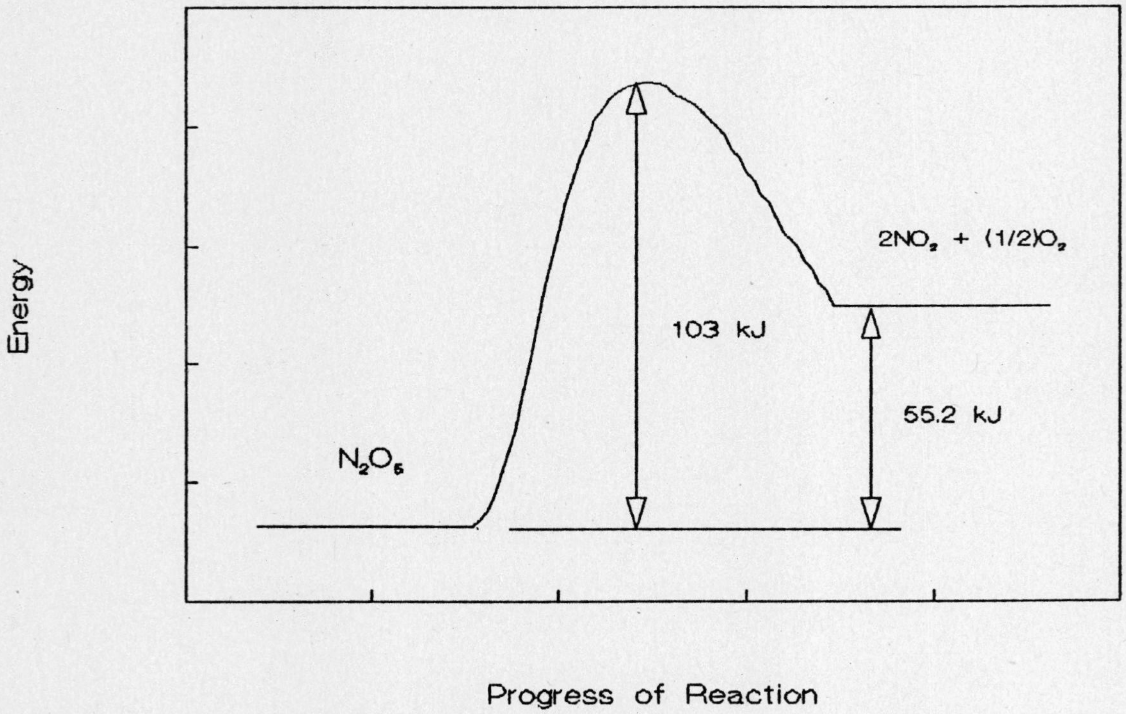

CHAPTER 19

CHEMICAL EQUILIBRIUM

CHAPTER OVERVIEW

This chapter reviews chemical equilibrium in a qualitative way and then presents a quantitative discussion. Equilibrium constant expressions and their formulation for homogeneous and heterogeneous systems are introduced. The formulation and interpretation of the reaction quotient is discussed. The influence that temperature, pressure, and concentration have on equilibrium is examined. A general method for solving equilibrium problems is presented.

COMPETENCIES

Definitions for the following terms should be learned:

- chromatography
- distribution coefficient
- equilibrium constant
- reaction quotient

General concepts and skills that should be learned include

- describing chemical equilibrium in terms of a dynamic state
- writing the expressions for the equilibrium constant or the reaction quotient for a given chemical equation
- interpreting the value of the equilibrium constant or the reaction quotient
- the general method for solving equilibrium problems

Numerical exercises that should be understood include

- calculating the value of the equilibrium constant from known values of concentration, pressure, etc.
- calculating K_p or K_c for reactions involving gases when one of the variables is given
- calculating the equilibrium constant for a chemical reaction from other chemical equations and the respective equilibrium constants

- evaluating and interpreting the value of the reaction quotient
- finding the concentration or partial pressure of a given reactant or product at equilibrium from either equilibrium concentration or pressure data or initial concentration or pressure data and the equilibrium constant

QUESTIONS

1. Complete the following sentences by filling in the various blanks with these terms: dynamic, equilibrium, equilibrium constant, kinetics, Le Chatelier's principle, and reaction quotient.

The study of the rate at which a chemical reaction reaches (a)_______ is the study of (b)_______. If given enough time, all reactions reach equilibrium conditions that may best be described as (c)_______. The product of the equilibrium concentration of the reaction products, each raised to the power equal to its coefficient in the balanced equation, divided by the product of the equilibrium concentration of reactants, each raised to the power equal to its coefficient in the balanced equation, is the (d)_______ expression. An expression that takes the same form as the equilibrium constant expression, but describes a reaction that is not at equilibrium, is the (e)_______. According to (f)_______, if a stress such as a pressure or concentration change is placed on a chemical reaction at equilibrium, the system will change so as to offset the stress.

2. Which of the following statements are true? Rewrite each false statement so that it is correct.

(a) The reaction quotient becomes the equilibrium constant at equilibrium.

(b) The law of chemical equilibrium states that if a system in equilibrium is subjected to a stress, the system will react in such a way as to tend to offset the stress.

(c) Once established, equilibrium conditions can be changed by changes in concentration or partial pressures, a temperature change, and by the addition of a catalyst.

(d) A system after reaching chemical equilibrium can be considered to be in a static state.

(e) Because units are usually dropped from values of equilibrium constants, K_c and K_p values are interchangeable.

(f) At equilibrium the rates of the forward and reverse reactions are zero.

(g) If any concentrations or partial pressures at equilibrium are altered by the addition or the removal of substances, the rates of the forward and reverse reactions will be temporarily different and one of the reactions will proceed at a greater rate than the other.

3. Choose the response that best completes each statement.

(a) Which is a true statement under equilibrium conditions?
  (i) The rates of the forward and reverse reactions are equal.
  (ii) All concentrations are equal.
  (iii) $Q > K$.
  (iv) All of these answers are true.

(b) The equilibrium constant for an equation written in reverse is related to the original constant by
  (i) $K^{1/2}$ (ii) $1/K$
  (iii) K^n (iv) K^2

(c) A reaction will proceed toward the products on the right side of an equation if
  (i) $Q = K$ (ii) $Q > K$
  (iii) $Q < K$ (iv) none of these answers

(d) Which is a factor that does not influence chemical equilibrium once it has been established?
  (i) temperature (ii) pressure
  (iii) concentration (iv) catalysis

(e) The equilibrium constant for the distribution of a solute between two immiscible solvents is called the
  (i) distribution coefficient (ii) chromatographic ratio
  (iii) overall rate constant (iv) reaction coefficient

(f) "If a system at equilibrium is subjected to a stress, the system will react in a way that tends to relieve the stress." This is a statement of
  (i) the law of mass action (ii) the first law of thermodynamics
  (iii) Le Chatelier's principle (iv) Avogadro's hypothesis

(g) The equilibrium constant for the chemical equation

$$aA + bB \rightleftharpoons cC$$

is given by

(i) $K = abc$ (ii) $K = [A]^a[B]^b/[C]^c$

(iii) $K = [A]^a[B]^b[C]^c$ (iv) none of these answers

(h) The equilibrium constant for the chemical equation

$$SnO_2(s) + 2CO(g) \rightleftharpoons Sn(s) + 2CO_2(g)$$

is given by

(i) $K_c = [Sn][CO_2]^2/[SnO_2][CO]^2$ (ii) $K_p = P_{CO_2}/P_{CO}$

(iii) $K_p = P^2_{CO_2}/P^2_{CO}$ (iv) $K_p = P_{CO}/P_{CO_2}$

(i) The equilibrium constant for the chemical equation

$$Cl_2(g) + H_2O(l) \rightarrow H^+ + Cl^- + HClO(aq)$$

is given by

(i) $K_c = [H^+][Cl^-][HClO]/[Cl_2]$ (ii) $K_c = [HCl][HClO]$

(iii) $K_p = P_{H^+}P_{Cl^-}P_{HClO}/P_{Cl_2}P_{H_2O}$ (iv) none of these answers

(j) Which "stress" will increase the partial pressure of O_2?

$$2H_2O(g) \rightleftharpoons 2H_2(g) + O_2(g) \qquad \Delta H° > 0$$

(i) increase the pressure of H_2

(ii) decrease the pressure of H_2O

(iii) increase the total pressure on the system

(iv) increase the temperature

(k) To obtain more reactant A in the chemical reaction

$$A(s) + B(s) \rightleftharpoons C(g) \qquad \Delta H° > 0$$

(i) increase the amount of B (ii) decrease the amount of C

(iii) increase the temperature (iv) none of these answers

4. Consider the simple reversible system represented by

$$H_2O(l) \rightleftharpoons H_2O(g)$$

What is the forward reaction in this case? What is the reverse reaction? What is the relationship between the rate of vaporization and the rate of condensation at equilibrium?

5. For the general chemical equation

$$aA + bB + \ldots \rightleftharpoons rR + sS + \ldots$$

write the expression for the concentration equilibrium constant.

6. In most equilibrium calculations, the units are usually omitted. What units are used for expressing the concentrations and partial pressures of substances in these calculations?

7. What is the relationship between K_p and K_c for gases? If there are more moles of gaseous products than of gaseous reactants, will K_p be larger, smaller, or equal to K_c?

8. Compare how equilibrium constant expressions are written for homogeneous and heterogeneous reactions. What terms are included and what terms are excluded in the expression for heterogeneous reactions?

9. Do the equilibrium constant expression and the value of the equilibrium constant depend on how the chemical equation describing a given reaction is written?

10. What is the relationship between the equilibrium constant for a given chemical equation and the equilibrium constant for the chemical equation written in reverse?

11. What is the relationship between the equilibrium constant for a given chemical equation and the equilibrium constant for the chemical equation written so that all of the stoichiometric coefficients are multiplied by a constant *n*?

12. How is the equilibrium constant found for a chemical equation that is the sum of several chemical equations for which the respective equilibrium constants are known?

13. How will a reaction relieve a stress brought about by (a) the addition of more of one of the reactants, (b) the removal of one of the products, (c) the addition of more of one of the products, or (d) the removal of one of the reactants?

14. Silver chloride is a very slightly soluble substance, but because of the equilibrium

$$AgCl(s) \rightleftharpoons Ag^+(aq) + Cl^-(aq)$$

small amounts of Ag^+ and Cl^- are present in the aqueous phase in contact with pure AgCl. Describe what happens as a soluble salt such as $AgNO_3$ is added to such a mixture and give a reason for your answer.

15. For the reaction between nitrogen and hydrogen to give ammonia,

$$N_2(g) + 3H_2(g) \rightleftharpoons 2NH_3(g) \qquad \Delta H^\circ < 0$$

identify what effect the following changes will have on $[H_2]$, $[NH_3]$, and *K*.

"Stress"	$[H_2]$	$[NH_3]$	*K*
(a) increase in temperature			
(b) increase in volume			
(c) removal of N_2			
(d) increase in the total pressure on the system			
(e) addition of a promoter			

16. For the reaction between carbon monoxide and chlorine to give phosgene,

$$CO(g) + Cl_2(g) \overset{C}{\rightleftarrows} COCl_2(g) \qquad \Delta H^\circ < 0$$

identify what effect the following changes will have on [CO, $[COCl_2]$, and K.

"Stress"	[CO]	$[COCl_2]$	*K*
(a) increase in $[Cl_2]$			
(b) increase in the temperature			
(c) addition of C			
(d) decrease in the total pressure on the system			
(e) decrease in [CO]			

17. Will increasing the temperature always increase the yield of products in a reaction at equilibrium? If not, what determines how the amounts of reactants and products will change?

18. What are the entries that appear in the table used to solve many equilibrium problems?

19. How can solving many algebraic equations in equilibrium problems be simplified? What are the general guidelines for this method?

Answers to questions

1. (a) equilibrium, (b) kinetics, (c) dynamic, (d) equilibrium constant, (e) reaction quotient, (f) Le Chatelier's principle.
2. (a) T; (b) F, replace *The law of chemical equilibrium* with *Le Chatelier's principle*; (c) F, replace *the addition of a catalyst* by *a change in total pressure*; (d) F, replace *static* by *dynamic*; (e) F, replace *are* with *are not* and add *are related by* $K_p = K_c(RT)^{\Delta n}$; (f) F, replace <u>zero</u> by <u>equal</u>; (g) T.
3. (a) i, (b) ii, (c) iii, (d) iv, (e) i, (f) iii, (g) iv, (h) iii, (i) i, (j) iv, (k) iv.
4. vaporization; condensation; the rates are equal to each other.
5. $K = [R]^r[S]^s \ldots/[A]^a[B]^b \ldots.$
6. The units are molarity and bars, respectively.

7. For gases

$$K_p = K_c(RT)^{\Delta n}$$

If there are more moles of gaeous products than of gaseous reactants, K_p will be larger than K_c.

8. Both equilibrium constant expressions have the same format--the products of the concentrations (or partial pressures) of the products each raised to an exponent corresponding to the stoichiometric coefficient in the balanced chemical equation divided by a similar product for the reactants. However, whenever the expressions for homogeneous reactions contain terms for all reactants and products, the expressions for heterogeneous reactions exclude terms for insoluble salts, pure liquids, and the solvent in dilute solutions.

9. yes.

10. The original expression and value are reciprocated.

11. The original expression and value are raised to the power of *n*.

12. The expressions and constants are multiplied together.

13. A reaction will relieve a stress brought about by
    (a) the addition of more of one of the reactants by the forward reaction's occurring to such an extent as to remove some of the reactants to give more products.
    (b) the removal of one of the products by the forward reaction's occurring to such an extent as to remove some of the reactants to give more products.
    (c) the addition of more of one of the products by the reverse reaction's occurring to such an extent as to remove some of the products to give more reactants.
    (d) the removal of one of the reactants by the reverse reaction's occurring to such an extent as to remove some of the products to give more reactants.

14. When a soluble salt such as $AgNO_3$ is added to a mixture of Ag^+, Cl^-, and AgCl, the Ag^+ concentration is too high, so more AgCl precipitates until $[Ag^+][Cl^-]$ in solution equals K.

15. (a) increase, decrease, decrease; (b) increase, decrease, no change; (c) increase, decrease, no change; (d) decrease, increase, no change; (e) no change, no change, no change.

16. (a) decrease, increase, no change; (b) increase, decrease, decrease; (c) no change, no change, no change; (d) increase, decrease, no change; (e) decrease, decrease, no change.
17. Increasing the temperature does not always increase the yield of products in a reaction at equilibrium. The heat of reaction determines how the reaction conditions will change.
18. The entries are the chemical equation, initial concentrations, changes in concentrations, and equilibrium concentrations.
19. Approximations that neglect the unknown in addition and/or subtraction terms are usually made. The guidelines are that K must be small compared to the concentrations and a 5 % error is usually allowed in the approximations.

SKILLS

Skill 19.1 Writing equilibrium constant and reaction quotient expressions

The general form for both the equilibrium constant expression (used for equilibrium conditions) and the reaction quotient expression (used for nonequilibrium conditions) is identical. It is given by the product of the molar concentrations of the chemical products, each raised to the power that corresponds to the stoichiometric coefficient in the balanced equation, divided by a similar product of concentrations raised to the appropriate powers for the chemical reactants. Thus for the equation

$$rR + sS \ldots \rightleftarrows xX + yY + \ldots$$

the form of K and Q would be

$$K \text{ or } Q = \frac{[X]^x[Y]^y \ldots}{[R]^r[S]^s \ldots}$$

For any reactant or product that is a pure solid, a pure liquid, or the solvent, the concentration terms do not appear in the expression for K or Q.

For example, for the reaction

$$PbCO_3(s) + 2H^+ + 2Cl^- \rightleftarrows PbCl_2(s) + CO_2(g) + H_2O(l)$$

the concentration equilibrium constant expression is

$$K_c = \frac{[CO_2]}{[H^+]^2[Cl^-]^2}$$

Skill Exercise: Write the concentration equilibrium expressions for the following equations:

(a) $H_2(g) + Cl_2(g) \rightleftharpoons 2HCl(g)$
(b) $2KClO_3(l) \rightleftharpoons 2KCl(s) + 3O_2(g)$
(c) $Ag^+ + Cl^- \rightleftharpoons AgCl(s)$
(d) $NH_3(aq) + H_2O(l) \rightleftharpoons NH_4^+ + OH^-$
(e) $HSO_4^- \rightleftharpoons H^+ + SO_4^{2-}$
(f) $[Cu(NH_3)_4]^{2+} \rightleftharpoons Cu^{2+} + 4NH_3(aq)$
(g) $Ag(s) + (1/2)Cl_2(g) \rightleftharpoons AgCl(l)$
(h) $BeSO_4(s) \rightleftharpoons BeO(s) + SO_3(g)$
(i) $2H_2(g) + O_2(g) \rightleftharpoons 2H_2O(g)$
(j) $2H_2(g) + O_2(g) \rightleftharpoons 2H_2O(l)$
(k) $H_2O(l) \rightleftharpoons H^+ + OH^-$
(l) $PCl_5(g) \rightleftharpoons PCl_3(g) + Cl_2(g)$
(m) $PCl_3(g) + Cl_2(g) \rightleftharpoons PCl_5(g)$
(n) $2Na(s) + 2H_2O(l) \rightleftharpoons 2Na^+ + 2OH^- + H_2(g)$
(o) $HNO_2(aq) \rightleftharpoons H^+ + NO_2^-$
(p) $HS^- + H_2O(l) \rightleftharpoons H_2S(aq) + OH^-$
(q) $ZnS(s) \rightleftharpoons Zn^{2+} + S^{2-}$
(r) $[Cu(NH_3)_5]^{2+} \rightleftharpoons [Cu(NH_3)_4]^{2+} + NH_3(aq)$
(s) $K_2Zn_3[Fe(CN)_6]_2(s) \rightleftharpoons 2K^+ + 3Zn^{2+} + 2[Fe(CN)_6]^{4-}$
(t) $(CH_3)_3N(aq) + H_2O(l) \rightleftharpoons (CN_3)_3NH^+ + OH^-$
(u) $H_2O(g) + C(s) \rightleftharpoons CO(g) + H_2(g)$

Answers:

(a) $K_c = \dfrac{[HCl]^2}{[H_2][Cl_2]}$
(b) $K_c = [O_2]^3$
(c) $K_c = \dfrac{1}{[Ag^+][Cl^-]}$
(d) $K_c = \dfrac{[NH_4^+][OH^-]}{[NH_3]}$
(e) $K_c = \dfrac{[H^+][SO_4^{2-}]}{[HSO_4^-]}$
(f) $K_c = \dfrac{[Cu^{2+}][NH_3]^4}{[Cu(NH_3)_4^{2+}]}$
(g) $K_c = \dfrac{1}{[Cl_2]^{1/2}}$
(h) $K_c = [SO_3]$
(i) $K_c = \dfrac{[H_2O]^2}{[H_2]^2[O_2]}$
(j) $K_c = \dfrac{1}{[H_2]^2[O_2]}$
(k) $K_c = [H^+][OH^-]$
(l) $K_c = \dfrac{[PCl_3][Cl_2]}{[PCl_5]}$
(m) $K_c = \dfrac{[PCl_5]}{[PCl_3][Cl_2]}$
(n) $K_c = [Na^+]^2[OH^-]^2[H_2]^2$
(o) $K_c = \dfrac{[H^+][NO_2^-]}{[HNO_2]}$
(p) $K_c = \dfrac{[H_2S][OH^-]}{[HS^-]}$
(q) $K_c = [Zn^{2+}][S^{2-}]$
(r) $K_c = \dfrac{[Cu(NH_3)_4^{2+}][NH_3]}{[Cu(NH_3)_5^{2+}]}$
(s) $K_c = [K^+]^2[Zn^{2+}]^3[Fe(CN)_6^{4-}]^2$
(t) $K_c = \dfrac{[(CH_3)_3NH^+][OH^-]}{[(CH_3)_3N]}$
(u) $K_c = \dfrac{[CO][H_2]}{[H_2O]}$

If the chemical reaction involves gases, partial pressures of the gases (in bars) often appear in the expression for Q or K instead of the concentrations. To distinguish between the two forms of Q or K, the subscripts c and p are used. The numerical values of Q_c or K_c and Q_p or K_p are related by

$$Q_p \text{ or } K_p = (Q_c \text{ or } K_c)(RT)^{\Delta n}$$

where Δn is given by

$$\Delta n = \text{(total number of moles of gaseous products)} - \text{(total number of moles of gaseous reactants)}$$

For example, in the lead carbonate/hydrochloric acid reaction given above, the equilibrium constant in terms of partial pressures is

$$K_p = \frac{P_{CO_2}}{[H^+]^2[Cl^-]^2}$$

and the value of Δn is +1 (the total number of moles of gaseous products is 1 and the total number of moles of gaseous reactants is 0).

Skill Exercise: Write the expression for K_p and determine Δn for reactions (a), (b), (g), (h), (i), (j), (l), (m), (n), and (u) given in the set of equations above. *Answers*:

(a) $\Delta n = 0$

$$K_p = \frac{P^2_{HCl}}{P_{H_2}\ P_{Cl_2}}$$

(b) $\Delta n = +3$

$$K_p = P^3_{O_3}$$

(g) $\Delta n = -½$

$$K_p = \frac{1}{P^{1/2}_{Cl_2}}$$

(h) $\Delta n = +1$

$$K_p = P_{SO_3}$$

(i) $\Delta n = -1$

$$K_p = \frac{P^2_{H_2O}}{P^2_{H_2}\ P_{O_2}}$$

(j) $\Delta n = -3$

$$K_p = \frac{1}{P^2_{H_2}\ P_{O_2}}$$

(l) $\Delta n = +1$

$$K_p = \frac{P_{PCl_3}\ P_{Cl_2}}{P_{PCl_5}}$$

(m) $\Delta n = -1$

$$K_p = \frac{P_{PCl_5}}{P_{PCl_3}\ P_{Cl_2}}$$

(m) $\Delta n = +1$

$$K_p = [Na^+]^2[OH^-]^2 P_{H_2}$$

(u) $\Delta n = +1$

$$K_p = \frac{P_{CO}\ P_{H_2}}{P_{H_2O}}$$

Comparison of the value of Q (the reaction quotient) to that of K (the equilibrium constant) indicates what will happen to a nonequilibrium mixture of reactants and products. If $Q < K$, then the system will reach equilibrium by increasing the concentrations (or partial pressures) of the products and decreasing the concentrations (or partial pressures) of the reactants--the forward reaction occurs. If $Q > K$, then the system will reach equilibrium by increasing the concentrations (or partial pressures) of the reactants and decreasing the concentrations (or partial pressures) of the products--the reverse reaction occurs.

Skill 19.2 Solving equilibrium problems

Many numerical problems involve finding concentrations of reactants and products under equilibrium conditions given values of equilibrium constants and initial, nonequilibrium concentrations. Summarized below is a method that is very useful in solving these kinds of problems.

Step 1. Write the balanced chemical equation.

Step 2. Write the corresponding equilibrium expression.

Step 3. Identify the unknown.

Step 4. Make a table consisting of the chemical equation, initial concentrations, changes in concentrations, and equilibrium concentrations.

Step 5. Substitute the equilibrium concentrations from the table into the equilibrium constant expression.

Step 6. Solve for the unknown quantity.

Step 7. If any approximations were made in Step 6 in order to simplify the calculations, check them for validity.

Step 8. Answer the questions in the problem.

Examples 19.6 - 19.8 illustrate using this method.

NUMERICAL EXERCISES

Example 19.1 Evaluation of K

The reaction

$$N_2(g) + O_2(g) \rightleftharpoons 2NO(g)$$

was studied at 4200 K, and at equilibrium the respective partial pressures were P_{N_2} = 0.51 bar, P_{O_2} = 0.75 bar, and P_{NO} = 0.069 bar. Calculate the value of the pressure equilibrium constant for this reaction.

Writing the equilibrium constant expression for the above equation and substituting the equilibrium data gives

$$K_p = \frac{P^2_{NO}}{P_{N_2}\, P_{O_2}} = \frac{(0.069)^2}{(0.51)(0.75)} = 0.012$$

The pressure equilibrium constant is 0.012.

Exercise 19.1 Calculate the value of the pressure equilibrium constant at 800 °C for the reaction

$$CO_2(g) + C(s) \rightleftharpoons 2CO(g)$$

given that the partial pressures at equilibrium are P_{CO_2} = 0.015 bar and P_{CO} = 4.30 bar. *Answer*: $K_p = 1.2 \times 10^3$.

Example 19.2 Evaluation of K

What is the value of K_c for the reaction described in *Example 19.1*? The relationship between K_p and K_c is

$$K_p = K_c\ (RT)^{\Delta n}$$

For the reaction under consideration

$$\Delta n = (2 \text{ mol NO}) - (1 \text{ mol } N_2 + 1 \text{ mol } O_2) = 0$$

Thus

$$K_c = \frac{K_p}{(RT)^{\Delta n}} = \frac{0.012}{[(0.08314)(4200)]^0} = 0.012$$

For this reaction, $K_c = K_p = 0.012$.

Exercise 19.2 What is the value of K_c for the reaction described in Exercise 19.1? *Answer*: 13.

Example 19.3 Evaluation of K*

A student wrote the equation for the reaction described in *Example 19.1* as

$$\frac{1}{2}N_2(g) + \frac{1}{2}O_2 \rightleftharpoons NO(g)$$

What is the value of the pressure equilibrium constant for this equation?

This problem is solved by using the comprehensive problem-solving approach presented in *Skill 2.7 Solving problems*.

1. Study the problem and be sure you understand it.
   (a) What is unknown?
   The value of the pressure equilibrium constant for the student's equation, K_2.
   (b) What is known?
   The value of the pressure equilibrium constant for the original equation, K_1, and the pressure equilibrium constant expressions for the two equations

$$K_1 = \frac{P^2_{NO}}{P_{N_2}\ P_{O_2}} \qquad K_2 = \frac{P_{NO}}{P^{1/2}_{N_2}\ P^{1/2}_{O_2}}$$

2. Decide how to solve the problem.
   (a) What is the connection between the known and the unknown?
   By inspection, K_1 is related to K_2 by

$$K_1 = \frac{P^2_{NO}}{P_{N_2}\ P_{O_2}} = \left[\frac{P_{NO}}{P^{1/2}_{N_2}\ P^{1/2}_{O_2}} \right]^2 = (K_2)^2$$

(b) What is necessary to make the connection?

To find K_2, find the square root of K_1.

3. Set up the problem and solve it.

The value of K_2 is

$$K_2 = (K_1)^{1/2} = (0.012)^{1/2} = 0.11$$

4. Check the result.

(a) Are significant figures and the location of the decimal point correct?

Yes.

(b) Did the answer come out in the correct units?

The units on equilibrium constants are usually omitted. However, all concentrations must be given in molarity and all partial pressures must be given in bars.

(c) Is the answer reasonable?

Yes.

Exercise 19.3 What is the value of the pressure equilibrium constant for the equation

$$CO(g) \rightleftharpoons \frac{1}{2}CO_2(g) + \frac{1}{2}C(s)$$

See Exercise 19.1 for further data. *Answer*: $K = (1/K)^{1/2} = 2.9 \times 10^{-2}$.

Example 19.4 Equilibrium concentrations

What is the partial pressure of O_2 in equilibrium at 4200 K with NO at a partial pressure of 0.15 bar and N_2 at a partial pressure of 10.5 bar? See *Example 19.1* for further data.

The partial pressure of O_2 can be found by rearranging the equilibrium expression and substituting the data to give

$$p_{O_2} = \frac{P^2_{NO}}{P_{N_2}\ K_p} = \frac{(0.15)^2}{(10.5)(0.012)} = 0.18 \text{ bar}$$

The partial pressure of O_2 is 0.18 bar.

Exercise 19.4 What is the partial pressure of CO_2 in equilibrium with 1.00 bar of CO? See Exercise 19.1 for further data. *Answer*: 8.3×10^{-4} bar.

Example 19.5 Reaction quotient

The partial pressures of N_2, O_2, and NO in a gaseous mixture at 4200 K were 0.25 bar each. Describe what happened in the gaseous mixture as it was allowed to come to equilibrium. See *Example 19.1* for the chemical equation and equilibrium data.

The value of the reaction quotient is

$$Q_p = \frac{P^2_{NO}}{P_{N_2}\,P_{O_2}} = \frac{(0.25)^2}{(0.25)(0.25)} = 1.0$$

Because $Q_p > K_p$, the reaction will proceed toward the reactants side of the equation. Thus the amount of NO will decrease and the amounts of N_2 and O_2 will increase until the partial pressures are such that $P^2_{NO}/P_{N_2}P_{O_2}$ equals 0.012.

Exercise 19.5 A gaseous mixture at 800 °C contained CO_2 at a pressure of 1.00 bar and CO at a pressure of 0.50 bar. The mixture was allowed to come to equilibrium in the presence of graphite. Describe what happened to the gaseous mixture. See Exercise 19.1 for the chemical equation and equilibrium data. *Answer*: Q_p = 0.25, CO_2 will react with C until equilibrium conditions are reached.

*Example 19.6 Equilibrium concentrations**

Bromine and chlorine dissolve in carbon tetrachloride and react to form BrCl. The concentration equilibrium constant at 25 °C is K_c = 7.0 for the reaction

$$Br_2(CCl_4) + Cl_2(CCl_4) \rightleftharpoons 2BrCl(CCl_4)$$

A reaction mixture at equilibrium contained $[Br_2] = [Cl_2]$ = 0.0043 mol/L and [BrCl] = 0.0114 mol/L. The addition of 0.0100 mol of bromine to 1.00 L of this mixture disturbed the equilibrium. Assuming a negligible volume change, calculate the concentrations of the three species present once equilibrium has been reestablished.

This problem is solved by using the comprehensive problem-solving approach presented in *Skill 2.7 Solving problems*.

1. Study the problem and be sure you understand it.
   (a) What is unknown?
   The concentrations of Cl_2, Br_2, and BrCl at the new equilibrium conditions.

(b) What is known?

The initial equilibrium concentrations of Cl_2, Br_2, and BrCl; the concentration of the additional Br_2; the chemical equation and corresponding equilibrium constant expression; and the value of the equilibrium constant.

2. <u>Decide how to solve the problem</u>.

(a) What is the connection between the known and the unknown?

The relationship between the concentrations is the expression for K,

$$K_c = \frac{[BrCl]^2}{[Br_2][Cl_2]}$$

(b) What is necessary to make the connection?

The steps necessary to solve this type of equilibrium problem are outlined in *Skill 19.2 Solving equilibrium problems.* Steps 1 and 2 appear above. Additional steps are shown below.

Step 3. Identify the unknown. According to Le Chatelier's principle, the addition of Br_2 will disturb the equilibrium and the equilibrium will be reestablished by some of the Br_2 reacting with some of the Cl_2 to produce more BrCl. This statement is borne out by the numerical value of

$$Q = (0.0114)^2/(0.0043 + 0.0100)(0.0043) = 2.1$$

which is less than K. Although there are several choices, for this problem let $x = [Br_2]$ that reacts.

Step 4. Prepare a table consisting of the chemical equation, initial concentrations, changes in concentrations, and equilibrium concentrations.

	Br_2	+	Cl_2	$\rightleftarrows$	2BrCl
initial	0.0043 + 0.0100		0.0043		0.0114
change	$-x$		$-x$		$+2x$
equilibrium	$0.0143 - x$		$0.0043 - x$		$0.0114 + 2x$

In the table above, the stoichiometry of the reaction indicates that for the reaction of x mol of Br_2, x mol of Cl_2 will also undergo reaction and $2x$ mol BrCl will form. The equilibrium concentrations are found by adding the initial concentrations and the changes in concentration.

It is possible to substitute the equilibrium concentration data in terms of the unknown quantity x into the equilibrium constant expression, solve for x, and find the concentrations after equilibrium has been reestablished.

3. Set up the problem and solve it.
Additional steps from *Skill 19.2 Solving equilibrium problems* are presented below.

Step 5. Substitute the equilibrium concentrations from the table into the equilibrium constant expression.

$$K_c = \frac{[BrCl]^2}{[Br_2][Cl_2]} = \frac{(0.0114 + 2x)^2}{(0.0143 - x)(0.0043 - x)} = 7.0$$

Step 6. Solve for the unknown quantity. Because K is not small compared to the concentrations, simplifying approximations cannot be used and the equation must be solved by using the quadratic formula.

$$(0.0114 + 2x)^2 = (7.0)(0.0143 - x)(0.0043 - x)$$

$$0.000130 + 0.0456x + 4x^2 = 0.00043 - 0.130x + 7.0x^2$$

$$3x^2 - 0.176x + 0.00030 = 0$$

$$x = \frac{-(-0.176) \pm \sqrt{(-0.176)^2 - (4)(3)(0.00030)}}{(2)(3)}$$

$$= 0.057 \text{ or } 0.0018$$

The value $x = 0.057$ is physically meaningless because it would give a negative concentration for Cl_2 at equilibrium:

$$[Cl_2] = 0.0043 - 0.057 = -0.053 \text{ mol/L}$$

Therefore $x = 0.0018$.

Step 8. Answer the questions in the problem. The equilibrium concentrations are

$$[Cl_2] = 0.0043 - 0.0018 = 0.0025 \text{ mol/L}$$

$$[Br_2] = 0.0143 - 0.0018 = 0.0125 \text{ mol/L}$$

$$[BrCl] = 0.0114 + 2(0.0018) = 0.0150 \text{ mol/L}$$

4. Check the result.
(a) Are significant figures and the location of the decimal point correct?
Yes.
(b) Did the answer come out in the correct units?
Yes.

(c) Is the answer reasonable?

Yes. The concentration of BrCl increased and the concentrations of Br_2 and Cl_2 decreased. In fact, the numerical values substituted into the equilibrium constant expression,

$$K_c = \frac{(0.0150)^2}{(0.0025)(0.0125)} = 7.2$$

confirm the accuracy of the results.

Exercise 19.6 A gaseous mixture originally contained $[Br_2]$ = 0.125 mol/L and $[F_2]$ = 1.25 mol/L. Determine the concentration of the gases after equilibrium was established for the reaction

$$Br_2(g) + F_2(g) \rightleftharpoons 2BrF(g)$$

K_C = 54.7 at 6000 K for this reaction. *Answer*: $[Br_2]$ = 0.002 mol/L, $[F_2]$ = 1.13 mol/L, [BrF] = 0.246 mol/L.

Example 19.7 Equilibrium concentrations

For the equation

$$N_2(g) + O_2(g) \rightleftharpoons 2NO(g)$$

the concentration equilibrium constant is 4.00×10^{-4} at 2000 K. Find the equilibrium concentration of NO if the original reaction mixture contained $[N_2]$ = 2.32×10^{-3} mol/L and $[O_2]$ = 5.8×10^{-4} mol/L.

For the equation above

$$K_c = \frac{[NO]^2}{[N_2][O_2]} = 4.00 \times 10^{-4}$$

and the unknown in this problem is [NO] = x. The table below lists the equation, the initial concentrations, the concentration changes, and the equilibrium concentrations.

	$N_2(g)$	+	O_2	$\rightleftharpoons$	$2NO(g)$
initial	2.32×10^{-3}		5.8×10^{-4}		0
change	$-0.5x$		$-0.5x$		$+x$
equilibrium	$(2.32 \times 10^{-3} - 0.5x)$		$(5.8 \times 10^{-4} - 0.5x)$		x

Substituting the expressions for the equilibrium concentrations into the equilibrium constant expression gives

$$K_c = \frac{x^2}{(2.32 \times 10^{-3} - 0.5x)(5.8 \times 10^{-4} - 0.5x)} = 4.00 \times 10^{-4}$$

Assuming $(2.32 \times 10^{-3} - 0.5x) \approx 2.32 \times 10^{-3}$ and $(5.8 \times 10^{-4} - 0.5x) \approx 5.8 \times 10^{-4}$ simplifies the calculations considerably and gives

$$x^2 = (4.00 \times 10^{-4})(2.32 \times 10^{-3})(5.8 \times 10^{-4}) = 5.4 \times 10^{-10}$$

$$x = 2.23 \times 10^{-5}$$

The approximation that 1.2×10^{-5} is negligible compared to 2.32×10^{-3} is only $1.2 \times 10^{-5}/2.32 \times 10^{-3} = 0.5$ % in error and the approximation that 1.2×10^{-5} is negligible compared to 5.8×10^{-4} is only $1.2 \times 10^{-5}/5.8 \times 10^{-4} = 2$ % in error, so the approximate solution is acceptable. The concentration of NO at equilibrium is 2.3×10^{-5} mol/L.

Exercise 19.7 Calculate the concentration of CH_4 formed at equilibrium conditions at 5000 K by reacting gaseous carbon at a concentration of 10.5 mol/L with hydrogen at a concentration of 35.6 mol/L. At this temperature, K_c = 1.47×10^{-5} for the equation

$$C(g) + 2H_2(g) \rightleftarrows CH_4(g)$$

Answer: 0.196 mol/L.

Example 19.8 Equilibrium concentrations

At 100 °C the concentration equilibrium constant for the reaction

$$CHCl_2COOC_5H_{11} \rightleftarrows CHCl_2COOH + C_5H_{10}$$

is 3.40. Calculate the concentration of C_5H_{10} in equilibrium in a solution originally containing $[CHCl_2COOC_5H_{11}] = 1.35$ mol/L and $[CHCl_2COOH] = 0.10$ mol/L.

For the foregoing equation, the equilibrium expression is

$$K_c = \frac{[CHCl_2COOH][C_5H_{10}]}{[CHCl_2COOC_5H_{11}]} = 3.40$$

and the unknown is $[C_5H_{10}] = x$. The table below lists the equation, the initial concentrations, the concentration changes, and the equilibrium concentrations.

	$CHCl_2COOC_5H_{11}$	$\rightleftarrows$	$CHCl_2COOH$ +	C_5H_{10}
initial	1.35		0.10	0
change	$-x$		$+x$	$+x$
equilibrium	$(1.35 - x)$		$(0.01 - x)$	x

Substituting the equilibrium concentrations into the equilibrium expression gives

$$\frac{(0.10 + x)(x)}{(1.35 - x)} = 3.40$$

Because the value of the equilibrium constant is not small compared to the concentrations, an approximate solution will not be valid. Thus

$$(0.10 + x)x = (3.40)(1.35 - x)$$

$$(0.10)x + x^2 = 4.59 - (3.40)x$$

$$x^2 + (3.50)x - 4.59 = 0$$

$$x = \frac{-(3.50) \pm \sqrt{(3.50)^2 - 4(1)(-4.59)}}{2(1)} = 1.02 \text{ or } -4.52$$

The concentration of C_5H_{10} at equilibrium is 1.02 mol/L.

Exercise 19.8 The concentration equilibrium constant for the reaction

$$C_2H_5OH(l) + CH_3COOH(l) \rightleftharpoons CH_3COOC_2H_5(l) + H_2O(l)$$

is 3.85. Find the concentration of $CH_3COOC_2H_5$ at equilibrium if the initial concentration of C_2H_5OH was 5.0 mol/L and of CH_3COOH was 6.5 mol/L. Note that in this case the concentration of water must be included in the equilibrium constant expression with the three other terms. *Answer*: 3.8 mol/L.

Example 19.9 Evaluation of K

Use the following equilibrium concentration data to determine the value of x in the chemical equation

$$M^{m+} + xX^- \rightarrow [MX_x]^{m-x}$$

Experiment	$[M^{m+}]$, mol/L	$[X^-]$, mol/L	$[MX_x]^{m-x}$, mol/L
1	0.10	0.25	5.1×10^{-4}
2	0.15	0.35	2.9×10^{-3}
3	0.20	0.30	2.1×10^{-3}

The concentration equilibrium constant expression for this reaction is

$$K_c = \frac{[MX_x^{m-x}]}{[M^{m+}][X^-]^x}$$

Instead of using a trial and error approach to find x, all that is required is to substitute the first two sets of data into the expression for K_c, equate the equations because K_c is a constant, and solve for x:

$$K_c = \frac{(5.1 \times 10^{-4})}{(0.10)(0.25)^x} = \frac{(2.9 \times 10^{-3})}{(0.15)(0.35)^x}$$

$$\frac{(0.35)^x}{(0.25)^x} = \left[\frac{0.35}{0.25}\right]^x = \frac{(2.9 \times 10^{-3})(0.10)}{(5.1 \times 10^{-4})(0.15)} = 3.8$$

$$(x)\log\left[\frac{0.35}{0.25}\right] = \log\ (3.8)$$

$$x = \frac{\log(3.8)}{\log[(0.35)/(0.25)]} = \frac{\log(3.8)}{\log(1.4)} = 4.0$$

The stoichiometric coefficient is 4. This value can be confirmed by checking how constant the value of K_c is by using this value for all the data given:

$$K_c = \frac{(5.1 \times 10^{-4})}{(0.10)(0.25)^4} = 1.3 \qquad K_c = \frac{(2.9 \times 10^{-3})}{(0.15)(0.35)^4} = 1.3 \qquad K_c = \frac{(2.1 \times 10^{-3})}{(0.20)(0.30)^4} = 1.3$$

Exercise 19.9 Find the value of x for the chemical equation

$$xA(g) \rightarrow A_x(g)$$

given the data in the following table.

Experiment	P_A, bar	P_{A_x}, bar
1	0.0063	0.10
2	0.0089	0.20
3	0.0110	0.30

Answer: $x = 2$, giving $K_p = 2.5 \times 10^3$.

PRACTICE TEST

1 (10 points). Write equilibrium constant expressions for the following chemical equations: (a) $H_2O(g) \rightleftarrows H_2(g) + \frac{1}{2}O_2(g)$, (b) $2ZnS_2(s) + 3O_2(g) \rightleftarrows 2ZnO(s) + 2SO_2(g)$, (c) $Ba^{2+} + SO_4^{2-} \rightleftarrows BaSO_4(s)$, (d) $[Fe(CN)_6]^{4-} \rightleftarrows Fe^{2+} + 6CN^-$, and (e) $CH_3COOH(aq) \rightleftarrows CH_3COO^- + H^+$.

2 (5 points). Choose the factors that will influence equilibrium conditions: (a) temperature, (b) concentration, (c) pressure, (d) inhibitor, (e) surfactant, (f) nature of substances.

3 (20 points). Choose the response that best completes each statement.

(a) The equilibrium constant expression

$$K_c = \frac{[CO_2]^2}{[CO]^2[O_2]}$$

represents which chemical equation?

(i) $2CO_2(g) \rightleftharpoons 2CO(g) + O_2(g)$ (ii) $CO(g) + \frac{1}{2}O_2(g) \rightleftharpoons CO_2(g)$

(iii) $2CO(g) \rightleftharpoons CO_2(g) + C(s)$ (iv) none of these answers

(b) Choose the equilibrium constant expression for the equation

$$2NH_3(g) + CO_2(g) + H_2O(l) \rightleftharpoons 2NH_4^+ + CO_3^{2-}$$

(i) $K_c = \frac{[(NH_4)_2CO_3]}{[NH_3]^2[CO_2]}$ (ii) $K_c = \frac{[NH_4^+]^2[CO_3^{2-}]}{[NH_3]^2[CO_2][H_2O]}$

(iii) $K_p = \frac{[2NH_4^+][CO_3^{2-}]}{2P_{NH_3}\ P_{CO_2}}$ (iv) $K_p = \frac{[NH_4^+]^2[CO_3^{2-}]}{P^2_{NH_3}\ P_{CO_2}}$

(c) For an exothermic reaction at equilibrium, the effect on the concentrations of the products of increasing the temperature will be

(i) an increase (ii) a decrease

(iii) no change (iv) none of these answers

(d) The effect on the concentration of N_2O_4 of increasing the total pressure on the system

$$2NO_2(g) \rightleftharpoons N_2O_4(g)$$

at equilibrium will be

(i) an increase (ii) a decrease

(iii) no change (iv) none of these answers

(e) The effect on the concentration of H^+ of adding CH_3COO^- to the system

$$CH_3COOH(aq) \rightleftharpoons H^+ + CH_3COO^-$$

at equilibrium will be

(i) an increase (ii) a decrease

(iii) no change (iv) none of these answers

(f) What will happen to the equilibrium system

$$C_2H_4(g) + Cl_2(g) \rightleftharpoons C_2H_4Cl(g)$$

as the total pressure is increased slightly (under constant temperature conditions)?

(i) $[C_2H_4Cl]$ increases. (ii) $P_{C_2H_4Cl}$ decreases.

(iii) $[Cl_2]$ increases. (iv) all of these answers

(g) What will happen to the equilibrium system

$$2N_2(g) + O_2(g) \rightleftharpoons 2N_2O(g)$$

as the partial pressure of N_2 is increased?

(i) P_{O_2} decreases. (ii) P_{N_2O} increases.

(iii) $[N_2O]$ increases. (iv) all of these answers

(h) For the endothermic reaction

$$3A(aq) \rightleftharpoons 2B(aq)$$

how can the yield of B be increased?

(i) increase the temperature (ii) increase the pressure
(iii) decrease the temperature (iv) add a catalyst

(i) To increase the yield of B

$$C(g) \rightarrow A(s) + B(s)$$

(i) increase [A] (ii) decrease [A]
(iii) decrease the total pressure (iv) increase the total pressure

(j) Which of the following changes will increase the amount of liquid water?

$$2H_2(g) + O_2(g) \rightleftharpoons 2H_2O(l) \qquad \Delta H° = -572 \text{ kJ}$$

(i) decrease the temperature (ii) increase P_{H_2}
(iii) increase $[O_2]$ (iv) all of these answers

4 (20 points). Choose the response that best completes each statement.

(a) At equilibrium $[N_2]$ = 0.02 mol/L, $[H_2]$ = 0.01 mol/L, and $[NH_3]$ = 0.10 mol/L for the reaction

$$N_2(g) + 3H_2(g) \rightleftharpoons 2NH_3(g)$$

What is the value of K_c for the equation?

(i) 5×10^2 (ii) 2×10^{-6}
(iii) 3×10^2 (iv) 5×10^5

(b) A reaction mixture consisted initially of [A] = 3 mol/L and [B] = 4 mol/L. At equilibrium [C] = 2 mol/L. Calculate K_c for the equation

$$A(aq) + B(aq) \rightleftharpoons C(aq) + D(aq)$$

(i) 8 (ii) 0.5
(iii) 2 (iv) none of these answers

(c) A reaction mixture consisted initially of $[Cl_2]$ = 0.50 mol/L, [NO] = 1.0 mol/L, and [NOCl] = 1.0 mol/L. At equilibrium, [NO] = 0.50 mol/L. Calculate the value of K_c for the equation

$$Cl_2(g) + 2NO(g) \rightleftharpoons 2NOCl(g)$$

(i) 36 (ii) 0.083
(iii) 12 (iv) 0.028

(d) For the equation

$$A(aq) + B(aq) \rightleftharpoons C(aq) + D(aq)$$

K_c = 3.0 at a given temperature. A 1.0 L flask contains 1.0 mol A, 1.0 mol B, 1.0 mol C, and 2.0 mol D. What will happen?

(i) Nothing, these are equilibrium concentrations.
(ii) The forward reaction is favored.
(iii) The reverse reaction is favored.
(iv) [A] will increase.

(e) For the equation

$$H_2(g) + I_2(g) \rightleftharpoons 2HI(g)$$

K_c = 1.0 at a given temperature. Starting with $[H_2] = [I_2] = 1.0$ mol/L, calculate [HI] at equilibrium.

(i) 0.33 mol/L (ii) 1.0 mol/L
(iii) 0.66 mol/L (iv) none of these answers

5 (10 points). The equilibrium constant for the reaction

$$H(g) + HCl(g) \rightleftharpoons H_2(g) + Cl(g)$$

at a certain temperature is $K_p = 3.25$. Calculate the partial pressure of Cl in equilibrium with $P_{H_2} = 1.00$ bar, $P_{HCl} = 3.25$ bar, and $P_H = 1.26 \times 10^{-6}$ bar.

6 (10 points). Calculate K_c for the equation in question 5. Assume T = 750 K.

7 (10 points). The concentration equilibrium constant for the equation given in question 5 is $K_c = 3.25$. Describe what will happen if a gaseous mixture containing $[H] = 1.2 \times 10^{-8}$ mol/L, [HCl] = 0.054 mol/L, $[H_2] = 0.057$ mol/L, and $[Cl] = 4.3 \times 10^{-9}$ mol/L is prepared.

8 (15 points). What will be the final concentration of Cl(g) at equilibrium with a mixture originally containing $[H] = 1.2 \times 10^{-8}$ mol/L, $[H_2] = 0.026$ mol/L, and [HCl] = 0.059 mol/L? $K_c = 3.25$.

Answers to practice test

1. (a) $K_p = \dfrac{P_{H_2}\, P_{O_2}^{1/2}}{P_{H_2O}}$ or $K_c = \dfrac{[H_2][O_2]^{1/2}}{[H_2O]}$

(b) $K_p = \dfrac{P_{SO_2}^2}{P_{O_2}^3}$ or $K_c = \dfrac{[SO_2]^2}{[O_2]^3}$

(c) $K_c = \dfrac{1}{[Ba^{2+}][SO_4^{2-}]}$ (d) $K_c = \dfrac{[Fe^{2+}][CN^-]^6}{[Fe(CN)_6^{4-}]}$ (e) $K_c = \dfrac{[H^+][CH_3COO^-]}{[CH_3COOH]}$

2. a, b, c, f.
3. (a) iv, (b) iv, (c) ii, (d) i, (e) ii, (f) i, (g) iv, (h) i, (i) iv, (j) iv.
4. (a) iv, (b) iii, (c) i, (d) ii, (e) iii.

5. 1.33×10^{-5} bar.
6. $\Delta n = 0$, $K_c = K_p = 3.25$.
7. $Q_c < K_c$, H will react with HCl to produce H_2 and Cl.
8. 1.0×10^{-8} mol/L.

ADDITIONAL NOTES

CHAPTER 20

ACIDS AND BASES

CHAPTER OVERVIEW

Two methods of defining acids and bases are introduced--the Brønsted-Lowry system (based on proton donation) and the Lewis system (based on electron-pair donation). The behavior of acids and bases in aqueous systems is discussed in detail. Structural factors that influence the strengths of acids and bases are examined, as are the equilibria in aqueous solutions of acids and bases.

COMPETENCIES

Definitions for the following terms should be learned:

- acid ionization constant
- acidic solution
- alkaline solution
- base ionization constant
- Brønsted-Lowry acid
- Brønsted-Lowry acid-base reaction
- Brønsted-Lowry base
- equivalent mass of an acid
- equivalent mass of a base
- ion product constant for water
- Lewis acid
- Lewis acid-base reaction
- Lewis base
- neutral solution
- normality
- oxoacid
- pH
- pOH

General concepts and skills that should be learned include

- defining acids, bases, and acid-base reactions in terms of the Brønsted-Lowry and Lewis systems
- the factors that influence the relative strengths of acids and bases
- the nomenclature of oxoacids and oxoanions
- the relationship between an equivalent and a mole for acids and bases
- expressing the concentration of an acid or base in normality
- writing and interpreting the equilibrium constant expressions for acids, bases, and water

- defining and using pH, pOH, and pK

Numerical exercises that should be understood include

- calculating $[H^+]$ or $[OH^-]$ when K_w and one of these variables are given
- calculating K_a or K_b when K_w and one of these variables are given
- calculating pH or $[H^+]$ when one of these variables is given
- calculating pOH or $[OH^-]$ when one of these variables is given
- calculating pH or pOH when one of these variables is given
- calculating equilibrium concentrations or the equilibrium constant for acidic or alkaline aqueous solutions when all variables except one are given
- calculating the fraction of ionization of a weak monoprotic acid
- calculating the equivalent mass or molar mass of a substance in a reaction when the chemical equation and one of the variables are given
- calculating the normality or molarity of a solution when the chemical equation and one of the variables are given

QUESTIONS

1. Complete the following sentences by filling in the various blanks with these terms: acidic aqueous solution, alkaline aqueous solution, amphoteric, pH, pOH, strong acid, and strong base.

   An (a)_______ is an aqueous solution containing $[H^+] > 1 \times 10^{-7}$ mol/L and $[OH^-] < 1 \times 10^{-7}$ mol/L. The (b)_______ of this solution would be less than 7 and for a concentrated solution of a (c)_______ the value could be very near 0 or even negative. Substances that act as either an acid or a base are called (d)_______. A solution in which $[OH^-] > 1 \times 10^{-7}$ mol/L and $[H^+] < 1 \times 10^{-7}$ mol/L is an (e)_______. If the pH is greater than 7, then the (f)_______ is less than 7. If the pH is significantly greater than 7, the substance is a (g)_______.

2. Match the correct term to each definition.

(a) donates OH^- ions	(i) Brønsted-Lowry acid
(b) accepts a pair of electrons	(ii) Brønsted-Lowry acid-base reaction
(c) a proton acceptor	(iii) Brønsted-Lowry base
(d) donates H^+ ions	(iv) Lewis acid
(e) acid + base → salt + H_2O	(v) Lewis acid-base reaction
(f) donates a pair of electrons	(vi) Lewis base

(g) two conjugate acid-base pairs in equilibrium
(h) formation of a coordinate covalent bond
(i) donates H^+ ions

(vii) water-ion acid
(viii) water-ion acid-base reaction
(ix) water-ion base

3. Which of the following statements are true? Rewrite each false statement so that it is correct.
   (a) Strong acids and bases are virtually 100% ionized or dissociated in dilute aqueous solutions.
   (b) The strength of an acid is related to the strength of its conjugate base with respect to water.
   (c) An acidic solution is characterized by pH > 7, a high value of $[H^+]$.
   (d) A conjugate acid is a molecule or ion formed by the addition of a proton to a base.
   (e) All water-ion acids can be categorized as Brønsted-Lowry acids.

4. Choose the response that best completes each statement.
   (a) Any substance that can lose a proton is
      (i) a Brønsted-Lowry base (ii) a water-ion acid
      (iii) amphoteric (iv) a solvent system acid
   (b) The ion product constant for water is given by
      (i) $[H^+][OH^-]$ (ii) $[H^+][OH^-]/[H_2O]$
      (iii) $K_w = K_a^2K_b$ (iv) $\text{acid}_1 + \text{base}_2 \rightleftharpoons \text{base}_1 + \text{acid}_2$
   (c) The ability of a substance to either gain or lose a proton is
      (i) an acid-base reaction (ii) the degree of ionization
      (iii) amphoterism (iv) oxidation-reduction
   (d) A weak base
      (i) produces H^+ (ii) has a weak conjugate acid
      (iii) is triprotic (iv) has a strong conjugate acid
   (e) The concentration of a solution expressed in the number of equivalents per liter of solution is the
      (i) molarity (ii) normality
      (iii) stoichiometric amount (iv) percent yield
   (f) The mass of a substance that will give one mole of OH^- is the
      (i) stoichiometric amount (ii) reducing agent
      (iii) equivalent mass of a base (iv) equivalent mass of an acid

(g) For binary acids, the acid strength increases with
(i) decreasing size of the nonhydrogen atom
(ii) decreasing atomic number of the nonhydrogen atom
(iii) increasing electronegativity of the nonhydrogen atom across a period
(iv) an increasing number of hydrogen atoms around the nonhydrogen atom

(h) The acid strength of binary acids increases with
(i) increasing atomic number within a family
(ii) decreasing atomic number within a family
(iii) decreasing electronegativity across a period
(iv) increasing electronegativity within a family

(i) A substance containing the X - O - H structure will be amphoteric if
(i) X is small and highly electronegative
(ii) X is intermediate in covalent radius and intermediate in oxidation state
(iii) the bond between the O and X is very polar
(iv) X has several oxygen atoms attached to it

(j) Acid strength of oxo acids increases with
(i) decreasing oxygen content
(ii) lower oxidation states
(iii) increasing electropositivity of the nonhydrogen atom
(iv) increasing electronegativity of the nonhydrogen atom

5. Name and write the formulas for the ions that are always produced (a) by a water-ion acid and (b) by a water-ion base. (c) Write the chemical equation illustrating a water-ion acid-base reaction.

6. What is the species formed when a Brønsted-Lowry acid loses a proton called? Write a "word" chemical equation illustrating this reaction.

7. What is the species formed when a Brønsted-Lowry base gains a proton called? Write a "word" chemical equation illustrating this reaction.

8. Write chemical equations showing (a) HSO_4^- acting as a base, (b) H_2O acting as an acid, (c) $Al(OH)_3(H_2O)_3$ acting as an acid, and (d) NH_4^+ acting as an acid.

9. Name and write the formulas for several strong bases. What limits the strength of the alkaline earth hydroxides?

10. Sort the following list of chemicals into (a) acidic, (b) alkaline, or (c) amphoteric species: (i) Cs_2O, (ii) N_2O_5, (iii) HCl, (iv) $SO_2(OH)_2$, (v) HNO_2, (vi) Al_2O_3, (vii) BaO, (viii) H_2O, and (ix) CO_2. Assume all oxides are dissolved in water. Do not be confused by the way the formula of the compound is written.

11. What is a polyprotic acid? Write chemical equations illustrating the stepwise ionization of a diprotic acid. Why are the intermediate ions of polyprotic acids amphoteric?

12. What is the difference between weak and strong acids? Is there a similar relationship between weak and strong bases?

13. What can be said about the strength of a conjugate base of a strong acid?

14. Briefly describe what happens during a Brønsted-Lowry acid-base reaction.

15. What is the general rule for naming a binary acid?

16. What is an oxoacid? What are the rules for naming oxoacids? Illustrate these rules by using the oxoacids of bromine. (Note: Not all of these acids exist.)

17. What is a binary acid? What are the two most important factors that influence the bond strength in a binary acid? Briefly explain why the acid strength of a binary acid generally increases as the electronegativity of the nonmetallic element increases.

18. Design an experiment which would correctly show that HI is a stronger acid than HBr.

19. Choose the statements that are valid in describing the acid strength of (a) binary acids and (b) oxoacids: (i) increases as atomic number increases within a periodic table family, (ii) increases as atomic number increases across a row in the periodic table, (iii) increases as atomic number decreases within a periodic table family for the same oxidation state of the elements, (iv) increases as oxidation number or oxygen content increases for the same element.

20. Why does NaOH behave as a base in water while ClOH behaves as an acid? Explain clearly this behavior and the general principles involved for any E - O - H compound.

21. Compare the acid strength of (a) $H_2Se(aq)$ to that of $H_2S(aq)$ and HBr(aq) and (b) HF(aq) to that of $H_2O(l)$ and HCl(aq).

22. Within each group of acids, rank the strongest acid first and the remainder in decreasing order: (a) HF, HCl, HBr, HI; (b) HClO, $HClO_2$, $HClO_3$, $HClO_4$; (c) $P(OH)_3$, $As(OH)_3$, $Sb(OH)_3$, $Bi(OH)_3$; (d) CH_4, NH_3, H_2O, HF; and (e) H_3PO_4, H_2SO_4, $HClO_4$.

23. Based on the relative strengths of the acid-base pairs given in Table 20.1 of the text, predict what will happen if an excess of (a) HCl(aq) is added to a solution containing S^{2-} and (b) formic acid is added to a solution containing trichloroacetate ion.

24. What is meant by the term "equivalent mass" of an acid or a base? How is this mass related to the molar mass?

25. Define the concentration unit of normality. What is the relationship between normality and molarity?

26. What is the relationship between the molarity and normality of the acids in each of the following reactions?

(a) $H_2SO_4(aq) + 2NaOH(aq) \rightarrow Na_2SO_4(aq) + 2H_2O(l)$

(b) $HCl(aq) + KOH(aq) \rightarrow KCl(aq) + H_2O(l)$

(c) $H_2SO_4(aq) + NaOH(aq) \rightarrow NaHSO_4(aq) + H_2O(l)$

(d) $2CH_3COOH(aq) + Ca(OH)_2(aq) \rightarrow Ca(CH_3COO)_2(aq) + 2H_2O(l)$

27. What is the pH of pure water at 25 °C? What is the pOH? Does a large value of pH mean a highly acidic solution?

28. Is the pH of a solution limited to values between 0 and 14? Why (or why not)?

29. If the pH of a solution is doubled from 2 to 4, does this mean that the $[H^+]$ has been doubled? Why (or why not)?

30. A teacher once mistakenly told a class that the $[H^+]$ increased as the pH increased. Explain what was wrong with the statement.

31. Explain why the pH of a solution of a weak acid increases as the concentration of the solution decreases, even though the fraction of molecules that are ionized increases.

32. Each of the following statements is a definition of (a) an acid, (b) a base, or (c) an acid-base reaction. Match each definition to one of these categories and identify which theory is (or theories are) being used: water-ion theory, Brønsted-Lowry theory, or Lewis theory.

(i) a molecule or ion capable of adding a proton
(ii) a reaction between an acid and a base in which a salt and water are formed
(iii) electron-pair donation--the formation of a coordinate covalent bond
(iv) a substance that produces OH^-
(v) a substance that acts as an electron-pair acceptor
(vi) a molecule or ion capable of losing a proton
(vii) a substance that acts as an electron-pair donor
(viii) a reaction between an acid and a base in which proton transfer takes place

33. Is the reaction

$$CaO + SiO_2 \rightarrow CaSiO_3$$

an acid-base reaction? If so, identify the acid.

34. Each of the following equations is a Lewis acid-base reaction.

(a) $Cu^{2+} + 4NH_3(aq) \rightarrow [Cu(NH_3)_4]^{2+}$
(b) $O^{2-} + SO_3 \rightarrow SO_4{}^{2-}$
(c) $FeCl_3 + Cl^- \rightarrow [FeCl_4]^-$

Identify the base in each case.

35. For each of the reactions described by the following equations, identify <u>all</u> of the acids.

(a) $NH_4I(NH_3) + KNH_2(NH_3) \rightarrow KI(NH_3) + 2NH_3(l)$
(b) $OCN^-(aq) + H_2O(l) \rightarrow HOCN(aq) + OH^-(aq)$
(c) $[Zn(H_2O)_4]^{2+}(aq) + H_2O(l) \rightleftharpoons [Zn(H_2O)_3(OH)]^+(aq) + H_3O^+(aq)$
(d) $K_2O(s) + CO_2(g) \rightleftharpoons K_2CO_3(s)$
(e) $Al^{3+}(aq) + 6F^-(aq) \rightleftharpoons [AlF_6]^{3-}(aq)$

Answers to questions

1. (a) acidic aqueous solution, (b) pH, (c) strong acid, (d) amphoteric, (e) alkaline aqueous solution, (f) pOH, (g) strong base.
2. (a) ix, (b) iv, (c) iii, (d) i or vii, (e) viii, (f) vi, (g) ii, (h) v, (i) vii or i.
3. (a) T; (b) T; (c) F, change $pH > 7$ to $pH < 7$; (d) T; (e) T.
4. (a) ii, (b) i, (c) iii, (d) iv, (e) ii, (f) iii, (g) iii, (h) i, (i) ii, (j) iv.

5. (a) hydrated proton or hydrogen ion, H^+; (b) hydroxide ion, OH^-; (c) acid + base $\rightarrow$ salt + water.
6. The species is the conjugate base of the acid; acid $\rightarrow$ proton + conjugate base.
7. The species is the conjugate acid of the base; base + proton $\rightarrow$ conjugate acid.
8. (a) $HSO_4^- + H^+ \rightarrow H_2SO_4(aq)$
   (b) $H_2O(l) \rightarrow H^+ + OH^-$
   (c) $Al(OH)_3(H_2O)_3(s) \rightarrow H^+ + [Al(OH)_4(H_2O)_2]^-$
   (d) $NH_4^+ \rightarrow H^+ + NH_3(aq)$
9. The list could include any hydroxide of the alkaline metal family (e.g., LiOH, lithium hydroxide) and the alkaline earth metal family, except Be (e.g., $Mg(OH)_2$, magnesium hydroxide). The strength of the alkaline earth hydroxides is limited by the low solubility of the base.
10. (a) ii, iii, iv, vi, ix; (b) i, vii; (c) vi, viii.
11. A polyprotic acid contains more than one ionizable proton; $H_2X(aq) + H_2O(l) \rightleftharpoons H_3O^+ + HX^-$, $HX^- + H_2O(l) \rightleftharpoons H_3O^+ + X^{2-}$. The intermediate ions can either gain or lose protons.
12. Weak acids are less than 100% ionized or dissociated and strong acids are nearly 100% ionized or dissociated; yes.
13. It is a very weak base.
14. The reaction is essentially a competition for the proton, which will end up with the stronger base.
15. Use the prefix "hydro," name the nonmetal modified with the suffix "ic," and add "acid."
16. Oxoacids contain H and O atoms and a central atom of a third element. The "most common" oxoacid is the "ic" acid and is assigned to contain n oxygen atoms. The oxoacid containing n+1 oxygen atoms is the "per...ic" acid, the oxoacid containing n-1 oxygen atoms is the "...ous" acid, and the oxoacid containing n-2 oxygen atoms is the "hypo...ous" acid. $HBrO_4$ is perbromic acid, $HBrO_3$ is bromic acid, $HBrO_2$ is bromous acid, and HBrO is hypobromous acid.
17. A binary acid is the hydride of a nonmetal. The factors are the radius and electronegativity of nonmetal atom. As the electronegativity of the nonmetal atom increases, the bond becomes more polar and the H^+ is easier to remove.
18. Dissolve NaBr in liquid HI and observe the formation of HBr as a result of $HI + Br^- \rightarrow HBr + I^-$.

19. (a) i, ii; (b) ii, iii, iv.
20. Acid strength for an oxoacid $E(OH)_x$ increases as the electronegativity of the element E increases. Because Cl is more electronegative than Na, ClOH is more acidic than NaOH.
21. (a) H_2Se is stronger than H_2S because acid strength increases down a family, HBr is a stronger acid than H_2Se because acid strength increases across a period, thus $H_2S < H_2Se <$ HBr; (b) HF is stronger than H_2O because acid strength increases across a period, HCl is a stronger acid than HF because acid strength increases down a family, thus $H_2O <$ HF $<$ HCl.
22. (a) HI > HBr > HCl > HF, (b) $HClO_4 > HClO_3 > HClO_2 >$ HClO, (c) $P(OH)_3 > As(OH)_3 > Sb(OH)_3 > Bi(OH)_3$, (d) HF $> H_2O > NH_3 > CH_4$, (e) $HClO_4 > H_2SO_4 > H_3PO_4$.
23. (a) Proton transfer to S^{2-} occurs to form HS^- and proton transfer to HS^- occurs to form $H_2S(aq)$; (b) only small concentrations of formate ion and trichloracetic acid will form.
24. The equivalent mass is the mass of acid that will give 1 mol H^+ or of base that will give 1 mol OH^-. The equivalent mass equals the molar mass divided by the number of ionizable H^+ ions per molecule for an acid or OH^- ions for a base.
25. Normality is the number of equivalents of solute per liter of solution. The normality equals the molarity multiplied by the number of ionizable H^+ ions per molecule for acids or OH^- ions for a base.
26. (a) normality = (2)(molarity), (b) normality = molarity, (c) normality = molarity, (d) normality = molarity.
27. pH = pOH = 7; no, the solution is alkaline if pH > 7.
28. No, pH < 0 for $[H^+] > 1$ mol/L and pH > 14 for $[OH^-] > 1$ mol/L.
29. No, [H+] has decreased by a factor of 100.
30. The teacher's statement was wrong because pH = $-\log [H^+]$, so that as the $[H^+]$ increases, the pH decreases.
31. Even though the fraction of molecules that are ionized increases as the concentration of the solution decreases, the net effect is still a lower $[H^+]$ and so the pH increases.
32. (a) v (Lewis); vi (water-ion and Brønsted-Lowry); (b) i (Brønsted-Lowry), iv (water-ion), vii (Lewis); (c) ii (water-ion), iii (Lewis), viii (Brønsted-Lowry).

33. Yes; according to the Lewis theory, O^{2-} ion from CaO donates a pair of electrons to the Si atom in SiO_2, so SiO_2 is the acid.
34. (a) NH_3, NH_3 donates an electron pair to the Cu^{2+} ion; (b) O^{2-}, O^{2-} donates an electron pair to the S atom in SO_3; (c) Cl^-, Cl^- donates an electron pair to the Fe^{3+} ion.
35. (a) NH_4I, NH_3; (b) H_2O, HOCN; (c) $[Zn(H_2O)_4]^{2+}$, H_3O^+; (d) CO_2; (e) Al^{3+}.

SKILLS

Skill 20.1 Classifying substances as acids and bases

Consider the following reaction between HCl and NaCN:

$$HCl(aq) + NaCN(aq) \rightarrow NaCl(aq) + HCN(aq)$$

Although HCl meets the definition of a water-ion acid (a proton donor), NaCN does not meet the water-ion definition of a base (OH^- donor) and so this is not an acid-base reaction according to the water-ion theory. However, NaCN is a proton acceptor and HCl is a proton donor, so the substances are acids and bases according to the Brønsted-Lowry theory and this reaction is an acid-base reaction. According to the Lewis theory, HCl contains the proton that will act as an electron-pair acceptor (an acid) and the carbon atom in the cyanide ion donates an electron pair (a base), so the reaction is considered to be a Lewis acid-base reaction.

Skill Exercise: Determine which of these reactions are acid-base reactions according to the (a) water-ion theory, (b) Brønsted-Lowry theory, and (c) Lewis theory: (i) $Al_2O_3(s) + 6HCl(aq) \rightarrow 2AlCl_3(aq) + 3H_2O(l)$, (ii) $H_2SO_4(aq) + NaOH(aq) \rightarrow NaHSO_4(aq) + H_2O(l)$, (iii) $NH_3(g) + BF_3(g) \rightarrow NH_3BF_3(g)$, (iv) $HF(aq) + H_2O(l) \rightarrow H_3O^+ + F^-$, (v) $H^+ + CO_3{}^{2-} \rightarrow HCO_3{}^-$, (vi) $2NH_3(l) \rightarrow NH_4{}^+ + NH_2{}^-$, and (vii) $CH_3COOH(aq) + NaOH(aq) \rightarrow CH_3COONa(aq) + H_2O(l)$. *Answers*: (a) ii, vii; (b) i, ii, iv, v, vi, vii; (c) i, ii, iii, iv, v, vi, vii.

Skill 20.2 Relative strengths of acids

When comparing binary acids, acid strength increases with increasing radius within a family and with increasing electronegativity of the nonmetal atom across a period. For example, HF would be predicted to be a stronger acid than H_2O, but HCl would be predicted to be stronger than HF. For oxoacids, acid strength increases with increasing oxygen content (or oxidation state) for a series of acids for a given element and with increasing electronegativity of the central atom for a series of acids having the same empirical formula. Therefore, $HClO_4$ is predicted to be a stronger acid than $HClO_3$, and $HClO_3$ is predicted to stronger than $HBrO_3$.

Skill Exercise: Choose the stronger acid from each pair: (a) H_2S and H_2O, (b) HBr and HCl, (c) HI and H_2O, (d) H_2Se and HBr, (e) $HClO_2$ and $HClO_3$, (f) H_2SO_4 and H_2SO_3, (g) $HClO_2$ and $HBrO_2$, (h) H_3PO_4 and H_3AsO_4, and (i) H_3PO_3 and H_3PO_4. *Answers*: (a) H_2S, (b) HBr, (c) HI, (d) HBr, (e) $HClO_3$, (f) H_2SO_4, (g) $HClO_2$, (h) H_3PO_4, (i) H_3PO_4.

Skill 20.3 Naming oxoacids and anions of oxoacids

The nomenclature of binary acids, salts of binary acids, and salts containing oxoacids has been presented in *Skill 4.2 Naming simple inorganic compounds*. Although the systematic approach presented in this skill will allow the writing of formulas and names of various oxoacids and oxoanions, not all of these species exist.

The oxoacids that carry the name of the central element and the suffix "ic" are those that at one time were the best known or most common. These acids are

H_3BO_3	boric acid	HNO_3	nitric acid	H_2SO_4	sulfuric acid
H_2AlO_3	aluminic acid	H_3PO_4	phosphoric acid	H_2SeO_4	selenic acid
H_3GaO_3	gallic acid	H_3AsO_4	arsenic acid	H_6TeO_6	telluric acid
		H_3SbO_4	antimonic acid		
H_2CO_3	carbonic acid	H_3BiO_3	bismuthic acid	$HClO_3$	chloric acid
H_2SiO_4	silicic acid			$HBrO_3$	bromic acid
H_4SnO_4	stannic acid			HIO_3	iodic acid
H_4PbO_4	plumbic acid				

The number of oxygen atoms in the "ic" acid is assigned the letter n.

The names of other acids of the central element are derived from the "ic" acid by changing the suffix and/or adding the prefixes shown in the table below, depending on the number of oxygen atoms present in the oxoacid to be named.

number of oxygen atoms	name
$n+1$	per...ic acid
n	...ic acid
$n-1$	...ous acid
$n-2$	hypo...ous acid

For example, H_3PO_4 is phosphoric acid. Thus H_3PO_3 is phosphorous acid and H_3PO_2 is hypophosphorous acid. According to the system, the formula of perphosphoric acid would be H_3PO_5, but this acid does not exist. Most oxoacids follow these rules, but there are some exceptions (for example, $H_2N_2O_2$ is hyponitrous acid and H_2TeO_3 is tellurous acid).

Skill Exercise: Name the following acids: (a) $HClO$, (b) $HBrO_4$, (c) H_2SeO_3, (d) H_3SbO_3, and (e) H_2SO_2 (which does not exist) and write the formulas for (f) periodic acid, (g) hypoiodous acid, and (h) hypophosphorous acid. *Answers*: (a) hypochlorous acid, (b) perbromic acid, (c) selenous acid, (d) antimonous acid, (e) hyposulfurous acid, (f) HIO_4, (g) HIO, (h) H_3PO_2.

Additional prefixes are sometimes used for the names of oxoacids. The prefix "ortho" represents the oxoacid that contains the greatest number of OH groups. Usually this prefix is not used except in some trade names and for orthophosphoric acid (H_3PO_4). A molecule of an "ortho" acid that has "lost" a water molecule (i.e., contains two less hydrogen atoms and one less oxygen atom) carries the prefix "meta." For example, the formula for metaphosphoric acid is HPO_3 ($H_3PO_4 - H_2O = HPO_3$). An "ortho" acid that has "lost" a water molecule from two molecules of the acid carries the prefix "pyro." For example, the name of $H_4P_2O_7$ is pyrophosphoric acid ($2H_3PO_4 - H_2O = H_4P_2O_7$). Acids containing sulfur atoms in place of oxygen atoms carry the prefix "thio" and acids containing an oxygen-oxygen bond carry the prefix "peroxo."

The names of the oxoanions are derived from the names of the oxoacids. The prefixes, if any, are unchanged. The "ic" suffix on the acid becomes "ate" for the anion and the "ous" suffix becomes "ite." When naming the anion alone, the word "ion" must be included. For example, IO_4^- is the anion of periodic acid, HIO_4. Thus the name of the ion is periodate ion and the name of $NaIO_4$ is sodium periodate.

Skill Exercise: Name the following species: (a) BrO_3^-, (b) $NaNO_2$, (c) IO_3^-, (d) NaH_2PO_4, (e) Na_2SO_3, (f) $Na_2S_2O_3$, and (g) $S_2O_7^{2-}$ and write the formulas for (h) aluminum metaphosphate, (i) ammonium sulfite, (j) hydrogen sulfate ion, (k) barium selenate, (l) calcium metaborate, and (m) cesium carbonate. *Answers*: (a) bromate ion, (b) sodium nitrite, (c) iodate ion, (d) sodium dihydrogen phosphate, (e) sodium sulfite, (f) sodium thiosulfate, (g) pyrosulfate ion, (h) $Al(PO_3)_3$, (i) $(NH_4)_2SO_3$, (j) HSO_4^-, (k) $BaSeO_4$, (l) $Ca(BO_2)_2$, (m) Cs_2CO_3.

NUMERICAL EXERCISES

Example 20.1 Ion product constant of water

A solution of dilute hydrochloric acid contains $[H^+]$ = 6 mol/L. What is the value of $[OH^-]$?

The ion product for water relates $[H^+]$ and $[OH^-]$:

$$K_w = [H^+][OH^-] = 1.00 \times 10^{-14}$$

Substituting $[H^+]$ = 6 mol/L gives

$$[OH^-] = \frac{1.00 \times 10^{-14}}{6} = 2 \times 10^{-15} \text{ mol/L}$$

The concentration of OH^- is 2×10^{-15} mol/L.

Exercise 20.1 A common concentration of aqueous NaOH used in the laboratory is 0.10 M. What is the value of $[H^+]$? *Answer*: 1.0×10^{-13} mol/L.

*Example 20.2 pH and pOH**

What is the pH of the solution described in *Example 20.1*?

This problem is solved by using the comprehensive problem-solving approach presented in *Skill 2.7 Solving problems*.

1. Study the problem and be sure you understand it.
   (a) What is unknown?
   The pH of the solution.
   (b) What is known?
   $[H^+]$ = 6 mol/L for the solution.
2. Decide how to solve the problem.
   (a) What is the connection between the known and the unknown?
   The pH of a solution is related to the concentration of H^+ by the equation
   $$pH = -\log [H^+]$$
   (b) What is necessary to make the connection?
   A simple substitution of the hydrogen ion concentration into the defining equation for pH will solve the problem.
3. Set up the problem and solve it.
   The pH of the solution is
   $$pH = -\log [H^+] = -\log(6) = -0.8$$
4. Check the result.
   (a) Are significant figures and the location of the decimal point correct?
   Yes. The concentration contains one significant figure and so the

mantissa of the logarithm should contain one significant figure (see Appendix I of the text).

(b) Did the answer come out in the correct units?

Yes. There are no units on pH.

(c) Is the answer reasonable?

Yes. Even though values of pH usually range from 0 to 14, the pH of a very strongly acid solution can be a negative number.

Exercise 20.2 What is the pH of the solution described in Exercise 20.1? *Answer*: 13.00.

Example 20.3 pH and pOH

What is the concentration of H^+ in a solution having a pH of 5.35?

Rearranging the definition for $[H^+]$ gives

$$\log [H^+] = -pH = -5.35$$

$$[H^+] = 4.5 \times 10^{-6} \text{ mol/L}$$

The concentration of H^+ is 4.5×10^{-6} mol/L.

Exercise 20.3 What is the concentration of H^+ in a solution having a pH of 12.35? *Answer*: 4.5×10^{-13} mol/L.

Example 20.4 pH and pOH

What is the pOH of a solution containing 0.015 M NaOH?

The pOH is defined as

$$pOH = -\log [OH^-]$$

Because NaOH is a strong base, $[OH^-] = 0.015$ mol/L so

$$pOH = -\log(0.015) = 1.82$$

The pOH of this solution is 1.82.

Exercise 20.4 What is the concentration of hydroxide ion in a solution having a pOH of 4.18? *Answer*: 6.6×10^{-5} mol/L.

Example 20.5 pH and pOH

What is the pOH of the solution described in *Example 20.3*?

The relation between pH and pOH is

$$pH + pOH = 14.00$$

Solving for pOH and pH = 5.35 gives

$$pOH = 14.00 - pH = 14.00 - 5.35 = 8.65$$

The pOH of the solution is 8.65.

Exercise 20.5 What is the pOH of the solution described in Exercise 20.3?
Answer: 1.65.

Example 20.6 Acid ionization constant*

Large amounts of lactic acid ($CH_3CHOHCOOH$, present in sour milk) are used for dyeing in the wool industry and for neutralizing lime in hides in the tanning industry. Find the pH of a 0.100 M solution of lactic acid.

$$CH_3CHOHCOOH(aq) + H_2O(l) \rightleftharpoons CH_3CHOHCOO^- + H_3O^+ \qquad K_a = 1.37 \times 10^{-4}$$

This problem is solved by using the comprehensive problem-solving approach presented in *Skill 2.7 Solving problems*.

1. <u>Study the problem and be sure you understand it</u>.
   (a) What is unknown?
   The pH of the solution.
   (b) What is known?
   The chemical equation for the ionization of the acid and the value of the equilibrium constant.
2. <u>Decide how to solve the problem</u>.
   (a) What is the connection between the known and the unknown?
   The pH is related to the concentration of H^+. The $[H^+]$ is related to K_a and the other concentrations of species at equilibrium by the equilibrium constant expression.
   $$K_a = \frac{[H^+][CH_3CHOHCOO^-]}{[CH_3CHOHCOOH]}$$
   (b) What is necessary to make the connection?
   Choosing $x = [H^+]$ at equilibrium, the usual table that relates the concentration of all of the species at equilibrium to the initial concentrations and x (see *Skill 19.2 Solving equilibrium problems*) is prepared. After substitution of these values into the equilibrium constant expression, the value of x is determined. Then the value of pH can be calculated.
3. <u>Set up the problem and solve it</u>.

	$CH_3CHOHCOOH(aq) + H_2O(l)$	$\rightleftharpoons$ $CH_3CHOHCOO^-$	+ H_3O^+
initial	0.100	0	0
change	$-x$	$+x$	$+x$
equilibrium	$0.100 - x$	x	x

$$K_a = \frac{(x)(x)}{(0.100 - x)} = 1.37 \times 10^{-4}$$

Assuming $0.100 - x \approx 0.100$ gives

$$x^2 = (1.37 \times 10^{-4})(0.100) = 1.37 \times 10^{-5}$$

$$x = 3.7 \times 10^{-3}$$

The approximation $0.100 - x \approx 0.100$ is valid to within the 5% error limit, $(0.0037/0.100) \times 100 = 3.7$ %, so $[H^+] = 3.7 \times 10^{-3}$ mol/L. From the definition of pH

$$pH = -\log [H^+] = -\log(3.7 \times 10^{-3}) = 2.43$$

The pH of the lactic acid solution is 2.43.

4. Check the result.
   (a) Are significant figures and the location of the decimal point correct?
       Yes.
   (b) Did the answer come out in the correct units?
       Yes.
   (c) Is the answer reasonable?
       Yes. The pH is between 0 and 7, which is the range for most aqueous acid solutions.

Exercise 20.6 Find the pH of a 0.107 M solution of benzoic acid. $K_a = 6.6 \times 10^{-5}$ for C_6H_5COOH. *Answer*: 2.57.

Example 20.7 Acid ionization constant

A weak acid HX is known to be 5.0 % ionized in a 0.1000 M solution. What is the value of the acid equilibrium constant?

For the equation

$$HX(aq) + H_2O(l) \rightarrow H_3O^+ + X^-$$

the equilibrium expression is

$$K_a = \frac{[H^+][X^-]}{[HX]}$$

The following table lists the equation, the initial concentrations, the concentration changes, and the equilibrium concentrations.

	$HX(aq) + H_2O(l) \rightarrow$	H_3O^+	+	X^-
initial	0.100	0		0
change	-(0.050)(0.100)	+(0.050)(0.100)		+(0.050)(0.100)
equilibrium	0.095	0.0050		0.0050

Substituting the equilibrium concentrations into the equilibrium expression gives

$$K_a = \frac{(0.0050)(0.0050)}{(0.095)} = 2.6 \times 10^{-4}$$

The acid dissociation constant is 2.6×10^{-4}.

Exercise 20.7 What is the percentage of ionization for a weak acid if K_a = 1.6×10^{-7} for a 0.010 M solution of HX? *Answer*: $[H^+] = [X^-] = 4.0 \times 10^{-5}$ mol/L, 0.40 %.

Example 20.8 Base ionization constant

Aniline, $C_6H_5NH_2$, is a weak base with $K_b = 4.2 \times 10^{-10}$. Find $[OH^-]$ for a 0.15 M aqueous solution of aniline.

For the equation

$$C_6H_5NH_2(aq) + H_2O(l) \rightarrow C_6H_5NH_3^+ + OH^-$$

the equilibrium expression is

$$K_b = \frac{[C_6H_5NH_3^+][OH^-]}{[C_6H_5NH_2]}$$

Preparing the usual table with $x = [OH^-]$ gives

	$C_6H_5NH_2(aq)$ + $H_2O(l)$	→	$C_6H_5NH_3^+$	+ OH^-
initial	0.15		0	0
change	$-x$		$+x$	$+x$
equilibrium	$0.15 - x$		x	x

Substituting the equilibrium concentrations into the equilibrium constant expression gives

$$K_b = \frac{(x)(x)}{(0.15 - x)} = 4.2 \times 10^{-10}$$

and solving gives $x = 7.9 \times 10^{-6}$. The concentration of OH^- from the aniline is 7.9×10^{-6} mol/L. [Note that the contribution to the $[OH^-]$ from the ionization water must also be considered for solutions of aniline that are more dilute. In this case, the initial $[OH^-]$ from the water is slightly more than 1 % of the total value.]

Exercise 20.8 Repeat the calculation in *Example 20.8* given that the solution also contains an initial concentration of $C_6H_5NH_3^+$ of 0.010 mol/L. *Answer*: $[OH^-] = 6.3 \times 10^{-9}$ mol/L.

Example 20.9 Base ionization constant

What is the percentage of ammonia molecules that undergo reaction with water in a 0.010 M aqueous solution of NH_3? $K_b = 1.6 \times 10^{-5}$ for NH_3.

For the equation

$$NH_3(aq) + H_2O(l) \rightarrow NH_4^+ + OH^-$$

the equilibrium expression is

$$K_b = \frac{[NH_4^+][OH^-]}{[NH_3]}$$

Preparing the usual table with $x = [OH^-]$ gives

	$NH_3(aq) + H_2O(l)$	$\rightarrow$	NH_4^+	$+ OH^-$
initial	0.010		0	0
change	$-x$		$+x$	$+x$
equilibrium	$0.010 - x$		x	x

and substituting the results into the equilibrium constant expression gives

$$\frac{(x)(x)}{(0.010 - x)} = 1.6 \times 10^{-5}$$

Assuming $0.010 - x \approx 0.010$ gives

$$x^2 = (0.010)(1.6 \times 10^{-5}) = 1.6 \times 10^{-7}$$

$$x = 4.0 \times 10^{-4}$$

The assumption is valid, so $[OH^-] = 4.0 \times 10^{-4}$ mol/L. The fraction of NH_3 molecules that reacts is

$$\frac{[OH^-]_{equilibrium}}{[NH_3]_{initial}} = \frac{4.0 \times 10^{-4} \text{ mol/L}}{0.010 \text{ mol/L}} = 0.040$$

Thus 4.0 % of the ammonia molecules react.

Exercise 20.9 A weak base X^- reacts with water to form OH^- to the extent of 8.3 % in a 0.35 M aqueous solution. What is K_b for this anion? *Answer*: 2.6×10^{-3}.

Example 20.10 Ion product constant of water

The acid ionization constant for benzoic acid, C_6H_5COOH, is 6.6×10^{-5}. What is K_b for the benzoate ion, $C_6H_5COO^-$?

The equation for the ionization of benzoic acid is

$$C_6H_5COOH(aq) + H_2O(l) \rightarrow C_6H_5COO^- + H_3O^+$$

The conjugate base of benzoic acid is the benzoate ion. Thus

$$K_b = \frac{K_w}{K_a} = \frac{1.00 \times 10^{-14}}{6.6 \times 10^{-5}} = 1.5 \times 10^{-10}$$

For the benzoate ion, $K_b = 1.5 \times 10^{-10}$.

Exercise 20.10 Find K_a for thiocyanic acid, HNCS, given that $K_b = 1.4 \times 10^{-16}$ for NCS^-. *Answer*: $K_a = 71$.

Example 20.11 Acid ionization constant

What is the pH of a 0.10 M solution of hypochlorous acid that also contains 0.010 M NaClO? $K_a = 2.90 \times 10^{-8}$ for HClO.

For the equation

$$HClO(aq) + H_2O(l) \rightarrow H_3O^+ + ClO^-(aq)$$

the equilibrium expression is

$$K_a = \frac{[H^+][ClO^-]}{[HClO]} = 2.90 \times 10^{-8}$$

Preparing the usual table with $x = [H^+]$ and substituting the equilibrium concentrations into the equilibrium expression gives

	$HClO(aq) + H_2O(l)$	$\rightarrow$ H_3O^+	+ ClO^-
initial	0.10	0	0.010
change	$-x$	$+x$	$+x$
equilibrium	$0.10 - x$	x	$0.010 + x$

$$K_a = \frac{(x)(0.010 + x)}{(0.10 - x)} = 2.90 \times 10^{-8}$$

Making the approximations that $(0.010 + x) \approx 0.010$ and $(0.10 - x) \approx 0.10$ gives

$$x = (2.90 \times 10^{-8})(0.10)/(0.010) = 2.9 \times 10^{-7}$$

The approximations are valid, so $[H^+] = 2.9 \times 10^{-7}$ mol/L and

$$pH = -\log [H^+] = -\log(2.9 \times 10^{-7}) = 6.54$$

The pH of the solution is 6.54.

Exercise 20.11 What is the (a) $[OH^-]$ and (b) pOH of a 0.15 M NH_3 solution that also contains 0.010 M NH_4NO_3? $K_b = 1.6 \times 10^{-5}$. *Answers*: (a) 2.4×10^{-4}, (b) 3.62.

Example 20.12 Equivalents

Two bottles of acid are marked 2.0 M. One of these bottles contains HCl and the other H_2SO_4. Which has the higher concentration expressed in units of normality?

Recalling that 2.0 M means that in each bottle there are 2.0 mol of acid for each liter of solution, all that is needed is the relationship between moles and equivalents of acid. For HCl, a mole of acid is equal to an equivalent of acid because there is only one available H^+ for each molecule. Hence

$$\text{normality} = \frac{\text{number of equiv}}{\text{solution volume in liters}} = \frac{(2.0\ \text{mol})(1\ \text{equiv}/1\ \text{mol})}{1\ \text{L}} = 2.0\ \text{equiv/L}$$

For the H_2SO_4, one mole is equivalent to two equivalents, hence

$$\text{normality} = \frac{(2.0\ \text{mol})(2\ \text{equiv}/1\ \text{mol})}{1\ \text{L}} = 4.0\ \text{equiv/L}$$

The H_2SO_4 has the higher concentration in terms of equivalents per liter or normality.

Exercise 20.12 Compare the number of equivalents of OH^- available from 1.0 L of 0.010 M NaOH, 2.0 L of 0.0050 M KOH, and 0.50 L of 0.010 M $Ba(OH)_2$. *Answer*: All three yield 0.010 equiv of OH^-.

Example 20.13 Equivalents*

What is the normality of an aqueous H_2SO_4 solution if 25.0 mL of the solution was neutralized by 16.3 mL of 0.107 N NaOH? The equation describing the neutralization process is

$$2NaOH(aq) + H_2SO_4(aq) \rightarrow Na_2SO_4(aq) + 2H_2O(l)$$

This problem is solved by using the comprehensive problem-solving approach presented in *Skill 2.7 Solving problems*.

1. Study the problem and be sure you understand it.
   (a) What is unknown?
   The concentration of the H_2SO_4 solution expressed in normality.
   (b) What is known?
   The concentration and volume of NaOH that reacts with a known volume of the acid and the chemical equation.
2. Decide how to solve the problem.
   (a) What is the connection between the known and the unknown?
   The chemical equation gives the molar relationship between the acid and base. The definition of molarity relates the number of moles of solute to the volume of the solution.
   (b) What is necessary to make the connection?
   The concentration of the NaOH solution can be changed from normality to

molarity, the number of moles of NaOH determined, the number of moles of H_2SO_4 calculated, the concentration of the H_2SO_4 solution determined, and the concentration changed from molarity to normality.

3. Set up the problem and solve it.

The chemical equation shows that there is 1 equiv/1 mol for NaOH and 2 equiv/1 mol for H_2SO_4. The molarity of the NaOH solution is

$$\text{molarity} = \left[\frac{0.107 \text{ equiv}}{1 \text{ L}}\right]\left[\frac{1 \text{ mol}}{1 \text{ equiv}}\right] = 1.107 \text{ mol NaOH/L}$$

Following the usual method to solve stoichiometric problems (see *Skill 5.5 Stoichiometry*) gives

$$\left[\frac{0.107 \text{ mol NaOH}}{1 \text{ L}}\right](16.3 \text{ mL})\left[\frac{1 \text{ L}}{1000 \text{ mL}}\right]\left[\frac{1 \text{ mol } H_2SO_4}{2 \text{ mol NaOH}}\right] = 8.72 \times 10^{-4} \text{ mol } H_2SO_4$$

The molarity of the solution is

$$\frac{8.72 \times 10^{-4} \text{ mol } H_2SO_4}{(25.0 \text{ mL})(1 \text{ L}/1000 \text{ mL})} = 0.0349 \text{ mol/L}$$

The normality of the solution is

$$\text{normlaity} = \left[\frac{0.0349 \text{ mol}}{1 \text{ L}}\right]\left[\frac{2 \text{ equiv}}{1 \text{ mol}}\right] = 0.0698 \text{ equiv/L}$$

4. Check the result.

(a) Are significant figures and the location of the decimal point correct?

Yes. Note that the factors such as (2 equiv/1 mol) are exact.

(b) Did the answer come out in the correct units?

Yes.

(c) Is the answer reasonable?

Yes.

Exercise 20.13 What is the normality of an aqueous solution of HNO_3 if 15.0 mL of the solution was neutralized by 20.2 mL of 0.107 N NaOH? *Answer*: 0.144 N.

PRACTICE TEST

1 (5 points). Identify which of the following reactions are acid-base reactions:

(a) $H_2(g) + Cl_2(g) \rightarrow 2HCl(g)$

(b) $CaCl_2(aq) + K_2C_2O_4(aq) \rightarrow 2KCl(aq) + CaC_2O_4(s)$

(c) $HNO_3(aq) + KOH(aq) \rightarrow KNO_3(aq) + H_2O(l)$

(d) $3Cu(s) + 8HNO_3(aq) \rightarrow 3Cu(NO_3)_2(aq) + 2NO(g) + 4H_2O(l)$

(e) $2H_2O(l) \rightarrow 2H_2(g) + O_2(g)$

2 (10 points). In the following reactions, choose the substances that are acting as (a) water-ion acids, (b) water-ion bases, (c) Brønsted-Lowry acids, and (d) Brønsted-Lowry bases.

$$NaHSO_4(aq) + NaOH(aq) \rightarrow Na_2SO_4(aq) + H_2O(l)$$
$$CaCO_3(s) + 2HCl(aq) \rightarrow CaCl_2(aq) + H_2O(l) + CO_2(g)$$
$$H_2O_2(aq) + H_2O(l) \rightarrow HO_2^- + H_3O^+$$
$$H_3O^+ + NH_3(aq) \rightarrow NH_4^+ + H_2O(l)$$
$$2HF(l) \rightarrow [H_2F^+] + F^-$$

3 (10 points). Show, by writing appropriate equations involving HCl(aq) and NaOH(aq), that $Al(OH)_3(s)$ is amphoteric.

4 (10 points). Identify which substance acts as the stronger acid: (a) HF or HI, (b) PH_3 or H_2S, (c) H_2SeO_3 or H_2SeO_4, (d) H_2SO_4 or H_2SeO_4, (e) $H_2PO_4^-$ or HPO_4^{2-}.

5 (10 points). Identify the conjugate acid-base pairs in the equation

$$HNO_3(aq) + H_2O(l) \rightleftharpoons H_3O^+ + NO_3^-$$

Which base is stronger? Which acid is stronger?

6 (15 points). Choose the response that best completes each statement.

(a) Which of the following statements is *not* true for the following reaction?

$$HSO_4^-(aq) + C_2O_4^{2-}(aq) \rightleftharpoons SO_4^{2-}(aq) + HC_2O_4^-(aq)$$

(i) $HSO_4^-(aq)$ is a water-ion acid.
(ii) $C_2O_4^{2-}$ is a Brønsted-Lowry base.
(iii) HSO_4^- is a Brønsted-Lowry acid.
(iv) HSO_4^- and $HC_2O_4^-$ are a conjugate acid-base pair.

(b) A Lewis base is defined as

(i) a proton acceptor
(ii) any substance that dissolves to give OH^-
(iii) an electron-pair acceptor
(iv) an electron-pair donor

(c) Which of the following can act as both an acid and a base according to the Brønsted-Lowry theory: (a) HCO_3^-, (b) $H_2PO_4^-$, (c) NH_3, (d) O^{2-}?

(i) only a and b (ii) all of them
(iii) only c and d (iv) none of these answers

(d) The hydrogen ion concentration of a solution of pH 4 is how many times as great as that of a solution of pH 8?

(i) 2 (ii) 3

(iii) 4000 (iv) none of these answers

(e) In any aqueous solution, the hydrogen ion concentration is equal to

(i) 1×10^{-14} (ii) 1×10^{-7}

(iii) $K_w - [OH^-]$ (iv) $K_w/[OH^-]$

(f) If pOH = 12.0, what does $[H^+]$ equal?

(i) 2.0 mol/L (ii) 12 mol/L

(iii) 1×10^{-2} mol/L (iv) 1×10^{-12} mol/L

(g) What is the pH of a 0.015 M HNO_3 solution?

(i) -1.82 (ii) 1.82

(iii) 2.18 (iv) 12.18

(h) Addition of NaOH to an aqueous solution causes the pH to

(i) increase (ii) decrease

(iii) remain the same (iv) none of the answers

(i) The pH of a 1×10^{-8} M aqueous solution of HNO_3 is

(i) 8.0 (ii) slightly less than 7

(iii) 6.0 (iv) slightly greater than 7

(j) The pH of a 0.01 M solution of a weak acid, HA, that is 1 % ionized is

(i) 7.0 (ii) 5.0

(iii) 4.0 (iv) 10.0

(k) What is the pH of a solution containing 0.0010 mol of KOH in enough water to make 10.0 L of solution?

(i) 10.00 (ii) 12.00

(iii) 11.00 (iv) 4.00

(l) For H_2S, $K_{a_1} = 1.0 \times 10^{-7}$ and $K_{a_2} = 3 \times 10^{-13}$ and for H_2SO_4, $K_{a_2} = 1.0 \times 10^{-2}$. Predict the major products formed by mixing HSO_4^- with HS^-.

(i) $HSO_4^- + HS^-$ (ii) $H_2SO_4(aq) + S^{2-}$

(iii) $SO_4^{2-} + H_2S(aq)$ (iv) $S_2O_3^{2-} + H_2S(aq) + O_2(g)$

(m) How many equivalents of OH^- are in 250 mL of 0.100 M $Ba(OH)_2$?

(i) 25 equiv (ii) 50. equiv

(iii) 0.025 equiv (iv) 0.050 equiv

(n) Which component of air will cause the pH of water to differ from 7?

(i) N_2 (ii) CO_2

(iii) O_2 (iv) H_2

(o) Which substance(s) is/are acting as a Lewis base in the following reaction?

$$F^- + BF_3 \rightarrow [BF_4]^-$$

(i) F^- (ii) $[BF_4]^-$

(iii) BF_3 (iv) BF_3 and $[BF_4]^-$

7 (10 points). A stockroom attendant prepares a bottle of 0.010 M HCl. What is the (a) $[H^+]$, (b) $[OH^-]$, (c) pH, and (d) pOH of this solution?

8 (15 points). The fraction of the weak base MOH that undergoes ionization is 1.3 % in a 0.025 M solution. What is K_b for the weak base?

9 (15 points). What is the pH of a 0.010 M solution of formic acid, HCOOH, that also contains 0.010 M HCOONa? $K_a = 1.772 \times 10^{-4}$ for HCOOH.

Answers to practice test

1. c
2. (a) $NaHSO_4$, HCl, H_2O_2, H_3O^+, HF; (b) NaOH; (c) $NaHSO_4$ and H_2O, HCl and H_2O, H_2O_2 and H_3O^+, H_3O^+ and NH_4^+, HF and $[H_2F^+]$; (d) NaOH and Na_2SO_4, $CaCO_3$ and CO_2, H_2O and HO_2^-, NH_3 and H_2O, HF and F^-.
3. $Al(OH)_3(s) + 3HCl(aq) \rightarrow AlCl_3(aq) + 3H_2O(l)$, $Al(OH)_3(s) + NaOH(aq) \rightarrow [Al(OH)_4]^- + Na^+$.
4. (a) HI, (b) H_2S, (c) H_2SeO_4, (d) H_2SO_4, (e) $H_2PO_4^-$.
5. HNO_3 and NO_3^-, H_2O and H_3O^+; H_2O; HNO_3.
6. (a) iv, (b) iv, (c) iv, (d) iv, (e) iv; (f) iii, (g) ii, (h) i, (i) ii, (j) iii, (k) i, (l) iii, (m) iv, (n) ii, (o) i.
7. (a) 0.010 mol/L, (b) 1.0×10^{-12} mol/L, (c) 2.00, (d) 12.00.
8. 4.3×10^{-6}.
9. 3.74.

ADDITIONAL NOTES

CHAPTER 21

IONS AND IONIC EQUILIBRIA: ACIDS AND BASES

CHAPTER OVERVIEW

This chapter shows the application of the theory of equilibria to acids, bases, and ions that react with water. Included are several topics of practical importance--hydrolysis of ions and salts, the common ion effect, the maintenance of constant pH by buffer solutions, polyprotic acids, and the titration of solutions of acids and bases.

COMPETENCIES

Definitions for the following terms should be learned:

- buffer solution
- common ion
- common ion effect
- endpoint
- equivalence point
- hydrolysis of an ion
- indicator (acid-base)
- titration
- titration curve (acid-base)

General concepts and skills that should be learned include

- the effect on the pH of a solution of the reaction of ions with water
- the common ion effect and buffer action
- describing the equilibrium in solutions of polyprotic acids
- preparing and interpreting titration curves for acid-base reactions

Numerical exercises that should be understood include

- calculating the equilibrium constant for the hydrolysis of an ion or salt from the K_a of the conjugate acid or the K_b of the conjugate base
- calculating the pH of a solution of a salt
- calculating the equilibrium concentrations in a solution containing ions common to the equilibrium

- calculating the pH of a buffer solution before and after small amounts of acid or base are added
- calculating the concentrations of the various species and pH of solutions of polyprotic acids
- using concentrations and volumes of acids and bases to calculate various stoichiometric quantities (volume, concentration, mass, pH, etc.)
- calculating titration curves for acid-base reactions

QUESTIONS

1. Complete the following sentences by filling in the various blanks with these terms: acid salts, endpoint, equivalence point, hydrolysis of an ion, indicator, and titration.

   Some salts, known as (a) ________, contain H^+ in addition to the normal cation and dissolve in water to give a solution having $pH < 7$. Even solutions of "normal salts" can result in a solution having a pH different from that of pure water because of the (b) ________. These effects are important in predicting the pH at the (c) ________ during an acid-base (d) ________. In many titrations, the (e) ________ is signaled by the color change of an (f) ________.

2. Which statements are true? Rewrite each false statement so that it is correct.
   (a) Most small, highly charged cations react with water to form acidic solutions.
   (b) A solution of a salt containing a cation that is a stronger acid than water and an anion that is a stronger base than water will always be neutral.
   (c) The H_3O^+ ion from the first ionization step of a polyprotic acid serves as a common ion to reduce the second ionization.
   (d) A buffer solution is one that resists changes in pH when small amounts of acid, base, or water are added and that illustrates the common ion effect.
   (e) A polyprotic acid will have more than one equivalence point.

3. Distinguish clearly between (a) hydrolysis and hydration and (b) the endpoint and equivalence point.

4. Choose the response that best completes each statement.

(a) A solution that maintains nearly a constant value of pH as small amounts of acid or base are added is

(i) an acid salt (ii) a buffer solution
(iii) a standard solution (iv) amphoteric

(b) A solution for which the exact concentration is known is

(i) standardized (ii) buffered
(iii) ideal (iv) titrated

(c) The equilibrium constant for the reaction of an ion with water is the

(i) hydrolysis constant (ii) ion product
(iii) solubility product (iv) water ionization constant

(d) The displacement of an ionic equilibrium by the addition of more of one of the ions involved is

(i) the hydrolysis of a salt (ii) an ion product
(iii) precipitation (iv) the common ion effect

(e) The pH of a solution of NaCN would be expected to be

(i) acidic (ii) alkaline
(iii) neutral (iv) slightly different from 7

(f) The pH of a solution of NH_4NO_3 would be expected to be

(i) acidic (ii) alkaline
(iii) neutral (iv) slightly different from 7

(g) The pH of a solution of $NaNO_3$ would be expected to be

(i) acidic (ii) alkaline
(iii) neutral (iv) slightly different from 7

(h) The pH of a solution of NH_4CN would be expected to be

(i) acidic (ii) alkaline
(iii) neutral (iv) slightly different from 7

(i) The equilibrium constant for any anion that reacts with water to form an alkaline solution is

(i) $K_aK_b = K_{sp}$ (ii) $K = K_w/K_aK_b$
(iii) $K_b = K_w/K_a$ (iv) $K_a = K_w/K_b$

(j) The equilibrium constant for any cation that reacts with water to form an acidic solution is

(i) $K_aK_b = K_{sp}$ (ii) $K = K_w/K_aK_b$
(iii) $K_b = K_w/K_a$ (iv) $K_a = K_w/K_b$

(k) Which statement is true about an acid-base titration at the equivalence point?

(i) pH = 7 (ii) pH = pK_a

(iii) pH = $pK_a \pm 1$ (iv) none of these answers

(l) Which of the following would be considered to be a salt that reacts with water to form an acidic solution?

(i) NH_4CN (ii) $NaHSO_4$

(iii) BiOCl (iv) $Na[Al(OH)_4]$

(m) Which of the following would be considered to be a salt that reacts with water to form an alkaline solution?

(i) $Pb(OH)(NO_3)$ (ii) KH_2PO_4

(iii) $NaCH_3COO$ (iv) all of these answers

(n) An organic acid or base that contains a group that reacts with H^+ or OH^- so that the color of the compound changes is a/an

(i) alkali (ii) water-ion base

(iii) amphoteric substance (iv) indicator

5. What happens to all ions as they dissolve in water? What is the intermolecular force that is primarily responsible for this effect?

6. Which of these solutions would be (a) acidic, (b) alkaline, or (c) nearly neutral: (i) an aqueous solution of a salt containing a cation that is a weaker base than water; (ii) an aqueous solution of a salt containing a cation that is a weaker acid than water and an anion that is a stronger base than water; (iii) an aqueous solution of a salt containing a cation that is a stronger acid than water and an anion that is a stronger base than water; and (iv) an aqueous solution of a salt containing a cation that is a stronger acid than water and an anion that is a weaker base than water?

7. Write the chemical equation for the hydrolysis of the anion A^-. The anion is a stronger base than water. Write the equilibrium constant expression for this equation. What is the relationship between this equilibrium constant and K_a for the weak conjugate acid?

8. Write the chemical equation for the hydrolysis of the cation BH^+. The cation is a stronger acid than water. Write the equilibrium constant expression for this equation. What is the relationship between this equilibrium constant and K_b for the weak conjugate base?

9. Write the chemical equation for the reaction with water of a salt containing a cation BH^+ that is a stronger acid than water and an anion A^- that is a stronger base than water. Write the equilibrium constant expression for this equation. What is the relationship among this equilibrium constant, K_a for the weak conjugate acid, and K_b for the weak conjugate base?

10. Briefly describe what is meant by the "common ion effect." How is this effect related to Le Chatelier's principle?

11. Consider the reaction of cyanide ion with water:

$$CN^- + H_2O(l) \rightleftarrows HCN(aq) + OH^-$$

Give some examples of substances that you could add that would illustrate the common ion effect. How would each of these substances affect the equilibrium?

12. What is a solution in which the common ion effect is used to control pH called? What ratio is important in determining the pH of this solution?

13. What is the general guideline in choosing a substance to prepare a buffer?

14. What is a polyprotic acid? Give an example.

15. Write chemical equations for the stepwise ionization of phosphoric acid. Which anion(s) of phosphoric acid will react with water to change the pH? Compare the base strengths of the various anions.

16. In a 0.01 M solution of H_2SO_4, the concentration of SO_4^{2-} is less than that found in a 0.01 M $NaHSO_4$ solution. Why?

17. Write the chemical equation and corresponding equilibrium constant expression for the overall ionization of hydrosulfuric acid. Under what conditions can this overall expression for K_a be used correctly in calculations?

18. What are the usual products of a reaction between a water-ion acid and a water-ion base?

19. Will a solution always be neutral when stoichiometric amounts of acid and base have been mixed? Explain your answer.

20. Briefly describe how an acid is titrated with a base. Name the pieces of apparatus needed. How could the endpoint be determined? How may the endpoint differ from the equivalence point?

21. Briefly describe why a weak acid should be titrated against a strong base rather than a weak base.

22. What is an acid-base indicator? Write the chemical equation showing the equilibrium involved for the indicator "HInd." Why do we limit the amount of indicator added to a titration solution?

23. Why is it necessary to have a supply of several indicators on hand in a general chemistry laboratory? Briefly describe how an indicator for a particular acid-base titration is chosen.

24. Why do most indicators have a useful range of about 2 pH units?

25. What is meant by the term "titration curve"? Briefly describe how laboratory data are collected to prepare a titration curve.

26. Make a rough sketch of the titration curve expected for the titration of a strong base with a strong acid. What determines the pH of the solution at the following points: (a) no titrant added, (b) half-equivalence point, (c) equivalence point, and (d) excess titrant?

27. Repeat Question 26 for the titration of a weak base with a strong acid.

Answers to questions

1. (a) acid salts, (b) hydrolysis of an ion, (c) equivalence point, (d) titration, (e) end point, (f) indicator.
2. (a) T; (b) F, change *always be* to *be nearly*; (c) T; (d) T; (e) T.
3. (a) *Hydrolysis* is the reaction of water with an ion to give H^+ or OH^- and hydroylsis products; *hydration* is the association of water molecules with an ion by intermolecular forces. (b) The *endpoint* is the point at which a chemical reaction such as a titration is stopped; the *equivalence point* refers to the point at which stoichiometric amounts of reactants have been added.
4. (a) ii, (b) i, (c) i, (d) iv, (e) ii, (f) i, (g) iii, (h) iv, (i) iii, (j) iv, (k) iv, (l) ii, (m) iv, (n) iv.
5. Hydration occurs; ion-dipole interactions.
6. (a) iv; (b) ii; (c) i, iii.
7. $A^- + H_2O(l) \rightleftharpoons HA(aq) + OH^-$, $K_b = [HA][OH^-]/[A^-] = K_w/K_a$.
8. $BH^+ + H_2O(l) \rightleftharpoons B(aq) + H_3O^+$, $K_a = [H^+][B]/[BH^+] = K_w/K_b$.
9. $BH^+ + A^- \rightleftharpoons B(aq) + HA(aq)$, $K = [HA][B]/[BH^+][A^-] = K_w/K_aK_b$.

10. The common ion effect is the displacement of an ionic equilibrium by an excess of one or more of the ions involved. This is a special application of Le Chatelier's principle to ionic equilibria.
11. Strong bases such as NaOH or KOH, or cyanide salts such as NaCN or KCN, could be added to illustrate the common ion effect. Strong bases would increase the $[OH^-]$, increasing the amount of reverse reaction; cyanide salts would increase the $[CN^-]$, increasing the amount of forward reaction.
12. The solution is a buffer solution. The important ratio is [anion]/[nonionized acid] or [cation]/[nonionized base].
13. Choose a weak acid or base with pK_a or pK_b within ±1 pH unit of the desired pH.
14. A polyprotic acid contains more than one replaceable H^+. Examples include H_2SO_4 and H_3PO_4.
15. $H_3PO_4(aq) + H_2O(l) \rightleftharpoons H_2PO_4^- + H_3O^+$
    $H_2PO_4^- + H_2O(l) \rightleftharpoons HPO_4^{2-} + H_3O^+$
    $HPO_4^{2-} + H_2O(l) \rightleftharpoons PO_4^{3-} + H_3O^+$
    All anions of phosphoric acid will react with water to change the pH because H_3PO_4, $H_2PO_4^-$, and HPO_4^{2-} are weak acids. The order of increasing base strengths of the anions is $H_2PO_4^- < HPO_4^{2-} < PO_4^{3-}$.
16. The H^+ produced from the first ionization of H_2SO_4 represses the second ionization step because of the common ion effect.
17. $H_2S(aq) + 2H_2O(l) \rightleftharpoons 2H_3O^+ + S^{2-}$, $K_a = [H^+]^2[S^{2-}]/[H_2S]$; the equation is accurate when $[H^+]$ is supplied from additional source.
18. The usual products are a salt and water.
19. A solution will not always be neutral when stoichiometric amounts of acid and base have been mixed because one or both of the ions of the salt that is formed may react with water to change the pH.
20. A known mass or volume of acid is placed in flask or beaker, the base is delivered from buret, and the reaction mixture is constantly stirred. The endpoint is determined either by using an indicator that changes color in the correct pH range or from titration curve data. These points may not be exactly the same, giving a systematic error.
21. The acid shows a buffer effect before the equivalence point and a weak base would show a buffer effect after the equivalence point, which would cause the titration curve to have very little vertical nature.

22. An acid-base indicator is an organic acid or base that has in its structure a group that reacts with hydrogen ion or hydroxide ion so that the color of the compound changes.

$$HInd(aq) + H_2O(l) \rightleftharpoons Ind^- + H_3O^+$$

The amount of indicator added to a titration solution is limited so only a very small amount of acid or base is needed to neutralize it.

23. All acid-base titrations have unique values of pH at the equivalence point. The choice depends on the indicator's changing color at a pH as near to the equivalence point as possible.

24. The human eye requires about a ±10-fold ratio of color to detect changes, which corresponds to a ±10-fold ratio of species in the equilibrium constant expression for an indicator. This ratio corresponds to ±1 pH unit.

25. A titration curve in the a plot of pH against the volume of titrant added for an acid-base reaction. The pH is measured after addition of various amounts of titrant--using smaller increments of volume near the equivalence point.

26.

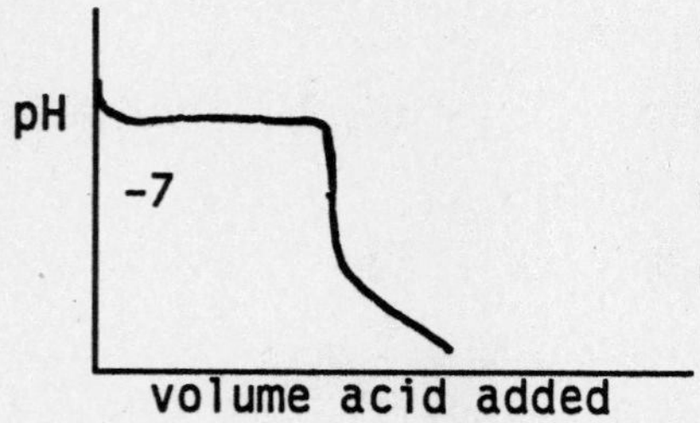

(a) concentration of base
(b) concentration of unreacted base
(c) pH = 7.00
(d) concentration of excess acid

27.

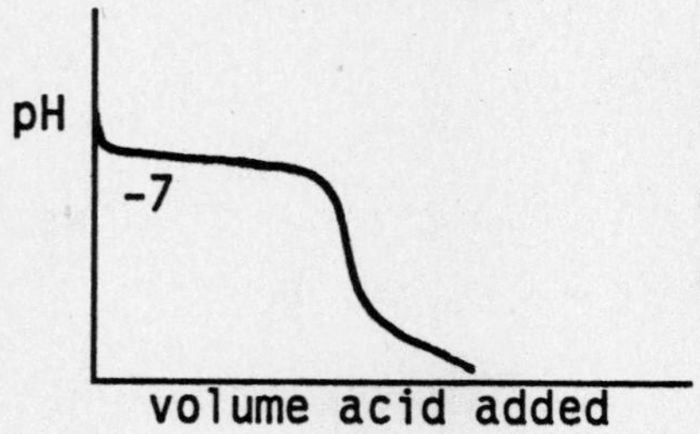

(a) K_b and concentration of base
(b) K_b, concentrations of base and anion (buffer)
(c) K_b, reaction of cation with water (hydrolysis)
(d) concentration of excess acid

NUMERICAL EXERCISES

Example 21.1 Reactions of ions with water

Find the value of the equilibrium constant for the hydrolysis of the NO_2^- ion. $K_a = 7.2 \times 10^{-4}$ for HNO_2.

For the reaction of the nitrite ion with water,

$$NO_2^- + H_2O(l) \rightleftharpoons HNO_2(aq) + OH^-$$

the equilibrium constant expression is

$$K_b = \frac{[HNO_2][OH^-]}{[NO_2^-]}$$

which is related to K_w and K_a for HNO_2 by

$$K_b = \frac{[HNO_2][OH^-]}{[NO_2^-]} \times \frac{[H^+]}{[H^+]} = \frac{[HNO_2]}{[NO_2^-][H^+]}[H^+][OH^-] = \frac{K_w}{K_a}$$

Thus the value of K_b is

$$K_b = \frac{K_w}{K_a} = \frac{(1.00 \times 10^{-14})}{(7.2 \times 10^{-4})} = 1.4 \times 10^{-11}$$

The hydrolysis constant of NO_2^- is 1.4×10^{-11}.

Exercise 21.1 Find the equilibrium constant for the reaction of the NH_4^+ ion with water. $K_b = 1.6 \times 10^{-5}$ for $NH_3(aq)$. *Answer*: 6.3×10^{-10}.

Example 21.2 Reactions of ions with water

What is the pH of a solution of 0.10 M $NaNO_2$? See *Example 21.1* for pertinent information.

The anion in sodium nitrite undergoes hydrolysis for which $K_b = 1.4 \times 10^{-11}$. Preparing the usual table with $x = [OH^-]$ gives

	NO_2^- + $H_2O(l)$	⇄ $HNO_2(aq)$	+ OH^-
initial	0.10	0	0
change	$-x$	$+x$	$+x$
equilibrium	$0.10 - x$	x	x

which upon substitution into the equilibrium expression gives

$$K_b = \frac{[HNO_2][OH^-]}{[NO_2^-]} = \frac{(x)(x)}{(0.10 - x)} = 1.4 \times 10^{-11}$$

Assuming $(0.10 - x) \approx 0.10$ simplifies the equation to

$$\frac{(x)(x)}{(0.10)} = 1.4 \times 10^{-11}$$

$$x = 1.2 \times 10^{-6} \text{ mol/L}$$

The approximation $(0.10 - x) \approx 0.10$ is valid, so

$$[H^+] = \frac{K_w}{[OH^-]} = \frac{1.00 \times 10^{-14}}{1.2 \times 10^{-6}} = 8.3 \times 10^{-9} \text{ mol/L}$$

$$pH = -\log [H^+] = -\log(8.3 \times 10^{-9}) = 8.08$$

The pH of the solution will be 8.08.

Exercise 21.2 What is the pH of a solution of 0.10 M NH_4Br? See Exercise 21.1 for pertinent information. *Answer*: 5.10.

Example 21.3 Common ion effect: buffers

What is the pH of a solution containing 0.100 M HBrO and 0.100 M NaBrO? $K_a = 2.2 \times 10^{-9}$ for HBrO.

Preparing the usual table with $x = [H^+]$ gives

	$HBrO(aq) + H_2O(l)$	$\rightleftharpoons$	H_3O^+	+ BrO^-
initial	0.100		0	0.100
change	$-x$		$+x$	$+x$
equilibrium	$0.100 - x$		x	$0.100 + x$

Substituting the equilibrium concentrations into the expression for K_a gives

$$K_a = \frac{[H^+][BrO^-]}{[HBrO]} = \frac{x(0.100 + x)}{(0.100 - x)} = 2.2 \times 10^{-9}$$

Assuming that x is negligible compared to 0.100, the solution to the equation is $x = 2.2 \times 10^{-9}$ mol/L. The approximation is valid, so the pH of the solution is

$$pH = -\log [H^+] = -\log(2.2 \times 10^{-9}) = 8.66$$

The pH of the solution is 8.66.

Exercise 21.3 What is the (a) $[OH^-]$ and (b) pH of a solution containing 0.015 M $NH_3(aq)$ and 0.0050 M NH_4Cl? $K_b = 1.6 \times 10^{-5}$ for $NH_3(aq)$. *Answers*: (a) 4.8×10^{-5} mol/L, (b) 9.68.

Example 21.4 Common ion effect: buffers*

What is the pH of a buffer solution prepared by mixing 0.100 mol of $NaCH_3COO$, 0.200 mol of acetic acid, and sufficient water to make 100.0 mL of solution? $K_a = 1.754 \times 10^{-5}$ for acetic acid.

This problem is solved by using the comprehensive problem-solving approach presented in *Skill 2.7 Solving problems*.

1. Study the problem and be sure you understand it.
   (a) What is unknown?
   The pH of the solution.
   (b) What is known?
   The number of moles of $NaCH_3COO$ and CH_3COOH, the value of K_a for CH_3COOH, and the volume of the solution.

2. Decide how to solve the problem.
   (a) What is the connection between the known and the unknown?
   The pH is related to the $[H^+]$ at equilibrium. The value of $[H^+]$ can be calculated by using the usual equilibrium calculation method or by using the Henderson-Hasselbach equation

$$pH = pK_a - \log \frac{[CH_3COOH]}{[CH_3COO^-]}$$

   [This equation is valid because K_a is considerably less than the concentrations of the acetate ion and nonionized acetic acid.]
   (b) What is necessary to make the connection?
   Calculate the pK_a, $[CH_3COOH]$, and $[CH_3COO^-]$ from the data provided and calculate the pH.
3. Set up the problem and solve it.

$$pK_a = -\log K_a = -\log(1.754 \times 10^{-5}) = 4.7560$$

$$[CH_3COOH] = \frac{0.200 \text{ mol } CH_3COOH}{(100.0 \text{ mL})(1 \text{ L}/1000 \text{ mL})} = 2.00 \text{ mol } CH_3COOH/L$$

$$[CH_3COO^-] = \frac{0.100 \text{ mol } CH_3COO^-}{(100.0 \text{ mL})(1 \text{ L}/1000 \text{ mL})} = 1.00 \text{ mol } CH_3COO^-/L$$

   The pH of the solution is

$$pH = (4.7560) - \log\left[\frac{2.00}{1.00}\right] = 4.455$$

4. Check the result.
   (a) Are significant figures and the location of the decimal point correct?
   Yes.
   (b) Did the answer come out in the correct units?
   Yes.
   (c) Is the answer reasonable?
   Yes. Most buffer solutions are designed to function between $pK_a \pm 1$.

Exercise 21.4 Repeat the calculations of *Example 21.4* assuming that formic acid, HCOOH, and NaHCOO are used instead of acetic acid and $NaCH_3COO$. K_a = 1.772×10^{-4} for formic acid. *Answer*: 3.451.

Example 21.5 Common ion effect: buffers

What would be the pH of the buffer solution described in *Example 21.3* if 0.005 mol of NaOH were added to 1.00 L of the buffer?

The base that is added reacts with the HBrO so that

$$[HBrO] = 0.100 - 0.005 = 0.095 \text{ mol/L}$$
$$[BrO^-] = 0.100 + 0.005 = 0.105 \text{ mol/L}$$

Thus the table of concentrations with $x = [H^+]$ becomes

	$HBrO(aq) + H_2O(l)$ ⇌	H_3O^+	+ BrO^-
initial	0.095	0	0.105
change	$-x$	$+x$	$+x$
equilibrium	$0.095 - x$	x	$0.105 + x$

Substituting the equilibrium concentrations into the expression for K_a gives

$$K_a = \frac{[H^+][BrO^-]}{[HBrO]} = \frac{x(0.105 + x)}{(0.095 - x)} = 2.2 \times 10^{-9}$$

Assuming that x is negligible compared to 0.095 and 0.105, the solution to the equation is $x = 2.0 \times 10^{-9}$ mol/L. The assumption is valid, so

$$pH = -\log [H^+] = -\log(2.0 \times 10^{-9}) = 8.70$$

The pH of the buffer has changed from 8.66 to 8.70.

Exercise 21.5 What is the pH of the buffer solution described in Exercise 21.3 after 0.0010 mol of HCl is added to 1.00 L of the buffer? *Answer*: 9.57.

Example 21.6 Polyprotic acids

Calculate $[H_2C_2O_4]$, $[HC_2O_4^-]$, $[H^+]$, and $[C_2O_4^{2-}]$ in a 1.50 M aqueous solution of oxalic acid. $K_{a1} = 5.6 \times 10^{-2}$ and $K_{a2} = 6.2 \times 10^{-5}$.

For the first ionization step

$$H_2C_2O_4(aq) + H_2O(l) \rightleftharpoons H_3O^+ + HC_2O_4^- \qquad K_{a1} = \frac{[H^+][HC_2O_4^-]}{[H_2C_2O_4]} = 5.6 \times 10^{-2}$$

Assuming the second ionization to produce negligible $[H^+]$ and to remove negligible $[HC_2O_4^-]$, the table for this step with $x = [H_2C_2O_4]$ that reacts is

	$H_2C_2O_4(aq) + H_2O(l)$ →	H_3O^+ +	$HC_2O_4^-$
initial	1.50	0	0
change	$-x$	$+x$	$+x$
equilibrium	$1.50 - x$	x	x

$$K_{a1} = \frac{(x)(x)}{(1.50 - x)} = 5.6 \times 10^{-2}$$

Because K_a is not considerably less than $[H_2C_2O_4]$, the usual approximations to simplify the calculations cannot be made.

$$x^2 = (5.6 \times 10^{-2})(1.50 - x) = 0.084 - (0.056)x$$

$$x^2 + (0.056)x - 0.084 = 0$$

$$x = \frac{-(0.056) \pm \sqrt{(0.056)^2 - (4)(1)(-0.084)}}{(2)(1)} = 0.26 \text{ mol/L}$$

At equilibrium,

$$[H_2C_2O_4] = 1.50 - 0.26 = 1.24 \text{ mol/L}$$

For the second ionization step

$$HC_2O_4^- + H_2O(l) \rightleftharpoons H_3O^+ + C_2O_4^{2-} \qquad K_{a2} = \frac{[H^+][C_2O_4^{2-}]}{[HC_2O_4^-]} = 6.2 \times 10^{-5}$$

The table describing this step where $x = [C_2O_4^{2-}]$ is

	$HC_2O_4^- + H_2O(l)$	$\rightarrow$ H_3O^+	+ $C_2O_4^{2-}$
initial	0.26	0.26	0
change	$-x$	$+x$	$+x$
equilibrium	$0.26 - x$	$0.26 + x$	x

$$K_{a2} = \frac{(0.026 + x)(x)}{(0.26 - x)} = 6.2 \times 10^{-5}$$

Assuming x to be small compared to 0.26, the solution to the equation is $x = 6.2 \times 10^{-5}$ mol/L. The approximation is valid and

$$[C_2O_4^{2-}] = 6.2 \times 10^{-5} \text{ mol/L}$$

The small value for $[C_2O_4^{2-}]$ indicates that the assumption that negligible $[H^+]$ is produced and negligible $[HC_2O_4^-]$ ionizes during the second step is valid. Thus $[HC_2O_4^-] = [H^+] = 0.26$ mol/L.

Exercise 21.6 Calculate $[H_2CO_3]$, $[HCO_3^-]$, $[CO_3^{2-}]$, and $[H^+]$ in a 0.0107 M solution of carbonic acid. $K_{a1} = 4.5 \times 10^{-7}$ and $K_{a2} = 4.8 \times 10^{-11}$. *Answers*: $[H_2CO_3] = 0.0107$ mol/L, $[H^+] = [HCO_3^-] = 6.9 \times 10^{-5}$ mol/L, $[CO_3^{2-}] = 4.8 \times 10^{-11}$ mol/L.

Example 21.7 Acid-base reactions

A 0.253 g sample of a mixture of Na_2CO_3 and Na_2SO_4 was dissolved in water and titrated with 0.1003 M HCl. If 12.36 mL of acid was required to reach the equivalence point, what is the percentage of Na_2CO_3 in the sample?

The acid-base reaction is given by the equation

$$Na_2CO_3(aq) + 2HCl(aq) \rightarrow 2NaCl(aq) + H_2O(l) + CO_2(g)$$

The number of moles of acid involved in the titration was

$$\left[\frac{0.1003 \text{ mol HCl}}{1 \text{ L}}\right](12.3 \text{ mL})\left[\frac{1 \text{ L}}{1000 \text{ mL}}\right] = 1.240 \times 10^{-3} \text{ mol HCl}$$

The number of moles of Na_2CO_3 is

$$(1.240 \times 10^{-3} \text{ mol HCl})\left[\frac{1 \text{ mol } Na_2CO_3}{2 \text{ mol HCl}}\right] = 6.200 \times 10^{-4} \text{ mol } Na_2CO_3$$

which is equivalent to

$$(6.200 \times 10^{-4} \text{ mol } Na_2CO_3)\left[\frac{105.99 \text{ g } Na_2CO_3}{1 \text{ mol } Na_2CO_3}\right] = 0.06571 \text{ g } Na_2CO_3$$

The fraction of Na_2CO_3 in the sample is

$$\frac{0.06571 \text{ g}}{0.253 \text{ g}} = 0.260$$

The mixture contained 26.0 mass % Na_2CO_3.

Exercise 21.7 A 0.1142 g sample of sulfamic acid, a monoprotic acid having the formula HSO_3NH_2, was used to standardize a solution of KOH. If 22.79 mL of the base were required to reach the equivalence point, what is the concentration of the base? *Answer*: 0.05161 mol/L.

Example 21.8 Acid-base reactions

Exactly 100 mL of a solution containing 0.0135 mol of perchloric acid, $HClO_4$, was titrated with 0.1039 M KOH. Calculate the pH of the solution (a) before the addition of any KOH and after the addition of (b) 5.0 mL, (c) 50.0 mL, (d) 100.0 mL, (e) 130.0 mL, and (f) 150.0 mL of base.

The reaction between the acid and base can be represented by

$$HClO_4(aq) + KOH(aq) \rightarrow H_2O(l) + KClO_4(aq)$$

(a) Perchloric acid is a strong acid, thus originally it is completely ionized. The concentration of H^+ initially present is

$$[H^+] = \frac{0.0135 \text{ mol}}{0.0100 \text{ L}} = 0.135 \text{ mol/L}$$

$$pH = -\log [H^+] = -\log(0.135) = 0.870$$

(b) The amount of $HClO_4$ left in solution before the equivalence point is

$$n_{HClO_4} = n_{HCl_4\text{,originally}} - n_{HClO_4\text{,reacted}}$$
$$= (0.0135 \text{ mol}) - n_{KOH\text{,added}}$$
$$= (0.0135 \text{ mol}) - \left[\frac{0.1039 \text{ mol}}{1 \text{ L}}\right](V_{KOH} \text{ in L})$$

giving the $[H^+]$ as

$$[H^+] = \frac{n_{HClO_4}}{(0.100) + (V_{KOH} \text{ in L})}$$

$$= \frac{(0.0135 \text{ mol}) - \left[\frac{0.1039 \text{ mol}}{1 \text{ L}}\right](V_{KOH} \text{ in L})}{(0.1000) + (V_{KOH} \text{ in L})}$$

Thus

$$[H^+] = \frac{(0.0135 \text{ mol}) - (0.139 \text{ mol/1 L})(0.0050 \text{ L})}{(0.1000 + 0.0050) \text{ L}} = 0.124 \text{ mol/L}$$

$$pH = -\log(0.124) = 0.907$$

(c) $$[H^+] = \frac{(0.0135) - (0.1039)(0.0500)}{(0.1000 + 0.0500)} = 0.055 \text{ mol/L}$$

$$pH = -\log(0.055) = 1.26$$

(d) $$[H^+] = \frac{(0.0135) - (0.1039)(0.1000)}{(0.1000 + 0.1000)} = 0.016 \text{ mol/L}$$

$$pH = -\log(0.016) = 1.80$$

(e) At the equivalence point where

n_{HClO_4} = (0.0135 mol) - (0.139 mol/1 L)(0.1300 L) = 0.0000 mol

the titration mixture is essentially a 200 mL aqueous solution of 0.0135 mol $KClO_4$. Because neither ion undergoes significant hydrolysis, the pH is 7.00.

(f) The amount of excess base is

$$n_{KOH} = n_{KOH,added} - n_{KOH,reacted}$$

$$= \left[\frac{0.1039 \text{ mol}}{1 \text{ L}}\right](V_{KOH} \text{ in L}) - n_{HClO_4,originally}$$

$$= \left[\frac{0.1039 \text{ mol}}{1 \text{ L}}\right](0.1500 \text{ L}) - (0.0135 \text{ mol})$$

$$= 0.0021 \text{ mol}$$

Because KOH is a strong base

$$[OH^-] = \left[\frac{0.0021 \text{ mol}}{100.0 \text{ mL} + 150.0 \text{ mL}}\right]\left[\frac{1000 \text{ mL}}{1 \text{ L}}\right] = 0.0084 \text{ mol/L}$$

the pOH is

$$pOH = -\log [OH^-] = -\log(0.0084) = 2.08$$

giving

$$pH = 14.00 - pOH = 14.00 - 7.08 = 11.92$$

Exercise 21.8 Exactly 10 mL of a solution containing 0.00361 mol of sodium hydroxide, NaOH, was titrated with 0.0963 M HCl. Calculate the pH of the solution (a) before the addition of any HCl and after the addition of (b) 10.0 mL, (c) 20.0 mL, (d) 30.0 mL, (e) 37.5 mL, and (f) 40.0 mL of acid. *Answers*: (a) 13.56, (b) 13.12, (c) 12.75, (d) 12.26, (e) 7.00, (f) 2.32.

Example 21.9 Acid-base reactions

Exactly 25 mL of a solution containing 0.0123 mol of NH_3 was titrated with 0.1077 M HCl. Calculate the pH of the solution (a) before the addition of any HCl and after the addition of (b) 25.0 mL, (c) 50.0 mL, (d) 75.0 mL, (e) 100.0 mL, (f) 114.2 mL, and (g) 125.0 mL of acid. $K_b = 1.6 \times 10^{-5}$ for NH_3.

The reaction between the acid and base is

$$NH_3(aq) + HCl(aq) \rightarrow NH_4^+ + Cl^-$$

(a) Ammonia is a weak base:

$$NH_3(aq) + H_2O(l) \rightleftharpoons NH_4^+ + OH^- \qquad K_b = \frac{[NH_4^+][OH^-]}{[NH_3]} = 1.6 \times 10^{-5}$$

The initial $[NH_3]$ is

$$[NH_3] = \frac{0.0123 \text{ mol}}{0.0250 \text{ L}} = 0.492 \text{ mol/L}$$

Preparing the usual table for the ionization of NH_3 with $x = [OH^-]$ gives

	$NH_3(aq)$ + $H_2O(l)$ →	NH_4^+ +	OH^-
initial	0.492	0	0
change	$-x$	$+x$	$+x$
equilibrium	$0.492 - x$	x	x

Substituting the expressions for the final equilibrium concentrations into the equilibrium expression and solving gives

$$K_b = \frac{(x)(x)}{(0.492 - x)} = 1.6 \times 10^{-5}$$

$$x = [OH^-] = 2.8 \times 10^{-3} \text{ mol/L}$$

$$pOH = -\log [OH^-] = -\log(2.8 \times 10^{-3}) = 2.55$$

$$pH = 14.00 - pOH = 14.00 - 2.55 = 11.45$$

(b) The amount of unreacted NH_3 left in solution before the equivalence point is

$$n_{NH_3} = n_{NH_3,\text{originally}} - n_{NH_3,\text{reacted}}$$

$$= (0.0123 \text{ mol}) - (n_{HCl,\text{added}})$$

$$= (0.0123 \text{ mol}) - (0.1077 \text{ mol/1 L})(V_{HCl} \text{ in L})$$

The amount of NH_4^+ produced is

$$n_{NH_4^+} = n_{NH_3,\text{reacted}} = n_{HCl,\text{added}} = (0.1077\ \text{mol/1 L})(V_{HCl}\ \text{in L})$$

Thus the concentrations of NH_3 and NH_4^+ are

$$[NH_3] = \frac{n_{NH_3}}{(0.0250\ \text{L}) + (V_{HCl}\ \text{in L})} - x$$

$$= \frac{(0.0123\ \text{mol}) - (0.1077\ \text{mol/1 L})(V_{HCl}\ \text{in L})}{(0.0250\ \text{L}) + (V_{HCl}\ \text{in L})} - x$$

$$[NH_4^+] = \frac{n_{NH_4^+}}{(0.0250\ \text{L}) + (V_{HCl}\ \text{in L})} + x = \frac{(0.1077\ \text{mol/1 L})(V_{HCl}\ \text{in L})}{(0.0250\ \text{L}) + (V_{HCl}\ \text{in L})} + x$$

where $x = [OH^-]$ produced by the hydrolysis of the unreacted NH_3. Using these expressions for $[NH_3]$ and $[NH_4^+]$, the value of x can be found from the equilibrium expression.

$$[NH_3] = \frac{(0.0123\ \text{mol}) - (0.1077\ \text{mol/1 L})(0.0250\ \text{L})}{(0.0250 + 0.0250)\ \text{L}} - x = (0.192 - x)\ \text{mol/L}$$

$$[NH_4^+] = \frac{(0.1077\ \text{mol/1 L})(0.0250\ \text{L})}{(0.0250 + 0.0250)\ \text{L}} + x = (0.05239 + x)\ \text{mol/L}$$

$$K_b = \frac{(x)(0.0539 + x)}{(0.192 - x)} = 1.6 \times 10^{-5}$$

$$x = [OH^-] = 5.7 \times 10^{-5}\ \text{mol/L}$$

$$\text{pOH} = -\log(5.7 \times 10^{-5}) = 4.24$$

$$\text{pH} = 14.00 - 4.24 = 9.76$$

Likewise, (c) $[NH_3] + (0.092 - x)$ mol/L, $[NH_4^+] = (0.072 + x)$ mol/L, $[OH^-] = 2.0 \times 10^{-5}$ mol/L, and pH = 9.31; (d) $[NH_3] = (0.042 - x)$ mol/L, $[NH_4^+] = (0.081 + x)$ mol/L, $[OH^-] = 8.3 \times 10^{-6}$ mol/L, and pH = 8.92; and (e) $[NH_3] = (0.012 - x)$ mol/L, $[NH_4^+] = (0.086 + x)$ mol/L, $[OH^-] = 2.2 \times 10^{-6}$ mol/L, and pOH = 8.35.

(f) At the equivalence point, the pH is determined by the hydrolysis of NH_4^+:

$$NH_4^+ + H_2O(l) \rightarrow NH_3(aq) + H_3O^+$$

with

$$K_a = \frac{[NH_3][H_3O^+]}{[NH_4^+]} = \frac{K_w}{K_b} = \frac{1.00 \times 10^{-14}}{1.6 \times 10^{-5}} = 6.3 \times 10^{-10}$$

The concentration of the NH_4Cl is

$$[NH_4Cl] = \frac{(0.0123\ \text{mol})}{(0.0250 + 0.1142)\ \text{L}} = 0.0884\ \text{mol/L}$$

Preparing the usual table for the hydrolysis with $x = [H_3O^+]$ gives

	$NH_4^+ + H_2O(l)$	$\rightarrow$	NH_3	$+ H_3O^+$
initial	0.0884		0	0
change	$+x$		$+x$	$+x$
equilibrium	$0.0884 - x$		x	x

Substituting the equilibrium concentrations into the K_a expression and solving for x gives

$$K_a = \frac{(x)(x)}{(0.0884 - x)} = 6.3 \times 10^{-10}$$

$$x = [H_3O^+] = 7.4 \times 10^{-6} \text{ mol/L}$$

$$pH = -\log(7.4 \times 10^{-6}) = 5.13$$

(g) The amount of excess acid is

$$n_{HCl} = n_{HCl,added} - n_{HCl,reacted}$$
$$= (0.1077 \text{ mol/1 L})(0.125 \text{ L}) - n_{NH_3,originally}$$
$$= 0.0135 \text{ mol} - 0.0123 \text{ mol} = 0.0012 \text{ mol}$$

which corresponds to

$$[H^+] = \frac{(0.0012 \text{ mol})}{(0.025 + 0.125) \text{ L}} = 0.0080 \text{ mol/L}$$

$$pH = -\log(0.0080) = 2.10$$

Exercise 21.9 Exactly 50 mL of a solution containing 0.0333 mol of formic acid, HCOOH, was titrated with 0.250 M NaOH. Calculate the pH of the solution (a) before the addition of base and after the addition of (b) 50.0 mL, (c) 100.0 mL, (d) 133.2 mL, and (e) 150.0 mL of base. $K_a = 1.772 \times 10^{-4}$ for HCOOH. *Answers*: (a) 1.963, (b) 3.530, (c) 4.24, (d) 8.505, (e) 12.32.

PRACTICE TEST

1 (20 points). Describe aqueous solutions of each of the following solutes as (a) acidic, (b) alkaline, or (c) nearly neutral: (i) KF, (ii) $CuCl_2$, (iii) HNO_3, (iv) NaOH, (v) $(NH_4)_2SO_4$, (vi) $NaHSO_4$, (vii) BiOCl, (viii) $(NH_4^+)(CH_3COO^-)$, (ix) HF, and (x) CsCl.

2 (10 points). Write the expression for the equilibrium constant describing the reaction with water (hydrolysis) of a salt containing (a) an anion that is a weaker base than water and a cation that is a stronger acid than water, (b) an anion that is a stronger base than water and a cation that is a weaker acid than water, and (c) an anion that is a stronger base than water and a cation that is a stronger acid than water in terms of K_w, K_a for the conjugate acid of the anion, and/or K_b for the conjugate base of the cation.

3 (30 points). Choose the response that best completes each statement.

(a) The anion of a weak acid reacts with water to form

(i) H_3O^+ ions and undissociated molecules of base

(ii) OH^- ions and undissociated molecules of acid

(iii) a weakly acidic solution

(iv) none of these answers

(b) In a 0.1 M solution the fraction of ionization of phosphoric acid is 0.32 for the first step and 0.0025 for the second step. Which of the following represents the ionization for the third step?

(i) 0.000067 (ii) 0.0055

(iii) 0.0025 (iv) 0.35

(c) A buffer contains 0.20 mol of NaA and 0.20 mol of HA in a liter of solution. What is the pH of the solution? $K_a = 1 \times 10^{-5}$ for HA.

(i) 1.0 (ii) 3.0

(iii) 5.0 (iv) 7.0

(d) Which of the following is true in a 0.01 M solution of sulfuric acid?

(i) $[H^+] = [SO_4^{2-}]$ (ii) $[H_2SO_4] = [HSO_4^-] = [SO_4^{2-}]$

(iii) $[H^+] = 2[SO_4^{3-}]$ (iv) $[HSO_4^-] > [SO_4^{2-}]$

(e) One liter of solution is prepared by adding 8.00 g NaOH and 3.65 g HCl to water. What is the pH of the solution?

(i) 0 (ii) 1.000

(iii) 13.000 (iv) 7.00

(f) How many moles of NaOH are needed to react completely with 0.50 mol H_2SO_4?

(i) 0.25 mol (ii) 0.50 mol

(iii) 1.0 mol (iv) 2.50 mol

(g) The diagram in Fig. 21-1 represents the titration of a

(i) strong acid with a strong base

(ii) weak base with a strong acid

(iii) strong acid with a weak base

(iv) weak acid with a strong base

(h) The region of best buffering action on the titration curve in Fig. 21-1 is between points

(i) 1 and 2 (ii) 4 and 6

(iii) 2 and 4 (iv) 6 and 7

Fig. 21-1

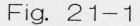

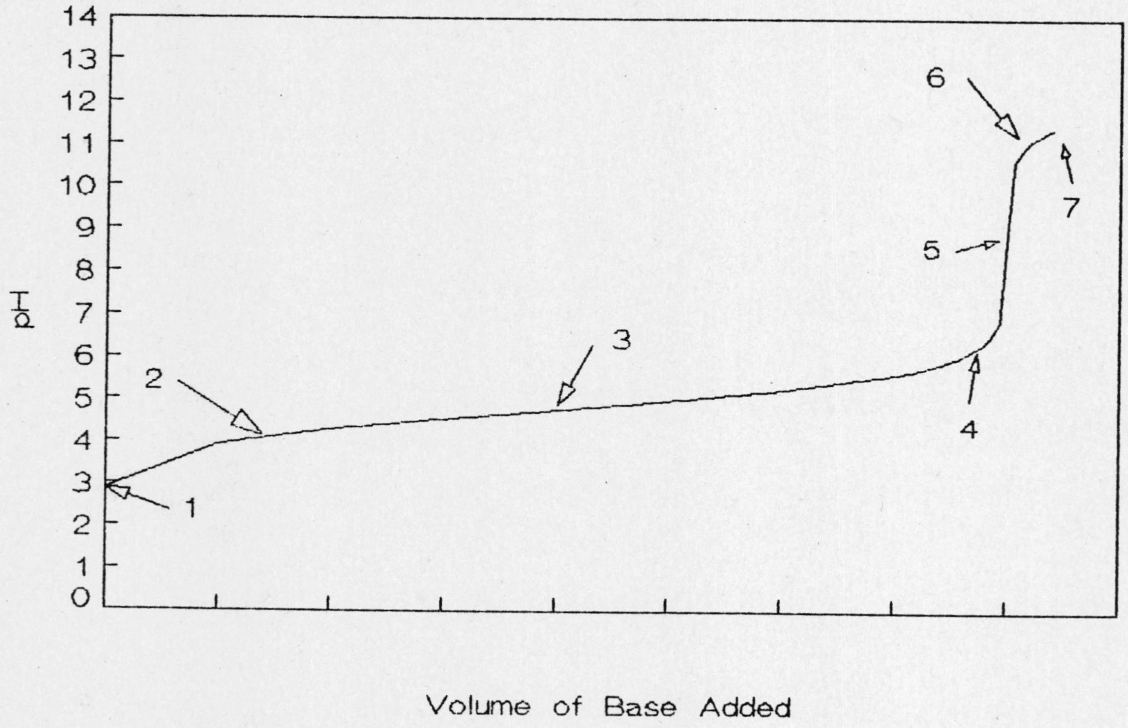

(i) The equivalence point in Fig. 21-1 is point number

(i) 3 (ii) 5

(iii) 1 (iv) 7

(j) Which of the following conditions characterizes the equivalence point of a neutralization titration?

(i) The pH is 7.00.

(ii) The indicator used changes color.

(iii) Equal volumes of acid and base have reacted.

(iv) Equivalent amounts of both acid and base have reacted.

(k) Which of the following indicators (pH intervals given) should be used for determining the end point of the titration of acetic acid by sodium hydroxide? $K_a = 1.754 \times 10^{-5}$ for acetic acid.

(i) thymol blue (1.2-2.8) (ii) methyl red (4.4-6.2)

(iii) phenolphthalein (8.3-10.0) (iv) malachite green (11.4-13.0)

4 (40 points). Calculate the pH of the following solutions: (a) 0.100 M $C_6H_5COOH(aq)$ with $K_a = 6.6 \times 10^{-5}$, (b) 0.100 M $CH_3NH_2(aq)$ with $K_b = 3.9 \times 10^{-4}$, (c) 0.100 M $C_6H_5COONa(aq)$, and (d) 0.100 M $CH_3NH_3Cl(aq)$.

Answers to practice test

1. (a) ii, iii, vi, ix; (b) i, iv, vii; (c) v, viii, x.
2. (a) $K_a = K_w/K_b$, (b) $K_b = K_w/K_a$, (c) $K = K_w/K_aK_b$.
3. (a) ii, (b) i, (c) iii, (d) iv, (e) iii, (f) iii, (g) iv, (h) iii, (i) ii, (j) iv, (k) iii.
4. (a) 2.59, (b) 11.79, (c) 8.59, (d) 5.80.

ADDITIONAL NOTES

CHAPTER 22

IONS AND IONIC EQUILIBRIA: COMPLEX IONS AND IONIC SOLIDS

CHAPTER OVERVIEW

This chapter shows the application of the theory of equilibrium to ions from slightly soluble ionic solutes and to complex ions in aqueous solution. Included are several topics of practical importance--overall dissociation of complexes, solubility of slightly soluble ionic solutes, and control of solubility by application of the principles of equilibria.

COMPETENCIES

Definitions for the following terms should be learned:

- dissociation constants (complex ions)
- formation constants
- inorganic qualitative analysis
- ion product
- molar solubility
- selective precipitation
- solubility product

General concepts and skills that should be learned include

- describing the equilibrium in solutions of complex ions
- describing the heterogeneous equilibria between slightly soluble salts and their saturated aqueous solutions

Numerical exercises that should be understood include

- calculating the various concentrations of species involved in the dissociation equilibria of a complex
- calculating the solubility of a salt or K_{sp} when one of these variables is given
- using the ion product to predict whether or not precipitation will occur in a solution
- using simultaneous equilibria to describe the selective precipitation of substances and the dissolution of ionic precipitates

QUESTIONS

1. Which statements are true? Rewrite each false statement so that it is correct.

(a) Precipitation of an ionic compound will occur if the ion product exceeds the solubility product.

(b) The values of the stepwise dissociation constants are widely different and so the equilibrium calculations can be treated like those for polyprotic acids.

(c) The product of a titration could be a precipitate.

(d) Salts containing anions that are strong conjugate bases of weak acids will be more soluble in acidic aqueous solutions than in water.

(e) For salts having the same general formula (e.g., MX), the solubility product values can be used to compare solubilities.

2. Distinguish clearly between (a) the solubility and solubility product, (b) the ion product and solubility product, (c) the stepwise and overall formation constants for a complex, and (d) the overall dissociation and formation constants for a complex.

3. Choose the response that best completes each statement.

(a) In a solution of $[Ag(NH_3)_2]^+$ complex ion, how many major chemical species are at equilibrium (excluding H^+, H_2O, and OH^-)?

(i) 2 (ii) 3
(iii) 4 (iv) 5

(b) For the $[Cu(NH_3)_4]^{2+}$ complex ion, how could the overall formation constant be calculated?

(i) $K_f = K_d$ (ii) $K_f = K_{d_1} K_{d_2} K_{d_3} K_{d_4}$
(iii) $K_f = K_d{}^4$ (iv) $K_f = K_{f_1} K_{f_2} K_{f_3} K_{f_4}$

(c) The product of the concentrations of the ions in a nonequilibrium solution is the

(i) hydrolysis constant (ii) solubility product
(iii) ion product (iv) solubility

(d) The product of the concentrations of the ions in equilibrium with a solid electrolyte is the

(i) common ion constant (ii) hydrolysis constant
(iii) ion product (iv) solubility product

(e) The presence of a common ion in a solution of a slightly soluble ionic compound
  (i) increases the solubility of the compound
  (ii) has no effect on the concentration of the compound
  (iii) decreases the solubility of the compound
  (iv) decreases the percent ionization of the compound

(f) The ion product can be used to
  (i) determine the pH of the solution
  (ii) calculate the percent dissociation of the substance
  (iii) estimate the forward rate of reaction
  (iv) predict whether or not precipitation will occur

(g) The expression for K_{sp} for the reaction $Ag_2CrO_4(s) \rightleftharpoons 2Ag^+ + CrO_4^{2-}$ is
  (i) $[Ag^+][CrO_4^{2-}]$ (ii) $[Ag^+]^2[CrO_4^{2-}]/[Ag_2CrO_4]$
  (iii) $[Ag^+]^2[CrO_4^{2-}]$ (iv) $[Ag^+]^2[CrO_4^{2-}][Ag_2CrO_4]$

(h) What is the relationship between the solubility, S, and the value of the solubility product, K_{sp}, for $Fe_2(CO_3)_3$?
  (i) $S = K_{sp}$ (ii) $S = \sqrt[5]{K_{sp}/108}$
  (iii) $S = \sqrt[5]{K_{sp}}$ (iv) $S = K_{sp}^5$

(i) The expression for the ion product for Ag_2SO_4 is
  (i) $[Ag^+][SO_4^{2-}]^2$ (ii) $[Ag^+]^2[SO_4^{2-}]$
  (iii) $[Ag^+]^2[SO_4^{2-}]/[Ag_2SO_4]$ (iv) $[Ag^+][SO_4^{2-}]$

(j) Which slightly soluble salt will be more soluble in an acidic solution than in an alkaline or neutral solution?
  (i) NaCl (ii) NH_4NO_3
  (iii) $NaHSO_4$ (iv) CdS

(k) Which slightly soluble salt will be more soluble in a solution of high Cl^- concentration than in a solution not containing Cl^- ion?
  (i) AgCl (ii) $AgNO_3$
  (iii) $Mg(OH)_2$ (iv) KCl

4. How is the ionization of a polyprotic acid similar to the dissociation of a complex? What is significantly different about the values of the equilibrium constants and how does this affect the ways in which the equilibrium calculations are done?

5. What is the relationship between the overall dissociation constant and the overall formation constant describing complex ion equilibria?

6. Write the chemical equations and the equilibrium constant expressions for the stepwise dissociation of the $[HgCl_4]^{2-}$ complex. What is the relationship between the overall dissociation constant and the stepwise constants?

7. Briefly describe the equilibrium in a saturated solution of a slightly soluble salt. How does this solution differ from that formed by a more soluble salt?

8. What concentration terms appear in the expression for the solubility produce of a salt?

9. What is the general relationship between the value of K_{sp} for a salt and its solubility? When comparing K_{sp} values for two salts to determine which salt is more soluble, what precautions must be taken?

10. How will the solubility of a slightly soluble salt be affected by (a) the presence of additional ions common to the equilibrium, (b) the hydrolysis of the ions formed upon dissolution, (c) the presence of H^+ if the salt contains an anion that is the conjugate base of a weak acid, and (d) the presence of OH^- if the salt contains a cation that is the conjugate acid of a weak base?

11. What are some ways by which the solubility of substances that are slightly soluble in water can be increased?

12. What is meant by the term "ion product"? What will happen in a solution in which the ion product is (a) larger than K_{sp}, (b) smaller than K_{sp}, or (c) equal to K_{sp}?

13. Compare the dissolution in strong acid of metal sulfides that have relatively large values of K_{sp} to those that have relatively small values of K_{sp}.

14. Write chemical equations for the dissolution of FeS and of SnS_2 in HCl solution.

Answers to questions

1. (a) T; (b) F, replace *widely different* to *similar* and *can* to *cannot*; (c) T; (d) T; (e) T.
2. (a) The *solubility* is the concentration of a saturated solution; the *solubility product* is the equilibrium constant expression describing the equilibrium between a slightly soluble substance and its aqueous solution. (b) The *ion product* describes a nonequilibrium system; the *solubility product* describes an equilibrium system. (c) The *stepwise formation constant*

describes each formation step; the *overall formation constant* describes the overall formation reaction and is related by $K_f = K_{f_1}K_{f_2} \cdots$. (d) The *overall dissociation constant* describes the overall dissociation of a complex; the *overall formation constant* describes the overall formation of the complex and is related by $K_d = 1/K_f$.

3. (a) iii, (b) iv, (c) iii, (d) iv, (e) iii, (f) iv, (g) iii, (h) ii, (i) ii, (j) iv, (k) i.
4. Both processes are stepwise processes, each step described by an equilibrium constant expression. However, successive K_a values are usually significantly different (which allows for the simplifying assumption that each step can be treated are at a time during equilibrium calculations) whereas successive K_d values are usually similar (which requires consideration of more different expressions to describe the simultaneous equilibria.
5. The constants are reciprocals of each other, i.e., $K_f = 1/K_d$.
6. $[HgCl_4]^{2-} \rightleftharpoons [HgCl_3]^- + Cl^-$, $K_{d_1} = [HgCl_3^-][Cl^-]/[HgCl_4^{2-}]$; $[HgCl_3]^- \rightleftharpoons HgCl_2(s) + Cl^-$, $K_{d_2} = [Cl^-]/[HgCl_3^-]$; $HgCl_2(s) \rightleftharpoons [HgCl]^+ + Cl^-$, $K_{d_3} = [HgCl^+][Cl^-]$; $[HgCl]^+ \rightleftharpoons Hg^{2+} + Cl^-$, $K_{d_4} = [Hg^{2+}][Cl^-]/[HgCl^+]$; $K_d = (K_{d_1})(K_{d_2})(K_{d_3})(K_{d_4})$.
7. In a saturated solution of a slightly soluble salt, the ions are dissociated, hydration or hydrolysis of ions occurs, and the ionic solid is in dynamic equilibrium with dissolved ions. The only difference is in the relative concentrations of ions.
8. Only the concentration terms for the soluble ions appear in the expression for K_{sp}.
9. Generally, more soluble salts have higher values of K_{sp}. Direct comparisons can be made only if the salts form exactly the same number of ions, otherwise terms involving the solubility might contain different multiplication factors (like 25, etc.) and be raised to different powers (like S^3, etc.)
10. (a) Because of the common ion effect, the solubility will be lower.
    (b) Because of the hydrolysis, the solubility will be greater.
    (c) Because of the acid-base reaction, the solubility will be greater.
    (d) Because of the acid-base reaction, the solubility will be greater.
11. The ways include changing temperature, adding strong acid, oxidizing the substance, and inducing a complexation reaction.
12. The ion product is the product of the concentrations of ions in a nonsaturated or nonequilibrium solution. In a solution in which the ion product

(b) Did the answer come out in the correct units?

Yes. Like equilibrium constants, units are not usually included on values of Q_i.

(c) Is the answer reasonable?

Yes.

Exercise 22.5 A 0.001 mol sample of Be^{2+} is added to 1.00 L of a solution having pH = 10.0. Will a precipitate form? $K_{sp} = 4 \times 10^{-13}$ for $Be(OH)_2$. *Answer*: $Q_i = 1 \times 10^{-11}$, yes.

Example 22.6 Controlling Solubility

Will FeS form in a solution containing 0.1 M $[Fe(CN)_6]^{4-}$, 0.1 M CN^-, and 0.1 M S^{2-}? $K_d = 1.3 \times 10^{-37}$ for $[Fe(CN)_6]^{4-}$ and $K_{sp} = 4.2 \times 10^{-17}$ for FeS.

The formation of FeS will occur if the ion product, $Q_i = [Fe^{2+}][S^{2-}]$, exceeds the value of the K_{sp}. The source of the Fe^{2+} will be from the complex ion. The concentration of Fe^{2+} can be calculated by using the usual procedures:

	$[Fe(CN)_6]^{4-}$	$\rightleftharpoons$	Fe^{2+}	+	$6CN^-$
initial	0.1		0		0.1
change	$-x$		$+x$		$+6x$
equilibrium	$0.1 - x$		x		$0.1 + 6x$

where $x = [Fe^{2+}]$.

$$K_d = \frac{[Fe^{2+}][CN^-]^6}{[Fe(CN)_6{}^{4-}]} = \frac{(x)(0.1 + 6x)^6}{(0.1 - x)} = 1.3 \times 10^{-37}$$

Assuming $(0.1 - x) \approx 0.1$ and $(0.1 + 6x) \approx 0.1$ gives

$$\frac{(x)(0.1)^6}{0.1} = 1.3 \times 10^{-37}$$

$$x = 1.3 \times 10^{-32} \text{ mol/L}$$

The approximate solution is valid, so $[Fe^{2+}] = 1.3 \times 10^{-32}$ mol/L. Thus the ion product for the precipitation is

$$Q_i = [Fe^{2+}][S^{2-}] = (1.3 \times 10^{-32})(0.1) = 1.3 \times 10^{-33}$$

Because the ion product is smaller than the K_{sp} precipitation of FeS will not occur.

Exercise 22.6 A solution that is 0.001 M in $MgCl_2$ and 0.01 M in NH_4Cl is made 0.01 M in NH_3. Will $Mg(OH)_2$ precipitate? K_b for NH_3 is 1.6×10^{-5} and K_{sp} for $Mg(OH)_2$ is 7.1×10^{-12}. *Answer*: $Q_i = 3 \times 10^{-13}$, $Q_i < K_{sp}$, no.

Example 22.7 Dissolution of Precipitates

A precipitate of AgBr forms as Br^- ion is added to a solution of Ag^+ ion. However, the silver(I) will redissolve as additional Br^- is added because of the formation of $[AgBr_2]^-$ complex ion. Find the range of Br^- ion concentration over which the concentration of silver(I) is less than 1×10^{-6} mol/L. $K_{sp} = 4 \times 10^{-13}$ for AgBr and $K_d = 5 \times 10^{-6}$ for $[AgBr_2]^-$.

At low Br^- ion concentrations, the formation of AgBr is important

$$AgBr(s) \rightleftharpoons Ag^+ + Br^- \qquad K_{sp} = [Ag^+][Br^-]$$

The $[Br^-]$ in equilibrium with $[Ag^+] = 1 \times 10^{-4}$ mol/L is

$$[Br^-] = \frac{K_{sp}}{[Ag^+]} = \frac{4 \times 10^{-13}}{1 \times 10^{-6}} = 4 \times 10^{-7} \text{ mol/L}$$

As the $[Br^-]$ increases, the complexation of AgBr becomes important. The solubility of silver(I) will be described by the equilibrium between the precipitate and the complex ion

$$AgBr \rightleftharpoons Ag^+ + Br^- \qquad K_{sp}$$

$$Ag^+ + 2Br^- \rightleftharpoons [AgBr_2]^- \qquad 1/K_d$$

$$AgBr(s) + Br^- \rightleftharpoons [AgBr_2]^- \qquad K = K_{sp}/K_d = \frac{[AgBr_2^-]}{[Br^-]}$$

The Br^- ion concentration in equilibrium with $[AgBr_2^-] = 1 \times 10^{-4}$ mol/L is

$$[Br^-] = \frac{[AgBr_2^-]}{K_{sp}/K_d} = \frac{1 \times 10^{-6}}{(4 \times 10^{-13})/(5 \times 10^{-6})} = 10 \text{ mol/L}$$

Exercise 22.7 A precipitate of $Pb(OH)_2$ forms as OH^- ion is added to an acidic solution of Pb^{2+} ion. However, the Pb(II) will redissolve as additional OH^- ion is added because of the formation of the $[Pb(OH)_4]^{2-}$ complex ion. Find the pH range over which the concentration of Pb(II) is less than 1×10^{-4} mol/L. $K_{sp} = 1.2 \times 10^{-15}$ for $Pb(OH)_2$ and $K_d = 5 \times 10^{-17}$ for $[Pb(OH)_4]^{2-}$. *Answer*: from pH 8.5 to pH 11.3.

PRACTICE TEST

1 (15 points). Choose the response that best completes each statement.

(a) The expression for the overall dissociation constant of the complex $[AlF_6]^{3-}$ is

(i) $[Al^{3+}][F^-]^6/[AlF_6^{3-}]$ (ii) $[AlF_6^{3-}]/[Al^{3+}][F^-]^6$

(iii) $[AlF_5^{2-}][F^-]/[AlF_6^{3-}]$ (iv) $[Al^{3+}][F^-]^6$

(b) Silver cyanide solution is used in silver plating. The equilibrium constant for the reaction

$$2AgCN(s) \rightarrow Ag^+ + [Ag(CN)_2]^-$$

is 4.0×10^{-12}. What is the equilibrium concentration of Ag^+ in a saturated solution?

(i) 4.0×10^{-12} mol/L (ii) 8.0×10^{-12} mol/L

(iii) 2.0×10^{-6} mol/L (iv) 1.0×10^{-6} M

(c) If the molar solubility of a slightly soluble hypothetical salt M_2X is given by S, the K_{sp} is equal to

(i) S^2 (ii) $4S^3$

(iii) $2S^2$ (iv) $4S^2$

(d) The K_{sp} of $Pb(OH)_2$ is 1.2×10^{-15}. In which of the following solvents will $Pb(OH)_2$ be most soluble?

(i) water (ii) a solution containing $PbNO_3$

(iii) an acidic solution (iv) an alkaline solution

(e) Which salt is the least soluble in water?

(i) $BaCO_3$ ($K_{sp} = 2.0 \times 10^{-9}$) (ii) $CaCO_3$ ($K_{sp} = 3.84 \times 10^{-9}$)

(iii) $CoCO_3$ ($K_{sp} = 1.0 \times 10^{-10}$) (iv) $CuCO_3$ ($K_{sp} = 2.3 \times 10^{-10}$)

2 (15 points). What is the concentration of Ni^{2+} in equilibrium with 6 M NH_3 in a solution having a total nickel(II) concentration of 0.10 mol/L? $K_d = 1 \times 10^{-9}$ for $[Ni(NH_3)_6]^{2+}$.

3 (15 points). The solubility product constant for MnS is 2.3×10^{-13}. If the concentration of the manganese ion, Mn^{2+}, in a solution is 2×10^{-4} mol/L, what is the highest concentration of sulfide ion that will not cause precipitation of MnS?

4 (15 points). The solubility product constant of $BaCrO_4$ is 1.2×10^{-10} at 25°C. If 10. mL of 0.010 M $BaCl_2$ is added to 10. mL of 0.0010 M K_2CrO_4 solution, will a precipitate of $BaCrO_4$ form?

5 (20 points). What is the solubility of $PbCrO_4$ in water if $K_{sp} = 2.8 \times 10^{-13}$? Repeat the calculation for $PbCrO_4$ in a solution containing $[Pb^{2+}] = 0.100$ mol/L.

6 (20 points). What is the concentration of Ag^+ in a solution containing 0.1 M SCN^- and 0.1 M $[Ag(SCN)_4]^{3-}$? $K_d = 2.1 \times 10^{-10}$ for $[Ag(SCN)_4]^{3-}$. Will Ag_2SO_4 precipitate if $[SO_4^{2-}] = 0.1$ mol/L? $K_{sp} = 1.5 \times 10^{-5}$ for Ag_2SO_4.

Answers to practice test

1. (a) i, (b) iii, (c) ii, (d) iii, (e) iii.
2. 2×10^{-15} mol/L.
3. 1×10^{-9} mol/L.
4. $Q_i = 2.5 \times 10^{-6}$, yes
5. 5.3×10^{-7} mol/L, 2.8×10^{-12} mol/L.
6. 2×10^{-7} mol/L; $Q_i = 4 \times 10^{-15}$, no.

ADDITIONAL NOTES

CHAPTER 23

THERMODYNAMICS

CHAPTER OVERVIEW

The thermodynamic concepts of entropy (S) and free energy (G) are introduced in this chapter. Entropy, entropy changes, and their influence on the spontaneity of a reaction are discussed first. Then the free energy change, a parameter based on enthalpy and entropy, is defined. Free energy can be used to predict unequivocally whether or not a reaction will be spontaneous. The relationship of ΔG to equilibrium is then discussed. Thermodynamic calculations of free energy changes of reaction and equilibrium conditions at temperatures other than 25 °C are considered.

COMPETENCIES

Definitions for the following terms should be learned:

- absolute entropy
- entropy
- free energy change
- second law of thermodynamics
- spontaneous chemical change
- third law of thermodynamics

General concepts and skills that should be learned include

- the physical meaning of entropy
- interpreting whether or not an entropy change is favorable
- predicting whether or not a process is spontaneous by using ΔG
- predicting the effect of temperature on spontaneity of a chemical reaction

Numerical exercises that should be understood include

- calculating ΔS, ΔH, or T for a phase change when two of these variables are given
- calculating $\Delta S°$ for a chemical reaction
- calculating $\Delta G°$, $\Delta S°$, $\Delta H°$, or T when three of these variables are given

- calculating $\Delta G°$ for a chemical reaction
- calculating $\Delta G°$ or K for a chemical reaction when one of these variables is given
- calculating ΔG, $\Delta G°$, or Q when two of the these variables are given
- calculating K or $\Delta G°$, $\Delta H°$, or T for a chemical reaction when two of these variables are given

QUESTIONS

1. Which of the following statements are true? Rewrite each false statement so that it is correct.

(a) The entropy content of a substance increases as its temperature increases.

(b) Both heat capacity and entropy have the units of J/K mol and hence are the same thermodynamic property.

(c) Unlike free energy and enthalpy, the absolute entropy of a substance can be determined.

(d) A chemical reaction in which the entropy decreases will never be spontaneous.

(e) For phase changes, $\Delta G°$ is zero because $\Delta S° = \Delta H°/T$ and thus these changes are examples of equilibrium.

2. Choose the response that best completes each statement.

(a) The study of the transformation of energy from any one form to another is called

(i) kinetics (ii) thermodynamics
(iii) stoichiometry (iv) equilibrium

(b) A measure of the randomness or disorder of a system is the

(i) enthalpy (ii) entropy
(iii) free energy (iv) equilibrium constant

(c) The change in the disorder accompanying any change in a system is given by

(i) ΔH (ii) ΔS
(iii) ΔG (iv) ΔT

(d) The unequivocal criterion for spontaneity of a process is given by the

(i) enthalpy change (ii) entropy change
(iii) free energy change (iv) rate constant

(e) The overall chemical driving force of a reaction is given by
   (i) the standard state free energy change
   (ii) the enthalpy change
   (iii) the free energy change
   (iv) the reverse rate constant

3. Clearly define enthalpy, entropy, and free energy. What are the symbols commonly used for these thermodynamic quantities? Give a suitable set of units for each of these quantities.

4. Choose all of the situations in which a reaction will occur: (a) $\Delta H < 0$, $\Delta S > 0$; (b) $\Delta H < 0$, $\Delta S < 0$, $\Delta G > 0$; (c) $\Delta H < 0$, $\Delta S < 0$, $\Delta G < 0$; (d) $\Delta H > 0$, $\Delta S < 0$; (e) $\Delta H > 0$, $\Delta S > 0$, $\Delta G > 0$; (f) $\Delta H > 0$, $\Delta S > 0$, $\Delta G < 0$; (g) $\Delta G^\circ < 0$, $\Delta G > 0$; (h) $\Delta G^\circ < 0$, $\Delta G < 0$; (i) $\Delta G^\circ > 0$, $\Delta G < 0$; (j) $\Delta G^\circ > 0$, $\Delta G > 0$; (k) $K > 1$, and (l) $K < 1$.

5. A flask containing nitrogen and a flask containing oxygen are connected by a small tube with a stopcock. After the stopcock is opened, the gases eventually mix so that a uniform composition is attained even though there is no enthalpy change. What is the driving force of this process?

6. What value do we assign to the entropy of a perfect crystalline substance at absolute zero? How does this value change for substances that do not form perfect crystals at absolute zero?

7. Why is the value of S° for Ne(g) at 25 °C greater than that for He(g)?

8. Does a favorable process produce an increase or decrease in the entropy of the universe?

9. Predict whether the entropy changes for the following reactions will be (i) large and negative, (ii) large and positive, or (iii) small.
   (a) $SO_2(g) + NO_2(g) \rightarrow SO_3(g) + NO(g)$
   (b) $2H_2S(g) + 3O_2(g) \rightarrow 2SO_2(g) + 2H_2O(g)$
   (c) $H_2(g) + \frac{1}{2}O_2(g) \rightarrow H_2O(g)$
   (d) $3Cu(s) + 8HNO_3(aq) \rightarrow 3Cu(NO_3)_2(aq) + 2NO(g) + 4H_2O(l)$

10. Repeat Question 9 for
   (a) $C(graphite) + 2Cl_2(g) \rightarrow CCl_4(l)$
   (b) $CCl_4(l) + H_2(g) \rightarrow HCl(g) + CHCl_3(l)$

(c) $2ZnS(s) + 3O_2(g) \xrightarrow{\Delta} 2ZnO(s) + 2SO_2(g)$

(d) $2N_2O(g) \xrightarrow{\Delta} 2N_2(g) + O_2(g)$

11. The standard enthalpy change at 25 °C for dissolving sodium chloride in water is 3880 J/mol,

$$NaCl(s) + nH_2O \rightarrow NaCl(aq, 1\ M)$$

which implies that the process is not spontaneous; yet we know that salt readily dissolves in water (think of the oceans!) (a) For this spontaneous process, is $\Delta G°$ positive or negative? (b) If the enthalpy change is unfavorable, what is the driving force for this reaction? (c) Is $\Delta S°$ positive or negative? (d) Does this sign represent an increase or decrease in randomness? (e) Does the formation of a mixture (in which the sodium and chloride ions are freely moving around in water) from a highly ordered crystal structure and a pure liquid confirm your answer to (d)?

12. Explain how $\Delta G°$ can be calculated for a chemical reaction from various data.

13. What is the relationship between $\Delta G°$ and the equilibrium constant for a chemical reaction?

14. A plot is made of log K against the inverse of the absolute temperature for a chemical reaction. What is the relationship between the slope of the straight line through the data and $\Delta H°$?

15. A chemical reaction has a negative entropy change. What will happen to the spontaneity of the reaction as the temperature increases? What would happen if the reaction had a positive entropy change?

Answers to questions

1. (a) T; (b) F, replace *and hence are* with *but are not*; (c) T; (d) F, change *will never* to *has an unfavorable entropy change, but may be spontaneous or nonspontaneous,* depending on the value of ΔG; (e) T.
2. (a) ii, (b) ii, (c) ii, (d) iii, (e) iii.
3. heat flow under constant pressure conditions, H, J/mol; measure of disorder or randomness, S, J/K mol; $H - TS$, G, J/mol.
4. a, c, f, h, i, k.
5. The driving force is entropy.
6. A perfect crystal has zero entropy at 0 K. The entropy is greater than zero.
7. Both gases have the same translational contribution to $S°$, but Ne has a higher electronic contribution.

8. increase.
9. (a) iii, (b) i, (c) i, (d) ii.
10. (a) i, (b) iii, (c) i, (d) ii.
11. (a) For the spontaneous process of sodium chloride dissolving in water, $\Delta G°$ is negative.
    (b) The driving force for this reaction is an increase in entropy.
    (c) The value of $\Delta S°$ is positive.
    (d) The positive sign represents an increase in randomness.
    (e) The formation of a mixture from a highly ordered crystal structure and a pure liquid does confirm an increase in randomness.
12. For a chemical reaction ΔG° can be calculated using $\Delta H°$, $\Delta S°$, and T; $\Delta G_f°$ for all reactants and products; a series of values of $\Delta G°$ for known reactions; K and T; and $E°$ (see Chapter 24).
13. $\Delta G° = -(2.303)RT \log K$.
14. $\Delta H° = -(2.303)R(\text{slope})$.
15. The spontaniety becomes less favorable as the temperature increases. The reaction would become more favorable.

SKILLS

Skill 23.1 Favorable and unfavorable entropy changes

The sign and magnitude of the entropy change can be used to decide whether or not a chemical process involves a favorable entropy change. Positive values of ΔS are favorable and negative values of ΔS are unfavorable. For example, the entropy changes for both of the following reactions are favorable.

$$C(s) + (1/2)O_2(g) \rightarrow CO(g) \qquad \Delta S°_{298} = 89.365 \text{ J/K}$$

$$C(s) + O_2(g) \rightarrow CO_2(g) \qquad \Delta S°_{298} = 2.86 \text{ J/K}$$

The above values of $\Delta S°$ can be confirmed semiquantitatively by realizing that the randomness of a substance at a given temperature generally decreases in the order of gas >> solution > liquid ≥ solid. In the CO reaction, 1 mol of gaseous CO is formed from 1 mol of solid carbon and 0.5 mol of gaseous O_2, which means that there is a net gain of randomness of about 0.5 mol of a gas (neglecting the solid). A typical value of $\Delta S°$ for an increase of 1 mol of gas is 150 J/K, so the value of 89.365 J/K agrees well with (0.5)(150 J/K) = 80 J/K. Likewise for the CO_2 reaction, 1 mol of gaseous CO_2 is formed from 1 mol of solid carbon and 1 mol of gaseous O_2, which means that there is no large net gain of randomness. So the small value of 2.86 J/K seems reasonable.

Skill Exercise: Indicate whether or not the following entropy changes are favorable: (a) ΔS°_{298} = 175.29 J/K for $AsI_3(s) \rightarrow AsI_3(g)$, (b) ΔS°_{298} = 165 J/K for $As_4(g) \rightarrow 2As_2(g)$, (c) ΔS°_{298} = -127.82 J/K for $2P(s) + 3Cl_2(g) \rightarrow 2PCl_3(g)$, and (d) ΔS°_{298} = -468.35 J/K for $2P(s) + 5Cl_2(g) \rightarrow 2PCl_5(g)$. Also semiquantitatively confirm the ΔS° values cited. *Answers*: (a) favorable, increase of 1 mol of gas; (b) favorable, increase of 1 mol of gas; (c) unfavorable, decrease of 1 mol of gas; (d) unfavorable, decrease of 3 mol of gas. (It is of interest to compare these answers with those determined in *Skill 7.1 Favorable and unfavorable energy changes*.)

Skill 23.2 Predicting reaction spontaneity by using ΔG

The sign and magnitude of the free energy change are the sole critera for predicting whether or not a chemical reaction will occur. Negative values of ΔG indicate spontaneous processes, positive values indicate nonspontaneous processes, and $\Delta G = 0$ indicates systems at equilibrium. For example, ΔG°_{298} = -137.168 kJ for the reaction

$$C(s) + \frac{1}{2}O_2(g) \rightarrow CO(g)$$

and the reaction will be spontaneous under standard state conditions. Likewise, ΔG°_{298}= -394.359 kJ for the reaction

$$C(s) + O_2(g) \rightarrow CO_2(g)$$

and this reaction will also be spontaneous under standard state conditions. Based on the values of ΔG°_{298}, the CO_2 reaction is predicted to be more favorable --which agrees with experimental fact. In *Skill 7.1 Favorable and unfavorable energy changes*, values of ΔH°_{298} were the bases on which both reactions were predicted to be favorable, and in *Skill 23.1 Favorable and unfavorable entropy changes*, values of ΔS°_{298} were the bases on which both reactions were predicted to be favorable. These predictions agree with the predictions based on ΔG°_{298} values.

Skill Exercise: Indicate whether or not the following reactions are spontaneous, and if two reactions are given, indicate which is more favorable: (a) ΔG°_{298} = 15 kJ for $AsI_3(s) \rightarrow AsI_3(g)$, (b) ΔG°_{298} = 251.4 kJ for $As_4(g) \rightarrow 2As_2(g)$, and (c) ΔG°_{298} = -535.6 kJ for $2P(s) + 3Cl_2(g) \rightarrow 2PCl_3(g)$ and ΔG°_{298} =

-610.0 kJ kcal for $2P(s) + 5Cl_2(g) \rightarrow 2PCl_5(g)$. *Answers*: (a) nonspontaneous; (b) nonspontaneous; (c) both spontaneous, formation of PCl_5 is more favorable.

NUMERICAL EXERCISES

Example 23.1 Entropy in physical changes

Bromine boils at 59.47 °C and has a heat of vaporization of 29.56 kJ/mol. Calculate the entropy change for this phase change.

The entropy change is given by

$$\Delta S° = \frac{\Delta H°}{T} = \frac{(29.56 \text{ kJ/mol})(10^3 \text{ J/1 kJ})}{(59.47 + 273.15) \text{ K}} = 88.87 \text{ J/K mol}$$

The entropy change is 88.87 J/K mol.

Exercise 23.1 Bromine melts at -7.25 °C and has a heat of fusion of 10.57 kJ/mol. Calculate the entropy change for this phase transition and compare the value for fusion to that of vaporization. *Answer*: 39.75 J/K mol, the fusion value is about half of the vaporization value.

Example 23.2 Entropy in chemical reactions

Calculate the standard state entropy change at 1000. K for the reaction

$$O_3(g) + NO(g) \rightarrow O_2(g) + NO_2(g)$$

given $S°_{1000}$ = 296.94 J/K mol for $O_3(g)$, 248.43 J/K mol for NO(g), 243.48 J/K mol for $O_2(g)$, and 293.78 J/K mol for $NO_2(g)$.

The entropy change for the reaction can be determined from absolute entropies of the reactants and products by

$$\begin{aligned} S° &= [\text{sum of } S°(\text{products})] - [\text{sum of } S°(\text{reactants})] \\ &= [(1 \text{ mol})S°(O_2) + (2 \text{ mol})S°(NO_2)] - [(1 \text{ mol})S°(O_3) + (1 \text{ mol})S°(NO)] \\ &= [(1 \text{ mol})(243.48 \text{ J/K mol}) + (1 \text{ mol})(293.78 \text{ J/K mol})] \\ &\quad - [(1 \text{ mol})(296.94 \text{ J/K mol}) + (1 \text{ mol})(248.43 \text{ J/K mol})] \\ &= -8.11 \text{ J/K} \end{aligned}$$

The entropy change is -8.11 J/K for the reaction.

Exercise 23.2 Calculate the standard state entropy change at 25 °C for the reaction

$$2H_2O_2(aq) \rightarrow 2H_2O(l) + O_2(g)$$

given $S°_{298}$ = 143.9 J/K mol for $H_2O_2(aq)$, 69.91 J/K mol for $H_2O(l)$, and 205.138 J/K mol for $O_2(g)$. *Answer*: 57.2 J/K.

Example 23.3 Entropy in chemical reactions*

Calculate the entropy change for the reaction given in *Example 23.2* by using the following data:

$O_3(g)$	$\rightarrow$	$3O(g)$	$\Delta S^\circ_{1000} = 263.12$ J/K
$NO(g)$	$\rightarrow$	$N(g) + O(g)$	$\Delta S^\circ_{1000} = 116.60$ J/K
$O_2(g)$	$\rightarrow$	$2O(g)$	$\Delta S^\circ_{1000} = 129.90$ J/K
$NO_2(g)$	$\rightarrow$	$N(g) + 2O(g)$	$\Delta S^\circ_{1000} = 257.94$ J/K

This problem is solved by using the comprehensive problem-solving approach presented in *Skill 2.7 Solving problems.*

1. Study the problem and be sure you understand it.
   (a) What is unknown?
   ΔS° for the reaction

$$O_3(g) + NO(g) \rightarrow O_2(g) + NO_2(g)$$

   (b) What is known?
   ΔS° for four chemical reactions involving the various reactants and products of the desired reaction.
2. Decide how to solve the problem.
   (a) What is the connection between the known and the unknown?
   The ΔS° for the desired equation can be calculated by using Hess's law (see *Skill 7.2 Using Hess's law*) if the given equations, upon proper combination, yield the desired equation.
   (b) What is necessary to make the connection?
   The desired ΔS°_{1000} can be found by adding the first two reactions (and values of ΔS°_{1000}) to the reverse of the last two reactions (and values of ΔS°_{1000} with sign changes).
3. Set up the problem and solve it.

O_3	$\rightarrow$	$3O(g)$	$\Delta S^\circ_{1000} = 263.12$ J/K
$NO(g)$	$\rightarrow$	$N(g) + O(g)$	$\Delta S^\circ_{1000} = 116.60$ J/K
$2O(g)$	$\rightarrow$	$O_2(g)$	$\Delta S^\circ_{1000} = -129.90$ J/K
$2O(g) + N(g)$	$\rightarrow$	$NO_2(g)$	$\Delta S^\circ_{1000} = -257.94$ J/K
$O_3(g) + NO_2(g)$	$\rightarrow$	$O_2(g) + NO(g)$	$\Delta S^\circ_{1000} = -8.12$ J/K

The standard state entropy change for the reaction is -8.12 J/K.

4. Check the result.
   (a) Are significant figures and the location of the decimal point correct?
   Yes.
   (b) Did the answer come out in the correct units?
   Yes.
   (c) Is the answer reasonable?
   Yes. Only a small value of ΔS would be expected because there is only a small change in randomness in this reaction (2 mol of gaseous reactants form 2 mol of gaseous products).

Exercise 23.3 Calculate the standard state entropy change for the reaction given in Exercise 23.2 by using the following data:

$$H_2O_2(aq) \rightarrow H_2O_2(g) \qquad \Delta S^\circ_{298} = 88.8 \text{ J/K}$$

$$H_2O_2(g) \rightarrow H_2(g) + O_2(g) \qquad \Delta S^\circ_{298} = 103.1 \text{ J/K}$$

$$2H_2O(l) \rightarrow 2H_2(g) + O_2(g) \qquad \Delta S^\circ_{298} = 326.69 \text{ J/K}$$

Answer: 57.1 J/K.

Example 23.4 ΔG° of reaction*

The standard state enthalpy change for the reaction given in *Example 23.2* is -201.87 kJ. What is the value of ΔG°_{1000}?

This problem is solved by using the comprehensive problem-solving approach presented in *Skill 2.7 Solving problems*.

1. Study the problem and be sure you understand it.
   (a) What is unknown?
   ΔG° for the reaction

$$O_3(g) + NO(g) \rightarrow O_2(g) + NO_2(g)$$

   (b) What is known?
   ΔH° = -201.87 kJ and ΔS° = -8.11 J/K for this reaction.
2. Decide how to solve the problem.
   (a) What is the connection between the known and the unknown?
   ΔG° for a chemical reaction is related to ΔH° and ΔS° by

$$\Delta G^\circ = \Delta H^\circ - T\,\Delta S^\circ$$

   (b) What is necessary to make the connection?
   Substitute the numerical values of ΔH°, ΔS°, and T into the equation.
3. Set up the problem and solve it.

$$\begin{aligned}\Delta G^\circ_{1000} &= \Delta H^\circ_{1000} - T\,\Delta S^\circ_{1000}\\ &= (-201.87\text{ kJ}) - (1000.\text{ K})(-8.11\text{ J/K})(1\text{ kJ/1000 J})\\ &= -193.76\text{ kJ}\end{aligned}$$

The standard state free energy change at 1000. K is -193.76 kJ.

4. Check the result.

   (a) Are significant figures and the location of the decimal point correct?
   Yes.

   (b) Did the answer come out in the correct units?
   Yes. The units of free energy are those of energy. Note that the units of ΔH° and $T\,\Delta S^\circ$ must agree and that the conversion from J to kJ for the $T\,\Delta S^\circ$ term was required.

   (c) Is the answer reasonable?
   Yes. Under standard state conditions, the reaction should be spontaneous because of the favorable ΔH° (large and negative).

Exercise 23.4 The value of ΔH°_{298} for the reaction in Exercise 23.2 is -189.32 kJ. Calculate ΔG°_{298}. *Answer*: -206.4 kJ.

Example 23.5 ΔG° of reaction

The respective values of ΔG° for the reactions given in *Example 23.3* are 351.56 kJ, 522.15 kJ, 375.51 kJ, and 691.95 kJ. Calculate ΔG°_{1000} by using these values for the reaction given in *Example 23.2*.

Using Hess's law gives

$$\begin{array}{rcll} O_3(g) & \rightarrow & 3O(g) & \Delta G^\circ_{1000} = 351.56\text{ kJ}\\ NO(g) & \rightarrow & N(g) + O(g) & \Delta G^\circ_{1000} = 522.15\text{ kJ}\\ 2O(g) & \rightarrow & O_2(g) & \Delta G^\circ_{1000} = -375.51\text{ kJ}\\ 2O(g) + N(g) & \rightarrow & NO_2(g) & \Delta G^\circ_{1000} = -691.95\text{ kJ}\\ \hline O_3(g) + NO(g) & \rightarrow & O_2(g) + NO_2(g) & \Delta G^\circ_{1000} = -193.75\text{ kJ}\end{array}$$

The standard state free energy change is -193.75 kJ.

Exercise 23.5 The respective values of ΔG° for the reactions given in Exercise 23.3 are 28.46 kJ, 105.57 kJ, and 474.258 kJ. Calculate ΔG°_{298} by using these values for the reaction given in Exercise 23.2. *Answer*: -206.20 kJ.

Example 23.6 ΔG° of reaction

The value of ΔG° at 1000. K is 211.71 kJ/mol for $O_3(g)$, 77.772 kJ/mol for $NO(g)$, 0 for $O_2(g)$, and 95.726 kJ/mol for $NO_2(g)$. Calculate ΔG°_{1000} for the reaction given in *Example 23.2*.

The free energy change for the reaction can be determined from values of free energy of formation by

$$\Delta G° = [(\text{sum of } \Delta G_f°(\text{products})] - [\text{sum of } \Delta G_f°(\text{reactants})]$$
$$= [(1\ \text{mol})\Delta G_f°(O_2) + (1\ \text{mol})\Delta G_f°(NO_2)]$$
$$- [(1\ \text{mol})\Delta G_f°(O_3) + (1\ \text{mol})\Delta G_f°(NO)]$$
$$= [(1\ \text{mol})(0) + (1\ \text{mol})(95.726\ \text{kJ/mol})]$$
$$- [(1\ \text{mol})(211.71\ \text{kJ/mol}) + (1\ \text{mol})(77.772\ \text{kJ/mol})]$$
$$= -193.76\ \text{kJ}$$

The standard state free energy change is -193.76 kJ.

Exercise 23.6 The value of $\Delta G_f°$ at 25 °C is -134.03 kJ/mol for H_2O_2(aq), -237.129 kJ/mol for H_2O(l), and 0 for O_2(g). Calculate $\Delta G_{298}°$ for the reaction given in Exercise 23.2. *Answer*: -206.20 kJ.

Example 23.7* ΔG *under nonstandard state conditions*

The conditions of the reaction

$$O_3(g) + NO(g) \rightarrow O_2(g) + NO_2(g) \qquad \Delta G_{1000}° = -193.75\ \text{kJ}$$

were such that P_{O_3} = 0.25 bar, P_{NO} = 0.50 bar, P_{O_2} = 0.20 bar, and P_{NO_2} = 0.001 bar. Calculate ΔG_{1000} and discuss the reaction spontaneity under these conditions.

This problem is solved by using the comprehensive problem-solving approach presented in *Skill 2.7 Solving problems*.

1. Study the problem and be sure you understand it.
   (a) What is unknown?
   The value of ΔG under nonstandard state conditions.
   (b) What is known?
   The value of $\Delta G°$ and the various partial pressures of the reactants and products.
2. Decide how to solve the problem.
   (a) What is the connection between the known and the unknown?
   ΔG is related to $\Delta G°$, T, and Q (the reaction quotient, see *Skill 19.1 Writing equilibrium constant and reaction quotient expressions*) by the equation
   $$\Delta G = \Delta G° + (2.303)RT \log Q$$
   (b) What is necessary to make the connection?

Substitute the partial pressures of the gases into the expression for the reaction quotient,

$$Q = \frac{P_{O_2}\ P_{NO_2}}{P_{O_3}\ P_{NO}}$$

and then substitute the numerical values of $\Delta G°$, Q, and T into the equation.

3. Set up the problem and solve it.

$$Q = \frac{(0.20)(0.001)}{(0.25)(0.50)} = 0.002$$

$$\Delta G = (-193.75 \text{ kJ}) + (2.303)(8.314 \text{ J/K mol})(1000 \text{ K})\left[\frac{1 \text{ kJ}}{1000 \text{ J}}\right] \log(0.002)$$

$$= (-193.75 \text{ kJ}) + (-52 \text{ kJ}) = -246 \text{ kJ}$$

The free energy change under these conditions is -246 kJ. The reaction is more favorable under these conditions than under standard state conditions.

4. Check the result.

(a) Are significant figures and the location of the decimal point correct? Yes. The number of significant figures in the value of Q limits the number of significant figures in the final answer.

(b) Did the answer come out in the correct units? Yes. Note that it was necessary to convert the units on the correction term for nonstandard state conditions from J to kJ.

(c) Is the answer reasonable? Yes. The equilibrium constant must be large because the reaction is spontaneous under standard state conditions. Because $Q < K$, the reaction must be more favorable under these conditions.

Exercise 23.7 The conditions of the reaction

$$2H_2O_2(aq) \rightarrow 2H_2O(l) + O_2(g) \qquad \Delta G°_{298} = -206.20 \text{ kJ}$$

were such that $[H_2O_2]$ = 0.65 mol/L and P_{O_2} = 0.20 bar. Calculate ΔG for this reaction and discuss the spontaneity under these conditions. *Answer*: -208.1 kJ, slightly more favorable.

Example 23.8 ΔG° and equilibrium

What is the value of the equilibrium constant for the reaction given in *Example 23.7*?

The equilibrium constant is related to $\Delta G°$ by

$$\log K_p = -\frac{\Delta G°}{(2.303)RT} = -\frac{(-193.75\text{ kJ})(1000\text{ J/1 kJ})}{(2.303)(8.314\text{ J/K mol})(1000\text{ K})}$$

$$= 10.12$$

$$K_p = 1.3 \times 10^{10}$$

The equilibrium constant at 1000 K is 1.3×10^{10}.

Exercise 23.8 Find the equilibrium constant for the reaction given in Exercise 23.7. *Answer*: $K_p = 1.3 \times 10^{36}$.

Example 23.9 ΔG° and equilibrium

In *Example 23.8*, K_p was found to be 1.3×10^{10} at 1000. K for the reaction given in *Example 23.7*. Calculate K_p at 1100. K given that $\Delta H° = -201.87$ kJ for this reaction.

The temperature dependence of the equilibrium constant is given by

$$\log\left[\frac{K_2}{K_1}\right] = \frac{-\Delta H°}{(2.303)R}\left[\frac{1}{T_2} - \frac{1}{T_1}\right]$$

Assigning T_1 = 1000. K, $K_1 = 1.3 \times 10^{10}$, and T_2 = 1100. K and substituting gives

$$\log\left[\frac{K_2}{1.3 \times 10^{10}}\right] = \frac{-(-201.87\text{ kJ})(1000\text{ J/1 kJ})}{(2.303)(8.314\text{ J/K mol})}\left[\frac{1}{1100.\text{ K}} - \frac{1}{1000.\text{ K}}\right]$$

$$= -0.96$$

$$\frac{K_2}{1.3 \times 10^{10}} = 0.11$$

$$K_2 = (0.11)(1.3 \times 10^{10}) = 1.4 \times 10^{9}$$

The equilibrium constant at this higher temperature is 1.4×10^{9}.

Exercise 23.9 Using the data and results in Exercises 23.7 and 23.8 and given $\Delta H° = -189.33$ kJ, calculate K_p at 20 °C. *Answer*: 4.8×10^{36}.

Example 23.10 ΔG° as a function of temperature

What is the value of $\Delta G°_{1100}$ for the reaction given in *Example 23.2*?

Using the value of $\Delta H°_{1000}$ from *Example 23.4* and the value of $\Delta S°_{1000}$ from *Example 23.2*,

$$\Delta G°_{1100} = \Delta H° - T\,\Delta S°$$

$$= (-201.87\text{ kJ}) - (1100\text{ K})(-8.11\text{ J/K})(1\text{ kJ/1000 J})$$

$$= -192.95\text{ kJ}$$

The standard state free energy change has decreased to -192.95 kJ.

Exercise 23.10 What is the value of ΔG°_{293} for the reaction given in Exercise 23.4? *Answer*: -206.1 kJ.

PRACTICE TEST

1 (10 points). Using the concept that the change in entropy represents a change in randomness, match the given values of ΔS° with the correct chemical equation:

(a) $H_2(g) + I_2(g) \rightarrow 2HI(g)$ (i) $\Delta S^\circ = -284.8$ J/K

(b) $MgCO_3(s) \xrightarrow{\Delta} MgO(s) + CO_2(g)$ (ii) $\Delta S^\circ_{600} = 170$ J/K

(c) $NH_3(g) + HCl(g) \rightarrow NH_4Cl(s)$ (iii) $\Delta S^\circ_{298} = 21.81$ J/K

2 (15 points). Which of the reactions given in the table (a) have favorable values of ΔH°, (b) have favorable values of ΔS°, and (c) are spontaneous?

Reaction	ΔH°_{298}, kJ	ΔG°_{298}, kJ	ΔS°_{298}, J/K
(i) $N_2(g) \rightarrow 2N(g)$	945.408	911.126	114.99
(ii) $N(g) + 3H(g) \rightarrow NH_3(g)$	-1172.71	-1081.75	-304.99
(iii) $H_2(g) \rightarrow 2H(g)$	435.930	406.494	98.742

3 (15 points). Using the data in question 2, calculate ΔG°_{298} and ΔS°_{298} for the reaction

$$N_2(g) + 3H_2(g) \rightarrow 2NH_3(g)$$

4 (10 points). Using the values of ΔG°_{298} and ΔS°_{298} obtained in question 3, calculate ΔH°_{298} for the production of ammonia.

5 (10 points). What is the value of the equilibrium constant for the reaction given in Question 3?

6 (10 points). What would be the value of ΔG for the reaction in Question 3 if $P_{N_2} = 15$ bar, $P_{H_2} = 25$ bar, and $P_{NH_3} = 0.3$ bar?

7 (20 points). Choose the response that best completes each statement.

(a) The absolute entropy of an element in its standard state is

(i) zero by definition

(ii) calculated by using Hess's law

(iii) calculated by using the third law of thermodynamics

(iv) impossible to determine

(b) Which state represents the largest entropy?

(i) $H_2(s)$ at 50 K and 1 bar (ii) $H_2(l)$ at 100 K and 1 bar

(iii) $H_2(g)$ at 200 K and 1 bar (iv) $H_2(g)$ at 200 K and 0.5 bar

(c) Calculate $\Delta S°$ for the reaction

$$Cu_2O(s) + H_2(g) \xrightarrow{\Delta} 2Cu(s) + H_2O(g)$$

given $S°$ = 94 J/K mol for $Cu_2O(s)$, 131 J/K mol for $H_2(g)$, 33 J/K mol for Cu(s), and 188.8 J/K mol for $H_2O(g)$.

(i) 29.7 J/K (ii) -3.3 J/K

(iii) 30. J/K (iv) 479.7 J/K

(d) For many liquids, $\Delta S°$ = 88.0 J/K mol for vaporization. The heat of vaporization of benzene is 30.7 kJ/mol at its boiling pint. The boiling point is approximately

(i) 0.35 K (ii) 2593 °C

(iii) 76 °C (iv) 2700 K

(e) The heat of fusion of lead iodide is 21.8 kJ/mol at 412 °C. The entropy change is

(i) 31.8 J/K mol (ii) 8.98 kJ/mol

(iii) 52.9 J/K mol (iv) none of these answers

(f) In order for a reaction to be spontaneous, the change in free energy must be

(i) positive (ii) negative

(iii) zero (iv) equal to ΔH

(g) Which combination of ΔH and ΔS will always be spontaneous?

(i) $\Delta H < 0$, $\Delta S > 0$ (ii) $\Delta H > 0$, $\Delta S < 0$

(iii) $\Delta H < 0$, $\Delta S < 0$ (iv) $\Delta H > 0$, $\Delta S > 0$

(h) At a given temperature ΔG = -50 kJ for the reaction

$$C(s) + O_2(g) \rightarrow CO_2(g)$$

Which statement is true?

(i) The system is at equilibrium.

(ii) CO_2 will decompose.

(iii) CO_2 will form from C(s) and $O_2(g)$.

(iv) none of these answers

(i) At equilibrium

(i) $\Delta G = 0$ (ii) $\Delta G < 0$

(iii) $\Delta H = \Delta S$ (iv) $\Delta H < 0$ and $\Delta S > 0$

(j) Which statement is true?
  (i) A reaction with a positive ΔG is spontaneous.
  (ii) The equilibrium constant of an endothermic reaction increases as the temperature increases.
  (iii) The absolute entropy of an element in its standard state at 25 °C is zero.
  (iv) none of these answers

8 (10 points). For the reaction

$$Na(s) + \frac{1}{2}Cl_2(g) \rightarrow NaCl(s)$$

ΔG°_f = -384.138 kJ and ΔH°_f = -411.153 kJ at 25 °C. What is the value of the standard state free energy change of formation of NaCl at 95 °C?

Answers to practice test

1. (a) iii, (b) ii, (c) i.
2. (a) ii; (b) i, iii; (c) ii.
3. -32.89 kJ, -198.76 J/K.
4. -92.15 kJ.
5. 5.77×10^5.
6. -700 kJ.
7. (a) iii, (b) iv, (c) iii, (d) iii, (e) i, (f) ii, (g) i, (h) iii, (i) i, (j) ii.
8. -377.795 kJ/mol.

ADDITIONAL NOTES

CHAPTER 24

OXIDATION-REDUCTION AND ELECTROCHEMISTRY

CHAPTER OVERVIEW

In this chapter the terms "oxidation" and "reduction" are defined in terms of loss and gain of electrons. The concept of half-reactions is presented and used as a method for balancing oxidation-reduction equations. Electrochemical cells are considered and the relative strengths of oxidizing and reducing agents are discussed in terms of standard reduction potentials. Faraday's law and stoichiometry are discussed. The Nernst equation is used to determine the spontaneity of a reaction under nonstandard state conditions, and equilibrium conditions for redox reactions are considered. The chapter concludes with a discussion on practical electrochemistry: common cells, storage cells, fuel cells, and corrosion.

COMPETENCIES

Definitions for the following terms should be learned:

- anode
- cathode
- cell potential, electromotive force, emf
- cell reaction
- electrochemical cell
- electrochemistry
- electrode potential
- electrode reaction
- electrolysis
- electrolytic cell
- faraday
- fuel cells
- half-cell
- half-reaction
- ion-electron equation
- overvoltage
- redox couple
- standard electrode potential, standard reduction potential
- storage cells
- voltaic cell, galvanic cell

General concepts and skills that should be learned include

- interpreting oxidation and reduction in terms of electron loss and gain

- writing ion-electron equations to describe oxidation and reduction processes
- balancing oxidation-reduction equations by the half-reaction method
- the types of reactions that can occur at the electrodes of voltaic and electrolytic cells
- using the shorthand notation to represent an electrochemical cell
- using the standard potential to predict reaction spontaneity under standard conditions
- using standard electrode potentials to describe the relative strengths of oxidizing or reducing agents
- predicting electrode reactions in electrolytic cells by using values of standard reduction potentials
- using cell potential to predict reaction spontaneity under nonstandard state conditions
- the chemistry of commercial dry cells, storage batteries, fuel cells, and corrosion

Numerical exercises that should be understood include

- calculating the electrical charge, electrical current, or time when two of these variables are given
- calculating masses, the volumes of gases, or the volumes and concentrations of solutes of reactants and products involved in an oxidation-reduction reaction
- calculating electrical work, cell potential, or electrical charge when two of these variables are given
- calculating $\Delta G°$ or $E°$ when one of these variables is given
- calculating K or $E°$ when one of these variables is given
- calculating a standard potential for an oxidation-reduction reaction from standard electrode potentials for each half-reaction
- calculating the potential for an ion-electron equation or a complete reaction at nonstandard state conditions

QUESTIONS

1. Complete the following sentences by filling in the various blanks with these terms: cell potential, electrochemical cell, electrode potentials, faradays,

fuel cell, half-cells, half-reaction, Nernst equation, primary cells, standard electrode potentials, standard state conditions, and storage cells.

A device in which an electrochemical reaction occurs is known as an (a)______. Such a device is constructed by using two (b)______ in which the respective oxidation and reduction processes occur. For each of these processes, we can write a (c)______. The voltage of the device is sometimes called the (d)______ and represents the chemical driving force of the cell reaction. This potential can be obtained by adding the values of the (e)______ or by adding the values of the (f)______, which represent the potentials of the respective oxidation and reduction reaction under (g)______, and then correcting the result to nonstandard state conditions by using the (h)______.

Voltaic cells may be classified into two groups: (i)______, in which the reactants are depleted irreversibly, and (j)______, in which the reactants may be regenerated by "charging" the cell. To overcome the drawbacks of discarding or charging cell, a third type of cell, the (k)______, was developed, in which the reactants are continuously fed to the cell and the products are usually removed from the cell. The amount of electricity that a cell will generate can be measured in units of (l)______, which is another word for a mole of electrons.

2. Match the correct term to each definition.

(a) the interchangeable oxidized and reduced forms of the same species
(b) the study of oxidation-reduction reactions that either produce or utilize electrical energy
(c) the gain of electrons
(d) the substance that brings about reduction
(e) the substance that is oxidized
(f) includes only the species involved in either the oxidation or the reduction of a single type of species
(g) the substance that is reduced
(h) the loss of electrons

(i) electrochemistry
(ii) ion-electron equation
(iii) oxidation
(iv) oxidizing agent
(v) redox couple
(vi) reducing agent
(vii) reduction

3. Match the correct equation to the reaction described for the various commercial electrochemical cells.

(a) lead storage cell cathode reaction during discharge
(b) nickel-cadmium cell cathode reaction during discharge
(c) hydrogen-oxygen fuel cell anode reaction
(d) Leclanché cell overall reaction
(e) mercury cell cathode reaction
(f) lead storage cell anode reaction during discharge

(i) $H_2(g) + 2OH^- \rightarrow 2H_2O(l) + 2e^-$
(ii) $Pb(s) + HSO_4^- \rightarrow PbSO_4(s) + H^+ + 2e^-$
(iii) $2MnO_2(s) + 2H_2O(l) + Zn(s) \rightarrow Zn(OH)_2(s) + 2MnOOH(s)$
(iv) $[NiOOH] + H_2O(l) + e^- \rightarrow Ni(OH)_2(s) + OH^-$
(v) $HgO(s) + H_2O(l) + 2e^- \rightarrow Hg(l) + 2OH^-$
(vi) $PbO_2(s) + 3H^+ + HSO_4^- + 2e^- \rightarrow PbSO_4(s) + 2H_2O(l)$

4. Which of the following statements are true? Rewrite each false statement so that it is correct.

(a) A reduction is characterized by a gain of electrons, a decrease in oxidation number, and possibly by the removal of oxygen from the substance undergoing reduction.
(b) The Nernst equation relates $E°$ to nonstandard state conditions.
(c) The electromotive force is the potential of a half-cell.
(d) A spontaneous oxidation-reduction reaction requires electrons from an external source.
(e) The standard reduction potential for a given couple has a positive value if the oxidizing agent in the couple has a greater tendency to gain electrons than does the hydrogen ion.
(f) A voltaic cell generates electrical energy from a spontaneous oxidation-reduction reaction.
(g) The anode is the electrode at which reduction occurs.
(h) Cations are always positively charged and move toward the cathode.
(i) The same number of coulombs can be delivered during a fast charge as during a slow charge of a lead storage battery.

5. Choose the response that best completes each statement.

(a) Reduction can be defined as the
(i) gain of electrons (ii) loss of electrons
(iii) increase in the oxidation number (iv) none of these answers

(b) The gain of electrons means

(i) oxidation (ii) reduction

(iii) an endothermic reaction (iv) a reducing agent

(c) Oxidation can be defined as the

(i) decrease in the oxidation number

(ii) gain of electrons

(iii) loss of neutrons

(iv) positive increase in the oxidation number

(d) The loss of electrons means

(i) reduction (ii) oxidation

(iii) an exothermic reaction (iv) an oxidizing agent

(e) Reducing agents

(i) lose electrons (ii) gain electrons

(iii) undergo reduction (iv) are short spies

(f) Metallic cadmium is placed in a solution of $CuSO_4$. At 25 °C, $E°$ = -0.4029 V for Cd^{2+}/Cd and 0.337 V for Cu^{2+}/Cu. Which statement is true?

(i) No reaction occurs.

(ii) Cd reduces Cu^{2+}.

(iii) Cu reduces any Cd^{2+} that is formed.

(iv) Cu^{2+} reduces the Cd.

(g) What is the quantity of electrical charge delivered by a current of one ampere in one second?

(i) a joule (ii) a coulomb

(iii) an ohm (iv) a faraday

(h) An Avogadro number of electrons is equal to

(i) an ampere (ii) a coulomb

(iii) a faraday (iv) a joule

(i) Which is *not* true about the shorthand notation used to represent electrochemical cells?

(i) The anode compartment is placed on the left side.

(ii) The symbol "|" represents a phase boundary.

(iii) The symbol "‖" represents the external load on the circuit.

(iv) The symbol "," represents two substances in the same phase.

(j) Which of the following statements is true?

(i) Oxidation occurs at the anode during electrolysis but at the cathode in a voltaic cell.

(ii) Oxidation always occurs at the cathode.

(iii) Reduction occurs at the cathode for cations and at the anode for anions.

(iv) none of these answers

(k) Consider the following reduction half-reaction to take place in the cathode compartment of a voltaic cell: $M^+(1\ M) + e^- \rightarrow M(s)$. The cathode half-cell would be symbolized by

(i) $Mn(s)|Mn^+(1\ M)$ (ii) $Mn^+(1\ M), Mn(s)$

(iii) $Mn^+(1\ M)\|Mn(s)$ (iv) $Mn^+(1\ M)|Mn(s)$

(l) A cell diagram is given as follows:

$$Ni(s)|Ni^{2+}(1\ M)\|Cu^{2+}(1\ M)|Cu(s)$$

The cell reaction corresponding to this diagram is

(i) $Ni(s) + Cu(s) \rightleftharpoons Ni^{2+} + Cu^{2+}$ (ii) $Ni^{2+} + Cu^{2+} \rightarrow Ni(s) + Cu(s)$

(iii) $Cu(s) + Ni^{2+} \rightleftharpoons Cu^{2+} + Ni(s)$ (iv) $Ni(s) + Cu^{2+} \rightleftharpoons Ni^{2+} + Cu(s)$

(m) A voltaic cell is one for which at standard conditions

(i) $E° < 0$ (ii) $E° > 0$

(iii) $E° = 0$ (iv) $E \neq E°$

(n) Which is *not* part of a functioning cell?

(i) oxidation occurring at one electrode and reduction at the other

(ii) equilibrium conditions

(iii) electron flow through an external conductor

(iv) ion flow in the electrolyte

(o) The anode is

(i) always positive

(ii) the side of oxidation

(iii) the electrode to which the cations move

(iv) the electrode that determines cell life

(p) A battery

(i) will deliver a potential that is proportional to its physical size

(ii) is two or more voltaic cells combined to increase voltage and/or current

(iii) is an accumulator

(iv) can generate electricity from an unfavorable chemical reaction

(q) The standard half-cell potential
(i) can be measured directly with a potentiometer
(ii) is 0.00 V
(iii) is assigned by reference to the standard hydrogen electrode
(iv) all of these answers

(r) Decomposition of a compound by using electrical energy is known as
(i) electrolysis
(ii) hydrolysis
(iii) galvanization
(iv) the Nernst effect

(s) In an electrolytic cell
(i) both electrons and positive ions flow toward the cathode
(ii) electrons flow toward the cathode and positive ions flow toward the anode
(iii) both electrons and negative ions flow toward the anode
(iv) both electrons and negative ions flow toward the cathode

(t) During electrolysis, the half-reaction that is predicted to occur
(i) at the anode is the one with the most positive standard reduction potential
(ii) at the cathode is the one with the most positive standard reduction potential
(iii) always involves the decomposition of the solvent
(iv) is the one involving the fewer number of electrons

(u) Which of the following equations represents the reaction occurring at the anode during electrolysis of a dilute solution of sulfuric acid?
(i) $2H^+ + 2e^- \rightarrow H_2(g)$
(ii) $2H_2O(l) \rightarrow O_2(g) + 4H^+ + 4e^-$
(iii) $2H_2O(l) + 2e^- \rightarrow 2OH^- + H_2(g)$
(iv) $HSO_4^- + 3H^+ \rightarrow SO_3(g) + 2H_2O(l) + 3e^-$

(v) Sodium metal is not obtained at the cathode during the electrolysis of an aqueous solution of NaCl because
(i) sodium metal reacts with the water as fast as it is formed
(ii) chloride ions are more readily oxidized than sodium
(iii) H_2O is oxidized more readily than Na^+
(iv) H_2O is reduced more readily than Na^+

(w) The electrolysis of a solution containing NaCl produces
(i) Na and Cl_2
(ii) H_2 and O_2
(iii) H_2 and ClO^-
(iv) H_2, Cl_2, and NaOH(aq)

(x) The electrolysis of aqueous $CuSO_4$ produces
   (i) $Cu(OH)_2$
   (ii) H_2 and O_2
   (iii) an acid solution near the cathode and an alkaline solution near the anode
   (iv) Cu and O_2

(y) When the surface of a tin can is scratched the exposed iron rusts very rapidly. The best explanation for this behavior is that
   (i) iron is a very active metal
   (ii) a galvanic cell is formed in which iron is the anode
   (iii) a galvanic cell is formed in which iron is the cathode
   (iv) tin is a less reactive metal than iron

6. Describe the operation of an electrochemical cell.

7. Whether we perform a specific redox reaction electrochemically or not, the same chemical change will occur. What is an advantage of carrying out the process electrochemically?

8. In writing the notation for an electrochemical cell, what does (a) a vertical line, (b) a comma, (c) the formula of a substance in parentheses, and (d) a double vertical line represent?

9. What is the purpose of a salt bridge?

10. What is Faraday's law? How do we calculate the number of moles of electrons from the change in the electrical charge? How do we calculate the amount of electrical charge from electrical current?

11. How would a list of standard oxidation potentials differ from a table of standard reduction potentials?

12. How does the numerical value of an electrode potential reflect the strength of a reducing agent?

13. When using $E°$ values to predict the course of a given oxidation-reduction reaction, what restrictions on reaction conditions must a chemist keep in mind?

14. How are standard reduction potentials for half reactions combined to find the potential of a given redox reaction?

15. How do we predict which of several reduction reactions will take place at the cathode during electrolysis? What about the oxidation processes at the anode? Why are such predictions not always correct?

16. Write the equation relating the potential of an electrochemical cell or half-reaction to the free energy change. Define all symbols.

17. What is the sign convention for potential used to describe the spontaneity of a reaction?

18. What is the mathematical relationship between the potential at nonstandard state conditions and $E°$ and the reaction quotient?

19. What happens to an electrochemical cell once equilibrium has been reached? How can the equilibrium constant be calculated from $E°$?

20. How are standard reduction potentials for half reactions combined to find the potential of a new half-reaction?

21. Why do we measure "open-circuit" potentials for electrochemical cells? What happens to the potential of a cell if an appreciable current is being drawn from the cell?

22. Distinguish between primary and secondary (storage) cells. Name a cell of each type that plays an important part of our everyday lives.

Answers to questions

1. (a) electrochemical cell, (b) half-cells, (c) half-reaction, (d) cell potential, (e) electrode potentials, (f) standard electrode potentials, (g) standard state conditions, (h) Nernst equation, (i) primary cells, (j) storage cells, (k) fuel cell, (l) faradays.
2. (a) v, (b) i, (c) vii, (d) vi, (e) vi, (f) ii, (g) iv, (h) iii.
3. (a) vi, (b) iv, (c) i, (d) iii, (e) v, (f) ii.
4. (a) T; (b) T; (c) F, replace *electromotive force* with *electrode potential*; (d) F, replace *spontaneous* with *nonspontaneous*; (e) T; (f) T; (g) F, replace *anode* with *cathode*; (h) T; (i) T.
5. (a) i, (b) ii, (c) iv, (d) ii, (e) i, (f) ii, (g) ii, (h) iii, (i) iii, (j) iv, (k) iv, (l) iv, (m) ii, (n) ii, (o) ii, (p) ii, (q) iii, (r) i, (s) i, (t) ii, (u) ii, (v) iv, (w) iv, (x) iv, (y) ii.
6. An electrochemical cell operates by (a) an oxidation reaction occurring at one electrode and a reduction reaction occurring at the other electrode,

(b) electrons flowing through an external circuit, and (c) ions flowing through the electrolyte.

7. The advantage is that it can serve as a source of electrical energy that can do work.
8. The symbols represent (a) physical contact between two different phases, (b) two substances in the same phase, (c) an inert electrode, and (d) a salt bridge or membrane.
9. A salt bridge permits ion flow while preventing the electrolytes from mixing.
10. Faraday's law states that the extent to which an electrochemical reaction occurs is related to the number of moles of electrons transferred. The total charge is divided by Faraday's constant. The current (expressed in A) is multiplied by the time (expressed in s).
11. Each half-reaction would be reversed and the values of $E°$ would be the same but with opposite sign.
12. At the top of the table the couples with large negative values will be strong reducing agents and at the bottom of the table the couples with large positive values will be weak reducing agents.
13. The reaction conditions must be standard state conditions.
14. The steps include using the reduction half-reaction and potential as found in the table and the oxidation half-reaction and potential obtained by reversing the corresponding reduction half-reaction and changing the sign on the potential, multiplying the two half-reactions by appropriate coefficients so that the number of electrons gained equals the number lost, and adding the two potentials.
15. The most probable of several reduction reactions that will take place at the cathode during electrolysis will be the half-reaction with the most positive reduction potential. The most probable of several oxidation reactions which will take place at the anode during electrolysis will be the half-reaction with the least positive reduction potential. The predictions may be error because of overvoltage effects, nonstandard state conditions, etc.
16. The equation is $\Delta G = -nFE$ where ΔG is free energy change, n is number of moles of electrons in the redox reaction, F is Faraday's constant, and E is electromotive force or potential.
17. The sign for the potential will be positive for all spontaneous reactions, negative for nonspontaneous reactions, and zero for systems at equilibrium.

18. $E = E° = - \frac{0.0592}{n} \log Q$

19. At equilibrium, the cell potential is zero and the flow of electrons ceases. The equilibrium constant can be calculated from $E°$ at 25 °C from

$$\log K = \frac{nE°}{0.0592}$$

20. The potential is obtained by multiplying each potential by the number of electrons in the respective half-reaction, adding the values algebraically, and dividing the sum by the number of electrons in the new half-reaction.

21. There is no voltage drop as a result of drawing current from the cell. The potential is less than an open-circuit potential.

22. Primary cells such as the Leclanché cell use up reactants irreversibly and secondary cells such as the lead storage cell can be recharged.

SKILLS

Skill 24.1 Balancing redox equations: half-reaction method

A redox equation can be balanced by writing two net ionic equations (see *Skill 5.3 Writing net ionic equations*), called ion-electron equations or half-reactions, one representing the oxidation process and the other representing the reduction process. The five steps in this method are as follows:

1. Write the overall unbalanced equation.
2. Identify the oxidized and reduced substances and write the unbalanced ion-electron equation.
3. Balance each ion-electron equation for atoms and charge.
   a. Balance for atoms other than O or H.
   b. Balance O atoms by adding H_2O on the side deficient in O atoms.
   c. Balance H atoms by adding H^+ for acidic solutions or OH^- and H_2O for alkaline solutions (add OH^- to same side as H_2O in Step 3b, add an equal number of H_2O as OH^- to the other side, cancel extra H_2O from both sides).
   d. Balance charge by adding electrons.
4. Multiply ion-electron equations by appropriate factors so that electrons gained equals electrons lost.
5. Add the ion-electron equations, cancelling where appropriate.

Three examples that illustrate this method are given:

Example (1) $I_2(s) + H_2S(aq) \rightarrow I^- + S(s)$

Step 1. Write the overall unbalanced equation.

$$H_2S + I_2 \rightarrow I^- + S$$

Step 2. Identify the oxidized and reduced substances and write the unbalanced ion-electron equations.

$$I_2 \rightarrow I^- \qquad H_2(s) \rightarrow S$$

Step 3. Balance each ion-electron equation for atom and charge.

a. Balance for atoms other than O or H.

$$I_2 \rightarrow 2I^- \qquad H_2S \rightarrow S$$

b. Balance O atoms by adding H_2O on side deficient in O atoms.

$$I_2 \rightarrow 2I^- \qquad H_2S \rightarrow S$$

c. Balance H atoms by adding H^+ for acidic solution.

$$I_2 \rightarrow 2I^- \qquad H_2S \rightarrow S + 2H^+$$

d. Balance charge by adding electrons.

$$I_2 + 2e^- \rightarrow 2I_2 \qquad H_2S \rightarrow S + 2H^+ + 2e^-$$

Step 4. Multiply ion-electron equations by appropriate factors so that electrons gained equals electrons lost.

$$I_2 + 2e^- \rightarrow 2I^- \qquad H_2S \rightarrow S + 2H^+ + 2e^-$$

Step 5. Add the ion-electron equations, cancelling where appropriate.

$$I_2 + 2e^- \rightarrow 2I^-$$

$$H_2S \rightarrow S + 2H^+ + 2e^-$$

$$I_2 + H_2S + 2e^- \rightarrow S + 2I^- + 2H^+ + 2e^-$$

$$I_2(s) + H_2S(aq) \rightarrow S(s) + 2H^+ + 2I^-$$

Example (2) $C_2H_2(g) + O_2(g) \xrightarrow{H^+} CO_2(aq)$

Step 1. Write the overall unbalanced equation.

$$C_2H_2 + O_2 \rightarrow CO_2$$

Step 2. Identify oxidized and reduced substances and write the unbalanced ion-electron equations.

$$C_2H_2 \rightarrow CO_2 \qquad O_2 \rightarrow$$

(The second half-reaction will involve H_2O as a product--see Step 3c.)

Step 3. Balance each ion-electron equation for atoms and charge.

a. Balance for atoms other than O or H.

$$C_2H_2 \rightarrow 2CO_2 \qquad O_2 \rightarrow$$

b. Balance O atoms by adding H_2O on the side deficient in O atoms.

$$C_2H_2 + 4H_2O \rightarrow 2CO_2 \qquad O_2 \rightarrow 2H_2O$$

c. Balance H atoms by adding H^+ for acidic solutions.

$$C_2H_2 + 4H_2O \rightarrow 2CO_2 + 10H^+ \qquad O_2 + 4H^+ \rightarrow 2H_2O$$

d. Balance charge by adding electrons.

$$C_2H_2 + 4H_2O \rightarrow 2CO_2 + 10H^+ + 10e^- \qquad O_2 + 4H^+ + 4e^- \rightarrow 2H_2O$$

4. Multiply ion-electron equations by appropriate factors so that electrons gained equals electrons lost.

$$(2)(C_2H_2 + 4H_2O \rightarrow 2CO_2 + 10H^+ + 10e^-) \quad (5)(O_2 + 4H^+ + 4e^- \rightarrow 2H_2O)$$

5. Add the ion-electron equations, cancelling where appropriate.

$$(2)(C_2H_2 + 4H_2O \rightarrow 2CO_2 + 10H^+ + 10e^-)$$

$$\underline{(5)(O_2 + 4H^+ + 4e^- \rightarrow 2H_2O)}$$

$$2C_2H_2 + 8H_2O + 5O_2 + 20H^+ + 20e^- \rightarrow 4CO_2 + 20H^+ + 20e^- + 10H_2O$$

$$2C_2H_2(g) + 5O_2(g) \rightarrow 4CO_2(aq) + 2H_2O(l)$$

Example (3) $Zn(s) + HgO(s) \xrightarrow{OH^-} Hg(l) + [Zn(OH)_4]^{2-}$

Step 1. Write the overall unbalanced equation.

$$Zn + HgO \rightarrow Hg + [Zn(OH)_4]^{2-}$$

Step 2. Identify the oxidized and reduced substances and write the unbalanced ion-electron equations.

$$Zn \rightarrow [Zn(OH)_4]_2^- \qquad HgO \rightarrow Hg$$

Step 3. Balance each ion-electron equation for atom and charge.

a. Balance for atoms other than O or H.

$$Zn \rightarrow [Zn(OH)_4]^{2-} \qquad HgO \rightarrow Hg$$

b. Balance O atoms by adding H_2O on side deficient in O atoms.

$$Zn + 4H_2O \rightarrow [Zn(OH)_4]^{2-} \qquad HgO \rightarrow Hg + H_2O$$

c. Balance H atoms by adding OH^- and H_2O for alkaline solutions.

Add OH^- to same side as H_2O.

$$Zn + 4H_2O + 4OH^- \rightarrow [Zn(OH)_4]^{2-} \qquad HgO \rightarrow Hg + H_2O + 2OH^-$$

Add equal number of H_2O as OH^-.

$$Zn + 4H_2O + 4OH^- \rightarrow [Zn(OH)_4]^{2-} + 4H_2O \qquad HgO + 2H_2O \rightarrow Hg + H_2O + 2OH^-$$

Cancel extra H_2O.

$$Zn + 4OH^- \rightarrow [Zn(OH)_4]^{2-} \qquad HgO + H_2O \rightarrow Hg + 2OH^-$$

d. Balance charge by adding electrons.

$$Zn + 4OH^- \rightarrow [Zn(OH)_4]^{2-} + 2e^- \qquad HgO + H_2O + 2e^- \rightarrow Hg + 2OH^-$$

Step 4. Multiply ion-electron equations by appropriate factors so that electrons gained equals electrons lost.

$$Zn + 4OH^- \rightarrow [Zn(OH)_4]^{2-} + 2e^- \qquad HgO + H_2O + 2e^- \rightarrow Hg + 2OH^-$$

Step 5. Add the ion-electron equations, cancelling where appropriate.

$$\begin{array}{rcl} Zn + 4OH^- & \rightarrow & [Zn(OH)_4]^{2-} + 2e^- \\ HgO + H_2O + 2e^- & \rightarrow & Hg + 2OH^- \\ \hline Zn + 4OH^- + HgO + H_2O + 2e^- & \rightarrow & [Zn(OH)_4]^{2-} + 2e^- + Hg + 2OH^- \\ Zn(s) + HgO(s) + 2OH^- + H_2O(l) & \rightarrow & [Zn(OH)_4]^{2-} + Hg(l) \end{array}$$

Once the stepwise process for balancing redox equations has been mastered, the steps are combined and the results look like those given in Step 5 only. Two examples will illustrate this abbreviated method:

Example (1) $C_6H_4(OH)_2(aq) + Cr_2O_7^{2-} \xrightarrow{H^+} C_6H_4O_2(aq) + Cr^{3+}$

$$\begin{array}{rcl} (3)(C_6H_4(OH)_2(aq) & \rightarrow & C_6H_4O_2 + 2H^+ + 2e^-) \\ Cr_2O_7^{2-} + 14H^+ + 6e^- & \rightarrow & 2Cr^{3+} + 7H_2O \\ \hline 3C_6H_4(OH)_2(aq) + Cr_2O_7^{2-} + 8H^+ & \rightarrow & 3C_6H_4O_2(aq) + 2Cr^{3+} + 7H_2O \end{array}$$

Example (2) $Cl_2(aq) + IO_3^- \xrightarrow{OH^-} IO_4^- + Cl^-$

$$\begin{array}{rcl} IO_3^- + 2OH^- & \rightarrow & IO_4^- + H_2O + 2e^- \\ Cl_2 + 2e^- & \rightarrow & 2Cl^- \\ \hline IO_3^- + Cl_2(aq) + 2OH^- & \rightarrow & IO_4^- + 2Cl^- + H_2O(l) \end{array}$$

Skill Exercise: Balance the following redox equations by the half-reaction method: (a) $Ag^+ + AsH_3(g) \rightarrow Ag(s) + H_3AsO_3(aq)$, (b) $MnO_4^- + HCl(aq) \rightarrow Mn^{2+} + Cl_2(g)$, (c) $H_2O(l) \rightarrow H_2(g) + O_2(g)$ under acidic conditions, (d) $F_2(g) + Cl^- \rightarrow Cl_2(g) + F^-$, and (e) $Cr(OH)_3(s) + Cl_2(g) \rightarrow CrO_4^{2-} + Cl^-$ under alkaline conditions. *Answers*: (a) $Ag^+ + e^- \rightarrow As$, $3H_2O + AsH_3 \rightarrow H_3AsO_3 + 6H^+ + 6e^-$, $6Ag^+ + 3H_2O(l) + AsH_3(g) \rightarrow H_3AsO_3(aq) + 6H^+ + 6As(s)$; (b) $MnO_4^- + 8H^+ + 5e^- \rightarrow Mn^{2+} + 4H_2O$, $2Cl^- \rightarrow Cl_2 + 2e^-$, $2MnO_4^- + 16H^+ + 10Cl^- \rightarrow 5Cl_2(g) + 2Mn^{2+} + 8H_2O(l)$; (c) $2H^+ + 2e^- \rightarrow H_2$, $2H_2O \rightarrow 4H^+ + O_2 + 4e^-$, $2H_2O(l) \rightarrow 2H_2(g) + O_2(g)$; (d) $F_2 + 2e^- \rightarrow 2F^-$, $2Cl^- \rightarrow Cl_2 + 2e^-$, $F_2(g) + 2Cl^- \rightarrow 2F^- + Cl_2(g)$; (e) $Cr(OH)_3 + 5OH^- \rightarrow CrO_4^{2-} + 4H_2O + 3e^-$, $Cl_2 + 2e^- \rightarrow 2Cl^-$, $2Cr(OH)_3(s) + 10OH^- + 3Cl_2(g) \rightarrow 6Cl^- + 2CrO_4^{2-} + 8H_2O(l)$.

Skill 24.2 Predicting reaction spontaneity by using E

Each ion-electron equation written in the form of a reduction half-reaction has a standard electrode potential which represents the ease or difficulty of the reduction process. For example, $E° = 0.5355$ V for the reduction of I_2 to I^- and $E° = 0.142$ V for the reduction of S to H_2S. Because the potential for the iodine reaction is more positive than that of the sulfur reaction, I_2 will undergo reduction and, consequently, oxidize the H_2S under standard conditions.

$$I_2(s) + H_2S(aq) \rightarrow S(s) + 2H^+ + 2I^-$$

Skill Exercise: Confirm that each of the reactions (a), (b), (d), and (e) in *Skill 24.1 Balancing redox equations: half-reaction method* will be spontaneous by using the following set of standard reduction potentials: 2.87 V for F_2/F^-, 1.51 V for MnO_4^-/Mn^{2+}, 1.3597 V for Cl_2/Cl^-, 1.229 V for O_2/H_2O, 0.7991 V for Ag^+/Ag, 0.0000 V for H^+/H_2, -0.13 V for $CrO_4^{2-}/Cr(OH)_3$ and -0.14 V for H_3AsO_3/AsH_3.

The sign and magnitude of the overall cell potential give the information necessary to predict whether or not an oxidation-reduction reaction will occur as written. For a spontaneous process, $E > 0$; for a nonspontaneous process, $E < 0$; and at equilibrium $E = 0$. For example, under standard conditions, $E° = 0.394$ V for the reaction between I_2 and H_2S and this reaction will occur as written. As a further example, if this reaction is carried out under reaction conditions such that $P_{H_2S} = 0.1$ bar, $[I^-] = 0.01$ mol/L, and $[H^+] = 1 \times 10^{-7}$ mol/L, then $E = 0.90$ V as determined by the Nernst equation, and this reaction will be even more favorable under these conditions.

Skill Exercise: Predict whether or not the reactions (a) - (e) given in *Skill 24.1 Balancing redox equations: half-reaction method* are spontaneous under standard state conditions by using the following overall cell potentials: (a) $E° = 0.94$ V, (b) $E° = 0.15$ V, (c) $E° = -1.229$ V, (d) $E° = 1.51$ V, (e) $E° = 1.49$ V. *Answers*: (a) spontaneous, (b) spontaneous, (c) nonspontaneous, (d) spontaneous, (e) spontaneous.

Skill 24.3 Interpreting electrochemical cell notation

Electrochemical cells are commonly represented by a short-hand notation in which the electrode materials are written at the extreme left for the anode and at the extreme right for the cathode, the species in solution are written in the center, single vertical lines indicate physical contact between species in different phases, commas separate species that are in the same phase, double vertical lines represent a salt bridge, and parentheses around a symbol or name of a material means that the material is an inert electrode. For example, the designation

$$(C)\ |\ Li(g)\ |\ LiCl(l)\ |\ Cl_2(g)\ |\ (C)$$

represents a cell in which gaseous lithium is undergoing oxidation at a graphite anode, gaseous chlorine is undergoing reduction at a graphite cathode, and the products of the half-reactions mix, giving molten lithium chloride. For the cell reaction given by the chemical equation

$$Hg(l) + H_2O(l) \xrightarrow{OH^-} H_2(g) + HgO(s)$$

the shorthand notation would be

$$(Pt)\ |\ Hg(l)\ |\ HgO(s)\ |\ OH^-\ |\ H_2\ |\ (Pt)$$

The five vertical lines represent the contact of each of the phases. The Hg and HgO are written on the left because these substances are involved in the oxidation process. The H_2 and OH^- are written on the right because these substances are involved in the reduction process. The additional inert electrode materials are necessary because of the construction of the half-cells.

Skill Exercise: Write the chemical equation for the following electrochemical cells: (a) (Pt) | Hg_2^{2+}, Hg^{2+}, H^+ | $H_2(g)$ | (Pt) and (b) (Pt) | Fe^{2+}, Fe^{3+} ‖ Ag^+ | Ag(s), and write the cell notation for (c) $Tl^+ + 2AgCl(s) \rightarrow 2Ag(s) + Tl^{3+} + 2Cl^-$ and (d) $Ag(s) + HgCl(s) \rightarrow AgCl(s) + Hg(l)$. *Answers*: (a) $Hg_2^{2+} + 2H^+ \rightarrow H_2(g) + 2Hg^{2+}$, (b) $Fe^{2+} + Ag^+ \rightarrow Fe^{3+} + Ag(s)$, (c) (Pt) | Tl^+, Tl^{3+}, Cl^- | AgCl(s) | Ag(s), (d) Ag | AgCl | Cl^- | HgCl | Hg | (Pt).

Skill 24.4 Predicting electrode reactions in electrolytic cells

The oxidation half-reaction predicted to occur at the anode during electrolysis is the one with the least positive standard reduction potential and the reduction half-reaction predicted to occur at the cathode is the one with the most positive standard reduction potential. For example, during the electrolysis of an aqueous solution of $CuCl_2$, two oxidation reactions that are reasonable at the anode involve the Cl^- ion and the solvent H_2O. Written as reduction half-reactions and voltages, these are

$$Cl_2 + 2e^- \rightarrow 2Cl^- \qquad E° = 1.3597\ V$$

$$O_2 + 4H^+ + 4e^- \rightarrow 2H_2O \qquad E° = 1.229\ V$$

Based on the values of $E°$, the anode reaction will be the one involving the decomposition of water because this half-reaction has the least positive $E°$. At the cathode the reasonable reduction reactions involve Cu^{2+}, H^+, and the solvent H_2O. The reactions and voltages are

$$Cu^{2+} + 2e^- \rightarrow Cu \qquad E° = 0.337\ V$$
$$2H^+ + 2e^- \rightarrow H_2 \qquad E° = 0.0000\ V$$
$$2H_2O + 2e^- \rightarrow H_2 + 2OH^- \qquad E° = -0.8281\ V$$

The favorable cathode reaction will be the one involving the plating out of Cu because this half-reaction has the most positive $E°$. Note that these predictions are based on standard state conditions and the assumption that there is no overvoltage.

Skill Exercise: Predict the electrode reactions taking place during the electrolysis of aqueous solutions of (a) $AuCl_3$, (b) $AuBr_3$, (c) $NiCl_2$, and (d) $NiBr_2$ given the following standard reduction potentials: $E°$ = 1.087 V for Br_2/Br^-; $E°$ = 1.229 V for $O_2/H^+,H_2O$; $E°$ = 1.3597 V for Cl_2/Cl^-; $E°$ = -0.8281 V for $H_2O,OH^-/H_2$; $E°$ = 0.0000 V for H^+/H_2; $E°$ = -0.250 V for Ni^{2+}/Ni; $E°$ = 1.498 V for Au^{3+}/Au. *Answers*: (a) reduction of Au^{3+} and formation of O_2, (b) reduction of Au^{3+} and generation of Br_2, (c) formation of H_2 and of O_2, (d) formation of H_2 and of Br_2.

NUMERICAL EXERCISES

*Example 24.1 Electrochemical stoichiometry**

An electrical current of 0.0100 A was passed between two copper electrodes placed in an aqueous solution of Cu^{2+} for a period of 24 h. What mass of copper metal was purified by being exchanged between the electrodes?

This problem is solved by using the comprehensive problem-solving approach presented in *Skill 2.7 Solving problems*.

1. Study the problem and be sure you understand it.
   (a) What is unknown?
   The mass of copper transferred between the electrodes.
   (b) What is known?
   The electrical current used in the process and the time.
2. Decide how to solve the problem.
   (a) What is the connection between the known and the unknown?
   The quantity of electrical charge that was passed through the cell is related to the current and the time. The number of moles of electrons is related to the charge by the faraday (see *Skill 4.4 Finding the number of moles*). The reaction

$$Cu^{2+} + 2e^- \rightarrow Cu(s)$$

shows that 1 mol of Cu is produced for 2 mol e^-.

(b) What is necessary to make the connection?

The amount of electrical charge is found by multiplying the electrical current in amperes by the time in seconds. The number of moles of electrons is found by multiplication of [1/faraday] and the usual stoichiometric calculations are performed.

3. Set up the problem and solve it.

The electrical charge is

$$(0.0100\ A)\left[\frac{1\ C}{1\ A\ s}\right](24\ h)\left[\frac{60\ min}{1\ h}\right]\left[\frac{60\ s}{1\ min}\right] = 864\ C$$

The number of moles of electrons is

$$(864\ C)\left[\frac{1\ mol\ e^-}{96{,}500\ C}\right] = 8.95 \times 10^{-3}\ mol\ e^-$$

The mass of Cu produced is

$$(8.95 \times 10^{-3}\ mol\ e^-)\left[\frac{1\ mol\ Cu}{2\ mol\ e^-}\right]\left[\frac{63.55\ g\ Cu}{1\ mol\ Cu}\right] = 0.284\ g\ Cu$$

4. Check the result.

(a) Are significant figures and the location of the decimal point correct?

Yes. For most calculations, the rounded value of 96,500 C/1 mol e^- for the faraday is valid.

(b) Did the answer come out in the correct units?

Yes.

(c) Is the answer reasonable?

Yes. Only a small amount of metal will be transferred between electrodes by using such a small electrical current.

Exercise 24.1 What must be the current (assuming 100% efficiency) to produce 1.00 g of Na each hour from molten NaCl? *Answer*: 1.17 A.

Example 24.2 Thermodynamics of electrochemical cells

Suppose an electrochemical cell was designed for the reaction

$$O_3(g) + NO(g) \rightarrow O_2(g) + NO_2(g) \qquad \Delta G° = -194\ kJ$$

What would be the potential of this cell?

The half-reactions and overall cell equation are

$$2e^- + 2H^+ + O_3 \rightarrow O_2 + H_2O$$

$$H_2O + NO \rightarrow NO_2 + 2H^+ + 2e^-$$

$$O_3 + NO + 2e^- \rightarrow O_2 + NO_2 + 2e^-$$

The relationship between cell potential and free energy is

$$\Delta G° = -nFE°$$

where n is the number of moles of electrons and F = 96.5 kJ/V mol e^-. For this reaction n = 2, giving

$$E° = \frac{-\Delta G°}{nF} = \frac{-(-194 \text{ kJ})}{(2 \text{ mol } e^-)(96.5 \text{ kJ/V mol } e^-)} = 1.01 \text{ V}$$

The electrochemical cell would generate 1.01 V under standard state conditions.

Exercise 24.2 The standard state cell potential for the reaction

$$2H_2O_2(aq) \rightarrow O_2(g) + 2H_2O(l)$$

is $E°$ = 0.547 V. Calculate the free energy change. *Answer*: n = 4, 211 kJ.

Example 24.3 Thermodynamics of electrochemical cells

For the reaction between Al and Fe^{2+}, $E°$ = 1.222 V at 25 °C.

$$2Al(s) + 3Fe^{2+} \rightarrow 3Fe(s) + 2Al^{3+}$$

What is the value of the equilibrium constant for this reaction?

The equilibrium constant is related to $E°$ by

$$\log K = \frac{n}{0.0592} E°$$

In this case

$$\begin{array}{rcl} (2)(Al & \rightarrow & Al^{3+} + 3e^-) \\ (3)(Fe^{2+} + 2e^- & \rightarrow & Fe) \\ \hline 2Al(s) + 3Fe^{2+} + 6e^- & \rightarrow & 3Fe(s) + 2Al^{3+} + 6e^- \end{array}$$

the value of n is 6, giving

$$\log K = \frac{n}{0.0592} E° = \frac{6}{0.0592}(1.222) = 124$$

and upon taking antilogarithms, $K = 10^{124}$.

Exercise 24.3 Calculate the value of $E°$ for the reaction

$$Zn(s) + Cu^{2+} \rightarrow Zn^{2+} + Cu(s)$$

given $K = 1.8 \times 10^{37}$. *Answer*: n = 2, 1.10 V.

Example 24.4 Thermodynamics of electrochemical cells

An electrochemical cell is capable of delivering a steady electrical current of 0.100 A for one week at a cell potential 1.35 V. Calculate the electrical work that the cell is designed to perform.

The electrical work is given by the product of the cell potential and the amount of electrical charge delivered:

$$w = -Eq$$

The electrical charge is

$$q = It$$
$$= (0.100\ \text{A})(1\ \text{wk})\left[\frac{7\ \text{day}}{1\ \text{wk}}\right]\left[\frac{24\ \text{h}}{1\ \text{day}}\right]\left[\frac{3600\ \text{s}}{1\ \text{h}}\right]\left[\frac{1\ \text{C}}{1\ \text{A s}}\right]$$
$$= 6.05 \times 10^4\ \text{C}$$

and the electrical work is

$$w = -(1.35\ \text{V})(6.05 \times 10^4\ \text{C})\left[\frac{1\ \text{J}}{1\ \text{V C}}\right] = -8.17 \times 10^4\ \text{J}$$

The cell is capable of delivery 8.17×10^4 J.

Exercise 24.4 An engineer required a battery capable of doing 6.0 J of work upon demand while only passing 0.50 C of electrical charge. What must be the voltage of this battery? *Answer*: 12 V.

Example 24.5 Calculating E°*

Consider the reaction between Al and Fe^{2+}

$$2Al(s) + 3Fe^{2+} \rightarrow 2Al^{3+} + 3Fe(s)$$

Given that the standard reduction potential for Al^{3+}/Al is -1.662 V and for Fe^{2+}/Fe is -0.4402 V, find $E°$ for the reaction.

This problem is solved by using the comprehensive problem-solving approach presented in *Skill 2.7 Solving problems*.

1. Study the problem and be sure you understand it.
   (a) What is unknown?
   The cell potential.
   (b) What is known?
   The $E°$ values for each reduction couple.
2. Decide how to solve the problem.
   (a) What is the connection between the known and the unknown?
   The cell potential is the sum of the potentials for each half-cell. The reduction half-cell potential will be the value given above. The oxidation half-cell potential will be the negative of the value given above for reduction.
   (b) What is necessary to make the connection?
   Each half-reaction (one for oxidation and one for reduction) and half-cell potential is written. The overall cell equation and potential are found by adding these half-reaction expressions together.

3. Set up the problem and solve it.

$$
\begin{array}{lll}
2(Al & \rightarrow & Al^{3+} + 3e^-) \qquad & E° = 1.662 \text{ V} \\
\underline{3(Fe^{2+} + 2e^-} & \underline{\rightarrow} & \underline{Fe)} \qquad & \underline{E° = -0.4402 \text{ V}} \\
2Al + 3Fe^{2+} & \rightarrow & 3Fe + 2Al^{3+} \qquad & E° = 1.222 \text{ V}
\end{array}
$$

The potential of the cell is 1.222 V under standard state conditions. Note that even though the half-reactions are multiplied by coefficients to make the number of electrons gained equal to the number lost, the half-cell voltages are not multiplied by these factors.

4. Check the result.
   (a) Are significant figures and the location of the decimal point correct?
   Yes. Quite often the number of significant figures for the electron couples are different. (Review *Skill 2.2 Using significant figures* if needed.)
   (b) Did the answer come out in the correct units?
   Yes.
   (c) Is the answer reasonable?
   Yes.

Exercise 24.5 $E°$ was measured as 1.100 V for the reaction

$$Zn(s) + Cu^{2+} \rightarrow Zn^{2+} + Cu(s)$$

Find $E°$ for the reduction of Cu^{2+} to Cu given that $E° = -0.7628$ V for reduction of Zn^{2+} to Zn. *Answer*: 0.337 V.

Example 24.6 Calculating **E°**

The standard reduction potential is 1.61 V for Ce^{4+}/Ce^{3+} and -2.483 V for Ce^{3+}/Ce. What is the standard reduction potential for Ce^{4+}/Ce?

The potential for a half-reaction such as

$$Ce^{4+} + 4e^- \rightarrow Ce$$

can be found from

$$
\begin{array}{lll}
Ce^{4+} + e^- \rightarrow Ce^{3+} & \qquad & E° = 1.61 \text{ V} \\
Ce^{3+} + 3e^- \rightarrow Ce & \qquad & E° = -2.483 \text{ V}
\end{array}
$$

by multiplying each known $E°$ value by the number of electrons involved in the half-reaction, adding these numbers, and dividing by the number of electrons in the desired half-reaction. Thus

$$E° = \frac{(1)(1.61 \text{ V}) + (3)(-2.483 \text{ V})}{4} = -1.46 \text{ V}$$

The standard reduction potential for Ce^{4+}/Ce is -1.46 V.

Exercise 24.6 The standard state reduction potential for Cr^{2+}/Cr is -0.913 V and -0.744 V for Cr^{3+}/Cr. Calculate $E°$ for Cr^{3+}/Cr^{2+}. *Answer*: -0.406 V.

Example 24.7 Calculating **E***

Find the cell potential for the cell described in *Example 24.5* given $[Al^{3+}] = 0.10$ mol/L and $[Fe^{2+}] = 0.25$ mol/L.

This problem is solved by using the comprehensive problem-solving approach presented in *Skill 2.7 Solving problems*.

1. Study the problem and be sure you understand it.
   (a) What is unknown?
   The cell potential under nonstandard state conditions.
   (b) What is known?
   $E° = 1.222$ V and the nonstandard state concentrations of the ions.
2. Decide how to solve the problem.
   (a) What is the connection between the known and the unknown?
   The cell potential under nonstandard state conditions is related to $E°$ and Q, the reaction quotient, by the Nernst equation

$$E = E° - \frac{0.0592}{n} \log Q$$

   where n is the number of electrons involved in the reaction.
   (b) What is necessary to make the connection?
   Write the expression for Q and evaluate it by using the known data. Substitute Q, $E°$, and n into the Nernst equation and solve for E.
3. Set up the problem and solve it.

$$Q = \frac{[Al^{3+}]^2}{[Fe^{2+}]^3} = \frac{(0.10)^2}{(0.25)^3} = 0.64$$

$$E = E° - \frac{0.0592}{n} \log Q = 1.222 - \frac{0.0592}{6} \log (0.64) = 1.224 \text{ V}$$

   The cell potential is 1.224 V.
4. Check the result.
   (a) Are significant figures and the location of the decimal point correct?
   Yes.
   (b) Did the answer come out in the correct units?
   Yes.
   (c) Is the answer reasonable?
   Yes.

Exercise 24.7 Calculate $E°$ for the copper-zinc reaction described in Exercise 24.5 given that the cell voltage is 1.175 V measured under conditions such that $[Zn^{2+}]$ = 0.010 mol/L and $[Cu^{2+}]$ = 3.5 mol/L. *Answer*: 1.100 V.

Example 24.8 Calculating **E**

If $[Al^{3+}]$ = 0.10 mol/L, find the electrode potential for the oxidation half-reaction

$$Al(s) \rightarrow Al^{3+} + 3e^- \qquad E° = 1.662 \text{ V}$$

The value of the electrode potential under these nonstandard state conditions is calculated by using the Nernst equation. For this equation

$$Q = [Al^{3+}] = 0.10$$

which gives

$$E = E° - \frac{0.0592}{n} \log Q = 1.662 - \frac{0.0592}{3} \log (0.10) = 1.682 \text{ V}$$

The electrode potential for the oxidation half-reaction is 1.682 V.

Exercise 24.8 Calculate the electrode potential for the reduction half-reaction

$$Fe^{2+} + 2e^- \rightarrow Fe(s) \qquad E° = -0.4402 \text{ V}$$

given $[Fe^{2+}]$ = 0.25 mol/L. *Answer*: -0.458 V.

PRACTICE TEST

1 (20 points). Choose the response that best completes each statement.

(a) A voltaic cell is characterized by

(i) a positive E (ii) a negative E

(iii) an overvoltage (iv) a nonspontaneous reaction

(b) A functioning electrochemical cell must have

(i) ion flow through an external conductor

(ii) electron flow through an electrolyte

(iii) oxidation occurring at one electrode and reduction at the other

(iv) all of these answers

(c) The purpose of making a battery by hooking two cells in series is to

(i) make a nonspontaneous reaction favorable

(ii) generate a higher current

(iii) overcome the overvoltage

(iv) generate a higher voltage

(d) Cells that can be recharged by using electrical energy from an external source to reverse the initial oxidation-reduction reaction are called
(i) primary cells (ii) storage cells
(iii) fuel cells (iv) conduction cells

(e) A faraday represents
(i) a mole of electrons (ii) voltage times time
(iii) current times time (iv) the internal resistance of a cell

(f) Oxidation is the
(i) gain of electrons (ii) loss of electrons
(iii) gain of hydrogen (iv) decrease in oxidation number

(g) The oxidation number of Cr in $K_2Cr_2O_7$ is
(i) +6 (ii) +4
(iii) -2 (iv) +12

(h) The process in which iron changes from +3 to +2 is
(i) catalysis (ii) redox
(iii) oxidation (iv) reduction

(i) The interchangeable oxidized and reduced form of the same element is the
(i) electrochemical cell (ii) redox couple
(iii) emf (iv) standard state condition

(j) The Nernst equation
(i) includes only the species directly involved in the oxidation process
(ii) is useful in balancing equations by the half-reaction method
(iii) relates the standard state potential to the potential under nonstandard state conditions
(iv) gives the potential of the cell under standard state conditions by adding the standard reduction potentials of the respective half cells

2 (10 points). Lead reacts with chlorate ion in acidic solution

$$Pb(s) + ClO_3^- \rightarrow PbCl_2(s)$$

(a) What is the oxidation state of Cl in ClO_3^-, Cl in $PbCl_2$, Pb in Pb, and Pb in $PbCl_2$? (b) Complete and balance the equation. (c) Identify the oxidizing agent, reducing agent, substance oxidized, and substance reduced.

3 (10 points). Complete and balance the following oxidation-reduction equations by the half-reaction method:

(a) $P_4(s) + NaOH(aq) \rightarrow PH_3(g) + NaH_2PO_4(aq)$ (in alkaline solution)

(b) $[AuCl_4]^- + H_2O_2(aq) \rightarrow Au(s) + O_2(g)$ (in acidic solution)

4 (10 points). The standard reduction potential of Ca^{2+} to Ca is -2.866 V and of Pb^{2+} to Pb is -0.126 V. Under standard state conditions, which of the following reactions will occur?

(a) $Pb^{2+} + Ca^{2+} \rightarrow Pb(s) + Ca(s)$ (b) $Pb(s) + Ca(s) \rightarrow Pb^{2+} + Ca^{2+}$

(c) $Pb^{2+} + Ca(s) \rightarrow Ca^{2+} + Pb(s)$ (d) $Pb(s) + Ca^{2+} \rightarrow Pb^{2+} + Ca(s)$

5 (10 points). Calculate the percentage of iron by mass in an iron ore sample if a 1.529 g sample of the ore was dissolved in acid, all the iron changed to Fe^{2+}, and 34.2 mL of 0.111 M $KMnO_4$ solution was required to titrate the Fe^{2+} to Fe^{3+}. In this reaction, the MnO_4^- is reduced to Mn^{2+}.

6 (10 points). An electrochemical cell is constructed by placing an iron electrode in an aqueous solution of $FeCl_3$ and by bubbling $Cl_2(g)$ over a graphite electrode that is also in contact with the $FeCl_3$ solution. (a) Write the half-reactions and overall cell equation for this electrochemical cell. (b) Calculate $E°$ for this cell given that the standard reduction potential for Cl_2 to Cl^- is 1.3597 V and for Fe^{3+} to Fe is -0.036 V. Is this reaction spontaneous under standard state conditions? (c) Calculate E for this cell if it is operating under nonstandard state conditions such that P_{Cl_2} = 0.1 bar, $[Fe^{3+}]$ = 0.001 mol/L, and $[Cl^-]$ = 0.003 mol/L. Is this reaction more or less favorable under these conditions than under standard state conditions? (d) Calculate the equilibrium constant for this reaction. (e) Find the pressure of $Cl_2(g)$ that will be in equilibrium with a 1.00 M solution of $FeCl_3$.

7 (10 points). Daniell cells were used for many years to provide electricity for telegraph wires. The shorthand notation for this cell is $Zn(s) \mid Zn^{2+} \parallel Cu^{2+} \mid Cu(s)$. Write the equations for the two half-reactions and the overall cell equation for this cell.

8 (10 points). Predict what will happen as aqueous solutions of (a) Na_2SO_4, (b) HCl, (c) NaOH, (d) H_2SO_4, and (e) NaCl are electrolyzed. Use the following standard reduction potentials: $E°$ = 1.229 V for $O_2,H^+/H_2O$; $E°$ = -0.8281 V for $H_2O/H_2,OH^-$; $E°$ = 0.0000 V for H^+/H_2; $E°$ = -2.714 V for Na^+/Na; $E°$ = 1.3597 V for Cl_2/Cl^-; $E°$ = 2.01 V for $S_2O_8^{2-}/SO_4^{2-}$.

9 (10 points). The nickel-cadmium alkaline storage cell is commonly found in most "rechargeable" appliances. The discharge reaction involves Cd(s) being oxidized to $Cd(OH)_2(s)$ and [NiOOH](s) being reduced to $Ni(OH)_2(s)$. (a) Write the two half-reactions for the discharge process and identify which electrode is the anode and cathode. What effect on the cell potential does the physical size of the cell have? (c) If a cell consists of 10.0 g of Cd, find the number of moles of electrons that the cell can deliver before it needs recharging. (d) What would be the lifetime of this cell if it can deliver an average of 0.010 A of current before becoming discharged?

Answers to practice test

1. (a) i, (b) iii, (c) iv, (d) ii, (e) i, (f) ii, (g) i, (h) iv, (i) ii, (j) iii.
2. (a) +5, -1, 0, +2; (b) $6Pb(s) + 12H^+ + 2ClO_3^- \rightarrow 5Pb^{2+} + PbCl_2(s) + 6H_2O(l)$; (c) ClO_3^-, Pb, Pb, ClO_3^-.
3. (a) $2P_4(s) + 3OH^- + 9H_2O(l) \rightarrow 3H_2PO_4^- + 5PH_3(g)$, (b) $2[AuCl_4]^- + 3H_2O_2(aq) \rightarrow 3O_2(g) + 6H^+ + 2Au(s) + 8Cl^-$.
4. C.
5. 69.3 mass %.
6. (a) $Fe \rightarrow Fe^{3+} + 3e^-$, $Cl_2 + 2e^- \rightarrow 2Cl^-$, $2Fe(s) + 3Cl_2(g) \rightarrow 2FeCl_3(aq)$; (b) 1.396 V, yes; (c) 1.575 V, more; (d) 10^{141}; (e) 10^{-46} bar.
7. $Zn(s) \rightarrow Zn^{2+} + 2e^-$, $Cu^{2+} + 2e^- \rightarrow Cu(s)$, $Zn(s) + Cu^{2+} \rightarrow Zn^{2+} + Cu(s)$.
8. H_2 and O_2 generated in all cells.
9. (a) $Cd(s) + 2OH^- \rightarrow Cd(OH)_2(s) + 2e^-$ at the anode and $[NiOOH](s) + H_2O(l) + e^- \rightarrow Ni(OH)_2(s) + 2OH^-$ at the cathode, (b) no change, (c) 0.178 mol e^-, (d) 1.7×10^6 s.

ADDITIONAL NOTES

CHAPTER 25

MOLECULAR ORBITAL THEORY

CHAPTER OVERVIEW

This chapter presents a brief introduction to molecular orbital theory. The types of molecular orbitals formed from the overlap of atomic orbitals and the order of filling of these orbitals by electrons in the molecule are discussed. The bonding in homonuclear diatomic molecules, and heteronuclear diatomic molecules, including those that contain multiple bonds, is discussed in terms of this theory.

COMPETENCIES

Definitions for the following terms should be learned:

- antibonding molecular orbital
- bond order
- bonding molecular orbital
- heteronuclear
- homonuclear
- molecular orbital
- molecular orbital theory

General concepts and skills that should be learned include

- describing the molecular orbitals formed by the overlap of atomic orbitals
- writing and interpreting molecular orbital electron configurations for diatomic molecules

Numerical exercises that should be understood include

- calculating the bond order for a diatomic molecule
- calculating bond dissociation energy

QUESTIONS

1. Complete the following sentences by filling in the various blanks with these terms: antibonding molecular orbitals, bond order, bonding molecular orbitals, molecular orbitals, and molecular orbital theory.

The (a)________ describes bonding in molecules in terms of electrons occupying (b)________ that exist for the molecule. The molecular orbitals that are favorable to bonding that are formed by the overlap of atomic orbitals and lie lower in energy than the atomic orbitals are known as (c)________. The molecular orbitals that are unfavorable to bonding and lie higher in energy than the atomic orbitals are known as (d)________. The (e)________ is related to the number of electrons in each of these orbitals.

2. Which of the following statements are true? Rewrite each false statement so that it is correct.

(a) Valence bond theory generates energy level diagrams that can be used to describe covalent bonding.

(b) The number of molecular orbitals formed from atomic orbitals is equal to the number of atomic orbitals used.

(c) Half of the molecular orbitals formed from atomic orbitals are bonding and half are antibonding.

(d) A molecular orbital is filled with a pair of electrons before electrons are placed in another molecular orbital of equal energy.

(e) The symbol π^4_{2p} means that there are four electrons in a π_{2p} molecular orbital.

(f) A bond order of 1 means that a σ bond exists.

3. Distinguish clearly between (a) bonding and antibonding molecular orbitals, (b) σ and π molecular orbitals, (c) homonuclear and heteronuclear diatomic molecules, and (d) multiple bonds and delocalized bonds.

4. Choose the response that best completes each statement.

(a) Which is not a requirement for successful bond formation?

(i) Orbitals forming the bond must be similar in energy.

(ii) One electron is donated by each atom.

(iii) Orbital overlap must occur for interaction.

(iv) Orbitals must have the same symmetry with respect to the internuclear axis.

(b) The symbol for an antibonding orbital formed by the parallel overlap of two p atomic orbitals is

(i) π_{2p} (ii) π^*_{2p}

(iii) σ_{2p} (iv) none of these answers

(c) The symbol for a bonding orbital formed by the head-on overlap of two p atomic orbitals is

(i) σ^*_{2p} (ii) π^*_{2p}

(iii) π_{2p} (iv) none of these answers

(d) The electron density is located between the nuclei of the bonded atoms in a/an

(i) antibonding molecular orbital (ii) bonding molecular orbital

(iii) delocalized orbital (iv) valence bond

(e) Which is *not* a rule for filling molecular orbitals?

(i) Lowest energy orbitals are used first.

(ii) Orbitals hold a maximum of two electrons.

(iii) Orbitals of equal energy are occupied by single electrons with parallel spins before electron pairing begins.

(iv) σ Orbitals are filled before π orbitals.

(f) A diatomic molecule that contains both atoms of the same element is called

(i) homonuclear (ii) heteronuclear

(iii) delocalized (iv) localized

(g) Which bonding description would account for a bond order of 2?

(i) one σ bond and one π bond

(ii) two π bonds

(iii) one σ bond, two π bonds, and one π antibond

(iv) all of these answers

(h) Which electron configuration would account for a bond order of 1?

(i) $1s^2\ 1s^2\ \sigma^2_{2s}$

(ii) $1s^2\ 1s^2\ \sigma^2_{2s}\ \sigma^{*2}_{2s}$

(iii) $1s^2\ 1s^2\ \sigma^2_{2s}\ \sigma^{*2}_{2s}\ \pi^2_{2p_y}\ \pi^2_{2p_z}$

(iv) $1s^2\ 1s^2\ \sigma^2_{2s}\ \sigma^{*2}_{2s}\ \pi^2_{2p_y}\ \pi^2_{2p_z}\ \sigma^2_{2p_x}$

(i) Choose the "molecule" that is not predicted to exist according to simple molecular orbital theory.

(i) Li_2 (ii) Be_2

(iii) B_2 (iv) C_2

(j) Choose the paramagnetic species.

(i) Li_2 (ii) N_2

(iii) O_2 (iv) F_2

(k) The bonding in CO most resembles that in

(i) Li_2 (ii) N_2

(iii) O_2 (iv) none of these answers

5. Briefly compare the four ways used to describe covalent bonding within a molecule.

6. What two types of molecular orbitals are formed by addition and subtraction of atomic orbitals? How do the electron waves interact with each other in these new orbitals?

7. Interpret the symbol σ_{1s}. Write the symbol for the corresponding antibonding molecular orbital.

8. Describe the two different types of molecular orbitals that are formed by the overlap of two $2p$ orbitals. Write the symbols for the bonding and antibonding orbitals of each type.

9. Write the order of filling of molecular orbitals for homonuclear diatomic molecules containing atoms with $Z \leq 7$. How does this order change for molecules containing atoms with $8 < Z \leq 10$?

10. Write the electron configuration for (a) N_2 and (b) O_2. Identify whether each molecule is paramagnetic or diamagnetic.

11. Write the electron configuration for $C_2(g)$ using the order of filling the molecular orbitals as (a) σ_{2p} before π_{2p} and (b) π_{2p} before σ_{2p}. (c) Which description correctly predicts C_2 to be diamagnetic? Although the σ_{2p} orbital is higher in energy than the π_{2p} orbitals, there is only a small energy difference between them in C_2, and a considerable number of excited molecules have the electron configuration $1s^2\ 1s^2\ \sigma_{2s}^2\ \sigma_{2s}^{*2}\ \sigma_{2p_x}^2\ \pi_{2p_y}^1\ \pi_{2p_z}^1$. (d) Is this excited state as stable as the ground state? (e) How does this excited state differ in magnetic properties from the ground state?

12. The bond energy of the O_2^+ ion is greater than that of O_2, but the bond energy of the N_2^+ ion is less than that of N_2. Explain why this is true.

13. We have assumed that the molecular orbital energy diagram given in Figure 25.6 can be used to describe gaseous heteronuclear diatomic molecules. How reasonable is this assumption for the molecule NO? How reasonable is this assumption for the molecule BeO?

Answers to questions

1. (a) molecular orbital theory, (b) molecular orbitals, (c) bonding molecular orbitals, (d) antibonding molecular orbitals, (e) bond order.
2. (a) F, replace *Valence bond* by *Molecular orbital*; (b) T; (c) T; (d) F, substitute *Molecular orbitals of equal energy are occupied by single electrons before electron pairing begins.* for the original statement; (e) F, replace *four electrons ... orbital* with *two electrons in each of two* π_{2p} *orbitals of equivalent energy*; (f) F, replace σ with *single* and add *which can be a σ or π bond* at the end of the statement.
3. (a) *Bonding molecular orbitals* are lower in energy than the atomic orbitals that were used to generate them; *antibonding molecular orbitals* are higher in energy. (b) In a σ *molecular orbital* the electron density is concentrated along the bond axis; in a π *molecular orbital* the electron density is formed by the parallel overlap of p atomic orbitals and does not lie along the bond axis. (c) *Homonuclear* refers to a species containing the same type of atoms; *heteronuclear* refers to a species containing different types of atoms. (d) *Multiple bonds* consist of localized π bonds; *delocalized bonds* consist of π bonds that are not restricted to two atoms.
4. (a) ii, (b) ii, (c) iv, (d) ii, (e) iv, (f) i, (g) iv, (h) i, (i) ii, (j) iii, (k) ii.
5. Lewis structures simply show the arrangement of valence electrons; VSEPR gives molecular geometry based on repulsion of valence electrons; valence bond theory gives molecular geometry based on overlap of unhybridized and/or hybridized atomic orbitals; molecular orbital theory gives molecular geometry and energy description based on orbitals characteristic of the molecule.
6. Molecular orbitals formed by addition of atomic orbitals are bonding orbitals and those formed by subtraction are antibonding orbitals. The waves reinforce each other (constructive interference) in a bonding orbital and cancel each other (destructive interference) in an antibonding orbital.
7. The symbol σ_{1s} represents the bonding molecular orbital formed by the overlap of two 1s atomic orbitals. The symbol for the corresponding antibonding molecular orbital is σ^*_{1s}.
8. Two pairs of *p* orbitals combine in parallel fashion to give four π molecular orbitals, two bonding (π_{2p_y} and π_{2p_z}) and two antibonding ($\pi^*_{2p_y}$

and $\pi^*_{2p_z}$). The third pair of *p* orbitals combine in end-on fashion to give two σ molecular orbitals, one bonding (σ_{2p_x}) and one antibonding ($\sigma^*_{2p_x}$).

9. $\sigma_{1s}\ \sigma^*_{1s}\ \sigma_{2s}\ \sigma^*_{2s}\ \pi_{2p_y}\ \pi_{2p_z}\ \sigma_{2p_x}\ \pi^*_{2p_y}\ \pi^*_{2p_z}\ \sigma^*_{2p_x}$; the σ_{2p_x} is lower in energy than the π_{2p_y}.
10. (a) $1s^2\ 1s^2\ \sigma^2_{2s}\ \sigma^{*2}_{2s}\ \pi^2_{2p_y}\ \pi^2_{2p_z}\ \sigma^2_{2p_x}$, diamagnetic;

    (b) $1s^2\ 1s^2\ \sigma^2_{2s}\ \sigma^{*2}_{2s}\ \sigma^2_{2p_x}\ \pi^2_{2p_y}\ \pi^2_{2p_z}\ \pi^{*1}_{2p_y}\ \pi^{*1}_{2p_z}$, paramagnetic.
11. (a) $1s^2\ 1s^2\ \sigma^2_{2s}\ \sigma^{*2}_{2s}\ \sigma^2_{2p_x}\ \pi^1_{2p_y}\ \pi^1_{2p_z}$

    (b) $1s^2\ 1s^2\ \sigma^2_{2s}\ \sigma^{*2}_{2s}\ \sigma^2_{2p_y}\ \pi^2_{2p_z}$

    (c) The configuration that has π_{2p} filling before σ_{2p}; configuration (b) correctly predicts C_2 to be diamagnetic.

    (d) The excited state is not as stable as the ground state.

    (e) The excited state is paramagnetic.
12. The removal of one electron from an antibonding π orbital in O_2 increases the bond order and the removal of one electron from a bonding σ orbital in N_2 decreases the bond order.
13. The assumption that the molecular orbital energy diagram given in Figure 25.6 of the text can be used for NO is reasonable because N and O do not differ greatly in atomic number, so orbitals should overlap properly. The assumption for BeO is not as good because of the large difference in atomic numbers.

SKILLS

Skill 25.1 Writing electron configurations for diatomic molecules

The general order of filling of molecular orbitals for homonuclear diatomic molecules containing atoms with atomic numbers

$1 < Z \leq 7$ is $\sigma_{1s}\ \sigma^*_{1s}\ \sigma_{2s}\ \sigma^*_{2s}\ \pi_{2p_y}\ \pi_{2p_z}\ \sigma_{2p_x}\ \pi^*_{2p_y}\ \pi^*_{2p_z}\ \sigma^*_{2p_x}$

$8 < Z \leq 10$ is $\sigma_{1s}\ \sigma^*_{1s}\ \sigma_{2s}\ \sigma^*_{2s}\ \sigma_{2p_x}\ \pi_{2p_y}\ \pi_{2p_z}\ \pi^*_{2p_y}\ \pi^*_{2p_z}\ \sigma^*_{2p_x}$

There are two ways of representing the $1s$ electrons when writing electron configurations for molecules in the second period: either $\sigma^2_{1s}\ \sigma^{*2}_{1s}$ or $1s^2\ 1s^2$. This order of filling is valid for heteronuclear diatomic molecules as long as the two atoms are close to each other in the periodic table.

Each molecular orbital can hold two electrons and electrons fill the lower energy orbitals before entering the higher energy orbitals. If two orbitals have the same energy, for example, π_{2p_y} and π_{2p_z}, they are each occupied by single electrons before electron pairing begins.

For example, the electron configurations for O_2^+, O_2, and O_2^- would be

O_2^+ 15 electrons $1s^2\ 1s^2\ \sigma_{2s}^2\ \sigma_{2s}^{*2}\ \sigma_{2p_x}^2\ \pi_{2p_y}^2\ \pi_{2p_z}^2\ \pi_{2p_y}^{*1}$

O_2 16 electrons $1s^2\ 1s^2\ \sigma_{2s}^2\ \sigma_{2s}^{*2}\ \sigma_{2p_x}^2\ \pi_{2p_y}^2\ \pi_{2p_z}^2\ \pi_{2p_y}^{*1}\ \pi_{2p_z}^{*1}$

O_2^- 17 electrons $1s^2\ 1s^2\ \sigma_{2s}^2\ \sigma_{2s}^{*2}\ \sigma_{2p_x}^2\ \pi_{2p_y}^2\ \pi_{2p_z}^2\ \pi_{2p_y}^{*2}\ \pi_{2p_z}^{*1}$

Skill Exercise: Write electron configurations for (a) B_2, (b) F_2, (c) Ne_2, (d) CN, (e) CN^+, and (f) CN^-. *Answers*: (a) $1s^2\ 1s^2\ \sigma_{2s}^2\ \sigma_{2s}^{*2}\ \pi_{2p_y}^1\ \pi_{2p_z}^1$, (b) $1s^2\ 1s^2\ \sigma_{2s}^2\ \sigma_{2s}^{*2}\ \sigma_{2p_x}^2\ \pi_{2p_y}^2\ \pi_{2p_z}^2\ \pi_{2p_y}^{*2}\ \pi_{2p_z}^{*2}$, (c) $1s^2\ 1s^2\ \sigma_{2s}^2\ \sigma_{2s}^{*2}\ \sigma_{2p_x}^2\ \pi^2_{2p_y}\ \pi^2_{2p_z}\ \pi_{2p_y}^{*2}\ \pi_{2p_z}^{*2}\ \sigma_{2p_x}^{*2}$, (d) $1s^2\ 1s^2\ \sigma_{2s}^2\ \sigma_{2s}^{*2}\ \pi_{2p_y}^2\ \pi_{2p_z}^2\ \sigma_{2p_x}^1$, (e) $1s^2\ 1s^2\ \sigma_{2s}^2\ \sigma_{2s}^{*2}\ \pi_{2p_y}^2\ \pi_{2p_z}^2$, (f) $1s^2\ 1s^2\ \sigma_{2s}^2\ \sigma_{2s}^{*2}\ \pi_{2p_y}^2\ \pi_{2p_z}^2\ \sigma_{2p_x}^2$.

The bond order is the number of effective bonding pairs of electrons shared between two atoms in a covalent bond.

$$\text{Bond order} = \frac{1}{2}[(\text{number of bonding electrons}) - (\text{number of antibonding electrons})]$$

Usually only the valence electrons are put into the equation. For example, the bond orders for the species O_2^+, O_2, and O_2^- are

$$\text{bond order} = \frac{1}{2}(6 - 1) = \frac{5}{2} \text{ for } O_2^+$$

$$\text{bond order} = \frac{1}{2}(6 - 2) = 2 \text{ for } O_2$$

$$\text{bond order} = \frac{1}{2}(6 - 3) = \frac{3}{2} \text{ for } O_2^-$$

(This decrease in bond order is reflected by the respective bond lengths of 0.11227 nm, 0.126 nm, and 0.149 nm.)

Skill Exercise: Determine the bond order in (a) B_2, (b) F_2, (c) Ne_2, (d) CN, (e) CN^+, and (f) CN^- by using the electron configurations written above. *Answers*: (a) 1, (b) 1, (c) 0, (d) 5/2, (e) 2, (f) 3.

NUMERICAL EXERCISES

Example 25.1 Bond order

The electron configuration of C_2 is $1s^2\ 1s^2\ \sigma^2_{2s}\ \sigma^{*2}_{2s}\ \pi^2_{2p_y}\ \pi^2_{2p_z}$ and of C_2^-, $1s^2\ 1s^2\ \sigma^2_{2s}\ \sigma^{*2}_{2s}\ \pi^2_{2p_y}\ \pi^2_{2p_z}\ \sigma^1_{2p_x}$. Calculate the bond order for each of these species.

The bond order is one half the difference between the number of bonding and the number of antibonding electrons. Usually only the valence electrons are considered in the calculation. For C_2, there are six bonding electrons ($\sigma^2_{2s}\ \pi^2_{2p_y}\ \pi^2_{2p_z}$) and two antibonding electrons (σ^{*2}_{2s}) giving

$$\text{bond order} = \frac{1}{2}[(\text{number of bonding electrons}) - (\text{number of antibonding electrons})]$$
$$= \frac{1}{2}(6 - 2) = 2$$

and for C_2^- there are seven bonding electrons ($\sigma^2_{2s}\ \pi^2_{2p_y}\ \pi^2_{2p_z}\ \sigma^1_{2p_x}$) and two antibonding electrons (σ^{*2}_{2s}), giving

$$\text{bond order} = \frac{1}{2}(7 - 2) = \frac{5}{2}$$

The bond order in C_2 is 2 and in C_2^- is $\frac{5}{2}$.

Exercise 25.1 The electron configuration of N_2^+ is $1s^2\ 1s^2\ \sigma^2_{2s}\ \sigma^{*2}_{2s}\ \pi^2_{2p_y}\ \pi^2_{2p_z}\ \sigma^1_{2p_x}$, of N_2 is $1s^2\ 1s^2\ \sigma^2_{2s}\ \sigma^{*2}_{2s}\ \pi^2_{2p_y}\ \pi^2_{2p_z}\ \sigma^2_{2p_x}$, and of N_2^- $1s^2\ 1s^2\ \sigma^2_{2s}\ \sigma^{*2}_{2s}\ \pi^2_{2p_y}\ \pi^2_{2p_z}\ \sigma^2_{2p_x}\ \pi^{*1}_{2p_y}$. Calculate the bond order for each of these species. *Answer*: 5/2 for N_2^+, 3 for N_2, 5/2 for N_2^-.

*Example 25.2 Bond energy**

The standard heat of formation is 443.5 kJ/mol for C_2^-(g), 587.9 kJ/mol for C^-(g), and 711.20 kJ/mol for C(g) at 0 K. Calculate the bond dissociation energy for C_2^-(g).

$$C_2^-(g) \rightarrow C(g) + C^-(g)$$

This problem is solved by using the comprehensive problem-solving approach presented in *Skill 2.7 Solving problems*.

1. <u>Study the problem and be sure you understand it.</u>
   (a) What is unknown?
   The bond dissociation energy.
   (b) What is known?

The chemical equation for the reaction and the heat of formation data for the various species involved.

2. Decide how to solve the problem.
   (a) What is the connection between the known and the unknown?
   The bond dissociation energy is equal to the $\Delta H°$ of the reaction, which can be calculated by using Hess's law (see *Skill 7.2 Using Hess's law* and *Examples 10.1* and *10.2*).
   (b) What is necessary to make the connection?
   Subtracting the heat of formation of $C_2^-(g)$ from the heats of formation of C(g) and $C^-(g)$ will give $\Delta H°$ for the reaction.
3. Set up the problem and solve it.
   For the reaction

$$\Delta H_0^\circ = (\text{sum of } \Delta H_f^\circ \text{ products}) - (\text{sum of } \Delta H_f^\circ \text{ reactants})$$
$$= [(1\ \text{mol})\Delta H_{f,0}^\circ(\text{C}) + (1\ \text{mol})\Delta H_{f,0}^\circ(\text{C}^-)] - [(1\ \text{mol})\Delta H_{f,0}^\circ(\text{C}_2^-)]$$
$$= [(1\ \text{mol})(711.20\ \text{kJ/mol}) + (1\ \text{mol})(587.9\ \text{kJ/mol})] - [(1\ \text{mol})(443.5\ \text{kJ/mol})]$$
$$= 855.6\ \text{kJ}$$

   The bond dissociation energy is 855.6 kJ/mol.
4. Check the result.
   (a) Are significant figures and the location of the decimal point correct?
   Yes.
   (b) Did the answer come out in the correct units?
   Yes.
   (c) Is the answer reasonable?
   Yes. A quick review of Table 10.5 of the text shows that bond energies are in the range of 150 to 1075 kJ/mol.

Exercise 25.2 Use the above data and ΔH_f° = 823.4 kJ/mol for $C_2(g)$ to calculate the bond dissociation energy for $C_2(g)$. Does this value confirm that the extra electron in $C_2^-(g)$ is in a bonding molecular orbital? *Answer*: 599.0 kJ/mol; yes, C_2^- has a higher bond dissociation energy.

PRACTICE TEST

1 (20 points). Write the electron configurations for (a) Li_2, (b) C_2, (c) N_2, (d) O_2, and (e) F_2. Determine the bond order for each molecule. Which of these are paramagnetic?

2 (20 points). Write the electron configurations for (a) H_2, (b) H_2^+, and (c) H_2^-. Which of these is the most stable?

3 (20 points). Choose the response that best completes each statement.

(a) Which is *not* a requirement for successful bond formation according to both the valence bond and molecular orbital theories?

(i) The overlapping orbitals must have the same symmetry with respect to the internuclear axis.

(ii) Orbitals of equal energy fill singly before electron pairing begins.

(iii) Orbitals must overlap sufficiently to allow for interaction.

(iv) Bonding orbitals must have similar energies.

(b) What is the bond order in B_2?

(i) 0 (ii) 1

(iii) 3/2 (iv) 2

(c) Compared to B_2, B_2^- would have a

(i) longer bond length and a higher bond dissociation energy

(ii) longer bond length and a smaller bond dissociation energy

(iii) shorter bond length and a higher bond dissociation energy

(iv) shorter bond length and a smaller bond dissociation energy

(d) Which has the longest carbon-nitrogen bond length?

(i) CN (ii) CN^-

(iii) CN^+ (iv) all identical

(e) The paramagnetism of molecular oxygen is best explained by the

(i) Lewis structure

(ii) valence-shell electron-pair repulsion theory

(iii) valence bond theory

(iv) molecular orbital theory

4 (20 points). The molecular orbital diagram at the right represents the energy levels in the H_4 molecule. In this diagram *b* represents a bonding orbital, *n* represents nonbonding orbitals, and *a* represents an antibonding orbital. Write the electron configuration for H_4. Based on this configuration, is this species predicted to exist? Is the molecule diamagnetic or paramagnetic?

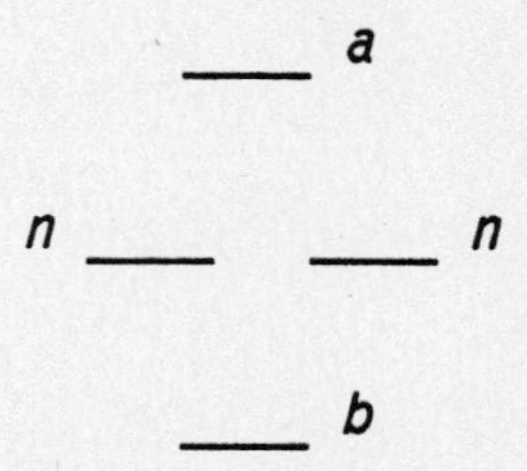

5 (20 points). Calculate the bond dissociation energy for CO at 0 K by using the following standard heat of formation data: $\Delta H^\circ_{f,0}$ = -113.801 kJ/mol for CO(g), 711.20 kJ/mol for C(g), and 246.785 kJ/mol for O(g).

Answers to practice test

1. (a) $1s^2\ 1s^2\ \sigma^2_{2s}$, 1, diamagnetic; (b) $1s^2\ 1s^2\ \sigma^2_{2s}\ \sigma^{*2}_{2s}\ \pi^2_{2p_y}\ \pi^2_{2p_z}$, 2, diamagnetic; (c) $1s^2\ 1s^2\ \sigma^2_{2s}\ \sigma^{*2}_{2s}\ \pi^2_{2p_y}\ \pi^2_{2p_z}\ \sigma^2_{2p_x}$, 3, diamagnetic; (d) $1s^2\ 1s^2\ \sigma^2_{2s}\ \sigma^{*2}_{2s}\ \sigma^2_{2p_x}\ \pi^2_{2p_y}\ \pi^2_{2p_z}\ \pi^{*1}_{2p_y}\ \pi^{*1}_{2p_z}$, 2, paramagnetic; (e) $1s^2\ 1s^2\ \sigma^2_{2s}\ \sigma^{*2}_{2s}\ \sigma^2_{2p_x}\ \pi^2_{2p_y}\ \pi^2_{2p_z}\ \pi^{*2}_{2p_y}\ \pi^{*2}_{2p_z}$, 1, diamagnetic.
2. (a) σ^2_{1s}, (b) σ^1_{1s}, (c) $\sigma^2_{1s}\ \sigma^{*1}_{1s}$; H_2.
3. (a) ii, (b) ii, (c) iii, (d) iii, (e) iv.
4. $b^2n^1n^1$, yes, paramagnetic.
5. 1071.79 kJ/mol.

ADDITIONAL NOTES

CHAPTER 26

METALS AND METALLURGY: THE *s*- AND *p*-BLOCK METALS

CHAPTER OVERVIEW

This chapter presents the chemical processes involved in producing metals from ores. To illustrate these processes, the metallurgy of sodium, aluminum, copper, zinc, and iron is discussed in detail. Properties of pure metals and of several alloys (including steel) are discussed.

The chemistry of the representative metals is presented in two convenient groupings: the *s*-block metals (the alkali and alkaline earth metals of Representative Groups I and II) and the *p*-block metals (the metals of Representative Groups III to V--aluminum, gallium, indium, thallium, tin, lead, and bismuth). Some of the most important aspects of the industrial chemistry and uses of these elements and their compounds are discussed.

COMPETENCIES

Definitions for the following terms should be learned:

- α-alumina
- alloy
- aluminothermic reactions
- calcining
- caustic soda
- electrometallurgy
- flux
- γ-alumina
- hydrated lime, slaked lime
- hydrometallurgy
- intermetallic compounds
- interstitial solid solution
- leaching
- lime
- metallurgy
- mineral
- ore
- pyrometallurgy
- quicklime, unslaked lime
- refining
- roasting
- sintering
- slag
- smelting
- soda ash
- solid solution
- steel

General concepts and skills that should be learned include

- the general properties of metals and alloys
- the occurrence of metals in minerals
- the general categories of procedures used to produce metals from ores
- the metallurgy of sodium, aluminum, copper, zinc, and iron
- the types, properties, and uses of alloys (including steel)
- the general physical and chemical properties, industrial preparations, and uses for the representative metals and some of their compounds

QUESTIONS

1. Match the correct term to each definition.

(a) all aspects of the science and technology of metals
(b) a naturally occurring substance with a characteristic range of chemical composition
(c) a mixture of minerals from which a particular metal (or several metals) can profitably be extracted
(d) unwanted rock originally in an ore
(e) the metal-bearing mineral is concentrated in a froth of bubbles that can be skimmed off water mixed with a detergent or oil
(f) washing a soluble compound from insoluble material with an appropriate solvent
(g) melting accompanied by a chemical change
(h) a mixture of gangue minerals produced during melting
(i) a substance that lowers the melting point of gangue
(j) heating an ore below its melting point in air or oxygen to produce oxides
(k) heating to drive off a gas
(l) heating a finely divided ore without melting to produce larger particles

(i) calcining
(ii) flotation
(iii) flux
(iv) gangue
(v) leaching
(vi) metallurgy
(vii) mineral
(viii) ore
(ix) roasting
(x) sintering
(xi) slag
(xii) smelting

2. Match the correct mineral to the element produced from it.

(a) Al	(i) bauxite
(b) Na and Mg	(ii) brines
(c) Pb	(iii) cassiterite
(d) Sn	(iv) galena

3. Distinguish clearly between (a) a mineral and an ore and (b) an acidic flux and a basic flux.

4. Which of the following statements are true? Rewrite each false statement so that it is correct.

(a) For representative elements that can form two or more oxidation states, the lower oxidation states are more favorable if the element is located near the bottom of the periodic table.

(b) Many salts having the empirical formula $M^{2+}X^{2-}$ are much less soluble than those having the empirical formula $M^{+}X^{-}$ because of the large increase in hydration energy.

(c) In many respects, compounds of Li^{+} are like those of Representative Group II, and those of Be^{2+} are like those of Representative Group III.

(d) A common name for NaOH is lye.

(e) Calcium oxide is an acidic anhydride.

5. Choose the response that best completes each statement.

(a) Which is *not* a general property of metals?

(i) good conductors of heat and electricity
(ii) electron donors in chemical reactions
(iii) malleable
(iv) form negative ions

(b) The transformation by controlled heating of a mass of small particles into a solid, dense, strong material is known as

(i) sintering (ii) condensation
(iii) superheating (iv) surface tension

(c) Which type of steel is malleable and ductile, so that it is used for wire, pipe, and sheet metal?

(i) low-carbon (ii) medium-carbon
(iii) high-carbon (iv) wrought iron

(d) Which is *not* an equation representing a step in the production of iron?

(i) $2C(s) + O_2(g) \rightarrow 2CO(g)$
(ii) $FeO(s) + 2CO(g) \rightarrow FeCO_3(s) + C(s)$
(iii) $CaCO_3(s) \rightarrow CaO(s) + CO_2(g)$
(iv) $C(s) + CO_2(g) \rightarrow 2CO(g)$

(e) Tempering of steel is

(i) plating the steel with a layer of zinc
(ii) coating the steel with a layer of cementite

(iii) heating and rapidly cooling the metal to "freeze" the C in the form of Fe_3C

(iv) adding an acidic flux

(f) Which flame test is correct?

(i) Rb^+ is red. (ii) Na^+ is yellow.

(iii) K^+ is green. (iv) Li^+ is violet.

(g) What are the product(s) formed as the metals of Representative Group I burn in air?

(i) M_2O, M_2O_2, or MO_2 (ii) MOH

(iii) M_2O (iv) MNO_3

(h) Which is *not* true about the metals in Representative Group I?

(i) All metals are extremely reactive.

(ii) Most crystalline compounds are water soluble.

(iii) The melting points, boiling points, hardness, and specific heats all increase with increasing atomic number.

(iv) $2M(s) + 2H_2O(l) \rightarrow H_2(g) + 2MOH(aq)$

(i) Which process is used for the commercial preparation of Na?

(i) $2NaH(s) + Ca(s) \rightarrow 2CaH_2(s)$

(ii) $2NaCl(l) \xrightarrow{\text{electrolysis}} 2Na(l) + Cl_2(g)$

(iii) $NaCl(l) + K(g) \rightarrow Na(g) + KCl(l)$

(iv) $2NaOH(aq) + Mg(s) \rightarrow 2Na(s) + Mg(OH)_2(aq)$

(j) Water reacts with sodium metal to form

(i) $H_2(g)$ (ii) $O_2(g)$

(iii) $Na_2O(s)$ (iv) $H_2(g)$ and NaOH(aq)

(k) Which process is used for preparing Be?

(i) $BeCl_2(l) \xrightarrow{\text{electrolysis}} Be(l) + Cl_2(g)$

(ii) $BeF_2(g) + Mg(s) \rightarrow Be(s) + MgF_2(g)$

(iii) $BeO(s) + 2Na(g) \rightarrow Be(g) + Na_2O(s)$

(iv) $Be_3Al_2(Si_6O_{18})(s) + 24F_2(g) \rightarrow 3Be(s) + 2AlF_6(s) + 6SiF_6(s) + 9O_2(g)$

(l) Which element(s) differ most from the rest of the metals in Representative Group II in behavior?

(i) Ca and Sr (ii) Be and Mg

(iii) Mg, Ca, and Ba (iv) Sr and Ba

(m) Which is *not* a common use for MgO?

(i) antacid (ii) preparation of O_2

(iii) dusting powder (iv) lining for furnaces

(n) Which reaction represents the slaking of lime?

(i) $CaCO_3(s) \xrightarrow{\Delta} CaO(s) + CO_2(g)$

(ii) $CaO(s) + H_2O(l) \rightarrow Ca(OH)_2(aq)$

(iii) $Ca(OH)_2(aq) + CO_2(l) \rightarrow CaCO_3(s) + H_2O(l)$

(iv) $CaH_2(s) + 2H_2O(l) \rightarrow Ca(OH)_2(aq) + 2H_2(g)$

(o) What should be remembered about compounds of Ba?

(i) explosive (ii) highly poisonous

(iii) radioactive (iv) hygroscopic

(p) Marble consists mainly of

(i) calcium chloride (ii) calcium sulfate

(iii) calcium phosphate (iv) calcium carbonate

(q) What are the common oxidation states for the Representative Group III elements?

(i) 0, +3 for lighter elements, and +1 for heavier elements

(ii) 0, ±3, and +5

(iii) 0, +2, and +3

(iv) none of these answers

(r) Why does Al seem rather inert?

(i) It is below H in the electromotive series.

(ii) It forms an adherent protective oxide coating.

(iii) It crystallizes in a diamond unit cell.

(iv) It is amphoteric.

(s) Which salt of Al is commonly used in the paper, tanning, and dyeing industries?

(i) $KAl(SO_4)_2 \cdot 12H_2O$ (ii) $LiAlH_4$

(iii) $Al_2(SO_4)_3$ (iv) $Al(OH)_3$

(t) Reactions between aluminum and metal oxides which produce considerable amounts of heat are known as

(i) aluminothermic reactions (ii) partner-exchange reactions

(iii) deliquescence (iv) flocculation

(u) Tin-plated iron is stable toward corrosion because

(i) the tin covers the metal

(ii) an alloy is formed

(iii) the Sn reacts preferentially to the Fe

(iv) none of these answers

6. First-year students in chemistry laboratories are usually surprised to find that metals in Group I such as Na are stored under oil; that metals in Group II such as Ca, when stored in the presence of air, become coated with white powders; and that Al reacts with H_2O and HCl much more slowly than expected. Explain these observations.

7. Name the three major types of rocks that are found in the earth's crust. How was each type formed? What is the most abundant kind of chemical compound in rocks?

8. Briefly describe a leaching process. Name some of the advantages of hydrometallurgy over other methods of extraction.

9. Why must alkali metals be produced by the electrolysis (in the absence of water) of their compounds or by redox reactions involving other very active metals? Briefly describe how metallic sodium is produced.

10. What metal sulfide is the common impurity in copper ores? How is this impurity removed during the production of copper metal?

11. Briefly describe the operation of a blast furnace. What are the raw materials and major products of the process?

12. What does the term "steel" mean? List a few types of steel and some of their properties and uses.

13. What are the common names given to the families of metals that make up Representative Groups I and II?

14. Table 26.2 of the text shows that as the atomic number increases within a family, the melting point generally decreases. Briefly explain this trend.

15. The metals in Groups I and II (represented by M and M', respectively) react vigorously with the halogens. Complete the following equations for the reactions with the hypothetical halogen X: (a) $M(s) + X_2 \rightarrow$ and (b) $M'(s) + X_2 \rightarrow$.

16. Lithium and its compounds resemble magnesium and its compounds in many respects. Why is this true? Is this kind of diagonal behavior shown by beryllium?

17. Write and discuss the chemical equations describing the formation of limestone caverns and the growth of stalagmites and stalactites in these caverns.

18. Within a family of *p*-block metals, how does the stability of the higher oxidation state for an element compare to the stability of the lower oxidation state? How does this trend affect the ionic and covalent properties of the substances that are formed?

19. A high school student wrote on an examination that bismuth is not a metal. The teacher marked this statement "wrong." Choose one of the points of view and defend your choice.

20. Identify the products of the following reactions:

(a) $M + O_2 \rightarrow$ M = Al, Ga, In, Tl, Bi

(b) $Sn + O_2 \rightarrow$

(c) $Pb + O_2 \rightarrow$

(d) $M + X_2 \rightarrow$ M = Al, Ga, In, Tl, and X = F, Br, Cl

(e) $Bi + F_2 \rightarrow$

21. In the "lead chamber" process for the manufacture of sulfuric acid (now obsolete), the acid was prepared in dilute form in large, lead-lined chambers. Why didn't the lead dissolve? Why was the acid concentration important?

22. Indium can be electrodeposited from an aqueous solution of In^{3+}, but aluminum cannot be produced by electroylsis of an aqueous solution of Al^{3+}. Explain this difference.

23. The characteristic oxidation state of gallium, indium, and thallium is +3. One of these elements is easily reduced to +1. Which element is this?

Answers to questions

1. (a) vi, (b) vii, (c) viii, (d) iv, (e) ii, (f) v, (g) xii, (h) xi, (i) iii, (j) ix, (k) i, (l) x.
2. (a) i, (b) ii, (c) iv, (d) iii.
3. (a) A *mineral* is a pure substance containing the desired element; an *ore* contains the mineral and impurities (which may be undesirable). (b) An *acidic flux* is the oxide of a nonmetal such as SiO_2; a *basic flux* is the oxide of a metal such as CaO.
4. (a) T; (b) F, replace *hydration* with *lattice*; (c) T; (d) T; (e) F; replace *an acidic* with *a basic*.

5. (a) iv, (b) i, (c) i, (d) ii, (e) iii, (f) ii, (g) i, (h) iii, (i) ii, (j) iv, (k) ii, (l) ii, (m) ii, (n) ii, (o) ii, (p) iv, (q) i, (r) ii, (s) iii, (t) i, (u) i.
6. Metals form oxides upon exposure to the atmosphere--oil protects the metals of Group I, the white powder is the oxide of the Group II metal, and Al forms a protective oxide coating which decreases its activity.
7. The three major types of rocks that are found in the earth's crust are igneous, formed by solidification of molten substances; sedimentary, formed by deposition of material by the oceans and rivers; and metamorphic, formed by the action of heat and pressure on existing rocks. The most abundant kind of chemical substance in rocks is silicates.
8. In leaching, the metal or compound of the metal is dissolved out of the ore by water or an aqueous solution. After suitable purification of the solution, a pure compound of the metal may be isolated from it, or reduction to the metal may be carried out directly in the solution. Some of the advantages of hydrometallurgy include elimination of concentration steps, large amounts of coal and coke are unnecessary, and some types of atmospheric pollution are avoided.
9. Alkali metals must be reduced by electrolysis of their compounds or by other very active metals because they are the most electropositive of all elements. Metallic sodium is produced by electrolysis of a molten mixture of NaCl and $CaCl_2$.
10. Iron sulfide is the common impurity in copper ores. Partial separation of iron and copper is accomplished by heating with a silica flux in a reverberatory furnace to convert iron oxides and some of the iron sulfide to slag. The remaining iron is separated by further smelting with silica and oxygen in a converter, where an iron silicate slag forms.
11. The raw materials for ironmaking (concentrated iron ore, coke, and limestone) are charged into the top of the blast furnace. Very hot air or oxygen entering at the botton converts the coke to CO, which reduces the ore to molten iron as it descends. The limestone reacts with impurities to produce the molten slag. The major products are crude iron, slag, CO, and N_2 (that was introduced in bottom air blast.)
12. Steel includes alloys of iron that contain up to 1.5 % of carbon, and often other metals. Low-carbon steel (up to 0.2 % C) is malleable and ductile and is used in making wire, pipe, and sheet steel. Medium steel

(0.2 - 0.6 % C) is strong and is used in rails, boiler plate, and structural pieces. High-carbon steel (0.6 - 1.5 % C) is hard but lacks ductility and flexibility, and is used for tools, springs, and cutlery.

13. The common names are the alkali metals and the alkaline earth metals.
14. The interatomic forces that hold atoms in the crystal structure decrease as the distance between the valence electrons and nucleus increases.
15. (a) $2M(s) + X_2 \rightarrow 2MX(s)$, (b) $M'(s) + X_2 \rightarrow M'X_2(s)$.
16. Lithium and its compounds resemble magnesium and its compounds because the charge-to-radius ratio of Li is quite similar to that of Mg. Beryllium also shows this kind of diagonal behavior with aluminum.
17. $CaCO_3 \cdot MgCO_3(s) + 2H_2O(l) + 2CO_2(aq) \rightleftharpoons Ca^{2+} + Mg^{2+} + 4HCO_3^-$
    Groundwater, always containing carbon dioxide, dissolves limestone and dolomite. A cavern is formed when the mineral is below the surface of the earth and is covered by rocks that do not dissolve. When groundwater reenters the cave, the lower partial pressure of the CO_2 allows the CO_2 to escape from solution and the insoluble calcium and magnesium carbonates are deposited as stalactites on the ceiling. Additional HCO_3^- decomposes as drops fall from the stalactites, forming deposits as stalagmites on the floor.
18. The lower oxidation states are preferred as the atomic mass increases within the family. Atoms with the lower oxidation states form more ionic compounds, and those with the higher oxidation states form more molecular compounds.
19. The metallic nature of Bi is shown by the following properties: +3 and +5 oxidation states, hard, reddish-white solid, reacts directly with O_2 and halogens, reacts with oxidizing acids, and Bi_2O_3 is basic oxide. The non-metallic nature of Bi is shown by the following properties: -3 oxidation state, brittle, and low electrical and thermal conductivity.
20. (a) M_2O_3, (b) SnO_2, (c) PbO or other oxides such as PbO_2 or Pb_3O_4, (d) MX_3, (e) BiF_5.
21. Pb is attacked only superficially by dilute H_2SO_4, forming an adherent $PbSO_4$ layer which protects the metal from further reaction. Concentrated H_2SO_4 will react readily with Pb.
22. $E° = -0.343$ V for $In^{3+} + 3e^- \rightarrow In(s)$, -0.8281 V for $2H_2O(l) + 2e^- \rightarrow H_2(g) + 2OH^-$, and -1.662 V for $Al^{3+} + 3e^- \rightarrow Al(s)$; In^{3+} will be reduced from aqueous solution, but water will be reduced instead of Al^{3+}.
23. Tl.

PRACTICE TEST

1 (10 points). Briefly discuss the metallurgy involved in making iron and in making steel.

2 (5 points). Write equations for the following reactions involved with the metallurgy of copper: (a) Cu_2S undergoing roasting, (b) copper(I) oxide reacting with Cu_2S, and (c) carbon reduction of Cu_2O.

3 (50 points). Choose the best answer for each statement.

(a) The mixture of minerals from which a particular metal can be profitably extracted is

(i) a flux (ii) an ore
(iii) a gangue (iv) a mineral

(b) The undesirable rock originally in an ore is the

(i) flux (ii) slag
(iii) gangue (iv) mineral

(c) The separation of a soluble compound from an insoluble substance by using an appropriate solvent is

(i) flotation (ii) leaching
(iii) sintering (iv) none of these answers

(d) The heating of a finely divided ore to produce larger particles without melting is

(i) leaching (ii) smelting
(iii) roasting (iv) none of these answers

(e) The processing of an ore in which reducing the ore produces the molten metal is

(i) roasting (ii) sintering
(iii) smelting (iv) calcining

(f) Before a metal can be extracted from its oxide, the material usually is

(i) roasted (ii) reduced
(iii) electrolyzed (iv) heated

(g) The usual production of Fe involves

(i) a thermite reaction
(ii) reduction of the oxide with C or CO
(iii) production of the volatile $Fe(CO)_4$
(iv) electrolysis of the oxide

(h) Steel is tempered to make it
(i) hard (ii) soft
(iii) malleable (iv) flexible

(i) Aluminum is prepared by
(i) reduction of $AlCl_3$ with CO
(ii) reduction of Al_2O_3 with H_2
(iii) electrolysis of a melt of Al_2O_3 dissolved in Na_3AlF_6
(iv) calcining an aluminum ore

(j) Why does Al seem rather inert?
(i) It is below H in the electromotive series.
(ii) It forms an adherent protective oxide coating.
(iii) It crystallizes in a diamond unit cell.
(iv) It is amphoteric.

(k) An automatic sprinkler system for fire protection must be equipped with an alloy that
(i) resists corrosion (ii) has a high electrical resistance
(iii) melts at a low temperature (iv) is inexpensive

(l) Which would *not* be a property of rubidium?
(i) good electrical and thermal conductivity
(ii) formation of +2 cation
(iii) metallic luster
(iv) vigorous reaction with halogens and water

(m) Which of the following reactions will *not* occur?
(i) $K(s) + O_2(g) \rightarrow KO_2(s)$
(ii) $BaO_2(s) + 2H_2O(l) \rightarrow H_2O_2(aq) + Ba(OH)_2(aq)$
(iii) $Mg(s) + O_2(g) \rightarrow MgO_2(s)$
(iv) $CaO(s) + H_2O(l) \rightarrow Ca(OH)_2(aq)$

(n) A major use for KO_2 is in
(i) abrasives (ii) a self-contained gas mask
(iii) insecticides (iv) none of these answers

(o) A major use for Al_2O_3 is in
(i) paint pigments (ii) water purification
(iii) abrasives (iv) all of these answers

(p) A natural source for Ca is
(i) trona (ii) saltpeter
(iii) bauxite (iv) none of these answers

(q) A natural source for Al is

(i) trona (ii) limestone

(iii) bauxite (iv) all of these answers

(r) A method for the preparation of Pb is

(i) bubbling calcium vapor through molten PbO

(ii) electroylsis of the molten halide

(iii) distillation of $[Pb(CO)_4]$

(iv) none of these answers

(s) A method for the preparation of Mg is

(i) decomposition of the oxide to oxygen and molten metal

(ii) sintering, roasting, and calcining MgO

(iii) reforming the halide

(iv) electrolysis of the molten halide

(t) Identify the correct set of products formed by the reaction of $MgCO_3$ with

(i) HCl(aq) to give $H_2(g)$, CO_2, and $MgCl_2(aq)$

(ii) NaOH(aq) to give $Na_2CO_3(aq)$ and $Mg(OH)_2(s)$

(iii) HCl(aq) to give $MgCl_2(aq)$, $H_2O(l)$, and a combustible gas

(iv) NaOH(aq) to give $Na_2[Mg(OH)_4]$, $CO_2(g)$, and $H_2O(l)$

(u) Identify the correct set of products formed by the reaction of PbO with

(i) HCl(aq) to give $H_2(g)$ and $Pb(OCl)_2(s)$

(ii) HCl(aq) to give $PbCl_2(s)$ and $H_2O(l)$

(iii) NaOH to give $H_2O_2(aq)$ and $Na_2[Pb(OH)](aq)$

(iv) none of these answers

(v) What mass of H_2 will be produced by the reaction of 1.00 g Mg with 25.0 mL of 6.0 M HCl?

(i) 1.00 g (ii) 0.151 g

(iii) 0.234 g (iv) 0.0829 g

(w) What mass of Pb_3O_4 will be needed to produce 15.1 g of lead(II) nitrate?

$$Pb_3O_4(s) + 4HNO_3(aq) \rightarrow PbO_2(s) + 2Pb(NO_3)_2(aq) + 2H_2O(l)$$

(i) 15.6 g (ii) 66.2 g

(iii) 31.3 g (iv) none of these answers

(x) What are the common oxidation states for the Representative Group III elements?

(i) 0, +3 for lighter elements, and +1 for heavier elements

(ii) 0, ±3, and +5

(iii) 0, +2, and +3

(iv) none of these answers

(y) Which is *not* an equation representing a step in the production of iron?

(i) $2C(s) + O_2(g) \rightarrow 2CO(g)$

(ii) $FeO(s) + 2CO(g) \rightarrow FeCO_3(s) + C(s)$

(iii) $CaCO_3(s) \rightarrow CaO(s) + CO_2(g)$

(iv) $C(s) + CO_2(g) \rightarrow 2CO(g)$

4 (10 points). How many coulombs of electricity is necessary to produce 100.0 kg of aluminum metal by the electrolysis of Al_2O_3? Assume that the efficiency of the process is 85 %.

5 (5 points). The following reaction takes place in a blast furnace at 800 K:

$$FeO(s) + CO(g) \rightarrow Fe(s) + CO_2(g)$$

The standard state free energy of formation at 800 K is -219.35 kJ/mol for FeO, -182.47 kJ/mol for CO, and -395.62 kJ/mol for CO_2. Calculate $\Delta G°$ for this reaction.

6 (5 points). An oxide of iron contained 72.4 mass % of iron. What is the formula of this oxide?

7 (5 points). Write chemical equations showing the amphoteric behavior of Al and Al_2O_3.

8 (5 points). Write chemical equations showing the reaction of the alkali metals with oxygen and name the compounds formed.

9 (5 points). Write the electron configuration of Pb and predict the formulas of the chlorides and oxides that it forms. Name these compounds.

Answers to practice test

1. See Section 26.7 of the text.
2. (a) $2Cu_2S + 3O_2 \xrightarrow{\Delta} 2Cu_2O + 2SO_2$, (b) $Cu_2S + 2Cu_2O \rightarrow 6Cu + SO_2$, (c) $Cu_2O + C \rightarrow 2Cu + CO$.
3. (a) ii, (b) iii, (c) ii, (d) iv, (e) iii, (f) ii, (g) ii, (h) i, (i) iii, (j) ii, (k) iii, (l) ii, (m) iii, (n) ii, (o) iii, (p) iv, (q) iii, (r) iv, (s) iv, (t) ii, (u) ii, (v) iv, (w) i, (x) i, (y) ii.
4. 1.3×10^9 C.
5. 6.20 kJ.
6. Fe_3O_4.

7. $2Al(s) + 6H^+ \rightarrow 2Al^{3+} + 3H_2(g)$, $2Al(s) + 2OH^- + 6H_2O(l) \rightarrow 2[Al(OH)_4]^- + 3H_2(g)$, $Al_2O_3(s) + 6H^+ \rightarrow 2Al^{3+} + 3H_2O(l)$, $Al_2O_3(s) + 3H_2O(l) + 2OH^- \rightarrow 2[Al(OH)_4]^-$.
8. $4Li(s) + O_2(g) \rightarrow 2Li_2O(s)$, lithium oxide; $2Na(s) + O_2(g) \rightarrow Na_2O_2(s)$, sodium peroxide; $M(s) + O_2(g) \rightarrow MO_2(s)$ where M = K, Rb, or Cs, potassium superoxide, rubidium superoxide, cesium superoxide.
9. $1s^22s^22p^63s^23p^63d^{10}4s^24p^64d^{10}4f^{14}5s^25p^65d^{10}6s^26p^2$; $PbCl_2$, lead(II) chloride; $PbCl_4$, lead(IV) chloride; PbO, lead(II) oxide; PbO_2, lead(IV) oxide.

ADDITIONAL NOTES

CHAPTER 27

NONMETALS: HALOGENS AND NOBLE GASES

CHAPTER OVERVIEW

In this chapter the descriptive chemistry of the halogens (the Representative Group VII elements) and the noble gases is presented. The properties of the halogens and their preparations and uses are described. The principal classes of halogen compounds are discussed--the hydrogen halides and their acidic aqueous solutions, the metal halides, the interhalogens, and the oxoacids of the halogens and their salts. The chapter concludes with a discussion of the noble gas family elements.

COMPETENCIES

Definitions for the following terms should be learned:

- halogenation
- interhalogens

General concepts and skills that should be learned include

- the names and symbols of the elements in the halogen family
- the general chemical and physical properties of the elements
- the sources, industrial and laboratory preparations, and uses of the elements
- the general descriptive chemistry (such as the properties, preparation, uses, and major reactions) of the following classes of compounds: hydrogen halides, metal halides, interhalogens, and the oxoacids of the halogens and their salts
- the sources, preparations, and uses of the noble gas elements
- the general physical and chemical properties of the noble gas elements

QUESTIONS

1. Match the correct term to each definition.

(a) a molecular compound formed between two different halogens
(b) a binary compound formed between a halogen and another element
(c) a binary acid of a halogen, HX(aq)
(d) the introduction of halogen atom(s) into a covalent compound
(e) the oxoacid of a halogen having the formula HXO(aq)

(i) halide
(ii) halogenation reaction
(iii) hydrohalic acid
(iv) hypohalous acid
(v) interhalogen

2. Which of the following statements are true? Rewrite each false statement so that it is correct.

(a) A halide is a binary compound of halogen atoms formed by direct union of two different halogens.
(b) All of the hydrogen halides form strong acids when dissolved in water.
(c) Fluorine is the strongest oxidizing agent among all the elements.
(d) The shape of molecular ClF_3 will be an isosceles triangle.
(e) The acid strengths of the oxochloro acids increase with oxygen content.
(f) Iodine and bromine are stronger oxidizing agents than chlorine and bromine.
(g) Chlorine is a colorless gas under room conditions.
(h) When chlorine is bubbled through water, a reaction takes place with the water, yielding HCl and HClO.
(i) HIO is called hypoiodous acid.
(j) Reactions involving pure perchlorates are usually explosive because they are good reducing agents.
(k) There is a relationship between the boiling point and atomic mass.
(l) Because neon is rather inert chemically, it is commonly used to fill incandescent light bulbs to keep the filament from burning out.

3. Choose the response that best completes each statement.

(a) Under room conditions, chlorine is a
(i) yellow-green gas (ii) red-brown liquid
(iii) violet solid (iv) silvery metal

(b) The halogen that sublimes under room conditions is
(i) F_2 (ii) Cl_2
(iii) Br_2 (iv) I_2

(c) The reaction represented by the chemical equation

$$F(g) + e^- \rightarrow F^-(g)$$

is called

(i) an electron affinity (ii) a coordinate covalent bond
(iii) an ionization process (iv) a heat of formation

(d) Which process represents the preparation of fluorine?

(i) $16HF + 2KMnO_4 \rightarrow 5F_2 + 2KF + 8H_2O + MnF_2$
(ii) electrolysis of brine
(iii) $2NaF + Cl_2 \rightarrow F_2 + 2NaCl$
(iv) electrolysis of molten KHF_2

(e) Which is a suitable laboratory means for preparing iodine?

(i) $PI_3 + 3H_2O \rightarrow 3HI + H_3PO_3$
(ii) $2HCl + 2NaI \rightarrow H_2 + I_2 + 2NaCl$
(iii) $2KI + 3H_2SO_4 + MnO_2 \rightarrow 2KHSO_4 + 2H_2O + I_2 + MnSO_4$
(iv) $2NaBr + Cl_2 \rightarrow 2NaCl + Br_2$

(f) The *least* active halogen listed is

(i) fluorine (ii) bromine
(iii) iodine (iv) chlorine

(g) What is the best representation for "chlorine water"?

(i) $HOCl + HCl$ (ii) $2HCl + \frac{1}{2}O_2$
(iii) $H^+ + OCl^-$ (iv) $HClO_2$

(h) Which is a test for Br^-?

(i) etching of glass (ii) white precipitate with Ag^+
(iii) pale yellow precipitate with Ag^+ (iv) blue starch test

(i) Which is the weakest binary acid?

(i) HI (ii) HBr
(iii) HF (iv) HClO

(j) Chlorine is added to water in order to

(i) make the water taste good
(ii) undergo reaction with water to product HClO, which oxidizes bacteria
(iii) prevent tooth decay
(iv) none of these answers

4. What are the common oxidation states of Cl and Br?

5. Draw the Lewis structure for a diatomic halogen molecule. What types of intermolecular forces will be present between halogen molecules?

6. The electronegativities of chlorine and bromine are 3.0 and 2.8, respectively. The standard reduction potentials are 1.3595 V and 1.087 V, respectively, which are not in the same ratio as the electronegativities. Why not?

7. Why must gaseous fluorine be prepared by electrolytic methods?

8. Write the formulas and name the binary acids of the halogens. Are any of these weak acids?

9. Suggest a structure for $(HF)_3$, a component of liquid HF.

10. Briefly explain why acid strength decreases in the order (a) HI > HBr > HCl > HF, (b) HClO > HBrO > HIO, and (c) $HClO_4$ > $HClO_3$ > $HClO_2$ > HClO.

11. Iron(II) chloride is an ionic compound, but iron(III) chloride is a volatile, molecular compound. Explain this difference.

12. Name three ways in which metal halides can be formed. Write a chemical equation illustrating each method.

13. What is the general trend for the thermal stabilities of the halide salts of the same cation?

14. List the general solubility rules for metal halides in water.

15. Name three types of interhalogens. Briefly describe the bonding in these species.

16. Draw the Lewis structure for IF_3. What is the geometry of this molecule? Is this a polar or a nonpolar molecule?

17. Describe the geometry of the (a) ClF_2^- and (b) ClF_4^- ions.

18. What are Freons, perfluorocarbons, and Teflon?

19. The compound SF_6, a remarkably thermally and chemically stable gaseous compound, is extensively used as an insulator in high-voltage transformers. Discuss the bonding in SF_6 and its molecular structure. Relate these to the chemical inertness of the compound.

20. The "highest" fluoride of sulfur is SF_6, the "highest" chloride is SCl_4, and the "highest" bromide is S_2Br_2. Iodine does not react with sulfur. Explain.

21. Complete and balance each of the following equations:
(a) $OCl^- + I^- \xrightarrow{OH^-} Cl^- + I_2(aq)$
(b) $ClO_3^- + Sn^{2+} \xrightarrow{H^+} Cl^- + Sn^{4+}$

(c) $I_2(aq) + S_2O_3^{2-} \rightarrow I^- + S_4O_6^{2-}$

(d) $IO_3^- + I^- \xrightarrow{H^+} I_2(aq)$

(e) $Cl_2(g) + H_2O(l) \rightarrow$

22. Briefly describe the discovery of each of the noble gases.

23. Write the chemical symbol and name for each of the noble gases. How do we isolate each of these elements? Identify one use for each of these gases.

24. What does the term "chemical stability" mean? Why are the noble gases the least reactive group of elements?

25. Complete the following table.

Element	Source	Preparation	Uses
He			
Ne			
Ar			
Kr			
Xe			________

26. Identify the noble gas (He, Ne, Ar, Kr, Xe, Rn) that has the highest/greatest (a) melting point, (b) boiling point, (c) density of gas at room conditions, (d) atomic radius, and (e) ionization energy.

27. Classify each of the following reactions as (i) combination, (ii) decomposition, (iii) displacement, or (iv) partner exchange:

(a) $Xe + F_2 \rightarrow XeF_2$

(b) $XeF_6 + RbF \rightarrow RbXeF_7$

(c) $2RbXeF_7 \rightarrow XeF_6 + Rb_2XeF_8$

(d) $XeF_6 + H_2O \rightarrow XeOF_4 + 2HF$

Answers to questions

1. (a) v, (b) i, (c) iii, (d) ii, (e) iv.
2. (a) F, replace *A halide* with *An interhalogen compound*; (b) F, add *except hydrogen fluoride*; (c) T; (d) F, replace *an isosceles triangle* with *T-shaped*; (e) T; (f) F, replace *stronger* with *weaker*; (g) F, replace *colorless* with *yellow-green*; (h) T; (i) T; (j) F, replace *reducing* with *oxidizing*; (k) T; (l) F, replace *neon* with *argon*.
3. (a) i, (b) iv, (c) i, (d) iv, (e) iii, (f) iii, (g) i, (h) iii, (i) iii, (j) ii.
4. +5, +7, -1, 0.

5. :X - X:, London forces.
6. The value of $E°$ includes bond energy, electron affinity, and heat of solution of ion. Electronegativity is simply a measure of the attraction of an e^- by a gaseous atom.
7. All natural sources of fluorine contain F^- and no usual chemical oxidizing agent can oxidize F^- to F_2 because fluorine itself is a very strong oxidizing agent.
8. HF, hydrofluoric acid, weak; HCl, hydrochloric acid; HBr, hydrobromic and; H_2^-, hydroiodic acid.
9. 

10. (a) Acid strength of binary acids decreases with decreasing size; (b) acid strength of a series of oxoacids having the general formula decreases with decreasing electronegativity of the central nonmetal atom; (c) acid strength of a series of oxoacids of the same element decreases with decreasing oxidation state of the central nonmetal atom.
11. Iron(II) chloride is an ionic compound but iron(III) chloride is a volatile molecular compound because Fe^{3+} is smaller than Fe^{2+}, causing covalent bonds to form for Fe^{3+}.
12. The three ways are direct combination of elements, $2Na(s) + Cl_2(g) \rightarrow 2NaCl(s)$; precipitation of ions, $Pb^{2+} + 2Br^- \rightarrow PbBr_2(s)$; and reaction of metal with HX, $Mg(s) + 2HCl(aq) \rightarrow MgCl_2(aq) + H_2(g)$.
13. The thermal stabilities decrease as the anionic sizes increase.
14. The general solubility rules for metal halides in water are that most fluorides are insoluble while most chlorides, bromides, and iodides are soluble, except for those of Ag^+, Hg_2^{2+}, Pb^{2+}, and Tl^+.
15. The types are diatomic molecules, polyatomic molecules, and polyhalide ions. The covalent bonding usually involves the unfilled *nd* subshells for the additional electrons.
16. 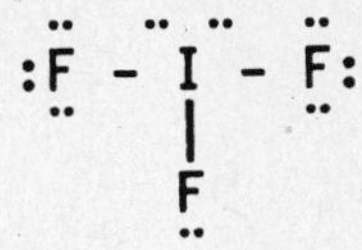

T-shaped (type AB_3E_2), polar.

17. (a) 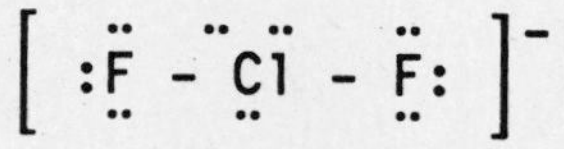 AB_2E_3 linear (b) 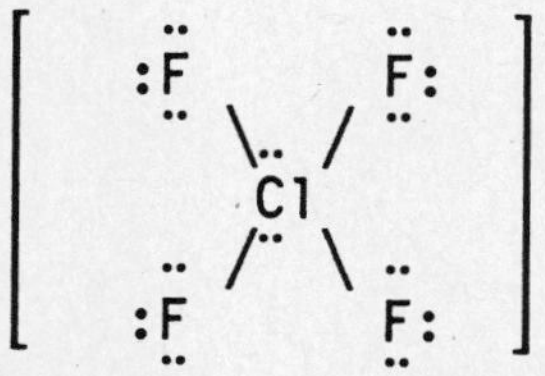 AB_4E_2 square planar

18. Freons are low molar mass fluorochlorocarbons, perfluorocarbons are organic compounds with all hydrogen atoms replaced by fluorine atoms, and Teflon is polytetrafluoroethylene.
19. There are strong polar covalent bonds between the S and F atoms. The shape is octahedral (type AB_6) and it is a nonpolar molecule.
20. Fluorine is the strongest oxidizing agent and its atoms are the smallest. There is not enough room around a sulfur atom for six chlorine atoms, and so the highest chloride has only four chlorine atoms. The still larger and less reactive bromine atoms react, but not vigorously enough to completely separate the sulfur atoms. Iodine is not a strong enough oxidizing agent to react with sulfur.
21. (a) $OCl^- + 2I^- + H_2O(l) \rightarrow Cl^- + I_2(aq) + 2OH^-$, (b) $ClO_3^- + 3Sn^{2+} + 6H^+ \rightarrow Cl^- + 3Sn^{4+} + 3H_2O(l)$, (c) $I_2(aq) + 2S_2O_3^{2-} \rightarrow 2I^- + S_4O_6^{2-}$, (d) $IO_3^- + 5I^- + 6H^+ \rightarrow 3I_2(aq) + 3H_2O(l)$, (e) $Cl_2(g) + H_2O(l) \rightarrow HCl(aq) + HClO(aq)$.
22. The existence of helium was recognized when sunlight was observed through the spectroscope. Argon was discovered as the residue from a sample of nitrogen from the atmosphere that did not react with magnesium. Krypton, neon, and xenon were obtained by separating the components of liquid air. Radon, the radioactive noble gas element, was discovered in 1900.
23. The chemical symbols and names of the noble gases are He, helium; Ne, neon; Ar, argon; Kr, krypton; Xe, xenon; and Rn, radon. Most helium is obtained from natural gas from some wells by liquefaction, followed by fractional distillation. Neon, argon, krypton, and xenon are obtained from liquefied air. Small amounts of radon are formed in radioactive processes. Helium is used in a variety of applications that take advantage of its unusual properties at low temperatures. Neon is used in "neon" signs. Argon is used with oxygen to help remove carbon impurities from molten steel. Krypton is used in expensive incandescent light bulbs.
24. Chemical stability is the resistance to chemical change. Noble gases have completely filled outermost energy levels.
25. source: He from natural gas, others from air

preparation: He from distillation of liquefied natural gas, others from distillation of liquefied air

uses: He for metallurgy, balloons, and neon signs; Ne for neon signs; Ar for metallurgy, light bulbs, and neon signs

26. (a) Rn, (b) Rn, (c) Rn, (d) Rn, (e) He.
27. (a) i, (b) i, (c) ii, (d) iv.

PRACTICE TEST

1 (10 points). Choose the response that best completes each statement.

(a) Which is a property of the halogens?
- (i) gain electrons to form monatomic cations with a -1 charge
- (ii) strong oxidizing agents
- (iii) high electronegativities
- (iv) all of these answers

(b) Which of the following statements is *not* true?
- (i) Bromine is a stronger oxidizing agent than iodine.
- (ii) Chlorine may be obtained by electrolyzing a solution of sodium chloride.
- (iii) Fluorine is the only halogen that exists as a gas under room conditions.
- (iv) None of the halogens are found free in nature.

(c) Which is the strongest oxidizing agent?
- (i) I_2
- (ii) Br_2
- (iii) Cl_2
- (iv) F_2

(d) Which of these reactions is *not* typical of the halogens?
- (i) $H_2 + H_2O(l) \rightarrow H^+ + X^- + HOX(aq)$
- (ii) $X_2 + Mg(s) \rightarrow MgX_2(s)$
- (iii) $X_2 + O_2(g) \rightarrow 2OX^-$
- (iv) $X_2 + H_2(g) \rightarrow 2HX(g)$

(e) The halogen compound used for etching of glass is
- (i) zonite
- (ii) freon
- (iii) hypochlorous acid
- (iv) hydrofluoric acid

(f) Which is a suitable means for preparing fluorine?
- (i) $H_2 + F_2 \rightarrow 2HF$
- (ii) $2KF + 3H_2SO_4 + MnO_2 \rightarrow 2KHSO_4 + MnSO_4 + 2H_2O + F_2$
- (iii) $2KHF_2 \rightarrow F_2 + 2KF + H_2$
- (iv) $2NaF + Cl_2 \rightarrow 2NaCl + F_2$

(g) The halogen that is a liquid under room conditions is

(i) chlorine (ii) iodine

(iii) fluorine (iv) bromine

(h) The most active halogen is

(i) chlorine (ii) bromine

(iii) fluorine (iv) iodine

(i) A test for the chloride ion is

(i) litmus paper (ii) odor and color

(iii) bromine water (iv) silver nitrate solution

(j) Identify the correct formula for an interhalogen.

(i) ClF (ii) XeF_4

(iii) FCl_4 (iv) Br_2

2 (10 points). For each of the halogens discussed in the text, (a) write the name, (b) give the symbol, and (c) describe the physical state (including color) for the diatomic molecule at room conditions.

3 (5 points). Write the electron configurations for (a) an atom of chlorine and (b) a chloride ion.

4 (10 points). Write a chemical equation for the reaction of $Cl_2(g)$ with (a) a metal such as Na, (b) a nonmetal such as P, and (c) a molecule such as $C_2H_4(g)$.

5 (10 points). From the compounds listed, identify which are (a) halides, (b) hydrohalic acids, (c) hypohalous acids, (d) interhalogens, and (e) halogen oxo acids: (a) $CCl_4(l)$ (ii) $HBrO_3(aq)$, (iii) $ICl_3(g)$, (iv) $UF_6(g)$, (v) HIO(aq), (vi) HCl(aq), and (vii) KF(s).

6 (10 points). The standard heat of formation at 25 °C is 121.679 kJ/mol for Cl(g), 111.884 kJ/mol for Br(s), 30.907 kJ/mol for $Br_2(g)$, 0 for $Cl_2(g)$, and 14.64 kJ/mol for BrCl(g). Calculate the bond dissociation energies for Cl-Cl, Br-Br, and Br-Cl. Which molecule contains the strongest bond?

7 (10 points). Metal halides can be produced by the reaction of active metals with the binary acids of the halogens. (a) Write the equation for the reaction between Mg(s) and HI(aq). (b) What mass of Mg will react with 25.0 mL of 0.102 M HI?

8 (10 points). The equilibrium constant for the ionization of chlorous acid is 1.1×10^{-2}. (a) Write the chemical equation for the ionization process. (b)

What is the $[H^+]$ in a 0.10 M solution of chlorous acid? (c) What is the pH of this solution?

9 (10 points). Chlorine can be prepared by the reaction of HCl with an acidic solution of MnO_4^-. (a) Write the half-reactions and (b) the overall equation. Using $E°$ as 1.3595 V for Cl_2/Cl^- and 1.51 V for MnO_4^-/Mn^{2+}, (c) find $E°$ for the reaction. (d) Calculate the potential for the reaction under nonstandard state conditions with $[H^+]$ = 6.0 mol/L, $[Cl^-]$ = 1.5 mol/L, $[MnO_4^-]$ = 0.021 mol/L = 0.021 mol/L, $[Mn^{2+}]$ = 0.0010 mol/L, and P_{Cl_2} = 0.0010 bar.

10 (15 points). Write the chemical equation for the combination reaction of $XeF_2(s)$ with $F_2(g)$ to give $XeF_4(s)$. Draw Lewis structures for these three substances. What is the oxidation number of each element in each substance?

Answers to practice test

1. (a) iv, (b) iii, (c) iv, (d) iii, (e) iv, (f) iii, (g) iv, (h) iii, (i) iv, (j) i.

2. (a) fluorine, chlorine, bromine, iodine; (b) F, Cl, Br, I; (c) yellow gas, yellow-green gas, red-brown liquid, violet-black solid.

3. (a) $1s^2 2s^2 2p^6 3s^2 3p^5$, (b) $1s^2 2s^2 2p^6 3s^2 3p^6$.

4. (a) $2Na(s) + Cl_2(g) \rightarrow 2NaCl(s)$, (b) $2P(s) + 5Cl_2(g) \rightarrow 2PCl_5(s)$, (c) $C_2H_4(g) + Cl_2(g) \rightarrow C_2H_4Cl_2(l)$.

5. (a) i, iii, iv, vi, vii; (b) vi; (c) v; (d) iii; (e) ii, v.

6. 243.358 kJ/mol, 192.861 kJ/mol, 218.92 kJ/mol; Cl_2.

7. (a) $Mg(s) + 2HI(aq) \rightarrow MgI_2(aq) + H_2(g)$, (b) 0.0310 g.

8. (a) $HClO_2(aq) + H_2O(l) \rightleftharpoons H_3O^+ + ClO_2^-$, (b) 2.8×10^{-2} mol/L, (c) 1.55.

9. (a) $2Cl^- \rightarrow Cl_2 + 2e^-$, $MnO_4^- + 8H^+ + 5e^- \rightarrow Mn^{2+} + 4H_2O$; (b) $2MnO_4^- + 16H^+ + 10Cl^- \rightarrow 5Cl_2(aq) + 2Mn^{2+} + 8H_2O(l)$; (c) 0.15 V; (d) 0.34 V.

10. $XeF_2(s) + F_2(g) \rightarrow XeF_4(s)$; F is -1 in XeF_2 and XeF_4 and 0 in F_2, Xe is +2 in XeF_2 and +4 in XeF_4.

$$:\ddot{F} - \ddot{X}e - \ddot{F}: \qquad :\ddot{F} - \ddot{F}: \qquad :\ddot{F} - \ddot{X}e - \ddot{F}:$$
$$\diagup \quad \diagdown$$
$$:\ddot{F}: \qquad :\ddot{F}:$$

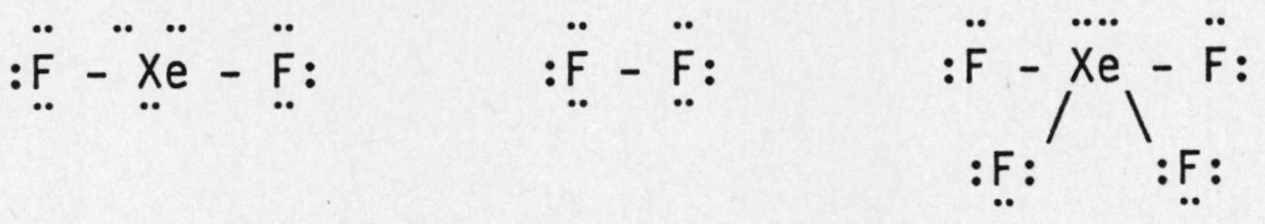

ADDITIONAL NOTES

CHAPTER 28

NONMETALS: NITROGEN, PHOSPHORUS, AND SULFUR

CHAPTER OVERVIEW

Nitrogen, phosphorus, and sulfur--three elements essential to both plant and animal life--are discussed in this chapter. The sources, preparation, and uses of each element are presented. Various chemical reactions of the elements are considered and the important industrial acids--sulfuric acid, phosphoric acid, and nitric acid--are discussed in detail. The chapter concludes with a discussion of several important compounds of nitrogen, phosphorus, and sulfur--nitrides, ammonia and ammonium salts, hydrazine, nitrogen oxides, nitrogen oxoacids and their salts, phosphides, phosphine, phosphorus halides, phosphorus oxides, phosphorus oxoacids and their salts, hydrogen sulfide and sulfides, sulfur oxides, sulfur halides, and sulfur oxoacids and their salts.

COMPETENCIES

Definitions for the following terms should be learned:

- aqua regia
- dimer
- London smog
- nitrogen fixation
- phosphate rock
- photochemical smog

General concepts and skills that should be learned include

- the sources, preparations, and uses of the elements
- the general physical and chemical properties of the elements
- the general physical and chemical properties of the major compounds of the elements, such as NH_3, the nitrogen oxides, HNO_2 and nitrites, HNO_3 and nitrates, H_3PO_4 and phosphates, H_2S and sulfides, the sulfur oxides, and H_2SO_4 and sulfates
- the basic parts of the nitrogen cycle

QUESTIONS

1. Match the correct term to each definition.
   (a) a mixture of 3 parts of HCl to 1 part of HNO_3 by volume
   (b) mineral containing $Ca_3(PO_4)_2$ and SiO_2
   (c) the combination of molecular nitrogen with other atoms
   (d) the conversion of NO_2^- and NO_3^- to N_2
   (e) the conversion of ammonia to NO_3^- and NO_2^-

   (i) aqua regia
   (ii) denitrification
   (iii) nitrification
   (iv) nitrogen fixation
   (v) phosphate rock

2. Match the correct name to each formula

(a) H_2NNH_2	(i) ammonia
(b) N_2O_5	(ii) ammonium ion
(c) NO^+	(iii) hydrazine
(d) N_2O	(iv) nitrate ion
(e) NO_2^-	(v) nitric acid
(f) N^{3-}	(vi) nitride ion
(g) HNO_2	(vii) nitrite ion
(h) NH_4^+	(viii) nitrogen
(i) NO_3^-	(ix) nitrogen(I) oxide or nitrous oxide
(j) NO_2 and N_2O_4	(x) nitrogen(II) oxide or nitric oxide
(k) N_2O_3	(xi) nitrogen(III) oxide or dinitrogen trioxide
(l) HNO_3	(xii) nitrogen(IV) oxides
(m) NH_3	(xiii) nitrogen(V) oxide or dinitrogen pentoxide
(n) NO	(xiv) nitrosyl ion
(o) N_2	(xv) nitrous acid

3. Match the correct name to each formula.

(a) PH_4^+	(i) dihydrogen phosphate ion
(b) P_4O_{10}	(ii) metaphosphoric acid
(c) HPO_4^{2-}	(iii) monohydrogen phosphate ion
(d) $(HPO_3)_n$	(iv) orthophosphoric or phosphoric acid
(e) P_4O_6	(v) phosphate ion
(f) PO_4^{3-}	(vi) phosphide ion
(g) P^{3-}	(vii) phosphine
(h) H_3PO_4	(viii) phosphonium ion
(i) PH_3	(ix) phosphorous acid

(j) $H_4P_2O_7$ (x) phosphorus(III) oxide

(k) H_2PO_3 (xi) phosphorus(V) oxide

(l) $H_2PO_4^-$ (xii) pyrophosphoric acid

4. Match the correct name to each formula.

(a) HS^-	(i) carbon disulfide
(b) SO_3	(ii) hydrogen sulfate ion
(c) H_2SO_4	(iii) hydrogen sulfide
(d) $S_2O_3^{2-}$	(iv) hydrogen sulfide ion
(e) S^{2-}	(v) hydrogen sulfite ion
(f) HSO_3^-	(vi) hydrosulfuric acid
(g) $H_2S_2O_7$	(vii) pyrosulfuric acid
(h) CH_3CSNH_2	(viii) sulfate ion
(i) HSO_4^-	(ix) sulfide ion
(j) $H_2S(aq)$	(x) sulfite ion
(k) SO_4^{2-}	(xi) sulfur dioxide
(l) H_2SO_3	(xii) sulfur trioxide
(m) SO_2	(xiii) sulfuric acid
(n) SO_3^{2-}	(xiv) sulfurous acid
(o) $H_2S(g)$	(xv) thioacetamide
(p) CS_2	(xvi) thiosulfate ion

5. Which of the following statements are true? Rewrite each false statement so that it is correct.

(a) Phosphorus differs in many respects from nitrogen, as does sulfur from oxygen, because of the respective sizes, electronegativities, and maximum number of orbitals available in the respective valence shells.

(b) Many nitrogen-containing compounds decompose to form N_2 because of the great stability of the nitrogen-nitrogen triple bond.

(c) Most ammonium salts are water soluble and produce highly alkaline solutions.

(d) Copper will not react with hydrochloric acid but will react with nitric acid.

(e) Most metal sulfides are more soluble in an acid solution than in water.

(f) Concentrated sulfuric acid is a strong dehydrating agent.

6. Choose the response that best completes each statement.

(a) What distinguishes nitrogen from sulfur and phosphorus?

(i) cannot use vacant *d* orbitals for bonding
(ii) nonmetal versus metal behavior
(iii) forms an ionic anion
(iv) forms covalent bonds

(b) Nitrogen is rather inert at ordinary temperatures because

(i) the outer shell can hold only eight electrons
(ii) the triple bond is very strong
(iii) nitrogen is a gas
(iv) nitrogen is nontoxic

(c) In which compound does the nitrogen atom have an oxidation number of -3?

(i) N_2O_3 (ii) HNO_3
(iii) NH_3 (iv) NCl_3

(d) Which is the strongest acid?

(i) $NH_3(aq)$ (ii) HNO_2
(iii) $HNO_3(aq)$ (iv) $H_2NNH_2(aq)$

(e) Which is *not* a chemical property of NH_3?

(i) acts as an electron-pair donor
(ii) forms alkaline aqueous solutions
(iii) forms complex ions with many metals
(iv) acts as an oxidizing agent

(f) What is the major commercial method for preparing NH_3?

(i) Frasch process
(ii) lead chamber process
(iii) fractional distillation of liquid air
(iv) Haber process

(g) Which is not a property of ammonium salts?

(i) The salts dissolve slightly in water.
(ii) The salts undergo hydrolysis.
(iii) Salts containing nonoxidizing anions decompose to form ammonia and the parent acid.
(iv) Salts containing oxidizing anions undergo internal oxidation-reduction.

(h) Which oxide is produced by the reaction of Cu with dilute HNO_3?

(i) N_2O (ii) NO

(iii) N_2O_3 (iv) NO_2

(i) Which oxide forms nitrous acid upon dissolving in water?

(i) N_2O (ii) NO

(iii) NO_2 (iv) N_2O_5

(j) Which is *not* a chemical property of HNO_2?

(i) decomposes to HNO_3 and NO (ii) acts as a strong acid

(iii) acts as an oxidizing agent (iv) acts as a reducing agent

(k) Which is *not* true about HNO_3?

(i) has three resonance structures in which all N-O bonds are identical

(ii) decomposes in sunlight to give NO_2

(iii) acts as a strong acid

(iv) acts as a strong oxidizing agent

(l) Which is *not* true about white phosphorus?

(i) is highly toxic

(ii) is moderately soluble in diethyl ether and benzene

(iii) is stable in air

(iv) reacts vigorously with metals and halogens

(m) Which is *not* true about phosphine?

(i) can be prepared by treating white phosphorus with warm aqueous solutions of strong bases

(ii) is extremely poisonous

(iii) ignites in air

(iv) is highly soluble in water because it undergoes decomposition

(n) Which halides are produced by reacting excess phosphorus with the free halogen?

(i) POCl (ii) PX_3

(iii) PX_5 (iv) P_4X_{10}

(o) Phosphoric and sulfuric acid

(i) are completely ionized in water

(ii) exhibit some π bonding between the oxygen atoms and the center atom

(iii) can act as reducing agents

(iv) form "pyro" acids by adding a water molecule to the acid molecule

(p) The various physical properties of the states of sulfur result from
(i) the length of sulfur atom chains
(ii) the fraction of S_4 tetrahedra
(iii) the amount of oxygen impurities present
(iv) the -2, +4, and +6 oxidation states of S

(q) Which is *not* true about H_2S?
(i) is highly soluble in water (ii) is a good reducing agent
(iii) is used in qualitative analysis (iv) is a poisonous, stinking gas

(r) Which is true about SO_2?
(i) decomposes readily
(ii) is a linear molecule
(iii) acts as a strong reducing agent
(iv) dissolves in water to form sulfuric acid

(s) Sulfuric acid
(i) is completely ionized in water
(ii) oxidizes nonmetals
(iii) reacts with very active metals to form SO_2
(iv) is mixed with HNO_3 to produce aqua regia

(t) In the contact process for preparing H_2SO_4
(i) NO is used as a catalyst to oxidize SO_2 to SO_3
(ii) the lead walls allow the SO_3 to mix with the water
(iii) very low temperatures are used to increase the yield
(iv) SO_3 mixes with H_2SO_4 to form $H_2S_2O_7$, which then reacts with water

(u) Which is *not* true about thiosulfates?
(i) act as fairly strong oxidizing agents
(ii) decompose in acidic solution
(iii) have structures that are similiar to SO_4^{2-} except one oxygen atom is replaced by a sulfur atom
(iv) are used in photography

7. Choose the response that best completes each statement.

(a) Which anions hydrolyze to give alkaline solutions?
(i) S^{2-} (ii) PO_4^{3-}
(iii) NO_2^- (iv) NO_3^-

(b) Which anions yield thermally unstable protonic acids that decompose to evolve gases?

(i) NO_3^- (ii) PO_4^{3-}

      (iii) SO_3^{2-} (iv) I^-

(c) Which anion is colored?

        (i) CO_3^{2-} (ii) CrO_4^{2-}

      (iii) I^- (iv) PO_4^{3-}

(d) Which anions are quite strong oxidizing agents?

        (i) I^- (ii) PO_4^{3-}

      (iii) NO_3^- (iv) CrO_4^{2-}

(e) Which anions are reducing agents?

        (i) I^- (ii) S^{2-}

      (iii) SO_3^{2-} (iv) NO_2^-

(f) Which anions form slightly soluble barium salts in acidic solution?

        (i) SO_3^{2-} (ii) CO_3^{2-}

      (iii) NO_3^- (iv) SO_4^{2-}

(g) Which ion does *not* form a precipitate with Ag^+ in neutral solution?

        (i) NO_3^- (ii) NO_2^-

      (iii) CrO_4^{2-} (iv) SO_3^{2-}

8. What are the undesirable effects on the environment produced by excessive amounts of NO and NO_2?

9. Write chemical equations illustrating the acidic nature of the oxides of nitrogen, phosphorus, and sulfur.

10. What do you think happens when a glowing wooden splint is inserted into a sample of nitrous oxide? Why?

11. The equation for reaction of PCl_3 with water can be written as

$$PCl_3(l) + 3H_2O(l) \rightarrow 3HCl(g) + P(OH)_3(aq)$$

which indicates that all of the oxygen and hydrogen atoms in the resulting phosphorus product are equivalent. The molecule $P(OH)_3$ probably exists for a very short time, but it rearranges rapidly to form

$$\begin{array}{c} O \\ \| \\ H - P - OH \\ | \\ OH \end{array}$$

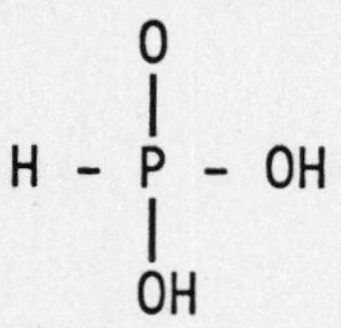

in which only two hydrogen atoms are ionizable. Why does this rearrangement take place?

12. Pyrophosphoric acid, $H_4P_2O_7$, readily forms the salts $Na_2H_2P_2O_7$ and $Na_4P_2O_7$, but it is difficult to get $NaH_3P_2O_7$ and $Na_3HP_2O_7$. Why?

13. What potential environmental hazard should be avoided when fluorapatite, $Ca_5(PO_4)_3F$, is converted to a phosphate fertilizer or phosphoric acid by reaction with sulfuric acid?

14. How is hydrogen sulfide generated under laboratory conditions for use in qualitative analysis? Write the chemical equation for this process.

15. When NH_4I is dissolved in water, and the solution is evaporated to dryness, solid NH_4I can be recovered. When PH_4I is dissolved in water, it dissociates into PH_3 and HI. Explain the difference.

16. Briefly explain why ΔH_f° is positive for the formation of the various nitrogen oxides.

17. Sulfur readily forms SF_6 and phosphorus forms $[PF_6]^-$. Why doesn't nitrogen form $[NF_6]^-$?

18. Draw Lewis structures for (a) HNO_2, (b) HNO_3, (c) NH_3, (d) PH_3, (e) H_3PO_3, (f) H_3PO_4, (g) $H_4P_2O_7$, (h) H_2S, (i) H_2SO_3, (j) H_2SO_4, (k) H_2SO_5, and (l) $H_2S_2O_3$.

19. Complete and balance the following oxidation-reduction equations:

(a) $NH_4NO_3(s) \xrightarrow{\Delta}$

(b) $N_2H_4(l) + O_2(g) \rightarrow$

(c) $N_2O(g) \rightarrow$

(d) $Cu(s) + H^+ + NO_3^-(dil) \rightarrow$

(e) $NO(g) + NO_2(g) \rightarrow$

(f) $NO_2(g) + H_2O(l) \rightarrow$

(g) $H_2S(g) + H_2SO_3(aq) \rightarrow$

(h) $SO_3^{2-} + S(s) \rightarrow$

Answers to questions

1. (a) i, (b) v, (c) iv, (d) ii, (e) iii.
2. (a) iii, (b) xiii, (c) xiv, (d) ix, (e) vii, (f) vi, (g) xv, (h) ii, (i) iv, (j) xii, (k) xi, (l) v, (m) i, (n) x, (o) viii.
3. (a) viii, (b) xi, (c) iii, (d) ii, (e) x, (f) v, (g) vi, (h) iv, (i) vii, (j) xii, (k) ix, (l) i.

4. (a) iv, (b) xii, (c) xiii, (d) xvi, (e) ix, (f) v, (g) vii, (h) xv, (i) ii,
    (j) vi, (k) viii, (l) xiv, (m) xi, (n) x, (o) iii, (p) i.

5. (a) T; (b) T; (c) F, replace *highly alkaline* with *acidic*; (d) T; (e) T; (f) T.

6. (a) i, (b) ii, (c) iii, (d) iii, (e) iv, (f) iv, (g) i, (h) ii, (i) iii, (j)
    ii, (k) i, (l) iii, (m) iv, (n) ii, (o) ii, (p) i, (q) i, (r) iii, (s) ii,
    (t) iv, (u) i.

7. (a) i, ii, iii; (b) iii; (c) ii; (d) iii, iv; (e) i, ii, iii, iv; (f) iv;
    (g) i.

8. The effects are photochemical smog and acid rain.

9. $3NO_2(g) + H_2O(l) \rightarrow 2HNO_3(aq) + NO(g)$

    $SO_3(g) + H_2O(l) \rightarrow H_2SO_4(aq)$

    $SO_2(g) + H_2O(l) \rightarrow H_2SO_3(aq)$

    $P_4O_6(s) + 6H_2O(l) \rightarrow 4H_3PO_3(aq)$

    $P_4O_{10}(s) + 6H_2O(l) \rightarrow 4H_3PO_4(aq)$

10. The splint will burst into flame. Nitrous oxide decomposes to generate O_2,
    $2N_2O(g) \xrightarrow{\Delta} 2N_2(g) + O_2(g)$.

11. A hydrogen ion can be formed from $P(OH)_3$, but it quickly forms a covalent
    bond with the remaining lone pair of electrons on the phosphorus atom to form
    a nonionizable hydrogen atom.

12. The ionization constants for the first proton at each end of the molecule
    are considerably greater than the ionization constants for the second proton
    at each end of the molecule.

13. Poisonous HF must not be released.

14. H_2S is formed by the decomposition of thioacetamide, $CH_3CSNH_2(aq) + 2H_2O(l)$
    $\xrightarrow{\Delta} CH_3COO^- + NH_4^+ + H_2S(aq)$.

15. NH_3 is a stronger base than PH_3. PH_4^+ reacts with water nearly completely to
    form PH_3, whereas NH_4^+ will not.

16. Considerable energy is needed to break the very strong $N \equiv N$ bond.

17. There are no *2d* orbitals available to hold the electrons from the additional
    bonds.

18. (a) $H - \ddot{O} - \ddot{N} = \ddot{O}:$ (b) $H - \ddot{O} - \overset{:O:}{\underset{\|}{N}} - \ddot{O}: \leftrightarrow H - \ddot{O} - \overset{:\ddot{O}:}{\underset{|}{N}} = \ddot{O}:$

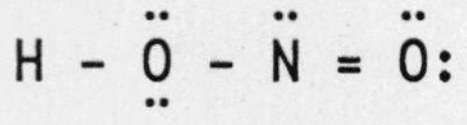

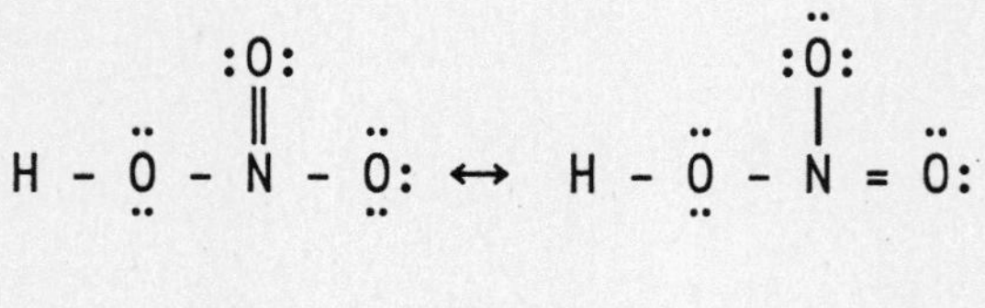

(c)
$$H - \overset{\displaystyle H}{\underset{\displaystyle ..}{N}} - H$$

(d)
$$H - \overset{\displaystyle H}{\underset{\displaystyle |}{P}} - H$$

(e)
$$H - \overset{..}{\underset{..}{O}} - \overset{\displaystyle H}{\underset{\displaystyle \overset{..}{O}:}{P}} - \overset{..}{\underset{..}{O}} - H$$
$$\overset{|}{H}$$

(f)
$$H - \overset{..}{\underset{..}{O}} - \overset{\displaystyle :\overset{..}{O}:}{\underset{\displaystyle :\overset{..}{O}:}{P}} - \overset{..}{\underset{..}{O}} - H$$
$$\overset{|}{H}$$

(g)
$$H - \overset{..}{\underset{..}{O}} - \overset{\displaystyle :\overset{..}{O}:}{\underset{\displaystyle :\overset{..}{O}:}{P}} - \overset{..}{\underset{..}{O}} - \overset{\displaystyle :\overset{..}{O}:}{\underset{\displaystyle :\overset{..}{O}:}{P}} - \overset{..}{\underset{..}{O}} - H$$
$$\overset{|}{H} \qquad \overset{|}{H}$$

(h) $H - \overset{..}{\underset{..}{S}} - H$

(i)
$$H - \overset{..}{\underset{..}{O}} - \overset{\displaystyle :\overset{..}{O}:}{S} - \overset{..}{\underset{..}{O}} - H$$

(j)
$$H - \overset{..}{\underset{..}{O}} - \overset{\displaystyle :\overset{..}{O}:}{\underset{\displaystyle :\overset{..}{O}:}{S}} - \overset{..}{\underset{..}{O}} - H$$

(k)
$$H - \overset{..}{\underset{..}{O}} - \overset{..}{\underset{..}{O}} - \overset{\displaystyle :\overset{..}{O}:}{\underset{\displaystyle :\overset{..}{O}:}{S}} - \overset{..}{\underset{..}{O}} - H$$

(l)
$$H - \overset{..}{\underset{..}{O}} - \overset{\displaystyle :\overset{..}{S}:}{\underset{\displaystyle :\overset{..}{O}:}{S}} - \overset{..}{\underset{..}{O}} - H$$

19. (a) $NH_4NO_3(s) \overset{\Delta}{\rightarrow} N_2O(g) + 2H_2O(g)$, (b) $N_2H_4(l) + O_2(g) \rightarrow N_2(g) + 2H_2O(l)$,
(c) $2N_2O(g) \rightarrow 2N_2(g) + O_2(g)$, (d) $3Cu(s) + 8H^+ + 2NO_3^- \rightarrow 3Cu^{2+} + 2NO(g)$
$+ 4H_2O(l)$, (e) $NO(g) + NO_2(g) \rightarrow N_2O_3(l)$, (f) $3NO_2(g) + H_2O(l) \rightarrow HNO_3(aq)$
$+ NO(g)$, (g) $2H_2S(g) + H_2SO_3(aq) \rightarrow 3H_2O(l) + 3S(s)$, (h) $SO_3^{2-} + S(s) \rightarrow$
$S_2O_3^{2-}$.

PRACTICE TEST

1 (10 points). Write the electron configurations for (a) N, (b) P, and (c) S.
Predict reasonable oxidation states for these elements based on the
configurations.

2 (10 points). Using the order of filling $1s < \sigma_{2s} < \sigma_{2s}^{\star} < \pi_{2p_y} =$
$\pi_{2p_z} < \sigma_{2p_x} < \pi_{2p_y}^{\star} = \pi_{2p_z}^{\star} < \sigma_{2p_x}^{\star} < \sigma_{3s} < \sigma_{3s}^{\star} < \pi_{3p_y} = \pi_{3p_z} < \sigma_{3p_x} < \pi_{3p_y}^{\star}$
$= \pi_{3p_z}^{\star} < \sigma_{3p_x}^{\star}$, write electron configurations for (a) S_2, (b) N_2, and (c) NO.
Discuss the bonding in each. Which are paramagnetic?

3 (30 points). Prepare Lewis structures for the following: (a) N_2, (b) SO_2,
(c) $S_2O_3^{2-}$, (d) N_2O, (e) PF_5, (f) PH_3, (g) PH_4^+, (h) P_4O_6, (i) HNO_3, and (j)

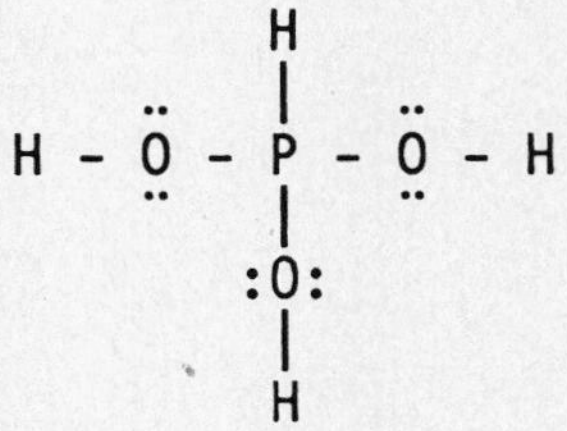

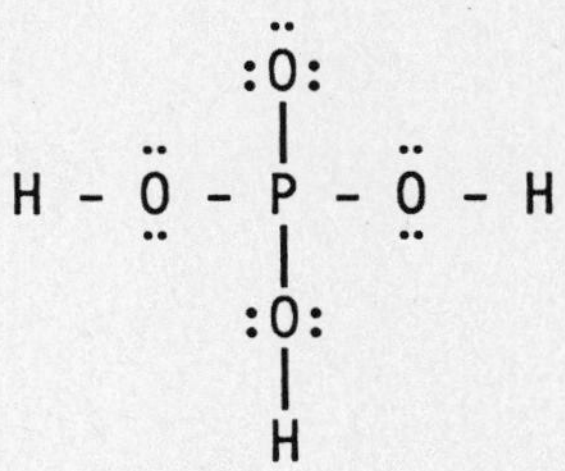

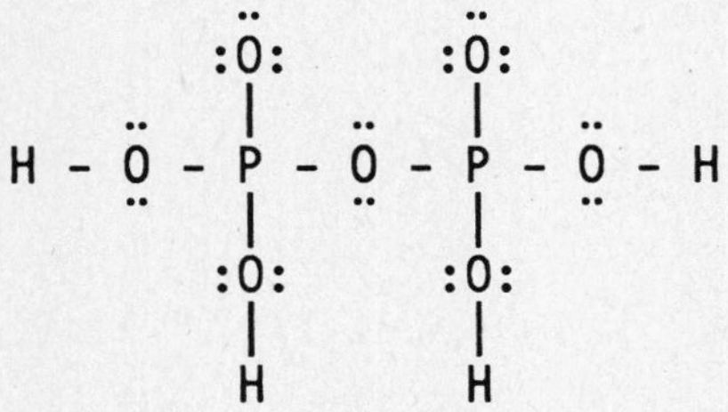

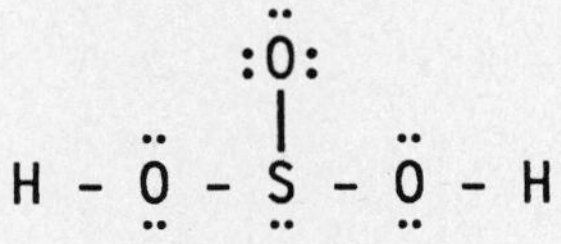

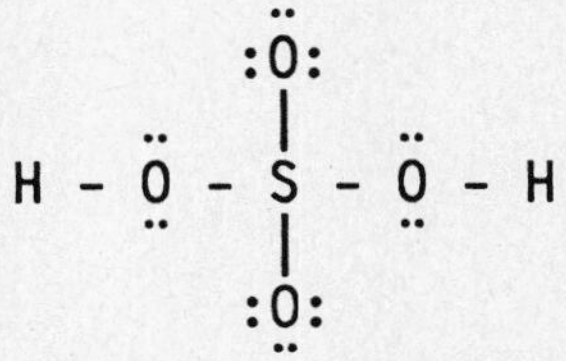

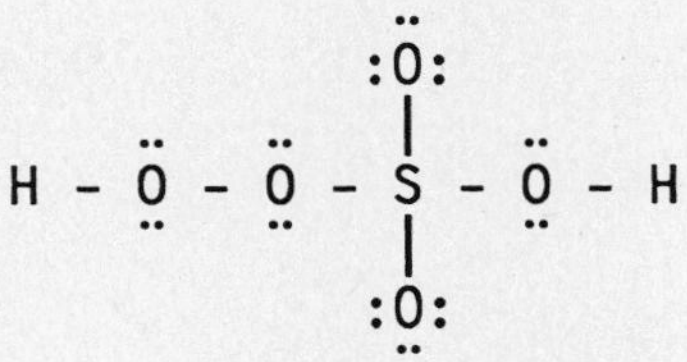

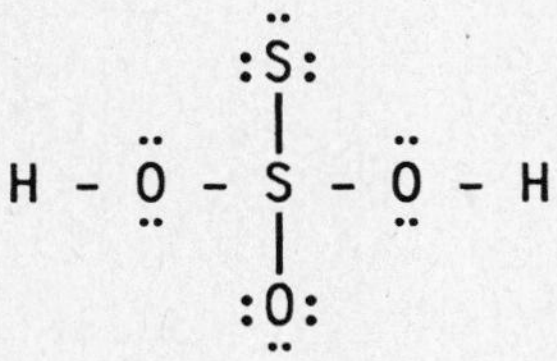

NO_3^-. Determine the type of hybridization for each element (except H) in each species and prepare three-dimensional sketches.

4 (10 points). List two major uses each for (a) N_2, (b) HNO_3, (c) P, (d) phosphates, (e) S, and (f) H_2SO_4.

5 (10 points). Write a balanced equation showing the preparation of (a) N_2O, (b) NO, (c) N_2O_3, (d) NO_2, (e) N_2O_4, and (f) N_2O_5. Name these compounds.

6 (10 points). Write a balanced chemical equation for each of the following reactions, outlining the contact method for producing H_2SO_4: (a) SO_2 being oxidized to SO_3 in the presence of V_2O_5, (b) producing pyrosulfuric acid, and (c) changing pyrosulfuric acid to H_2SO_4.

7 (20 points). Choose the response that best completes each statement.

(a) Which is *not* a physical property of nitrogen?

(i) slightly soluble in water (ii) odorless
(iii) colored (iv) less dense than air

(b) Elemental phosphorus at room temperature is

(i) a monatomic gas
(ii) a diatomic gas
(iii) tetraatomic with P atoms at the corners of a tetrahedron
(iv) a liquid like mercury

(c) Laughing gas is

(i) N_2O_4 (ii) N_2O_3
(iii) N_2O (iv) NO

(d) Decaying animal wastes give off

(i) oxygen (ii) hydrogen
(iii) ammonia (iv) nitric oxide

(e) What type of hybridization occurs for the sulfur atom in SF_6?

(i) sp (ii) sp^3d^2
(iii) sp^3 (iv) none of these answers

(f) Which statement is *not* true?

(i) Nitrogen is very soluble in water.
(ii) Nitrogen exists as an extremely stable diatomic molecule.
(iii) Most of the ammonia produced in the United States is used in fertilizers.
(iv) Nitrogen dioxide is a brown toxic gas.

(g) Which statement is *not* true?
  (i) Phosphorus is essential for the growth of plants and animals.
  (ii) White phosphorus is highly reactive and must be stored under water or in an air-tight container.
  (iii) Phosphine is a colorless, nontoxic, nonflammable gas.
  (iv) Phosphorus trichloride reacts with water to give phosphorous acid.

(h) Which statement is *not* true?
  (i) Sulfur trioxide, SO_3, is an acid anhydride.
  (ii) Sulfur in moist air is slowly oxidized to sulfuric acid.
  (iii) Sulfuric acid is a good reducing agent.
  (iv) Hydrogen sulfide is an extemely poisonous, colorless gas having the odor of rotten eggs.

(i) Photochemical smog requires the presence of
  (i) sunlight (ii) hydrocarbons
  (iii) nitrogen oxides (iv) all of the above

(j) Which of the following contributes significantly to the formation of acid rain?
  (i) N_2 (ii) SO_3
  (iii) O_3 (iv) CO_2

Answers to practice test

1. (a) $1s^22s^22p^3$, ±3, +5, and 0; (b) $1s^22s^22p^63s^23p^3$, ±3 +5, and 0; (c) $1s^22s^2 2p^63s^23p^4$, +4, +6, 0, -2.
2. (a) $1s^2\ 1s^2\ \sigma_{2s}^2\ \sigma_{2s}^{*2}\ \pi_{2p_y}^2\ \pi_{2p_z}^2\ \sigma_{2p_x}^2\ \pi_{2p_y}^{*2}\ \pi_{2p_z}^{*2}\ \sigma_{2p_x}^{*2}\ \sigma_{3s}^2\ \sigma_{3s}^{*2}\ \pi_{3p_y}^2\ \pi_{3p_z}^2\ \sigma_{3p_x}^2\ \pi_{3p_y}^{*1}\ \pi_{3p_z}^{*1}$, one σ and one π bond, paramagnetic; (b) $1s^2\ 1s^2\ \sigma_{2s}^2\ \sigma_{2s}^{*2}\ \pi_{2p_y}^2\ \pi_{2p_z}^2\ \sigma_{2p_x}^2$, one σ and two π bonds; (c) $1s^2\ 1s^2\ \sigma_{2s}^2\ \sigma_{2s}^{*2}\ \pi_{2p_y}^2\ \pi_{2p_z}^2\ \sigma_{2p_x}^2\ \pi_{2p_y}^{*1}$, one σ and (3/2)π bonds, paramagnetic.
3. (a) *sp*, linear (b) sp^2 for all, angular (c) sp^3 for all, trigonal pyramidal

   :N ≡ N: :Ö = S̈ - Ö: ↔ :Ö - S̈ = Ö:

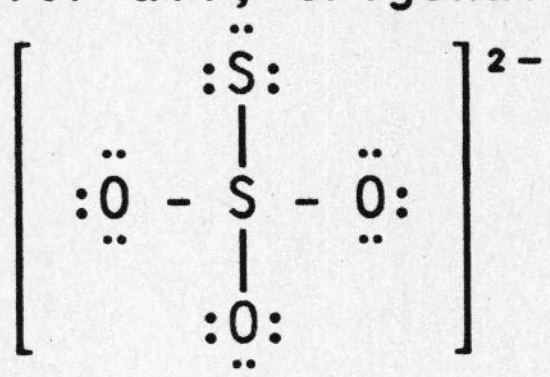

   (d) *sp* for both nitrogens, sp^2 for oxygen, linear

   :N̈ = N = Ö: ↔ :N ≡ N - Ö:

(e) sp^3 for fluorine, sp^3d for phosphorus,   (f) sp^3, trigonal pyramidal
trigonal bipyramidal

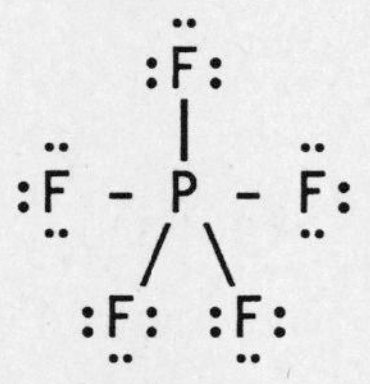

$$:\ddot{F}: \\ | \\ :\ddot{F} - P - \ddot{F}: \\ / \backslash \\ :\ddot{F}: :\ddot{F}:$$

$$H \\ | \\ H - P - H \\ ..$$

(g) sp^3, tetrahedral

$$\left[\begin{array}{c} H \\ | \\ H - P - H \\ | \\ H \end{array} \right]^+$$

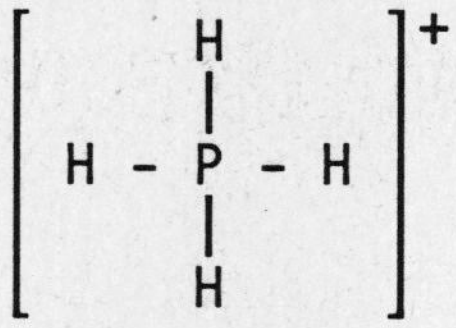

(h) sp^3 for all, tetrahedral

$$P \\ /\ \backslash \\ \ddot{O}\ \ |\ \ \ddot{O} \\ / \quad :\ddot{O}: \quad \backslash \\ :P{-}O \quad\quad P: \\ \backslash \quad \quad / \\ \ddot{O}\ \ |\ \ \ddot{O} \\ \backslash\ / \\ P \\ ..$$

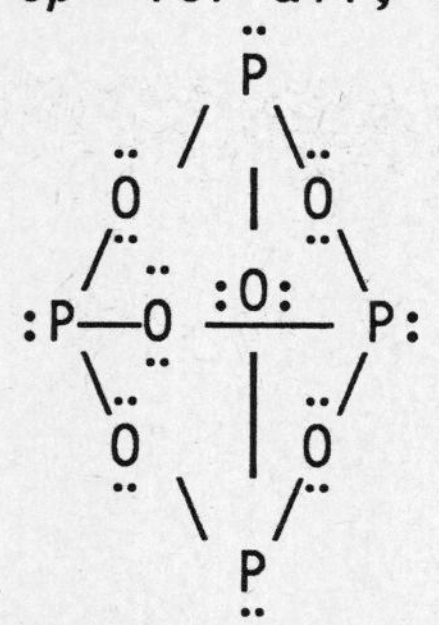

(i) sp^3 for oxygen connected to H, sp^2 for rest, the NO_3^- group in the shape of an isosceles triangle with the H attached to the longer bond

$$:\ddot{O}: \quad\quad\quad :\ddot{O}: \\ \parallel \quad\quad\quad\quad | \\ H - \ddot{O} - N - \ddot{O}: \ \leftrightarrow\ H - \ddot{O} - N = \ddot{O}$$

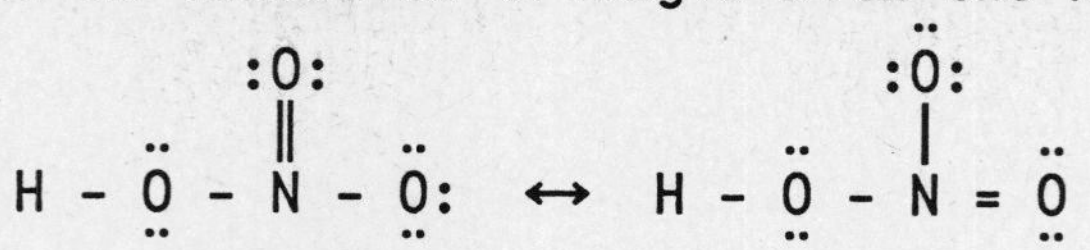

(j) sp^2 for all, equilateral triangle

$$\left[\begin{array}{c} :\ddot{O}: \\ \parallel \\ :\ddot{O} - N - \ddot{O}: \end{array}\right]^- \leftrightarrow \left[\begin{array}{c} :\ddot{O}: \\ | \\ :\ddot{O} - N = \ddot{O}: \end{array}\right]^- \leftrightarrow \left[\begin{array}{c} :\ddot{O}: \\ | \\ :\ddot{O} = N - \ddot{O}: \end{array}\right]^-$$

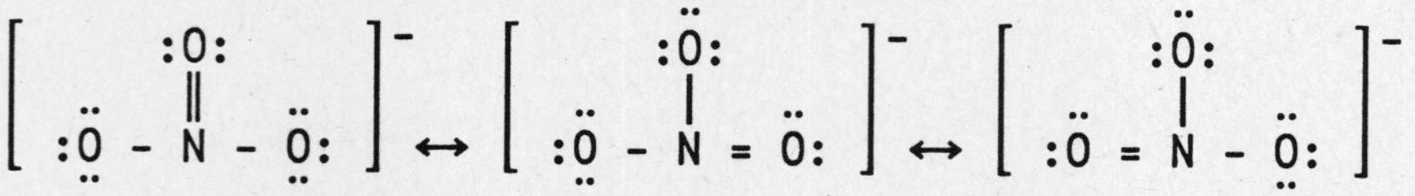

The three-dimensional sketches are shown in Fig. 18-1.

4. (a) See Table 28.3 of the text; (b) synthesis of NH_4NO_3, explosives and nitrocellulose, industrial oxidation and nitration reactions; (c) see Table 28.4 of the text; (d) see Table 28.14 of the text; (e) see Table 28.5 of the text; (f) see Table 28.6 of the text.

5. (a) $NH_4NO_3(s) \overset{\Delta}{\to} N_2O(g) + 2H_2O(g)$, nitrous oxide or nitrogen(I) oxide; (b) $3Cu(s) + 8H^+ + 2NO_3^- \to 3Cu^{2+} + 2NO(g) + 4H_2O(l)$, nitric oxide or nitrogen(II) oxide; (c) $NO(g) + NO_2(g) \to N_2O_3(l)$, nitrogen(III) oxide or dinitrogen trioxide; (d) $2NO(g) + O_2(g) \to 2NO_2(g)$, nitrogen dioxide; (e) $2NO_2(g) \to N_2O_4(g)$, dinitrogen tetroxide; (f) $4HNO_3(l) + P_4O_{10}(s) \to 2N_2O_5(s) + 4HPO_3(s)$, nitrogen(V) oxide or dinitrogen pentoxide.

6. (a) $2SO_2(g) + O_2(g) \overset{V_2O_5}{\to} 2SO_3(g)$, (b) $SO_3(g) + H_2SO_4(l) \to H_2S_2O_7(l)$, (c) $H_2S_2O_7(l) + H_2O(l) \to 2H_2SO_4(aq)$.

Fig. 28-1

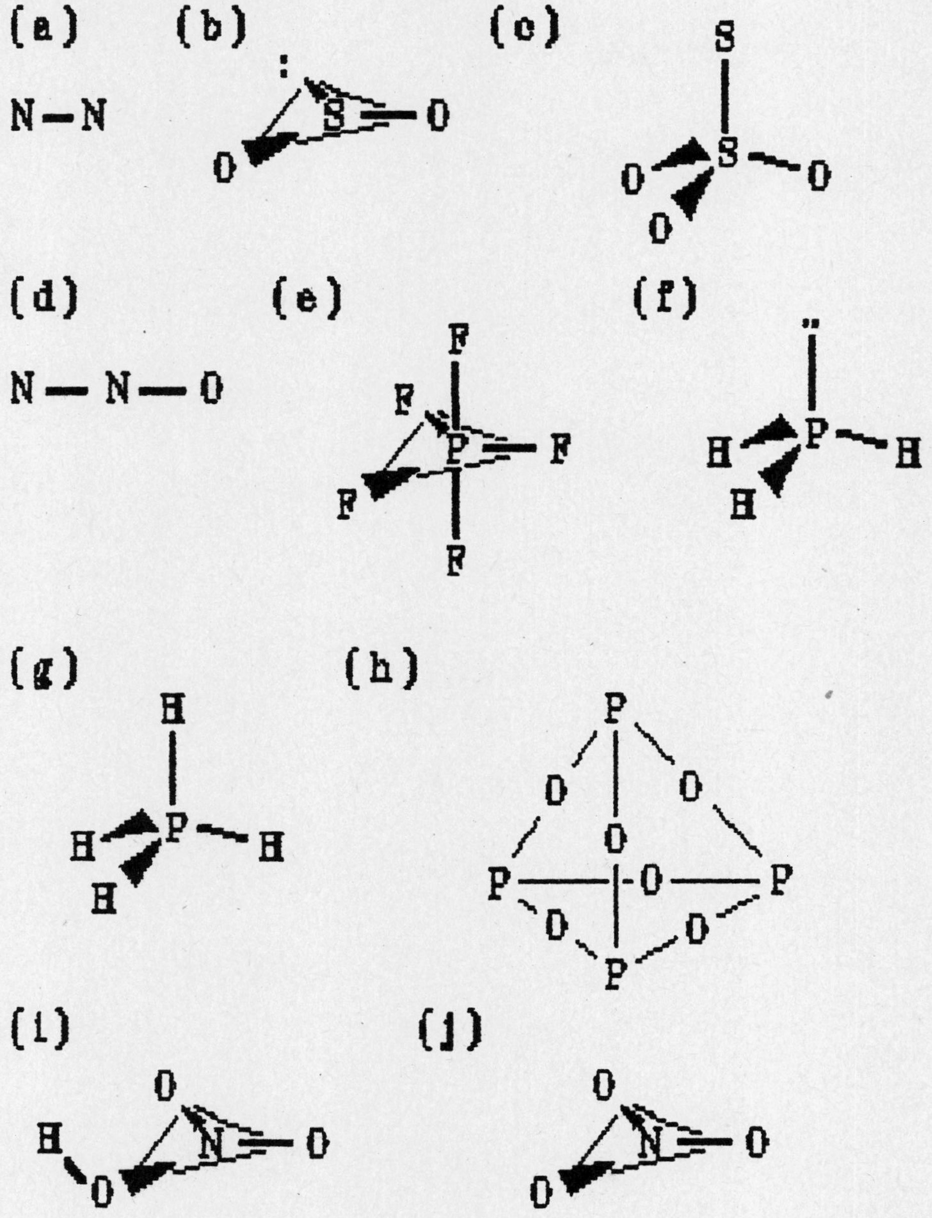

7. (a) iii, (b) iii, (c) iii, (d) iii, (e) ii, (f) i, (g) iii, (h) iii, (i) iv, (j) ii.

ADDITIONAL NOTES

CHAPTER 29

CARBON AND THE SEMICONDUCTING ELEMENTS

CHAPTER OVERVIEW

This chapter is divided into two major parts. The first considers the descriptive chemistry of carbon and some important inorganic compounds of carbon. The second part of the chapter presents the properties of the semiconducting elements and their hydrides, oxygen compounds, and halides. Three-center bonding in boron compounds is described. Some of the natural and manmade silicon-oxygen compounds and their uses are discussed. The nature of bonding and conductivity in semiconductors is explained. Brief descriptions are given for the preparation of ultrapure silicon and germanium for the electronics industry and for the nature of semiconducting devices.

COMPETENCIES

Definitions for the following terms should be learned:

- acceptor impurity
- acceptor level
- conduction band
- donor impurity
- donor level
- dopant
- doping
- electron-deficient compounds
- energy band
- extrinsic semiconductor
- forbidden energy gap, energy gap
- insulator
- intrinsic semiconductor
- *n*-type semiconductor
- *p-n* junction
- *p*-type semiconductor
- semiconductor
- silicates
- silicones
- three-center bond
- valence band
- zeolites
- zone refining

General concepts and skills that should be learned include

- the properties and uses of elemental carbon and the important inorganic compounds of carbon (CO, CO_2, the carbonates, CS_2, CCl_4, the carbides, and the cyanides)
- the general properties, preparations, and uses of the seven semiconducting elements and of their compounds
- how a semiconductor works
- the general properties of the silicates and silicones

QUESTIONS

1. Complete the following sentences by filling in the various blanks with these terms: acceptor impurities, acceptor level, donor impurities, donor level, doping, extrinsic semiconductors, insulator, integrated circuit, intrinsic semiconductors, *n*-type semiconductors, *p-n* junction, *p*-type semiconductors, rectifier, transistors, and zone refining.

   The band theory is based on the concept of energy bands being formed by the combination of many atomic orbitals. In a metal, the valence band (the highest fully occupied band) and the conduction band (the band above the valence band) are either very close or the conduction band is half-filled, so that electrons can move and current can flow. If there is an energy gap between the bands so large that electrons cannot cross, the substance is an (a)_______ and no current will flow. In (b)_______, some electrons can move across the energy gap and some current can flow as a result of equal numbers of electrons and "holes."

   (c)_______ is the addition of controlled amounts of impurities to ultrapure semiconductors like Si and Ge (which are produced by (d)_______) to give semiconductors which have different numbers of electrons and holes, known as (e)_______. In (f)_______, the impurities are usually elements from Representative Group V, which act as (g)_______. These impurities contribute electrons to the conduction band without leaving holes in the valence band because the extra electron occupies the (h)_______, a valence band that usually lies slightly below the conduction band of the host semiconductors. These extra electrons are easily promoted, so that electrical current can flow. In (i)_______, the impurities are usually elements from Representative Group III, which act as (j)_______. In this case the (k)_______ is located slightly above the valence band, so that electrons are easily promoted to this level and positive holes conduct the electrical current.

The joining of the two types of semiconducting materials, a (l)_______, produces a (m)_______, which converts alternating current to direct current. (n)_______ contain three sections of semiconductor materials and are used to amplify and control electrical current. Modern technology has produced the (o)_______, which is a combination of various electronic elements of an electrical circuit (transitors, resistors, capacitors, etc.) on one piece of silicon.

2. Match the correct term to each definition.

(a) contain As^{3-}	(i) arsenides
(b) contain B and H	(ii) boranes
(c) Ge_nH_{2n+2}	(iii) germanes
(d) contain Sb^{3-}	(iv) metalloids
(e) contain Se^{2-}	(v) selenides
(f) semiconducting elements	(vi) silanes
(g) SiO_2	(vii) silica
(h) contain Si, C, H, and O	(viii) silicates
(i) contain Si, O, and metals	(ix) silicones
(j) Si_nH_{2n+2}	(x) stibides
(k) contains Te^{2-}	(xi) tellurides

3. Which of the following statements are true? Rewrite each false statement so that it is correct.
   (a) The semiconducting elements generally look like metals and act like metals in most of their chemical behavior.
   (b) At very low temperatures the semiconducting elements approach the nonmetal elements in behavior, and at high temperatures they approach the metallic elements in behavior.
   (c) The addition of controlled amounts of impurities to semiconducting elements is known as sintering.
   (d) During zone refining, an ultrapure zone of liquid is formed and is then used for making electronic parts.
   (e) The electrical resistance of a semiconducting element decreases with increasing temperature while that of metals increases.

4. Distinguish clearly between/among (a) insulators, semiconductors, and metals; (b) intrinsic and extrinsic semiconductors; (c) *n*-type and *p*-type semiconductors; (d) "ane" and "ate" endings as in borane and borate or silane and silicate; (e) valence and conduction bands; and (f) donor and acceptor impurities.

5. Choose the response that best completes each statement.
   (a) Which type of bonds are not formed by carbon?
      (i) ionic with C^{4+}
      (ii) ionic with C^{4-} and C_2^{2-}
      (iii) covalent single bonds with sp^3 hybridization
      (iv) covalent multiple bonds with sp^2 and sp hybridization
   (b) Which form of carbon is used as a lubricant?
      (i) diamond (ii) graphite
      (iii) amorphous forms (iv) carbon black
   (c) Which is *not* a method for preparing CO?
      (i) burning carbon and hydrocarbons in a limited amount of air
      (ii) the reduction of CO_2 at high temperature
      (iii) the reaction of an acid with a carbonate
      (iv) burning carbon at high temperatures
   (d) Which is *not* a chemical property of CO?
      (i) reacts with water to form carbonates
      (ii) combines with Cl_2 to form phosgene
      (iii) forms metal carbonyls with many transition metals and some of their salts
      (iv) reduces metal oxides at high temperatures
   (e) The geometry and bonding in the carbonate ion can be described as
      (i) linear with double bonds
      (ii) tetrahedral with sp^3 hybridization for the C
      (iii) trigonal pyramidal with sp^3 hybridization for the C
      (iv) equilateral triangular with delocalized π bonding
   (f) Aqueous solutions of carbonic acid are
      (i) highly acidic
      (ii) weakly acidic
      (iii) mildly alkaline because of hydrolysis of CO_3^{2-}
      (iv) very weakly alkaline

(g) Which substance is not considered toxic?
(i) CO (ii) CO_2
(iii) CS_2 (iv) CCl_4

(h) Which species is used to extract Ag and Au from ores?
(i) C_2^{2-} (ii) CS_2
(iii) CN^- (iv) CCl_4

(i) Compounds that possess too few valence electrons for ordinary covalent bonds are called
(i) ionic (ii) saturated
(iii) electron deficient (iv) rectified

(j) A bond in which a single pair of electrons bonds three atoms covalently is a
(i) three-center bond (ii) hydrogen bond
(iii) triple bond (iv) multiple bond

(k) Which is *not* a property of the semiconducting elements?
(i) generally resemble metals in appearance
(ii) conduct electricity well
(iii) form molecular, volatile halides
(iv) form acidic oxides except for As_2O_3 and Sb_2O_3, which are amphoteric

(l) At high temperatures, elemental boron will
(i) undergo reaction with only F_2 and HNO_3
(ii) add "spring" to steel
(iii) form ionic borides with many metals
(iv) undergo reaction with molten alkali or alkaline earth hydroxides to form H_2 and borates

(m) Which is *not* a step in the metallurgy of Ge?
(i) roasting of the sulfide
(ii) extraction using HCl to form $GeCl_4$
(iii) hydrolysis forming GeO_2
(iv) C or H reduction

(n) When heated in air, As and Sb form
(i) arsenides and stibides (ii) As_4O_{10} and Sb_4O_{10}
(iii) As_2O_3 and Sb_2O_3 (iv) arsenates and stibates

(o) For which element does the electrical conductivity depend markedly on the intensity of incident light?

(i) Se (ii) Te
(iii) Si (iv) B

(p) Which is *not* a property of boric acid?
(i) white, crystalline substance
(ii) strong triprotic acid
(iii) neutralization gives tetraborates
(iv) metaboric acid can be formed from it

(q) Boron hydrides
(i) readily dimerize
(ii) use extensive three-center bonding
(iii) are slightly soluble in water
(iv) act as strong Lewis bases

(r) Arsine and stibine
(i) are basic
(ii) form cations similar to NH_4^+
(iii) are poisonous
(iv) dimerize through three-center bonding

(s) Arsenic(V) oxide is
(i) produced by burning As in air
(ii) formed by oxidizing As_2O_3 with concentrated HNO_3
(iii) forms alkaline solutions in water
(iv) has the structure corresponding to the formula As_4O_{10}

(t) Which is *not* a property of silica?
(i) inert toward all halogens except F_2 and all acids except HF
(ii) oxidized at high temperatures by active metals
(iii) dissolves in NaOH to form a mixture of soluble silicates
(iv) form three-dimensional polymeric substances for all crystal modifications

(u) Natural silicates do *not* contain
(i) ring structures
(ii) chain and double chain structures
(iii) sheets and three-dimensional networks
(iv) multiple bonding

(v) Common glass
(i) is a mixture of the silicates of Na and Ca
(ii) is inert to strong bases

(iii) is an aluminosilicate

(iv) has a definite crystalline structure

(w) An energy band in which electrons are free to move is the

(i) valence band (ii) forbidden energy gap

(iii) delocalized band (iv) conduction band

6. Name some of the uses of graphite. Briefly describe why graphite can be used as lubricant.

7. What are the products of the reactions involving the oxoanions of carbon (a) with acids and (b) when heated?

8. Name three classes of carbides. Briefly describe the bonding characteristic of each. Which of these will produce acetylene when water is added?

9. Before silicon carbide was first prepared, would it have been possible to predict that it would be very hard? Why?

10. Complete and balance the following equations:

(a) $CaCO_3(s) \xrightarrow{\Delta}$ (b) $NaHCO_3(s) + H^+ \rightarrow$

(c) $CS_2(g) + Cl_2(g) \rightarrow$ (d) $Fe_2O_3(s) + CO(g) \xrightarrow{\Delta}$

(e) $CO_2(g) + C(s) \xrightarrow{\Delta}$ (f) $CaO(s) + C(s) \xrightarrow{\Delta}$

11. Classify each of the following reactions according to the reaction types listed in Tables 17.2 and 17.7 of the text.

(a) $CaO(s) + 3C(s) \xrightarrow{\Delta} CaC_2(s) + CO(g)$

(b) $H_2C_2O_2(s) \xrightarrow{H_2SO_4(conc)} CO_2(g) + CO(g) + H_2O(l)$

(c) $COCl_2(g) + 2NH_3(g) \xrightarrow{toluene} CO(NH_2)_2(s) + 2HCl(g)$

12. Which of the semiconducting elements would you predict to have the highest melting points? Explain your answer.

13. List some of the uses of boron.

14. Write the chemical equation describing the preparation of elemental silicon from sand using coke. What are some of the uses for the silicon prepared in this way?

15. Aluminum, silicon, and phosphorus occupy adjacent positions on the periodic table. List similarities and differences in the properties of these elements.

16. Briefly discuss the allotropic forms of elemental arsenic and antimony.

17. Write the chemical formulas for the oxides of arsenic and antimony. How do they react with water?

18. Write the chemical formulas for arsenic acid, pyroarsenic acid, and metaarsenic acid.

19. What is the common commercial source for selenium? What are some of the uses for this element?

20. Sulfur dioxide is a gas, but selenium dioxide is a solid. Explain the difference in terms of molecular structure.

21. Write the formula for each of the following anions: (a) tetraborate ion, (b) arsenate ion, and (c) hydrogen selenite ion.

22. What properties does a glass have in common with a crystalline solid and with a liquid?

Answers to questions

1. (a) insulator, (b) intrinsic semiconductors, (c) Doping, (d) zone refining, (e) extrinsic semiconductors, (f) *n*-type semiconductors, (g) donor impurities, (h) donor level, (i) *p*-type semiconductors, (j) acceptor impurities, (k) acceptor level, (l) *p-n* junction, (m) rectifier, (n) Transistors, (o) integrated circuit.
2. (a) i, (b) ii, (c) iii, (d) x, (e) v, (f) iv, (g) vii, (h) ix, (i) viii, (j) vi, (k) xi.
3. (a) F, replace *and act like metals* with *but act like nonmetals*; (b) T; (c) F, replace *sintering* by *doping*; (d) F, replace *ultrapure zone ... parts* with *impure zone of liquid is formed which is discarded*; (e) T.
4. (a) *Insulators* have completely filled valence bands with a large energy gap to the conduction band; *semiconductors* have completely filled valence bands with a small energy gap to the conduction band; *metals* have partially filled valence bands or overlapping conduction bands with completely filled valence bands. (b) *Intrinsic semiconductors* conduct electricity as a result of the promotion of an electron over an energy gap by thermal energy, thereby producing an equal number of electrons and holes; *extrinsic* semiconductors conduct because of an unequal number of electrons or holes caused by the presence of impurities. (c) *n*-Type semiconductors conduct by the flow of electrons; *p*-type semiconductors conduct by the flow of holes (the movement of vacancies caused by the dopants). (d) The suffix *ane* refers to the

binary compounds of the element and hydrogen; the suffix *ate* refers to the oxo anions of the element. (e) The *valence band* is the highest completely filled energy band; the *conduction band* is a band in which electrons can move. (f) A *donor impurity* contributes electrons to the conduction band; an *acceptor* impurity accepts electrons to leave holes.

5. (a) i, (b) ii, (c) iii, (d) i, (e) iv, (f) ii, (g) ii, (h) iii, (i) iii, (j) i, (k) ii, (l) iv, (m) i, (n) iii, (o) i, (p) ii, (q) ii, (r) iii, (s) ii, (t) ii, (u) iv, (v) i, (w) iv.
6. Uses include lubricant, electrodes, molds and crucibles for hot metals, pencil "lead," and fabricated parts. The planes of atoms slip over each other aided by presence of moisture or gas molecules adsorbed on surface of layers.
7. (a) CO_2, (b) CO_2.
8. Saltlike carbides contain C_2^{2-} or C^{4-} ions, metallic carbides contain interstitial C atoms in the metal crystal structure, and covalent carbides are three-dimensional networks of atoms. Saltlike carbides containing C_2^{2-} will form acetylene.
9. The prediction that silicon carbide would be very hard is reasonable because it would likely have the same crystal form and bonding as diamond.
10. (a) $CaCO_3(s) \xrightarrow{\Delta} CaO(s) + CO_2(g)$, (b) $NaHCO_3(s) + H^+ \rightarrow Na^+ + H_2O(l) + CO_2(g)$, (c) $CS_2(g) + 3Cl_2(g) \rightarrow CCl_4(g) + S_2Cl_2(g)$, (d) $Fe_2O_3(s) + 3CO(g) \xrightarrow{\Delta} 2Fe(s) + 3CO_2(g)$, (e) $CO_2(g) + C(s) \xrightarrow{\Delta} 2CO(g)$, (f) $CaO(s) + 3C(s) \xrightarrow{\Delta} CaC_2(s) + CO(g)$.
11. (a) redox--disproportionation reaction in which the carbon has been reduced to the C_2^{2-} ion and oxidized to CO; (b) redox--disproportionation reaction in which oxalic acid decomposes to form CO_2, CO, and H_2O; (c) nonredox--partner-exchange reaction in which amine groups are exchanged for chloro groups.
12. The elements are B and Si. The small radii of these atoms produce strong covalent bonds in the crystal structure.
13. Uses include aluminum alloys, steel, oxygen scavenger, fiber-reinforced materials, nuclear reactors, and dopant in electronics.
14. $SiO_2(s) + 2C(s) \xrightarrow{\Delta} Si(l) + 2CO(g)$
    The silicon prepared in this way is used in alloys, in silicones, and for the preparation of ultrapure silicon for semiconductor devices.
15. A comparison of the properties of aluminum, silicon, and phosphorus, shows that their sizes decrease, their electronegativities increase, their

tendencies to form negative oxidation states increase, and their number of positive oxidation states increase in order of increasing atomic number.

16. Crystalline metallic forms are gray, lustrous, and semiconducting. Nonmetallic allotropes are yellow and readily revert to the crystalline forms.
17. The oxides are As_2O_3--amphoteric, more acidic; Sb_2O_3--amphoteric, more alkaline; As_2O_5--acidic; and Sb_2O_5--relatively insoluble, but the oxide is an acidic oxide.
18. H_3AsO_4, $H_4As_2O_7$, $HAsO_3$.
19. The source is the anode mud formed in electrolytic copper refining. Uses include the preparation of glasses and the vulcanizing agent for rubber.
20. Sulfur dioxide is a gas because it exists as discrete molecules. Selenium dioxide is a solid because it exists in polymeric chains.
21. (a) $B_4O_7^{2-}$, (b) AsO_4^{3-}, (c) HSO_3^-.
22. Like a solid, glass is hard and retains its form for long periods. Like a liquid, however, it is highly viscous and does flow in time.

PRACTICE TEST

1 (20 points). Choose the response that best completes each statement.

(a) Which is *not* a natural source of carbon?

(i) cassiterite (ii) petroleum and natural gas

(iii) carbonate minerals (iv) animal and plant material

(b) Which of the following substances does *not* have the properties listed?

(i) graphite--a soft, black solid which conducts electricity

(ii) diamond--a clear, colorless, hard crystalline solid with a very high melting point

(iii) carbon monoxide--a colorless, odorless, toxic gas

(iv) carbon disulfide--a colorless, noncombustible gas that has a disagreeable odor

(c) Which is a chemical property of carbon?

(i) Only fluorine reacts directly with diamond without heating.

(ii) Strong acids react with carbon to form diamonic acid.

(iii) Carbon dissolves in several inorganic solvents at room temperature.

(iv) C exists as diamond, graphite, and in amorphous forms.

(d) Which is a chemical property of CO_2?

(i) reacts with H_2O to form carbonates

(ii) combines with Cl_2 to form phosgene

(iii) forms metal carbonyls with many transition metals and some of their salts

(iv) all of the above answers

(e) Which compound is used for an abrasive?

(i) CS_2 (ii) SiC

(iii) CaC_2 (iv) CCl_4

(f) Which compound reacts with water to produce a hydrocarbon?

(i) SiC (ii) CaC_2

(iii) CS_2 (iv) $C_{10}H_{22}$

(g) An *n*-type semiconductor contains

(i) a donor impurity

(ii) an acceptor impurity

(iii) equal numbers of electrons and holes

(iv) a scavenger

(h) Compounds such as the boron hydrides which do not contain enough electrons for ordinary covalent bonds are said to be

(i) electron deficient (ii) an insulator

(iii) an extrinsic semiconductor (iv) saturated

(i) Compounds containing Si, H, C, and O are called

(i) silanes (ii) silicas

(iii) silicates (iv) silicones

(j) Silicon is the second most abundant element in the earth's crust and is found as

(i) the free element (ii) sulfide ores

(iii) silica and silicates (iv) silicones

(k) Which is *not* a property of boranes?

(i) decompose in the absence of air to give B and H_2

(ii) stable in water solution

(iii) many contain three-center bonds

(iv) spontaneously flammable in air

(l) Germanium doped with gallium is

(i) an insulator (ii) a *p*-type semiconductor

(iii) an *n*-type semiconductor (iv) none of these answers

(m) In which of the following compounds is the oxidation state of arsenic equal to +5?

(i) arsenite ion (ii) arsenic trioxide
(iii) arsenous acid (iv) potassium arsenate

(n) What is a major use for Ge and Si?
(i) electronics (ii) pesticides
(iii) fertilizers (iv) paint pigments

(o) What is the formula for "borax"?
(i) BN (ii) H_3BO_3
(iii) B_2H_6 (iv) $Na_2B_4O_7 \cdot 10H_2O$

2 (15 points). Describe the bonding in (a) HCN, (b) CO_3^{2-}, (c) C_2^{2-}, and (d) CCl_4 by preparing Lewis structures, preparing three-dimensional sketches of the structures, discussing the hybridization used by the atoms, and identifying the geometries of the species.

3 (10 points). What will be the pressure of acetylene produced by 10.0 g of calcium carbide reacting with an excess of water in a closed 1.03 L container at 23 °C?

4 (15 points). What are the formulas of the products of the respective reactions of arsenic and antimony with (a) hot dilute nitric acid and (b) hot concentrated nitric acid? Name these compounds.

5 (5 points). Draw the Lewis structure for diborane.

6 (15 points). Find the empirical formula of the compound containing 5.2 mass % H, 71.4 mass % O, 11.3 mass % B, and 12.1 mass % Na. Name this compound.

7 (15 points). Write the chemical equation for (a) arsenous acid acting as a weak monoprotic acid, (b) arsenous acid acting as a base, (c) arsenic(III) oxide reacting with water, (d) arsenic(V) oxide reacting with water, and (e) $AsCl_3$ undergoing hydrolysis.

8 (5 points). Explain how an aluminosilicate can be used to soften water.

9 (15 points). The standard state heat of formation at 25 °C is 34.3 kJ/mol for $SiH_4(g)$, -910.94 kJ/mol for $SiO_2(s)$, and -241.818 kJ/mol for $H_2O(g)$. What is the enthalpy change for 1.00 g of $SiH_4(g)$ igniting in air?

Answers to practice test

1. (1) i, (b) iv, (c) i, (d) i, (e) ii, (f) ii, (g) i, (h) i, (i) iv, (j) iii, (k) ii, (l) ii, (m) iv, (n) i, (o) iv.

Fig. 29-1

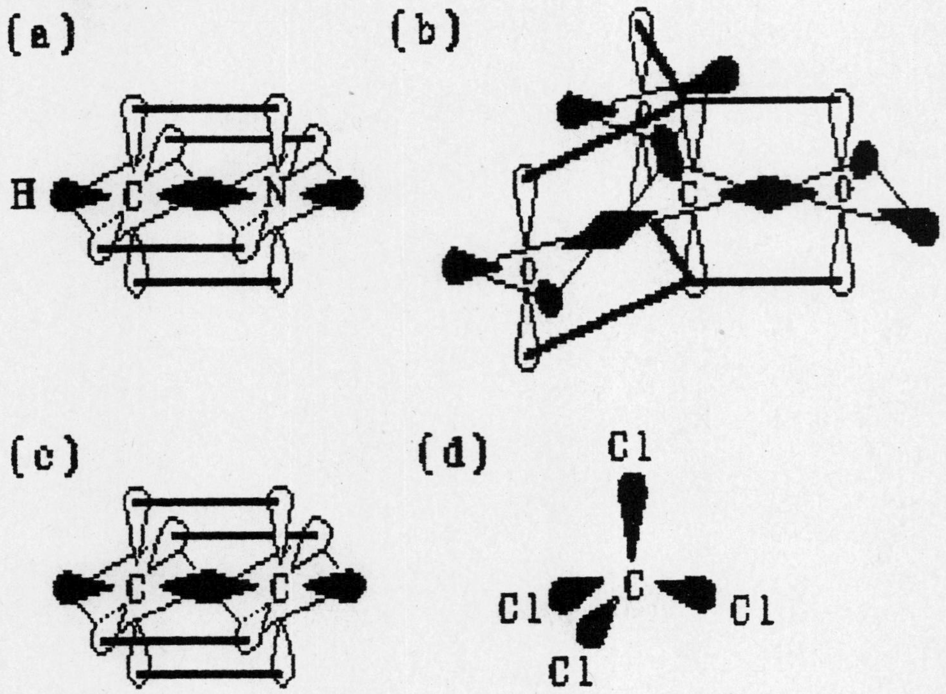

2. (a) *sp* hybridization for C and for N, linear; H-C≡N:
   (b) sp^2 hybridization for all oxygens and for carbon, equilateral triangle;

```
              ..                        ..
    :O:     ]2-         :O:     ]2-          :O:     ]2-
 ..  ||  ..         ..   |   ..         ..    |
[:O - C - O:]  ↔  [:O = C - O:]   ↔   [:O - C = O]
 ..      ..                  ..         ..         ..
```

   (c) *sp* hybridization for both carbons, linear; $[:C \equiv C:]^{2-}$
   (d) sp^3 hybridization for carbon, tetrahedral;

```
        ..
       :Cl:
    ..  |   ..
   :Cl - C - Cl:
    ..  |   ..
       :Cl:
        ..
```

   The three-dimensional sketches are shown in Fig. 29-1.
3. 3.68 atm.
4. (a) H_3AsO_3, arsenous acid; Sb_4O_6, antimony(III) oxide; (b) H_3AsO_4, arsenic acid; Sb_2O_5, antimony(V) oxide.
5.

```
H     H     H
 \ .    . /
  B       B
 / .    . \
H     H     H
```

6. $Na_2B_4O_7 \cdot 10H_2O$, sodium tetraborate decahydrate.
7. (a) $H_3AsO_3(aq) \rightarrow H^+ + H_2As_2O_3^-$, (b) $H_3AsO_3(aq) \rightarrow AsO^+ + OH^- + H_2O(l)$, (c) $As_2O_3(s) + 3H_2O(l) \rightarrow 2H_3AsO_3(aq)$, (d) $As_2O_5(s) + 3H_2O(l) \rightarrow 2H_3AsO_4(aq)$, (e) $AsCl_3(l) + 3H_2O(l) \rightarrow H_3AsO_3(aq) + 3HCl(aq)$.
8. See Section 29.13 of the text.
9. -44.5 kJ.

ADDITIONAL NOTES

CHAPTER 30

d- AND *f*-BLOCK ELEMENTS

CHAPTER OVERVIEW

The chemistry of the *d*- and *f*-block metals is considered in this chapter. The descriptive chemistry of chromium, manganese, iron, cobalt, nickel, copper, silver, gold, zinc, cadmium, and mercury is presented in separate sections. A brief look is taken at some of the most interesting aspects of the chemistry and uses of the other transition metals.

COMPETENCIES

Definitions for the following terms should be learned:

- amalgams
- coinage metals
- ferromagnetic
- platinum metals

General concepts and skills that should be learned include

- the general physical and chemical properties, industrial preparations, uses, and reactions with nonmetals and compounds for the transition metals and some of their compounds

QUESTIONS

1. Match the correct term to each definition.

(a) the significant decrease that occurs as a result of filling the f level for the radii of elements

(b) the electrolytic deposition of a layer of Cr on another metal

(c) elements having atomic numbers of 57 through 71

(d) elements having atomic numbers 89 through 103

(i) actinides
(ii) amalgam
(iii) chrome plating
(iv) coinage metals
(v) ferromagnetism
(vi) lanthanide contraction
(vii) lanthanides
(viii) platinum metals
(ix) transuranium elements

(e) the property of exhibiting magnetism in the absence of an external magnetic field
(f) elements having atomic numbers 29, 47, and 79
(g) elements having atomic numbers of 44 through 46 and 76 through 78
(h) elements with atomic numbers greater than 92
(i) an alloy containing mercury

2. Which of the following statements are true? Rewrite each false statement so that it is correct.

(a) Similarity in electron configuration and size causes the first and second members of each transition metal family to more closely resemble each other in chemical behavior than to resemble the third member of the family.
(b) Ionic compounds are most likely to form between transition metals and the representative nonmetals when the transition metal is in its higher oxidation state(s).
(c) Copper is an excellent "sacrificial anode" for protecting iron and steel structures because of its high electrical conductivity.
(d) Aqueous solutions of Fe^{3+} are rather unstable because the Fe^{3+} can be air-oxidized to Fe^{2+}.
(e) The compound $[Fe(H_2O)_6]^{3+}[Fe(CN)_6]^{3-}$ provides an example of interaction absorption.
(f) A sample of frozen mercury would be expected to be malleable instead of brittle like ice.

3. Choose the response that best completes each statement.

(a) Which is *not* a general property of *d*-block metal atoms?
 (i) act as metals
 (ii) contain one or two *s* electrons (except for Pd)
 (iii) contain partially filled *d* subshells (except near the end of each series)
 (iv) contain partially filled *f* subshells

(b) Which is a property characteristic of transition metals?
 (i) increasing atomic radii from left to right in the periodic table
 (ii) form colorless compounds
 (iii) make good catalysts because of the ease with which electrons are lost and gained or promoted
 (iv) do not form coordination compounds

(c) Which is *not* a general trend in atomic and ionic radii for transition metals?

(i) general, regular increase from left to right in the periodic table

(ii) general increase down a family

(iii) lanthanide contraction nearly counteracting the expected increase between the second and third members of a family

(iv) similar trends for ions of the same charge to those for atomic radii

(d) All of the *d*-block metals exhibit multiple oxidation states

(i) except Sc

(ii) for which the most stable is the maximum

(iii) in which ionic compounds involve the highest values

(iv) which illustrate ferromagnetism

(e) Chromium(III) hydroxide

(i) is the anhydride of chromic acid

(ii) forms polymeric forms of chromium(VI) in strong acids

(iii) is a strong reducing agent

(iv) is an amphoteric highly polymeric, hydrated material

(f) Manganese is

(i) the mineral rutile

(ii) very stable in moist air

(iii) prepared by sulfate oxidation

(iv) an essential trace element in plants and animals

(g) Manganese(IV) oxide is a conductor of electricity because

(i) it forms polymeric anions

(ii) it has a nonstoichiometric crystal lattice

(iii) the manganese is in an intermediate oxidation state

(iv) it is ferromagnetic

(h) Which statement is *not* true concerning the reduction of the permanganate ion?

(i) forms MnO_4^- in strongly alkaline solution

(ii) forms MnO_2 in a neutral or slightly alkaline solution

(iii) forms $Mn(OH)_2$ in slightly alkaline solutions or Mn^{2+} in weakly acidic solutions

(iv) forms MnO_4^{2-} in strongly alkaline solution

(i) The phenomenon in which a molecule containing atoms of the same metal in two different oxidation states is deeply colored is called

(i) interaction absorption (ii) auto oxidation-reduction
(iii) hydrate isomerism (iv) disproportionation

(j) Which is a correct oxidation state for iron?

(i) +2 in Fe_2O_3 (ii) +2 and +3 in Fe_3O_4
(iii) +2 in FeO_2 (iv) +4 in FeS_2

(k) The formula for fool's gold is

(i) $FeSO_4$ (ii) FeS_2
(iii) $Fe_4[Fe(CN)_6]_3$ (iv) $FeSO_4 \cdot 7H_2O$

(l) Which complex is blue and is reversible with a pink complex?

(i) $[Co(H_2O)_6]^2$ (ii) $[Co(CN)_4]^{2+}$
(iii) $[Co(CN)_6]^{3-}$ (iv) $[Co(NH_3)_6]Cl_3$

(m) The confirmatory test used for Ni^{2+} in qualitative analysis schemes is often

(i) the formation of $[Ni(NH_3)_6]^{2+}$ from $Ni(OH)_2$
(ii) the smelting of $[Ni(CO)_4]$
(iii) the production of the dimethylglyoxime complex
(iv) none of these answers

(n) The coinage metals

(i) all show oxidation states of +1, +2, and +3
(ii) react with nonoxidizing acids
(iii) form stable oxides at high temperatures
(iv) none of these answers

(o) Which metal is usually found naturally in the metallic state?

(i) Cu (ii) Ag
(iii) Pt (iv) Au

(p) The green color of the Statue of Liberty indicates that it contains

(i) zinc (ii) iron
(iii) lead (iv) copper

(q) Which is *not* true about the elements in the zinc family?

(i) The outer electron configurations are $(n-1)d^{10}ns^2$.
(ii) They form only +2 ions.
(iii) They form complex ions extensively.
(iv) Oxides become more deeply colored and less stable from Zn to Cd to Hg.

(r) Naturally occurring mercury is found as
   (i) Hg_2Cl_2
   (ii) cinnabar
   (iii) in flue deposits of zinc plants
   (iv) calamine

(s) Alloys of mercury are called
   (i) amalgams (ii) fluxes
   (iii) couples (iv) scavengers

(t) The density of lead is 11.29 g/cm³ and that of mercury is 13.59 g/cm³. Thus
   (i) lead sinks in a pool of mercury
   (ii) lead floats on a pool of mercury
   (iii) an exothermic reaction between the metals takes place
   (iv) none of the above answers

(u) A general characteristic of the actinides is
   (i) radioactivity
   (ii) their great stability in air
   (iii) their great number of oxidation states, i.e., +1 to +8
   (iv) that their sizes generally increase from left to right in the periodic table

4. Why are many ions of the *d*-block metals colored?

5. Briefly describe the reactivities of the *d*-block elements with some nonmetals.

6. Briefly describe the reactivities of the *d*-block elements with acids.

7. What are some significant types of reactions in which the ions of the *d*-block elements participate?

8. Discuss what happens as OH^- is added to a solution containing dichromate ion.

9. The outer-shell electron configurations of the Ti, V, Cr, and Mn atoms are $3d^24s^2$, $3d^34s^2$, $3d^54s^1$, $3d^54s^2$, respectively. Why is there a discontinuity between V and Cr?

10. Pretend that both manganic acid, H_2MnO_4, and permanganic acid, $HMnO_4$, exist. Which would you predict to be the stronger acid? Why is this only a hypothetical question?

11. Is the Mn-O bond in MnO_4^- more or less ionic than in MnO? Why?

12. What is the most common water-soluble iron(II) salt? How is it prepared? For what is it used?

13. A common laboratory experiment in general chemistry is to prepare FeS(s) by strongly heating a mixture of Fe and S. The iron is carefully weighed, excess sulfur is thoroughly mixed with the iron, the mixture is heated and cooled, the iron sulfide is carefully weighed, and the mass of combined sulfur is found by difference. From the masses of Fe and of S combined in the compound, the empirical formula of the compound can be determined. A usual result is something like $Fe_{0.83}S$. Some students will report the result as FeS, Fe_2S_3, or FeS_2 rather than the fractional formula. Why was the first answer probably correct?

14. Potassium cyanide, KCN, is extremely poisonous, but $K_3[Fe(CN)_6]$ is only slightly so. Explain.

15. What types of ores are common for nickel? From what source is cobalt obtained?

16. What are the common oxidation states of nickel and cobalt? Under what conditions is the +3 oxidation state of cobalt rather stable?

17. Name the coinage metals. Why are these metals used for this purpose?

18. Name the platinum metals. Why are these metals grouped together even though they are not in the same periodic table family?

19. Briefly describe what happens as silver tarnishes.

20. Name some of the classes of stable copper(I) compounds. What happens to Cu_2SO_4 in aqueous solution?

21. As $CuCl_2$ in HCl is boiled with metallic copper, the solution, which was originally yellow, becomes very dark in color and then becomes colorless. What is the explanation for this?

22. Compare the behavior of Zn^{2+}, Cd^{2+}, Hg^{2+}, Hg_2^{2+} as sodium hydroxide is added until an excess is present.

23. Which of the metal hydroxides in the zinc family is amphoteric? Write chemical equations showing the hydroxide in the role as an acid and as a

base. Write chemical equations showing the reactions between this element and strong acids and bases.

24. How would you explain the lanthanide contraction to your younger brother who is taking a high school chemistry course? How would you answer his question about whether you would predict an "actinide" contraction?

25. Write balanced equations for the following chemical reactions: (a) chromate-dichromate equilibrium in aqueous solution, (b) preparation of $K_2MnO_4(s)$ from $MnO_2(s)$ at elevated temperature, (c) reaction of $Fe^{3+}(aq)$ with excess $CN^-(aq)$, and (d) reaction of neodymium(III) oxide(s) with aqueous HCl (net ionic).

Answers to questions

1. (a) vi, (b) iii, (c) vii, (d) i, (e) v, (f) iv, (g) viii, (h) ix, (i) ii.
2. (a) F, replace *first and second* and *third* with *second and third* and *first*; (b) F, replace *higher* by *lower*; (c) F, replace *Copper* by *Magnesium* and *electrical conductivity* with *chemical reactivity*; (d) F, change Fe^{3+} to Fe^{2+} *in both places and* Fe^{2+} to Fe^{3+}; (e) F, replace $[Fe(H_2O)_6]^{3+}[Fe(CN)_6]^{3-}$ by $[Fe(H_2O)_6{}^{2+}]_3[Fe(CN)_6{}^{3-}]_2$; (f) T.
3. (a) iv, (b) iii, (c) i, (d) i, (e) iv, (f) iv, (g) ii, (h) i, (i) i, (j) ii, (k) ii, (l) ii, (m) iii, (n) iv, (o) iv, (p) iv, (q) ii, (r) ii, (s) i, (t) ii, (u) i.
4. The adsorption of light by the *d* electrons occurs in the visible range of the spectrum.
5. The reactivity is related to temperature and state of division. Reactions include the formation of halides, oxides, and nitrides. Often the formation of an oxide coating can render the metal inactive with oxygen.
6. Many metals will react with nonoxidizing acids, several metals will react only with oxidizing acids, and a few metals will react only with aqua regia. The reactivity is inhibited in many cases by the formation of protective coatings.
7. The significant types of reactions include complexation, redox, hydration, hydrolysis and precipitation reactions.
8. The dichromate ion is in equilibrium with hydrogen chromate and chromate ions.

$$Cr_2O_7{}^{2-} + H_2O(l) \rightleftharpoons 2HCrO_4{}^- \rightleftharpoons 2H^+ + 2CrO_4{}^{2-}$$

The added OH^- reacts with H^+, disrupting the equilibrium (Le Chatelier's principle) by displacing the equilibrium to the right and causing more chromate ion to form. The originally deep orange solution turns yellow as $Cr_2O_7^{2-}$ is converted to CrO_4^{2-}.

9. The 4*s* and 3*d* levels are close in energy and half-completion of the 3*d* subshell is energetically favorable, so in Cr a 4*s* electron enters the 3*d* subshell to make five electrons in that subshell.
10. The stronger acid should be $HMnO_4$. The acid strength is related to the ease of ionization of the O-H bond--in $HMnO_4$ the O-H bond is more ionic because the strength of the Mn-O bonds are greater (because of the higher oxidation state for Mn). The acid H_2MnO_4 is only hypothetical because MnO_4^{2-} in the presence of H^+ forms MnO_4^- and MnO_2.
11. The bond is less ionic. The higher oxidation state of Mn in MnO_4^- results in a more covalent Mn-O bond.
12. The salt is $FeSO_4$. It is produced as a by-product of pickling steel with dilute H_2SO_4 and used as a starting material for production of most iron compounds.
13. FeS is a nonstoichiometric compound.
14. The CN^- ions lose their usual chemical properties because they are bonded strongly to Fe^{3+} in the complex.
15. The common ores are the oxides and sulfides. Cobolt is obtained as a by-product of nickel metallurgy.
16. The oxidation states are +2 for Ni and +2 and +3 for Co. The +3 state is stable in complexes formed by strong ligands that can force the odd electron out to an outer energy level so that it can be easily removed.
17. The coinage metals include copper, silver, and gold. These metals are used for this purpose because they are somewhat rare and relatively unreactive.
18. Ruthenium, osmium, rhodium, iridium, palladium, and platinum are collectively called the platinum metals. Even though they are not in the same periodic table family, these metals are grouped together because they are all quite alike and occur together in various combinations in nature.
19. H_2S in air reacts with Ag to form $Ag_2S(s)$.
20. The classes include halides, sulfides, and cyanides. Disproportionation of Cu_2SO_4 occurs.
21. The Cu^+ that forms generates a dark colored complex with some of the remaining Cu^{2+} (interaction absorption). The complex eventually disappears as the remainder of the Cu^{2+} is reduced.

22. $Zn^{2+} + 2OH^- \rightarrow Zn(OH)_2(s) \xrightarrow{OH^-} [Zn(OH)_4]^{2-}$
$Cd^{2+} + 2OH^- \rightarrow Cd(OH)_2(s)$
$Hg^{2+} + 2OH^- \rightarrow HgO(s) + H_2O(l)$
$Hg_2{}^{2+} + 2OH^- \rightarrow Hg(l) + HgO(s) + H_2O(l)$
23. Only $Zn(OH)_2$ is amphoteric. $Zn(OH)_2(s) + 2H^+ \rightarrow Zn^{2+} + 2H_2O(l)$, $Zn(OH)_2(s) + 2OH^- \rightarrow [Zn(OH)_4]^{2-}$; $Zn(s) + 2H^+ \rightarrow Zn^{2+} + H_2(g)$, $Zn(s) + 2OH^- + 2H_2O(l) \rightarrow [Zn(OH)_4]^{2-} + H_2(g)$.
24. Across a period, the general size decreases for atoms. After the 14 lanthanides intervene, the increase in size from the increase of the principal quantum number is nearly offset by this decrease, and so the last two elements in each of the families of *d*-block elements are close in size. Because of similar size and electron configurations, the elements have very similiar chemical and physical properties. There would probably be the same effect for actinides.
25. (a) $2CrO_4{}^{2-} + 2H^+ \rightleftharpoons Cr_2O_7{}^{2-} + H_2O(l)$
(b) $MnO_2(s) + KNO_3(l) + 2KOH(l) \xrightarrow{\Delta} K_2MnO_4(s) + KNO_2(s) + H_2O(g)$
(c) $Fe^{2+} + 6CN^- \rightarrow [Fe(CN)_6]^{4-}$
(d) $Nd_2O_3(s) + 6H^+ \rightarrow 2Nd^{3+} + 3H_2O(l)$

PRACTICE TEST

1 (5 points). Write the electron configuration for the ground state of uranium. Predict reasonable oxidation numbers for uranium in combination with other elements.

2 (10 points). The standard reduction potentials are 0.153 V for Cu^{2+}/Cu^+ and 0.521 V for Cu^+/Cu. Show that Cu^+ undergoes disproportionation.

3 (10 points). Write the half-reactions for $MnO_4{}^{2-}$ in acidic solution forming (a) $MnO_4{}^-$ and (b) MnO_2. (c) Combine these half-reactions to obtain the overall equation for the disproportionation of $MnO_4{}^{2-}$.

4 (10 points). There are three compounds having the empirical formula $CrCl_3{\cdot}6H_2O$. Careful study shows that the chromium is complexed to 4, 5, or 6 water molecules in these compounds. (a) Write structural formulas for these compounds. (b) Which compound would require the largest amount of $AgNO_3$ to precipitate any available Cl^- in solution?

5 (15 points). Exactly 0.100 A for 1.00 h was passed through electrolysis cells containing Cu^{2+}, Ag^{+}, and Au^{3+}. In which cell was the largest mass of metal transferred between electrodes and what mass of the metal was transferred?

6 (10 points). The standard state enthalpy of formation at 25 °C is -213.8 kJ/mol for freshly precipitated MnS(s, pink) and -214.2 kJ/mol for MnS(s, green). (a) Calculate the enthalpy change for the reaction

$$\text{MnS(s, pink)} \rightarrow \text{MsS(s, green)}$$

(b) Based on the enthalpy change, which form is the more stable at 25 °C? The pink solid is amorphous and the green form is crystalline. (c) Which form has the greater amount of randomness and thus is the more stable based on entropy content? (d) Discuss these opposite driving forces in light of the knowledge that the reaction as written is spontaneous.

7 (10 points). Write chemical equations showing the reaction (if any) of mercury with (a) a nonoxidizing acid such as HCl, (b) an excess of an oxidizing acid such as H_2SO_4, and (c) a limited amount of an oxidizing acid H_2SO_4.

8 (30 points). Choose the response that best completes each statement.

(a) The electron configuration for the elements in the coinage group is

(i) $...(n\text{-}1)d^9ns^2$ (ii) $...(n\text{-}1)d^{10}ns^1$

(iii) $...ns^2np^7$ (iv) $...(n\text{-}1)d^8ns^2$

(b) The predominant oxidation state for the actinide elements is

(i) +3 (ii) +2

(iii) +6 (iv) +4

(c) The relative sizes of the atoms and ions in the vanadium family are

(i) V > Nb > Ta (ii) Nb > Ta > V

(iii) Ta ≈ Nb > V (iv) $X^{5+} > X^{3+} > X$

(d) To make ^{240}Pu, an isotope with a half-life of 6580 years,

(i) unstable ^{239}U is bombarded with neutrons

(ii) ^{239}Np undergoes positron emission

(iii) ^{238}U is bombarded with alpha particles

(iv) $^{234}Pu^{6+}$ is reduced

(e) The hemoglobin from the red corpuscles of blood contains approximately 0.33 mass % iron. The molecular mass is 68,000 u. How many iron atoms are there in each molecule of the hemoglobin?

(i) 1 (ii) 2
(iii) 3 (iv) 4

(f) Which of the following will form a precipitate when added to 0.1 M $AgNO_3$?
(i) HCl(aq) (ii) NaBr(aq)
(iii) H_2SO_4(aq) (iv) all of the above

(g) The hydroxides of transition elements having the formula $M(OH)_3$
(i) are amphoteric
(ii) react with excess base to form soluble complex ions
(iii) decompose upon heating to form MO(OH) and then M_2O_3
(iv) are highly soluble in water and act as strong bases

(h) Chromates are
(i) red (ii) yellow
(iii) green (iv) deep blue

(i) In acidic solution, manganates
(i) explode to give MnO_2 and O_2
(ii) form permanganic acid
(iii) disproportionate into MnO_2 and MnO_4^-
(iv) produce manganites

(j) What happens to anhydrous cobalt compounds in the presence of water?
(i) become green (ii) change from blue to pink
(iii) form insoluble hydroxides (iv) become passive

(k) An alloy of mercury is called
(i) an amalgam (ii) a flux
(iii) a mordant (iv) a chalcogen

(l) Which of the following reactions will *not* occur?
(i) $Hg(l) + 2H^+ \rightarrow Hg^{2+} + H_2(g)$
(ii) $2ZnS(s) + 3O_2(g) \xrightarrow{\Delta} 2ZnO(s) + 2SO_2(g)$
(iii) $Zn(s) + 2NaOH(aq) + 2H_2O(l) \rightarrow Na_2[Zn(OH)_4](aq) + H_2(g)$
(iv) $2HgO(s) \xrightarrow{\Delta} 2Hg(l) + O_2(g)$

(m) A major use for ZnO is in
(i) paint pigments (ii) water purification
(iii) abrasives (iv) a self-contained gas mask

Answers to practice test

1. $1s^2 2s^2 2p^6 3s^2 3p^6 3d^{10} 4s^2 4p^6 4d^{10} 4f^{14} 5s^2 5p^6 5d^{10} 5f^4 6s^2 6p^6 7s^2$; 0, +2, +6.

2.

$Cu^+ \rightarrow Cu^{2+} + e^-$		$E° = -0.153$ V
$Cu^+ + e^- \rightarrow Cu$		$E° = 0.521$ V
$2Cu^+ \rightarrow Cu(s) + Cu^{2+}$		$E° = 0.368$ V

3. (a) $MnO_4^{2-} \rightarrow MnO_4^- + e^-$, (b) $MnO_4^{2-} + 4H^+ + 2e^- \rightarrow MnO_2(s) + 2H_2O(l)$, (c) $3MnO_4^{2-} + 4H^+ \rightarrow 2MnO_4^- + MnO_2(s) + 2H_2O(l)$.
4. (a) $[Cr(H_2O)_6]Cl_3$, $[Cr(H_2O)_5Cl]Cl_2 \cdot H_2O$, $[Cr(H_2O)_4Cl_2]Cl \cdot 2H_2O$; (b) $[Cr(H_2O)_6]Cl_3$.
5. 0.402 g Ag.
6. (a) -0.4 kJ, (b) green, (c) pink, (d) enthalpy change is more important.
7. (a) $Hg(l) + HCl(aq) \rightarrow$ no reaction, (b) $Hg(l) + 2H_2SO_4(aq) \rightarrow HgSO_4(s) + SO_2(g) + 2H_2O(l)$, (c) $2Hg(l) + 2H_2SO_4(aq) \rightarrow Hg_2SO_4(s) + SO_2(g) + 2H_2O(l)$.
8. (a) ii, (b) i, (c) iii, (d) iii, (e) iv, (f) iv, (g) iii, (h) ii, (i) iii, (j) ii, (k) i, (l) i, (m) i.

ADDITIONAL NOTES

CHAPTER 31

COORDINATION CHEMISTRY

CHAPTER OVERVIEW

This chapter presents the nomenclature of complexes and describes their geometry, isomerism, and general properties. Three different approaches to the explanation of bonding in complexes are introduced. Two of these--valence bond theory and molecular orbital theory--have been used before. The third--crystal field theory--is presented here for the first time. Several practical applications of complexes are mentioned.

COMPETENCIES

Definitions for the following terms should be learned:

- bidentate
- chelate ring
- chelation
- chirality
- *cis* isomer
- *cis-trans* isomerism
- coordination number (complex)
- crystal field stabilization energy
- crystal field theory
- geometrical isomerism
- high-spin complex
- inert complex
- isomers
- labile complex
- low-spin complex
- optical isomerism
- optical isomers, enantiomers
- optically active
- plane-polarized light
- strong-field ligand
- structural isomerism
- *trans* isomer
- weak field ligand

General concepts and skills that should be learned include

- naming complexes
- preparing Lewis structures and discussing the geometries of complexes
- identifying and discussing isomers of complexes
- discussing the relative stabilities of complexes

- explaining the bonding in complexes by using the valence bond, molecular orbital, and crystal field theories

Numerical exercises that should be understood include

- calculating the crystal field stabilization energy of a complex

QUESTIONS

1. Match the correct term to each definition.

(a) a species formed between a metal atom or ion (by accepting an electron pair) and ions or neutral molecules (that donate electron pairs)
(b) a molecule or ion that contains the donor atom
(c) the number of nonmetal atoms surrounding the central metal atom or ion in a complex
(d) positively or negatively charged complexes
(e) a neutral compound formed between a complex ion and other ions or molecules
(f) the formation of a ring in a complex by a ligand that coordinates to the metal ion by using two or more donor atoms
(g) a complex in which the d electrons remain unpaired
(h) a ligand that will not force pairing of d electrons
(i) a complex containing paired d electrons
(j) a ligand that will force pairing of d electrons

(i) chelation
(ii) complex
(iii) complex ion
(iv) coordination compound
(v) coordination number
(vi) high-spin complex
(vii) ligand
(viii) low-spin complex
(ix) strong-field ligand
(x) weak-field ligand

2. Match the correct term to each definition.

(a) the existence of compounds with the same molecular formula but with the atoms joined in a different order
(b) isomerism that can occur about a double bond and in other molecules with a rigid structure
(c) an isomer having identical groups on the same side of the molecule
(d) an isomer having identical groups on opposite sides of the molecule

(i) chirality
(ii) *cis* isomer
(iii) *cis-trans* isomerism
(iv) optical isomers
(v) structural isomerism
(vi) *trans* isomer

(e) molecules having the same molecular formula that rotate plane-polarized light differently
(f) the property of having nonsuperimposable mirror images

3. Which of the following statements are true? Rewrite each false statement so that it is correct.
   (a) Those metal ions with small radius and high charge and those that have vacant *d* orbitals readily form complexes.
   (b) Chelation greatly increases the stability of a complex compared to a similar complex not having chelation.
   (c) Valence bond theory describes metal-ligand bonding as coordinate covalent bonding in terms of overlap of atomic orbitals. The atomic orbitals on the metal that are filled by the ligand electrons are hybridized.
   (d) Crystal field theory views the ions and groups in complexes in terms of the spatial relationships of the orbitals and the interactions of the ions and groups based on their electrostatic attraction and repulsion.
   (e) Molecular orbital theory describes the bonding in terms of bonding, antibonding, and nonbonding molecular orbitals formed by the combination of orbitals on the metal and ligands.
   (f) Geometric isomers are compounds that have the same formula and the same atoms attached to each other but in different spatial arrangements.
   (g) Optical isomers have the same molecular formula, but the atoms are joined in a different order.

4. Distinguish clearly between (a) labile and inert complexes, (b) high- and low-spin complexes, (c) strong- and weak-field ligands, and (d) *cis* and *trans* isomers.

5. Choose the response that best completes each statement.
   (a) A ring formed between the ligand and the central metal or ion in a complex
      (i) is a chelate ring
      (ii) is an inner orbital complex
      (iii) is a low-spin complex
      (iv) increases the value of the dissociation constant of the complex
   (b) The equilibrium constant for the dissociation of complex ions into their components is the

(i) coordination number   (ii) crystal field stabilization energy
(iii) dissociation constant   (iv) valence bond constant

(c) The amount of stabilization provided by the splitting of d orbitals is
    known as the

      (i) back-bonding

     (ii) coordination number

    (iii) crystal field stabilization energy

     (iv) dissociation constant

(d) Which set of compounds form a colored complex upon mixing equal volumes
    of their dilute solutions?

      (i) $NaCl + CuSO_4$ (ii) $CuSO_4 + NH_4Cl$

    (iii) $CuSO_4 + NH_3$ (iv) $NH_3 + Ca(NO_3)_2$

(e) What is the color of the complex formed when aqueous Fe^{3+} reacts with
    SCN^-?

      (i) green   (ii) dark blue

    (iii) yellow   (iv) blood red

(f) Which of the following complexes has a coordination number of 4?

      (i) $[Pt(NH_3)_2Cl_4]$ (ii) $[Ag(NH_3)_2]Cl$

    (iii) $[Cu(NH_3)_4]Cl_2$ (iv) $[Co(NH_3)_6]Br_3$

(g) The geometry of a complex having a coordination number of 6 is

      (i) square planar   (ii) tetrahedral

    (iii) hexagonal   (iv) octahedral

(h) In which complex is there chelation?

(i)
$$\begin{bmatrix} & NH_3 & \\ & | & \\ H_3N & - Ni - & NH_3 \\ & | & \\ & NH_3 & \end{bmatrix}^{2+}$$

(ii)
$$\begin{bmatrix} H_2N & - C_2H_4 \\ | & | \\ H_2N - Ni - & NH_2 \\ | & | \\ H_4C_2 - NH_2 & \end{bmatrix}^{2+}$$

(iii)
$$\begin{bmatrix} & Br & \\ & | & \\ H_3N & - Ni - & OH \\ & | & \\ & NH_3 & \end{bmatrix}$$

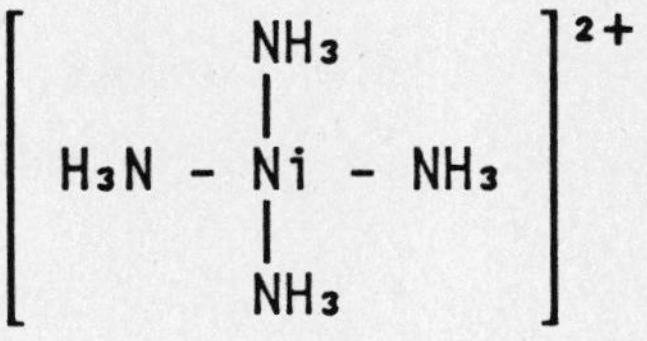

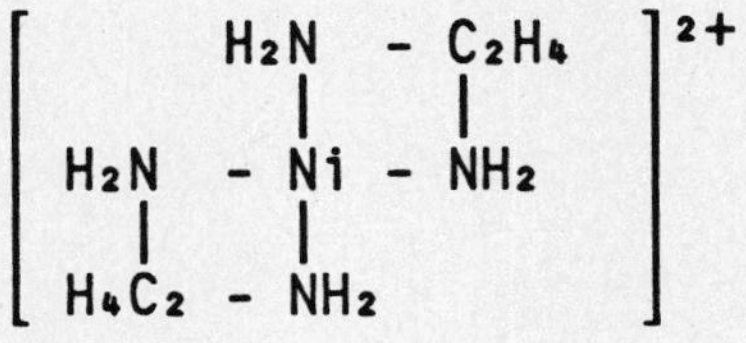

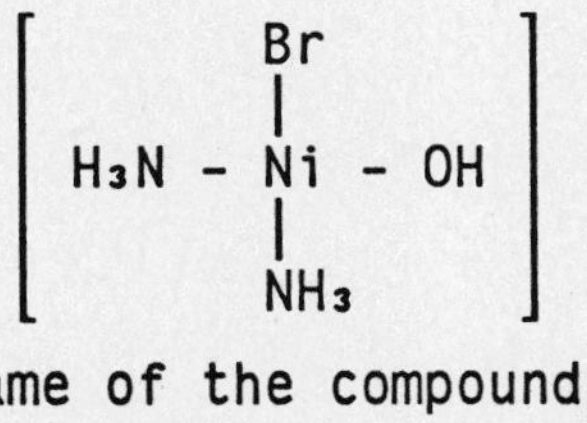

(iv) none of these

(i) The name of the compound $[Cu(NH_3)_4]Cl_2$ is

      (i) copper ammonium chloride

     (ii) tetraamminedichlorocopper(II)

    (iii) tetraamminecopper(II) chloride

     (iv) tetraamminecuprate(II) chloride

(j) The name of the compound $K_3[Fe(CN)_6]$ is

(i) potassium ferrocyanide

(ii) potassium hexacyanatoferrate(III)

(iii) potassium hexocyanoiron(III)

(iv) none of these answers

(k) The name of the complex ion shown is

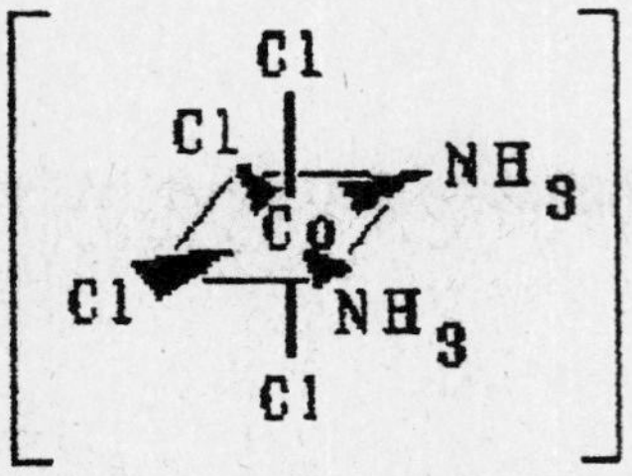

(i) *cis*-diamminetetrachlorocobaltate(III) ion

(ii) *trans*-diamminetetrachlorobaltate(III) ion

(iii) *cis*-diamminetetrachlorocobalt(III)

(iv) *trans*-bisamminetetrakischlorocobaltatoate(III) ion

(l) The formula of potassium hexachloroplatinate(IV) is

(i) $KCl \cdot PtCl_6$ (ii) $K_2[PtCl_6]$

(iii) $[KPtCl_5]Cl$ (iv) $K_4[PtCl_6]$

(m) The formula of the pentaaquamonocyanoiron(III) ion is

(i) $[Fe(CN)(H_2O)_5]^{2+}$ (ii) $[Fe(CN)_5(H_2O)]^{2-}$

(iii) $[Fe(CN)(H_2O_5)]^{2+}$ (iv) none of these answers

(n) *Cis* and *trans* isomers are often found in compounds containing

(i) single bonds (ii) triple bonds

(iii) double bonds (iv) tetrahedral geometry

(o) Pairs of isomers that rotate the direction of plane-polarized light are known as

(i) *cis-trans* isomers (ii) optical isomers

(iii) structural isomers (iv) secondary carbon atoms

(p) Light rays that vibrate in one plane are

(i) resolved (ii) steric hindered

(iii) optically active (iv) plane polarized

(q) A molecule that has the ability to rotate the plane of polarized light is

(i) optically active (ii) electrophilic

(iii) aliphatic (iv) resolvable

6. How does a coordination compound differ from a complex ion?

7. Write the formulas for (a) tetraamminediaquachromium(III) chloride, (b) pentaamminenitritocobalt(III) sulfate, (c) pentaamminenitrocobalt(III) sulfate, (d) hexacyanoplatinate(IV) ion, and (e) diamminebis(ethylenediamine)-platinum(IV) chloride.

8. Name the following: (a) $[Mn(H_2O)_6]SO_4$, (b) $[Ni(NH_3)_4](NO_3)_2$, (c) $[Ni(CO)_4]$, (d) $Na[FeCl_4]$, and (e) $[Co(NH_3)_4(NO_2)_2]NO_3$.

9. A useful empirical notation for representing complexes is

M = central metal atom or ion

a,b,c,... = monodentate ligands

(ab),... = bidentate ligands

Prepare three-dimensional sketches for the following complexes: (a) Ma_3, (b) Ma_2b, (c) Mabc, and (d) M(ab)c. Do any geometric or optical isomers of these complexes exist?

10. Repeat question 9 for square planar and tetrahedral (a) Ma_4, (b) Ma_3b , (c) Ma_2b_2, (d) Ma_2bc, and (e) Mabcd.

11. Repeat question 9 for square planar and tetrahedral (a) $M(aa)_2$ and (b) $M(ab)_2$.

12. What does the term "isomer" mean? What are the two types of geometric isomers?

13. Define "structural isomerism."

14. There are two geometric isomers of $[Pt(NH_3)_2Cl_2]$ that are known. (One of these, the *cis* isomer, is one of the most important anticancer drugs to be discovered in recent years.) Prepare sketches of the $[Pt(NH_3)_2Cl_2]$ molecule using sp^3 (tetrahedral) and sp^2d (square planar) hybridization for the Pt atom. For which type of hybridization can two isomers be drawn?

15. Using the valence bond theory, describe the bonding in (a) $[Co(NH_3)_6]^{2+}$, (b) $[Fe(CN)_6]^{4-}$, (c) $[Cr(NH_3)_6]^{3+}$, and (d) $[Ni(H_2O)_6]^{2+}$.

16. How many t_{2g} electrons and how many e_g electrons are there in the complexes listed in question 15?

17. How is the crystal field stabilization energy (CFSE) for a complex calculated? What does the symbol 10 Dq represent? What does the symbol P represent? Calculate the CFSE for the complexes listed in question 15.

Fig. 31-1

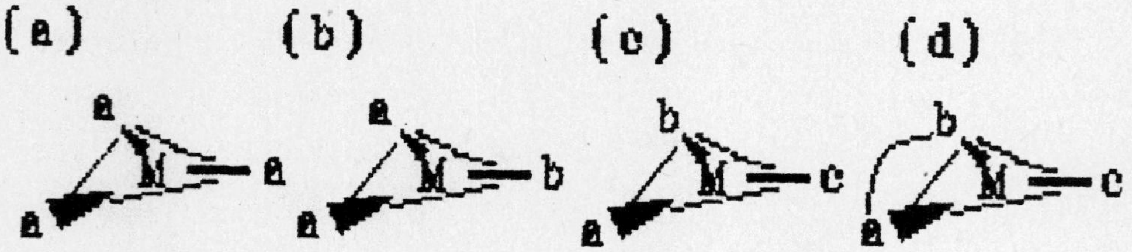

Answers to questions

1. (a) ii, (b) vii, (c) v, (d) iii, (e) iv, (f) i, (g) vi, (h) x, (i) viii, (j) ix.
2. (a) v, (b) iii, (c) ii, (d) vi, (e) iv, (f) i.
3. (a) T; (b) T; (c) T; (d) T; (e) T; (f) T, (g) F, replace *optical* with *structural*.
4. (a) *Labile complexes* exchange ligands rapidly; *inert complexes* exchange ligands slowly. (b) *High-spin complexes* contain several unpaired electrons; *low-spin complexes* have as many electrons paired as possible. (c) *Strong-field ligands* cause large separations in the *d* orbital splitting and usually force pairing of unpaired electrons; *weak-field ligands* cause small separations in the splitting and usually do not cause electron pairing to occur. (d) Cis *isomers* are geometric isomers that have the groups under consideration on the same side of the molecule; trans *isomers* have the groups on opposite sides.
5. (a) i, (b) iii, (c) iii, (d) iii, (e) iv, (f) iii, (g) iv, (h) ii, (i) iii, (j) iv, (k) i, (l) ii, (m) i, (n) iii, (o) ii, (p) iv, (q) i.
6. A complex ion has a charge; a coordination compound is neutral.
7. (a) $[Cr(NH_3)_4(H_2O)_2]Cl_3$, (b) $[Co(NH_3)_5(NO_2)]SO_4$, (c) $[Co(NH_3)_5(NO_2)]SO_4$, (d) $[Pt(CN)_6]^{2-}$, (e) $[Pt(NH_3)_2(en)_2]Cl_4$.
8. (a) hexaaquamanganese(II) sulfate, (b) tetraamminenickel(II) nitrate, (c) tetracarbonylnickel(0), (d) sodium tetrachloroferrate(III), (e) tetraamminedinitrocobalt(III) nitrate or tetraamminedinitritocobalt(III) nitrate.
9. See Fig. 31-1. No geometric or optical isomers are possible.
10. The sketches are shown in Fig. 31-2. (a) No geometric or optical isomers are possible. (b) No geometric or optical isomers are possible. (c) *Cis-trans* isomers of square planar are possible. (d) *Cis-trans* isomers of

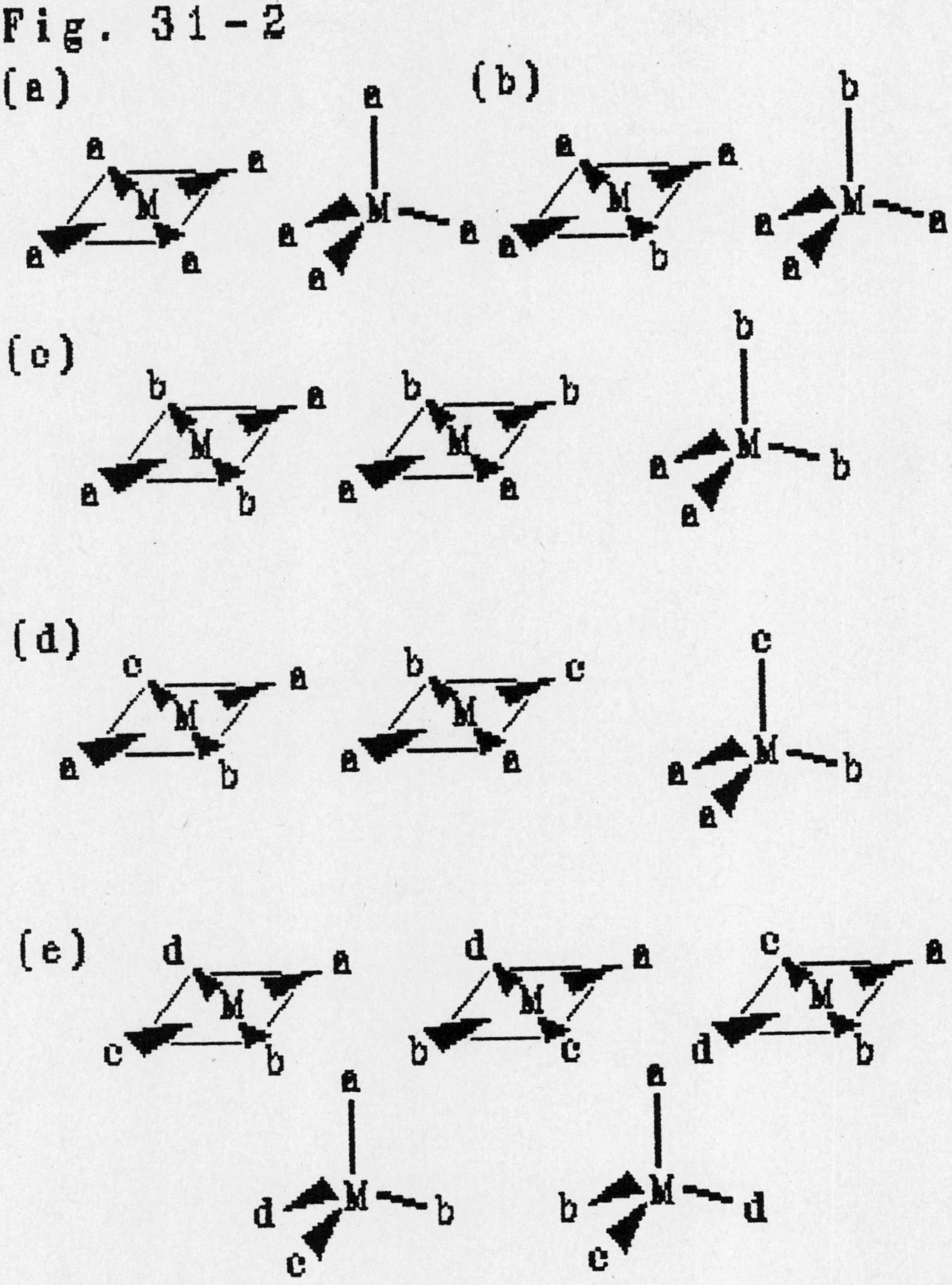

square planar are possible. (e) Three geometric isomers of square planar are possible; optical isomers of tetrahedral are possible.

11. The sketches are shown in Fig. 31-3. (a) No geometric or optical isomers are possible. (b) *Cis-trans* isomers of square planar are possible; optical isomers of tetrahedral are possible.

Fig. 31-3

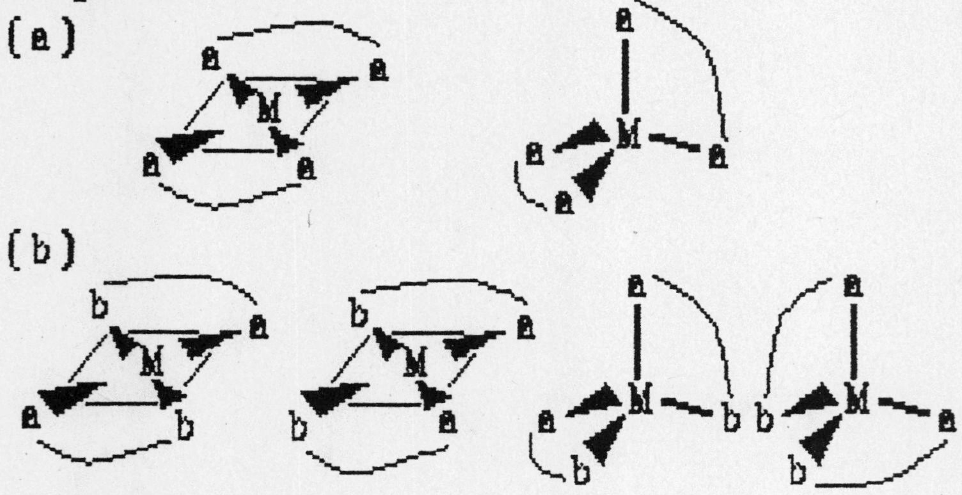

12. Isomers are compounds having the same molecular formula, but differing in structure. The two types are called *cis* and *trans*.
13. Structural isomers are compounds having the same formula, but a different arrangement of atoms.
14. The sketches are shown in Fig. 31-4. The sp^2d hybridization predicts isomerism.
15. (a) ligands in d^2sp^3, 1 electron promoted to higher *d* orbital, one unpaired electron; (b) ligands in d^2sp^3, no unpaired electrons; (c) ligands in d^2sp^3, three unpaired electrons; (d) ligands in sp^3d^2, two unpaired electrons.
16. (a) 6, 0; (b) 6, 0; (c) 3, 0; (d) 6, 2.
17. CFSE is the algebraic sum of energy gained and lost by splitting the *d* orbitals, 10 Dq is the energy difference in the *d* orbitals, and P is the energy consumed in pairing two electrons. The values of CFSE are (a) (6)(-4 Dq) + 3 P = -24 Dq + 3 P, (b) (6)(-4 Dq) + 3 P = -24 Dq + 3 P, (c) (3)(-4 Dq) = -12 Dq, (d) (6)(-4 Dq) + (2)(6 Dq) + 3 P = -12 Dq + 3 P.

SKILLS

Skill 31.1 Naming complexes

Although the name of a complex often looks long and complicated, the name is systematically determined:

(a) The ligands are named first, using prefixes to indicate the number of each ligand present.

Fig. 31-4

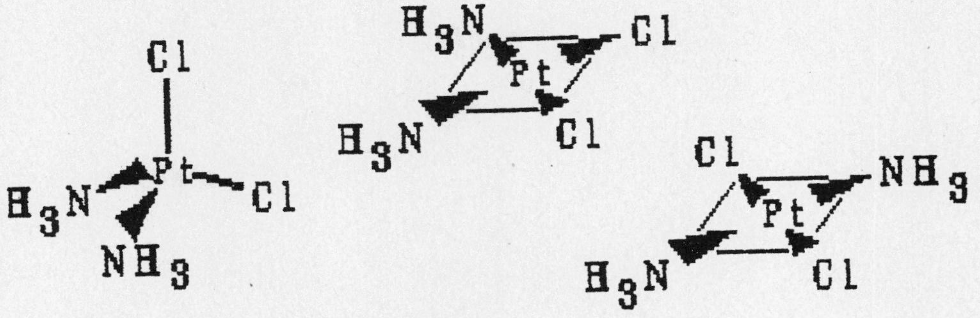

(b) The ligands are named in alphabetical order. The names of the negative ligands end in "o" and two common ligands carry special names--H_2O is "aqua" and NH_3 is "ammine."

(c) The name of the central metal atom is given after the ligand names. The oxidation state of the metal is given in parentheses. Use the word "ion" for complex ions.

(d) The name of a negative complex ion ends in "ate" and the Latin, Greek, etc., name of the element is used for those elements having historical symbols.

The following examples illustrate the foregoing rules:

$[Ag(H_2O)_2]^+$	diaquasilver(I) ion
$[Co(NH_3)_6]^{3+}$	hexaamminecobalt(III) ion
$[CoF_6]^{3-}$	hexafluorocobaltate(III) ion
$[Fe(CN)_6]^{3-}$	hexacyanoferrate(III) ion
$[Pt(NH_3)_2Cl_2]$	diamminedichloroplatinum(II)
$K_2[Ni(CN)_4]$	potassium tetracyanonickelate(II)
$[Co(NH_3)_5Br]SO_4$	pentaamminemonobromocobalt(III) sulfate
$[Co(NH_3)_6][Cr(C_2O_4)_3]$	hexaamminecobalt(III) trioxalatochromate(III)

Skill Exercise: Name the following complexes: (a) $[Ni(NH_3)_6]^{2+}$, (b) $[SiF_6]^{2-}$, (c) $[Ag(CN)_2]^-$, (d) $[Ni(CN)_4]^{2-}$, (e) $[Fe(H_2O)_6]^{3+}$, (f) $[Ir(NH_3)_3Cl_3]$, (g) $K_2[Ni(CN)_4]$, (h) $[Co(NH_3)_5(SO_4)]Br$, (i) $[Cr(NH_3)_6][Co(C_2O_4)_3]$, and (j) $[Pt(NH_3)_4][PtCl_6]$. *Answers*: (a) hexaamminenickel(II) ion, (b) hexafluorosilicate(IV) ion, (c) dicyanoargentate(I) ion, (d) tetracyanonickelate(II), (e) hexaaquairon(III) ion, (f) triamminetrichloroiridium(III), (g) potassium tetracyanonickelate(II), (h) pentaamminemonosulfatocobalt(III) bromide, (i) hexaamminechromium(III) trioxalatocobaltate(III), (j) tetraammineplatinum(II) hexachloroplatinate(IV).

Skill 31.2 Understanding geometric isomerism

Various geometric isomeric forms of a compound exist because of the symmetry or the lack of symmetry about a double bond or in some other molecular structures that do not permit rotation of the parts of a molecule. For example, $C_2H_2Cl_2$ has two geometric isomers

To distinguish between these, the first is called the *cis* isomer (meaning on the same side) and the second is called the *trans* isomer (meaning on the opposite side). The *cis-trans* isomers of $C_2H_2Cl_2$ have somewhat different properties because of the difference in structures, for example, different rates of reaction, a 13 °C difference in boiling points, a 30 °C difference in melting points, and one molecule is polar and the other is not. (Strictly speaking, there is a third isomer of $C_2H_2Cl_2$:

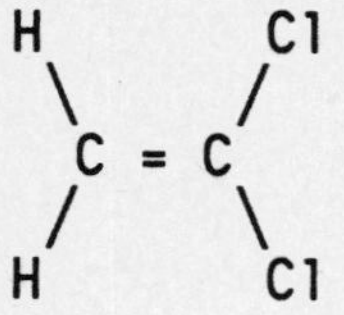

but this is an example of structural isomerism, not geometric isomerism, and is not considered further in this skill.)

Skill Exercise: Prepare sketches of the geometric isomers of diimide (HNNH) and name them. Likewise prepare sketches of the geometric isomers for $(NH_3)_2PtCl_2$, in which the platinum is in the center of a square planar molecule, and name the isomers. *Answers*:

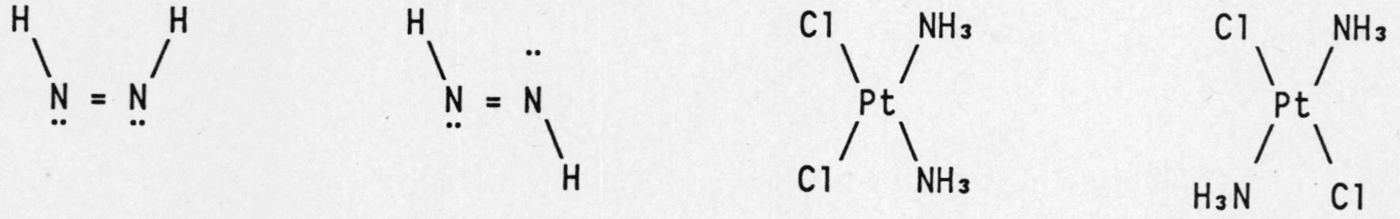

cis *trans* *cis* *trans*

Fig. 31-5

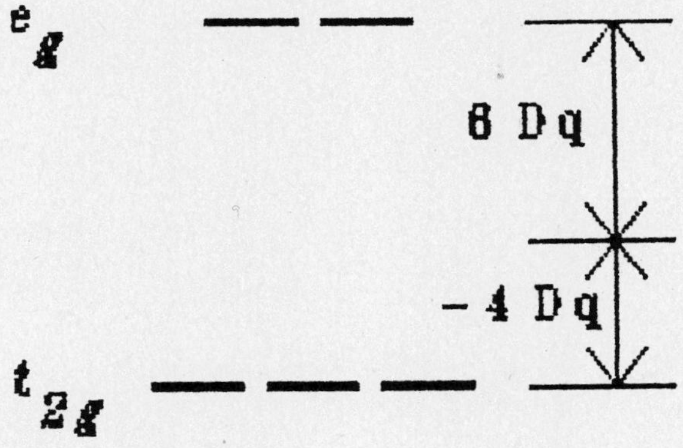

NUMERICAL EXERCISES

Example 31.1 Crystal field stabilization energy

Using the octahedral crystal field splitting shown in Fig. 31-5, determine the crystal field stabilization energy for a d^5 system in which there is (a) high spin as in $[Mn(H_2O)_6]^{2+}$ and (b) low spin as in $[Mn(SCN)_6]^{4-}$.

In the high-spin complex for a d^5 system each orbital would contain 1 electron corresponding to the configuration $(t_{2g})^3(e_g)^2$. The crystal field stabilization energy, CSFE, is found by adding the energy lost by the t_{2g} electrons to that gained by the e_g electrons.

$$\text{CFSE} = 3(-4\text{ Dq}) + 2(6\text{ Dq}) = 0$$

For the low-spin complex, the orbitals would be filled as $(t_{2g})^5$, giving

$$\text{CFSE} = 5(-4\text{ Dq}) + 2\text{ P} = -20\text{ Dq} + 2\text{ P}$$

where P is the energy required for pairing electrons.

Exercise 31.1 Using the octahedral crystal field splitting shown in Fig. 31-5, determine the CFSE for a d^7 system in which there is (a) high spin as in $[Co(H_2O)_6]^{2+}$ and (b) low spin as in $[Co(NH_2)_6]^{4-}$. *Answers*: (a) $(t_{2g})^5(e_g)^2$ gives -8 Dq + 2 P, (b) $(t_{2g})^6(e_g)^1$ gives -18 Dq + 3 P.

PRACTICE TEST

1 (30 points). Choose the response that best completes each statement.

(a) A complex that has either a positive or negative charge is

(i) a chelate (ii) a complex ion

(iii) a high-spin complex (iv) an outer orbital complex

(b) The formation of a ring in a complex by a ligand is
   (i) back-bonding
   (ii) chelation
   (iii) the crystal field stabilization energy
   (iv) an outer orbital complex

(c) Which type of metal ions forms stable complexes?
   (i) large radius and high charge   (ii) small radius and high charge
   (iii) small radius and low charge   (iv) large radius and low charge

(d) The energy difference between the orbitals of lower and higher energy is called
   (i) back-bonding
   (ii) the crystal field stabilization energy
   (iii) the coordination number
   (iv) the dissociation constant

(e) Which is *not* true about the stability of complexes?
   (i) The halide complexes usually fall in the order $I^- > Br^- > Cl^-$.
   (ii) CN^- resembles the halide ions in many respects and forms less stable complexes than the halide ions.
   (iii) Chelate ring formation increases the stability.
   (iv) The higher the oxidation state of the metal, the more stable are the complexes it forms with a given ligand.

(f) The name of the compound $K_3[Fe(CN)_6]$ is
   (i) potassium hexacyanoferrate(III)   (ii) potassium hexacyanoferrate(II)
   (iii) potassium ferrocyanide   (iv) potassium iron(III) cyanide

(g) The name of the compound $[Co(NO_2)_3(NH_3)_3]$ is
   (i) cobalt triamminenitrite   (ii) triamminetrinitratocobalt(III)
   (iii) triaminetrinitrocobalt(III)   (iv) none of these answers

(h) The name of the compound $[Co(H_2O)_4Cl_2]Br$ is
   (i) tetraaquamonobromodichlorocobalt(III)
   (ii) dichlorotetraaquacobaltate(III) bromide
   (iii) tetraaquadichlorocobalt(III) bromide
   (iv) none of these answers

(i) Which formula-name pair is *incorrect*?
   (i) $K_6[Ni(SO_4)_4]$ potassium tetrasulfatonickelate(II)
   (ii) $K_2[Ni(OH)_4]$ potassium tetraaquanickelate(II)

(iii) $[Ni(NO_2)_4]^{2-}$ tetranitronickelate(II) ion

(iv) $K_2[NiCl_4]$ potassium tetrachloronickelate(II)

(j) Which complex has a coordination number of 5?

(i) $[Zn(CN)_4]^{2-}$ (ii) $K_3[Al(C_2O_4)_2(OH)_2]$

(iii) $[Ni(CO)_4]$ (iv) $[NiBr_3(P(CH_3)_3)_2]$

(k) What is the coordination number of nickel in $[Ni(H_2O)_6]^{2+}$?

(i) 3 (ii) 4

(iii) 8 (iv) none of these answers

(l) In an octahedral splitting, the d_{xy}, d_{xz}, and d_{yz} orbitals are designated

(i) e_g (ii) t_{2g}

(iii) Dq (iv) antibonding

(m) The hybridization of the nickel ion in $[Ni(CN)_4]^{2-}$ is dsp^2. The geometry of this complex ion is

(i) tetrahedral (ii) square planar

(iii) octahedral (iv) square antiprismatic

(n) An iron(III) ion has a $3d^5$ electron configuration. The number of unpaired electrons in the $[FeF_6]^{3-}$ complex ion is

(i) 1 (ii) 3

(iii) 5 (iv) none of these answers

(o) An iron(III) ion has a $3d^5$ electron configuration. The number of unpaired electrons in the $[Fe(CN)_6]^{3-}$ complex ion is

(i) 1 (ii) 3

(iii) 5 (iv) none of these answers

2 (15 points). Name the following complexes: (a) $[AuCl_4]^-$, (b) $[Co(NH_3)_5(OH)]^{2+}$, (c) $[Fe(CO)_4I_2]$, (d) $[Ag(NH_3)_2]Cl$, (e) $[Co(NH_3)_6][Co(NO_2)_6]$.

3 (15 points). Prepare sketches of three different isomers of dichlorodiammineplatinum(II). Discuss the molecular geometry of this compound if it is known that two different geometric isomers exist.

4 (15 points). Using the octahedral crystal field splitting shown in Fig. 31-5, determine the crystal field stabilization energy for a d^6 system in which there is (a) high spin as in $[CoF_6]^{3-}$ and (b) low spin as in $[Co(NH_3)_6]^{3+}$.

5 (15 points). Calculate the concentration of Fe^{3+} in a solution in which the initial oxalate ion concentration is 0.15 mol/L and the initial iron concentration is 0.01 mol/L. $K_d = 3 \times 10^{-21}$ for $[Fe(C_2O_4)_3]^{3-}$.

6 (10 points). What is the density of gaseous $[Ni(CO)_4]$ at 100.0 °C and 1.00 atm?

Answers to practice test

1. (a) ii, (b) ii, (c) ii, (d) ii, (e) ii, (f) i, (g) iv, (h) iii, (i) ii, (j) iv, (k) iv, (l) ii, (m) ii, (n) iii, (o) i.
2. (a) tetrachloroaurate(III) ion, (b) pentaamminemonohydroxocobalt(III) ion, (c) tetracarbonyldiiodoiron(II), (d) diamminesilver(I) chloride, (e) hexaamminecobalt(III) hexanitrocobaltate(III).
3. The sketches are shown in Fig. 31-4. The existence of two isomers implies that the structure is square planar.
4. (a) -4 Dq + P, (b) -24 Dq + 3 P.
5. 2×10^{-20} mol/L.
6. 5.57 g/L.

ADDITIONAL NOTES

CHAPTER 32

ORGANIC CHEMISTRY

CHAPTER OVERVIEW

The chapter introduces the simplest of organic compounds, the hydrocarbons--compounds composed only of hydrogen and carbon atoms. The nomenclature for saturated, unsaturated, and aromatic hydrocarbons is presented. Fossil fuels, the major sources of hydrocarbons, are also discussed. The chapter concludes by introducing several functional groups and discusses the nomenclature and some common reactions of compounds containing these groups.

COMPETENCIES

Definitions for the following terms should be learned:

- addition reactions
- alcohols
- aldehyde
- aliphatic hydrocarbons
- alkanes
- alkenes, olefins
- alkyl groups
- alkyl halides
- alkynes, acetylenes
- amine
- aromatic hydrocarbons
- aryl halides
- asymmetric atom
- carbonyl group
- carboxylic acid
- cycloalkanes
- electrophile
- ester
- ether
- functional group
- hydroxyl group
- ketone
- nucleophile
- phenol
- racemic mixture
- saturated hydrocarbons
- substituents
- substitution reactions
- unsaturated hydrocarbons

General concepts and skills that should be learned include

- the properties and reactions of the hydrocarbons
- the nomenclature of the hydrocarbons
- the descriptive chemistry of fossil fuels and synfuels
- the names, significant reactions, properties, and uses for the various functional groups of organic molecules
- the fundamentals of organic nomenclature for compounds containing functional groups

QUESTIONS

1. Complete the following sentences by filling in the various blanks with these terms: aliphatic hydrocarbons, alkanes, alkenes, alkynes, aromatic hydrocarbons, cycloalkanes, homologous series, saturated hydrocarbon, and unsaturated hydrocarbons.

A (a)______ contains only covalent single bonds, whereas (b)______ contain covalent double or triple bonds between carbon atoms. A series of compounds that can be represented by a general molecular formula is called a (c)______. The series of compounds that can be represented by the general molecular formula C_nH_{2n+2} is the (d)______, by C_2H_{2n} and containing no multiple bonds is the (e)______, by C_nH_{2n} and containing double bonds is the (f)______, and by C_nH_{2n-2} is the (g)______. Hydrocarbons containing resonance-stabilized ring systems are known as (h)______ and those that do not contain such rings are known as (i)______.

2. Match the correct term to each definition.

(a) hydrocarbons containing an unbranched carbon chain

(b) a carbon atom joined to one other carbon atom

(c) double bonds that alternate with single bonds

(d) a carbon atom joined to three other carbon atoms

(e) the interaction of large groups attached to a double bond so that the *trans* form is more stable

(i) asymmetric atom

(ii) conjugated double bonds

(iii) normal hydrocarbons

(iv) primary carbon atom

(v) secondary carbon atom

(vi) steric hindrance

(vii) tertiary carbon atom

(f) an atom bonded to four different atoms or groups, which causes a molecule to be optically active
(g) a carbon atom joined to two other carbon atoms

3. Match the correct term to each definition.

(a) replacement of one or more hydrogen atoms by other atoms	(i) addition reactions
(b) the addition of atoms or groups of atoms to the carbon atoms in a multiple bond	(ii) petroleum alkylation
(c) converting straight-chain alkanes into branched-chain alkanes	(iii) petroleum cracking
(d) breaking large molecules into small molecules	(iv) petroleum isomerization
(e) converting noncyclic hydrocarbons into aromatic compounds	(v) petroleum reforming
(f) combining lower molecular mass alkanes and alkenes to form larger molecules	(vi) substitution reactions

4. Match the correct name to the general formula.

(a) R-X	(i) acid anhydride
(b) Ar-X	(ii) acyl group
(c) -OH	(iii) acyl halide
(d) R-OH	(iv) alcohol
(e) Ar-OH	(v) aldehyde
(f) R-O-R	(vi) alkoxide
(g) R-O-	(vii) alkoxy group
(h) $R_xH_{(3-x)}N$	(viii) alkyl halide
(i) $-\underset{\vert}{C}=O$	(ix) amide
	(x) amine
(j) RCHO	(xi) aryl halide
(k) R_2CO	(xii) carbonyl group
(l) RCOOH	(xiii) carboxylate ion
(m) $RCOO^-$	(xiv) carboxylic acid
(n) RCOOR'	(xv) ester
(o) RCO-X	(xvi) ether
(p) $RCONH_2$	(xvii) hydroxyl group
(q) RO^-M^+	(xviii) ketone
(r) RCO^-	(xix) phenol
(s) RCOO-OCR'	

5. Which of the following statements are true? Rewrite each false statement so that it is correct.
   (a) The cycloalkanes are cyclic saturated hydrocarbons with the general formula C_nH_{2n+2}.
   (b) An alkyl group contains one less hydrogen atom than the alkane from which it was formed.
   (c) Acetylides readily react with water to produce acetylene.
   (d) The stability of a carbon ring to oxidation and to the preference of substitution reactions instead of addition reactions is known as aromatic character.
   (e) The covalent bond between the carbon atom and the halogen atom in an alkyl halide is polar because of the larger electronegativity of the carbon atom.
   (f) The boiling point of ethyl alcohol is significantly greater than that of dimethyl ether because there is considerable hydrogen bonding in the alcohol.
   (g) Denatured alcohol contains various toxic or objectionable materials that are difficult to remove so that it cannot be used for internal consumption.
   (h) Phenols act as weak acids, alcohols as weak acids and bases, and amines as weak bases in many reactions.
   (i) All alcohols can be oxidized easily to carboxylic acids.

6. Distinguish clearly between/among (a) saturated and unsaturated hydrocarbons; (b) alkanes, alkenes, and alkynes; and (c) aromatic and aliphatic hydrocarbons; (d) alkyl, aryl, and acyl halides; (e) alcohols and phenols; (f) simple (symmetrical) and mixed (unsymmetrical) ethers; (g) primary, secondary, and tertiary amines; (h) amines and amides; and (i) aldehydes and ketones.

7. Choose the response that best completes each statement.
   (a) The hydrocarbon group that contains one less hydrogen atom than the parent alkane is known as
      (i) a secondary carbon (ii) an unsaturated hydrocarbon
      (iii) a primary carbon (iv) an alkyl group
   (b) Acetylene can be produced by reacting water with certain
      (i) carbides (ii) cyanides
      (iii) alkanes (iv) phenyl groups

(c) The stability of the ring in aromatic compounds to addition reactions is

(i) structural isomerism (ii) aromatic character

(iii) steric hindrance (iv) disproportionation

(d) A mixture containing equal amounts of levorotatory and dextrorotatory isomers is

(i) a racemic mixture (ii) an example of chirality

(iii) conjugated (iv) saturated

(e) The C_6H_5- group is

(i) an alkyl group (ii) a homologous series

(iii) the phenyl group (iv) a paraffin

(f) An alcohol containing a saturated hydrocarbon chain is

(i) an alkanol (ii) an alkenol

(iii) an alkynol (iv) a phenol

(g) An alcohol which has the hydroxyl group attached to a carbon atom that is attached to two other carbon atoms is a

(i) primary alcohol (ii) secondary alcohol

(iii) tertiary alcohol (iv) all of these answers

(h) Which can act as the strongest acid?

(i) an alcohol (ii) a phenol

(iii) an ether (iv) a carboxylic acid

(i) Which is *not* true about ethers?

(i) Mixed (unsymmetrical) ethers contain different hydrocarbon groups.

(ii) Ethers are produced from alcohols by intermolecular dehydration.

(iii) Ethers have considerable hydrogen bonding between molecules.

(iv) none of these answers

(j) The formula R_4N^+ represents a

(i) primary amine (ii) secondary amine

(iii) tertiary amine (iv) quaternary ammonium ion

(j) Which is *not* true?

(i) Oxidation of primary alcohols gives aldehydes.

(ii) Oxidation of secondary alcohols gives ketones.

(iii) Ketones are easily oxidized.

(iv) Aldehydes and ketones are commercially prepared by dehydrogenation reactions.

(k) The reaction between a carboxylic acid and an alcohol is

(i) an esterification reaction (ii) a dehydrogenation

(iii) a hemiacetal formation (iv) saponification

(l) The hydrolysis of an ester by hot aqueous NaOH is

(i) dehydrogenation (ii) esterification

(iii) saponification (iv) peptidization

8. Write the structural formula of a saturated hydrocarbon that contains a primary carbon atom, a secondary carbon atom, a tertiary carbon atom, and a quaternary carbon atom. Label each of these types of carbon atoms.

9. What is the hybridization of the carbon atoms that form multiple covalent bonds in an alkene and in an alkyne? Why are alkenes and alkynes said to be unsaturated?

10. Indicate the type of hybridization on each carbon atom in the following compounds.

(a) $CH_3 - CH = CH_2$

(b) $CH_2 = CH - C \equiv CH$

(c) $CH_3 - \underset{\underset{\displaystyle H}{|}}{C} = O$

(d) 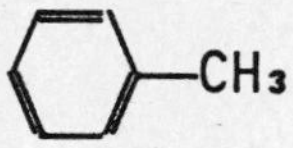

(e) $HC \equiv N$

(f) 

(g) $CH_3 - \overset{\overset{\displaystyle H}{|}}{\underset{\underset{\displaystyle Br}{|}}{C}} - \overset{\overset{\displaystyle H}{|}}{\underset{\underset{\displaystyle Br}{|}}{C}} - CH_3$

(h) 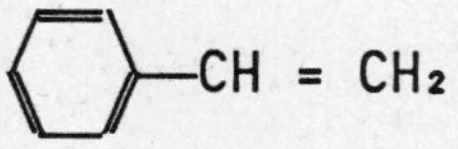

(i) $CH_2 = C = CH_2$

(j) $O = C = O$

11. Do the physical properties (melting point, boiling point, solubility, etc.) usually differ for (a) geometric isomers, (b) structural isomers, and (c) optical isomers containing one asymmetric carbon atom?

12. What is a racemic mixture? Could a polarimeter by used to distinguish between a racemic mixture and a compound that is not optically active?

13. What types of isomerism are possible for (a) 1,2-dibromoethene, (b) trimethylbenzene, and (c) 2-chlorobutane? Write the structural formulas showing the possible isomers.

14. Write structural formulas for all of the isomeric structures that could result from the substitution of one chlorine atom for one hydrogen atom in each of the following compounds.

```
(a) CH3 - CH2 - CH3

(b)        CH3
           |
     CH3 - C - CH3
           |
           CH3

(c) CH3 - CH2 - CH2 - CH2 - CH3

(d)        CH3
           |
     CH3 - CH - CH3

(e) H2C - CH2
       \ /
       CH2

(f) CH3 - CH = CH2

(g) CH3 - CH - CH - CH3
          |    |
          CH3  CH3

(h) CH3 - CH - CH2 - CH3
          |
          CH3
```

15. Which of the following compounds can have geometric isomers? Write structural formulas for the *cis* and *trans* isomers of each.

```
(a) CH3 - CH = CH - CH3

(b) CH3 - CH2 - CH = CH2

(c) CH3 - CH2 - C ≡ CH

(d) CH3 - HC - CH - CH3
           \ /
           CH2

(e) CH3 - CH = CH - CH = CH2

(f) CH2 = C - CH = CH2
          |
          CH3

(g) CH3 - CH = CHCl
```

(h) 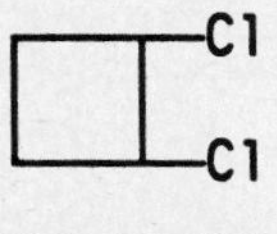

16. Draw all of the possible molecular structures corresponding to each of the following molecular formulas.

(a) C_5H_{12} (3 structures)
(b) C_2H_6O (2 structures)
(c) C_3H_6 (2 structures)
(d) $C_3H_6Cl_2$ (4 structures)
(e) C_6H_{14} (5 structures)
(f) C_3H_4 (3 structures)
(g) C_3H_8O (3 structures)
(h) C_4H_6 (8 structures)

17. Write structural formulas for each of the following compounds: (a) *n*-pentane, (b) methylcyclobutane, (c) 2,4-hexadiene, (d) 2,3-dibromobutane, (e) *m*-dinitrobenzene, (f) phenylacetylene, and (g) 1,2,4,5-tetraethylbenzene.

18. Name the compounds given in Question 14.

19. Name the compounds given in Question 15.

20. Name each of the following groups.

(a) $CH_3 - CH_2 -$

(b) $CH_3 - \underset{\displaystyle |}{CH} - CH_3$

(c) 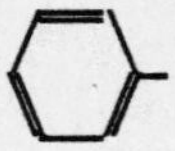

(d)
$$\begin{array}{c} CH_3 \\ | \\ CH_3 - C - \\ | \\ CH_3 \end{array}$$

(e) 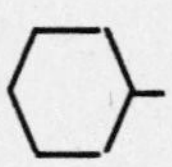

(f) $CH_3 - \underset{\displaystyle CH_3}{\underset{\displaystyle |}{CH}} - CH_2 -$

(g) $CH_3 - CH_2 - \underset{\displaystyle |}{CH} - CH_3$

(h) 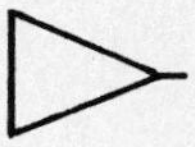

21. Identify the products of the following reactions.

(a)  $+ Br_2 \rightarrow$

(b) $CH_3 - \underset{\displaystyle CH_3}{\underset{\displaystyle |}{C}} = CH_2 + H_2SO_4 \rightarrow$

(c) $CH_3 - C \equiv CH + 2HBr \rightarrow$

(d)  $\xrightarrow[H_2O]{KMnO_4\,(excess)}$

(e) H_3C 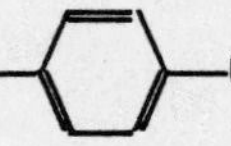 CH_3 $\xrightarrow[H_2O,\Delta]{KMnO_4}$

(f) 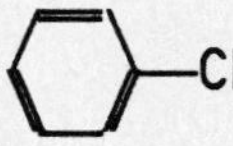 $CH_3 + Cl_2 \xrightarrow{h\nu}$

(g) 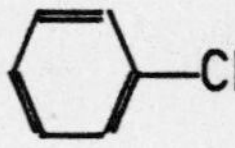 $CH_3 + Cl_2 \xrightarrow{Fe}$

(h)  $+ CH_3 - \underset{\displaystyle Cl}{\underset{\displaystyle |}{CH}} - CH_3 \xrightarrow{AlCl_3}$

22. Gasoline is given an "octane number" on the basis of matching its performance to a synthetic gasoline made by mixing isooctane (octane rating = 100) with normal heptane (octane rating = 0). For example, a rating of 85 for a gasoline sample indicates that it produces the same amount of "knocking" in a test engine as a mixture containing 85% isooctane and 15% *n*-heptane. Assuming that a gasoline sample contains only a mixture of the various structural isomers of octane and heptane, write the structural formulas for as many of the isomers as you can and name each compound.

23. What is "water gas"? How can it be prepared? Write the chemical equations describing the use of water gas as a fuel.

24. What does the term "combustion" of fossil fuels mean? What are the products of combustion reactions of compounds containing hydrogen, carbon, and oxygen with excess oxygen?

25. Write chemical reactions describing the complete combustion of (a) CH_4(g), (b) CH_3CH_2OH(l), and (c) $C_{12}H_{22}O_{11}$(s).

26. Consider the diagram shown in Fig. 32-1, which represents some of the reactions involving the "CH_3C" group. Name the seven compounds in this diagram.

27. Classify each molecule (alcohol, amine, etc.) and name the functional group(s) present.

   (a)

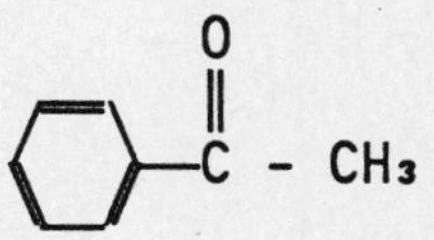

$$\bigcirc\!\!\!\!- \overset{\overset{O}{\|}}{C} - CH_3$$

   (b)

$$CH_3 - \overset{\overset{CH_3}{|}}{\underset{\underset{CH_3}{|}}{C}} - OH$$

   (c)

$$H - \overset{\overset{O}{\|}}{C} -\bigcirc\!\!\!\!- \overset{\overset{O}{\|}}{C} - O - \overset{\overset{O}{\|}}{C} - CH_2 - CH_2 - \overset{\overset{O}{\|}}{C} - Br$$

   (d)

$$H - \overset{\overset{H}{|}}{\underset{\underset{Cl}{|}}{C}} - Cl$$

   (e)

$$CH_3 - CH_2 - \overset{\overset{O}{\|}}{C} - NH_2$$

   (f)

$$H_2N - CH_2 - \overset{\overset{O}{\|}}{C} - NH - CH_2 - \overset{\overset{O}{\|}}{C} - O - CH_3$$

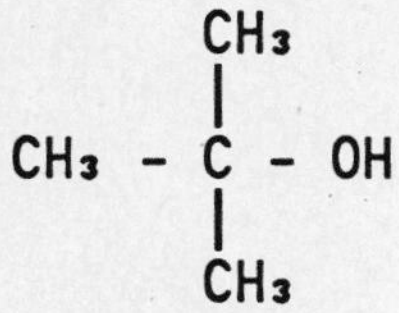

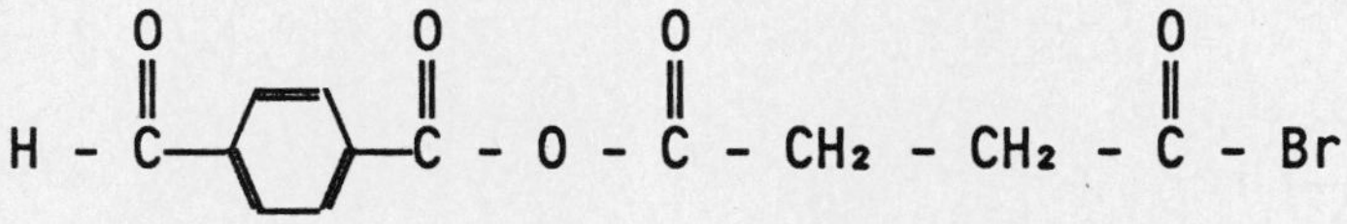

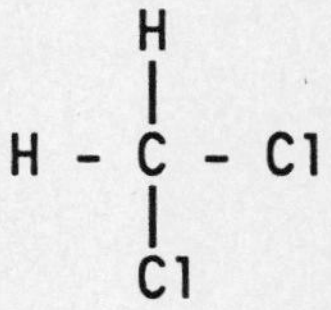

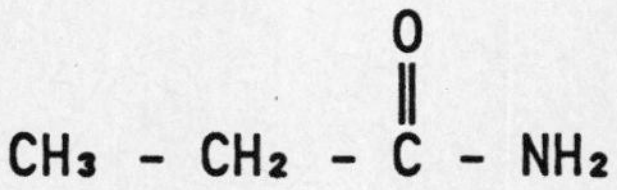

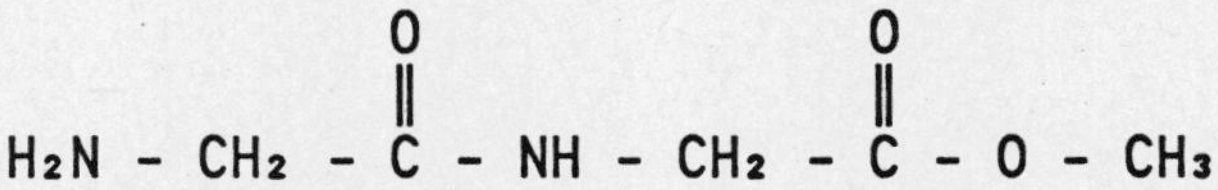

28. Write the molecular structure of a tertiary alcohol that possesses one asymmetric carbon atom.

29. The value of K_b for CH_3NH_2 is less than that for $C_2H_5NH_2$ and for $(CH_3)_2NH$ it is less than that for $(C_2H_5)_2NH$. Predict the relative order for the values for $(CH_3)_3N$ and $(C_2H_5)_3N$.

30. Complete the following equations:

   (a) $CH_3 - CH_2 - I + NaOH \overset{H_2O}{\rightarrow}$

   (b) $H_3C -\bigcirc\!\!\!\!- OH + NaOH \rightarrow$

   (c) $\bigcirc\!\!\!\!- NH_3{}^+Cl^- + NaOH \overset{H_2O}{\rightarrow}$

   (d) $\bigcirc\!\!\!\!- OH \overset{Cr_2O_7{}^{2-}}{\underset{H^+}{\rightarrow}}$

   (e) $CH_3 - COOH + Ba(OH)_2 \rightarrow$

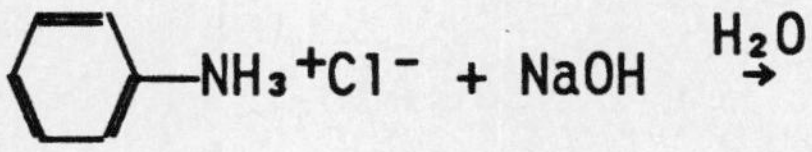

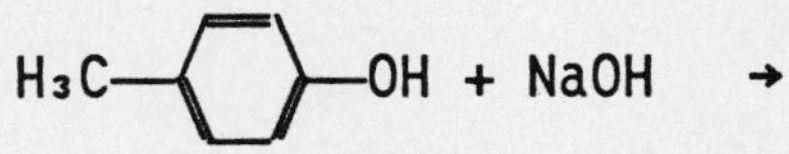

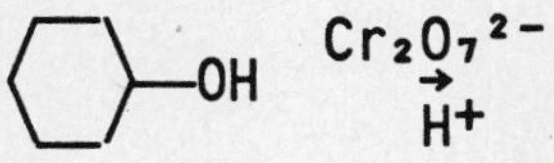

Fig. 32-1

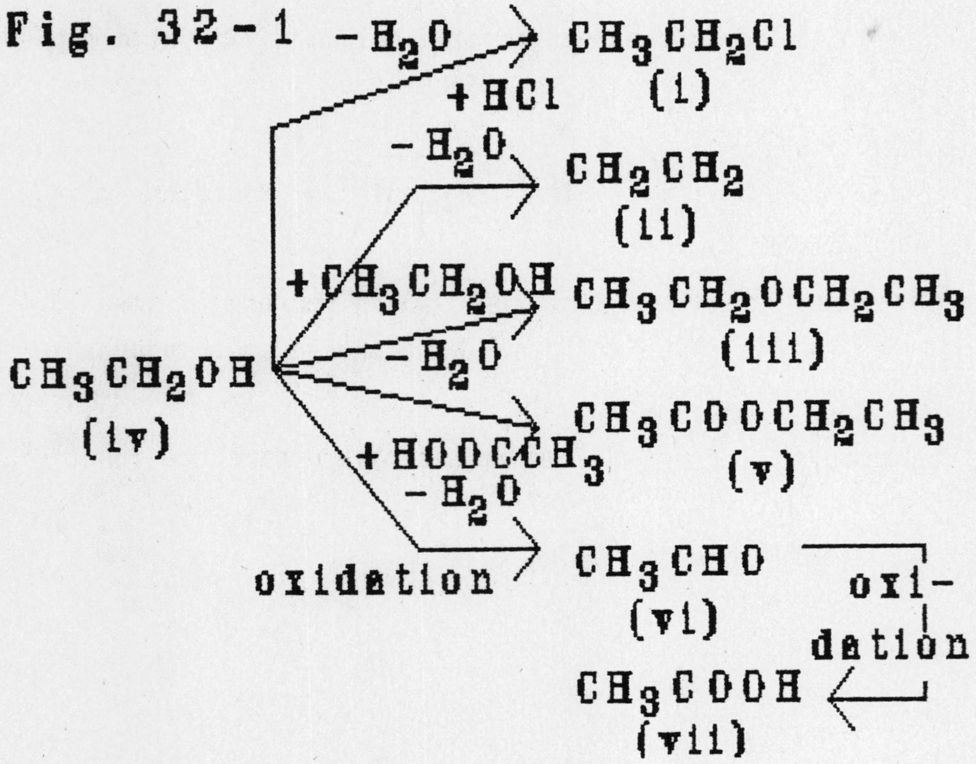

31. Complete the following equations:

(a) $CH_3 - CH_2 - NH_2 + HBr \rightarrow$

(b) $CH_3 - CH_2 - Br + NH_3 \rightarrow$

(c) $CH_3 - \underset{\displaystyle OH}{\underset{|}{CH}} - CH_2 - CH_3 + Na \rightarrow$

32. Show by equations how the following conversions might be made by using other organic or inorganic compounds. (More than one step may be necessary.)

(a) $CH_3 - \underset{\displaystyle OH}{\underset{|}{CH}} - CH_3$ to $CH_3 - \underset{\displaystyle Br}{\underset{|}{CH}} - CH_3$

(b) $CH_3 - CH_2 - CH_2 - Br$ to $CH_3 - \overset{\displaystyle Br}{\overset{|}{CH}} - \underset{\displaystyle Br}{\underset{|}{CH_2}}$

(c) $CH_3 - CH_2 - OH$ to $CH_3 - CH_2 - I$

(d) $CH_3 - CH_2 - OH$ to $(CH_3 - CH_2)_4N^+Br^-$

33. The organic compound 3,5,7,8-tridecatetraene-10,12-diynoic acid is rather unusual in that it possesses covalent triple bonds between carbon atoms, a feature not commonly found in most naturally occurring compounds. Draw the molecular structure of this compound. Hint: The name of the unbranched saturated hydrocarbon $C_{13}H_{28}$ is tridecane.

34. Write the chemical equation for the saponification of ethyl formate by sodium hydroxide.

35. Give the general formula for (a) a carbonyl group, (b) a carboxyl group, (c) an acyl group, and (d) a carboxylate ion.

36. Write the structural formula for each of the following: (a) propionaldehyde, (b) cyclobutanone, (c) 2,2-dimethylpropanoic acid, (d) ammonium benzoate, (e) acetic propionic anhydride, and (f) isopropyl formate.

37. Name each of the following:

(a) $CH_3 - CH_2 - CHO$

(b)
$$\text{Ph} - O - \overset{\overset{\displaystyle O}{\|}}{C} - CH_3$$

(c)
$$\text{Ph} - \overset{\overset{\displaystyle O}{\|}}{C} - CH_2 - CH_3$$

(d) $CH_3 - CH_2 - \overset{\overset{\displaystyle O}{\|}}{C} - NH_2$

(e)
$$CH_3 - \overset{\overset{\displaystyle CH_3}{|}}{CH} - \overset{\overset{\displaystyle O}{\|}}{C} - OH$$

(f)
$$\text{Ph} - \overset{\overset{\displaystyle O}{\|}}{C} - NH - CH_3$$

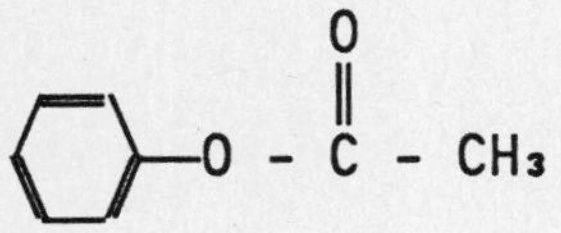

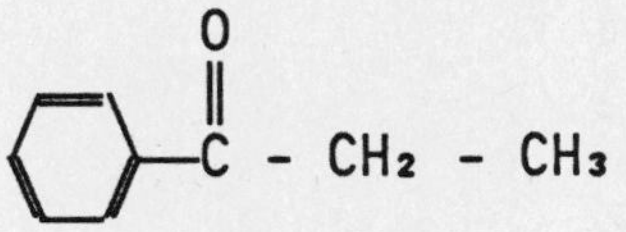

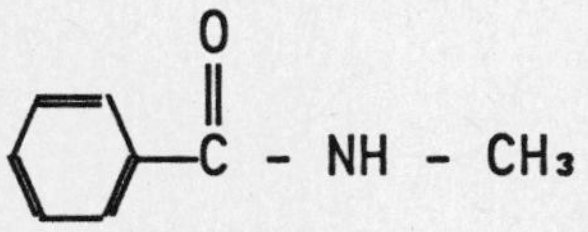

38. Give a structural formula of an organic compound that is an example of (a) an alkyl halide, (b) an ether, (c) an ester of a carboxylic acid, (d) an aldehyde, (e) a ketone, (f) a carboxylic acid, (g) an alcohol, (h) an amide, (i) an acyl halide, (j) an anhydride of a carboxylic acid, (k) an amine, (l) a phenol, and (m) an alkoxide.

39. Name the class of organic compound (aldehyde, ketone, ester, etc.) that is indicated by each of the following suffixes in the IUPAC system of nomenclature: (a) ol, (b) oic, (c) al, (d) one, and (e) ate.

40. Name the structural feature that is indicated by the following: (a) dial, (b) diol, (c) diethoxy, (d) trione, (e) dioic, and (f) ynol.

Answers to questions

1. (a) saturated hydrocarbon, (b) unsaturated hydrocarbons, (c) homologous series, (d) alkanes, (e) cycloalkanes, (f) alkenes, (g) alkynes, (h) aromatic hydrocarbons, (i) aliphatic hydrocarbons.

2. (a) iii, (b) iv, (c) ii, (d) vii, (e) vi, (f) i, (g) v.
3. (a) vi, (b) i, (c) iv, (d) iii, (e) v, (f) ii.
4. (a) viii, (b) xi, (c) xvii, (d) iv, (e) xix, (f) xvi, (g) vii, (h) x, (i) xii, (j) v, (k) xviii, (l) xiv, (m) xiii, (n) xv, (o) iii, (p) ix, (q) vi, (r) ii, (s) i.
5. (a) F, replace C_nH_{2n+2} with C_nH_{2n}; (b) T; (c) T; (d) T; (e) F, replace *larger* by *smaller*; (f) T; (g) T; (h) T; (i) F, replace *All* by *Primary*.
6. (a) *Saturated hydrocarbons* do not contain any multiple covalent bonds; *unsaturated hydrocarbons* contain double and/or triple bonds. (b) *Alkanes* contain only single covalent bonds; *alkenes* contain covalent double bonds; *alkynes* contain covalent triple bonds. (c) *Aromatic hydrocarbons* contain an aromatic (resonance-stabilized) ring of atoms; *aliphatic hydrocarbons* do not contain an aromatic ring. (d) *Alkyl halides* are organic halides that contain saturated hydrocarbon groups (R-X); *aryl halides* contain an aromatic group (Ar-X); *acyl halides* contain an acyl group (RCO-X). (e) *Alcohols* contain nonaromatic hydrocarbon groups (R-OH); *phenols* contain aromatic hydrocarbon groups (Ar-OH). (f) *Simple ethers* have identical hydrocarbon groups attached to the oxygen atom (ROR); *mixed ethers* have different hydrocarbon groups (ROR'). (g) *Primary amines* contain one hydrocarbon group (RNH_2); *secondary amines* contain two hydrocarbon groups (R_2NH); *tertiary amines* contain three hydrocarbon groups (R_3N). (h) *Amines* contain hydrocarbon groups attached to a nitrogen atom (RNH_2, R_2NH, and R_3N); *amides* contain an amino group attached to an acyl group ($RCONH_2$, RCONHR', and RCONR'R"). (i) *Aldehydes* have a hydrocarbon group and a hydrogen atom attached to the carbonyl group (RCOH); *ketones* have two hydrocarbon group attached (RCOR).
7. (a) iv, (b) i, (c) ii, (d) i, (e) iii, (f) i, (g) ii, (h) iv, (i) iii, (j) iv, (k) i, (l) iii.
8.
```
          1°
          CH3
          |    3°
    CH3 - C - CH - CH2 - CH3
    1°  4°|    |   2°    1°
          CH3 CH3
          1°  1°
```
9. sp^2, sp; multiple bonds can undergo addition reactions.
10. (a) sp^3, sp^2, sp^2; (b) sp^2, sp^2, sp, sp; (c) sp^3, sp^2; (d) sp^2 in the ring, sp^3; (e) sp; (f) sp^2 for C atoms involved in the double bond, sp^3 for the others; (g) all are sp^3; (h) all are sp^2; (i) sp^2, sp, sp^2; (j) sp.

11. (a) yes, (b) yes, (c) no.
12. A racemic mixture contains equal amounts of levorotatory and dextrorotatory isomers; no.
13. (a) *cis-trans* isomerism

```
 Br      Br          H       Br
   \    /             \     /
    C = C              C = C
   /    \             /     \
 H       H          Br       H
```

(b) structural isomerism

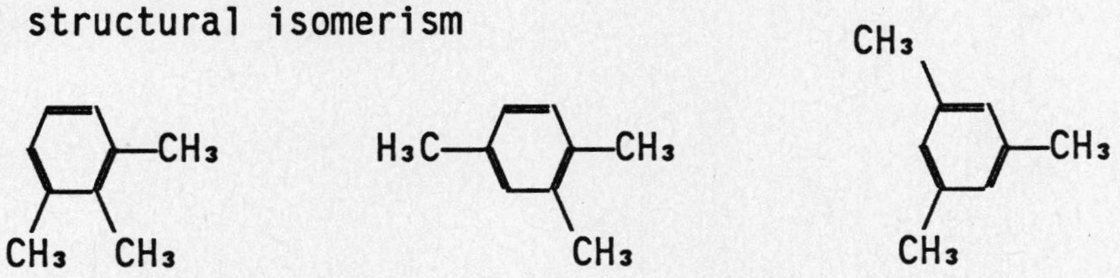

(c) optical isomerism

```
     CH3                      CH3
     Δ                        |
H  - C - Cl                   CH2
     Δ                        ∇
     CH2                 H  - C - Cl
     |                        Δ
     CH3                      CH3
```

14. (a) $CH_3 - CH_2 - CH_2Cl$

$CH_3 - CHCl - CH_3$

(b)

```
         CH3
         |
 CH3 -   C - CH2Cl
         |
         CH3
```

(c) $CH_3 - CH_2 - CH_2 - CH_2 - CH_2Cl$

$CH_3 - CH_2 - CH_2 - CHCl - CH_3$ (optical isomers)

$CH_3 - CH_2 - CHCl - CH_2 - CH_3$

(d)

```
        CH3                     CH3
        |                       |
 CH3 -  CH - CH2Cl       CH3 -  CCl - CH3
```

(e)

```
 H2C — CHCl
    \ /
    CH2
```

(f) $CH_3 - CH = CHCl$ (*cis-trans* isomers)

$CH_3 - CCl = CH_2$

$CH_2Cl - CH = CH_2$

(g)

```
 CH3 - CH - CH - CH2Cl
       |    |
       CH3  CH3  (optical isomers)

 CH3 - CH - CCl - CH3
       |    |
       H3   CH3
```

(h)

```
 CH3 - CH - CH2 - CH2Cl
       |
       CH3

 CH3 - CH - CHCl - CH3
       |
       CH3   (optical isomers)
```

```
CH3 - CCl - CH2 - CH3
      |
      CH3

CH3 - CH - CH2 - CH3
      |
      CH2Cl   (optical isomers)
```

15. (a)

```
H3C         CH3
   \       /
    C  =  C
   /       \
  H         H

  H         CH3
   \       /
    C  =  C
   /       \
H3C         H
```

(d)

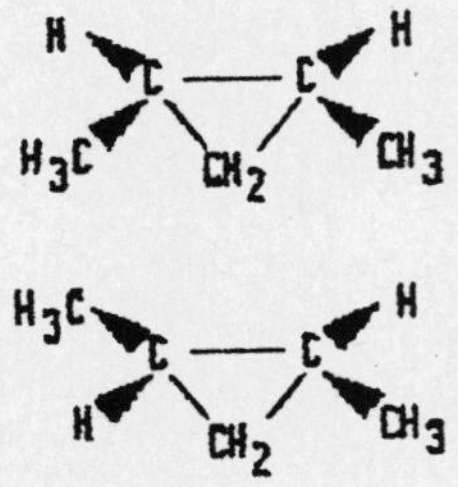

(e)

```
H3C         H
   \       /
    C  =  C
   /       \
  H         CH = CH2

H3C         CH = CH2
   \       /
    C  =  C
   /       \
  H         H
```

(g)

```
H3C         H
   \       /
    C  =  C
   /       \
  H         Cl

H3C         Cl
   \       /
    C  =  C
   /       \
  H         H
```

(h)

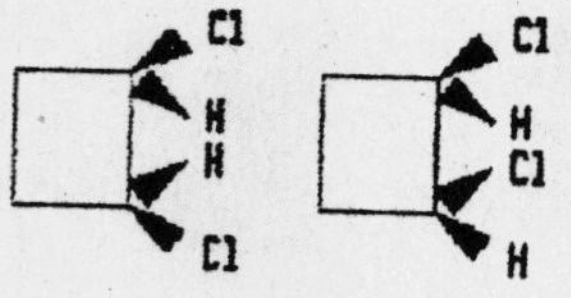

16. (a)

```
CH3 - CH2 - CH2 - CH2 - CH3

CH3 - CH2 - CH - CH3
            |
            CH3
      CH3
      |
CH3 - C - CH3
      |
      CH3
```

(b)

```
CH3 - O - CH3

      H
      |
CH3 - C - OH
      |
      H
```

(c) $CH_3 - CH = CH_2$

$$\begin{array}{ccc} H_2C & \!\!\!-\!\!\! & CH_2 \\ & \diagdown \;\; \diagup & \\ & CH_2 & \end{array}$$

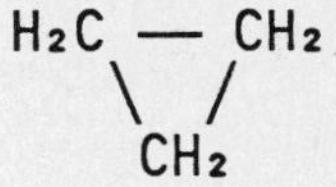

(d) $CH_3 - CH_2 - CHCl_2$ $CH_3 - CHCl - CH_2Cl$

$CH_2Cl - CH_2 - CH_2Cl$ $CH_3 - CCl_2 - CH_3$

(e) $CH_3 - CH_2 - CH_2 - CH_2 - CH_2 - CH_3$ $\begin{array}{c} CH_3 - CH - CH - CH_3 \\ | | \\ CH_3 \;\; CH_3 \end{array}$

$\begin{array}{c} CH_3 - CH_2 - CH - CH_2 - CH_3 \\ | \\ CH_3 \end{array}$

$\begin{array}{c} CH_3 \\ | \\ CH_3 - C - CH_2 - CH_3 \\ | \\ CH_3 \end{array}$

$\begin{array}{c} CH_3 - CH_2 - CH_2 - CH - CH_3 \\ | \\ CH_3 \end{array}$

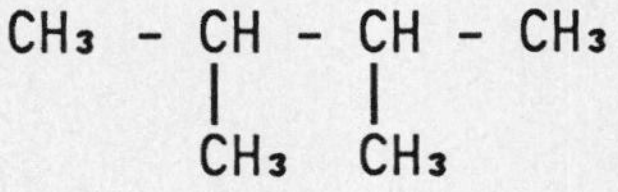

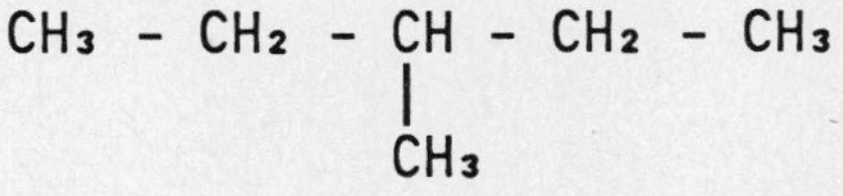

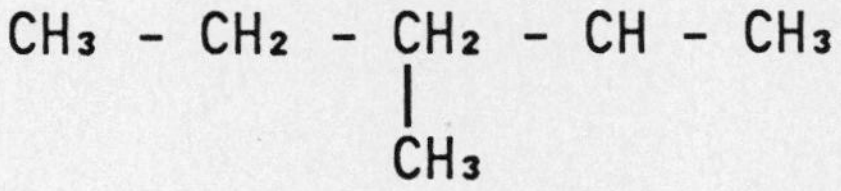

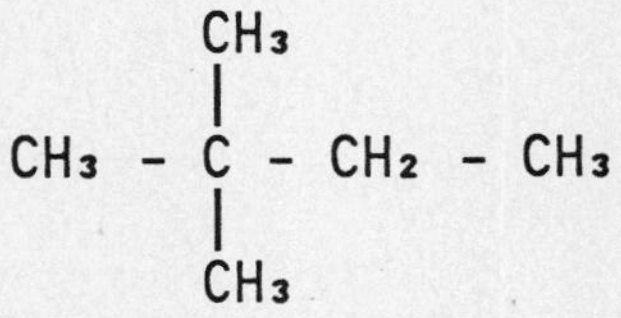

(f) $CH_3 - C \equiv CH$ $CH_2 = C = CH_2$

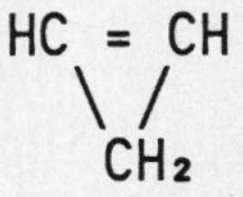

$$\begin{array}{ccc} HC & \!\!\!=\!\!\! & CH \\ & \diagdown \;\; \diagup & \\ & CH_2 & \end{array}$$

(g) $CH_3 - O - CH_2 - CH_3$ $\begin{array}{c} H \\ | \\ CH_3 - CH_2 - C - OH \\ | \\ H \end{array}$ $\begin{array}{c} CH_3 - CH - CH_3 \\ | \\ OH \end{array}$

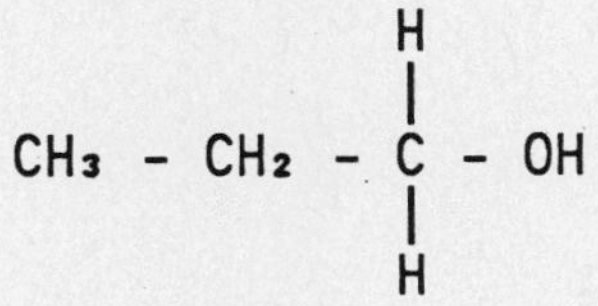

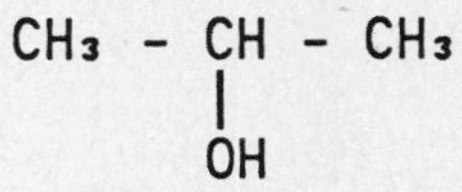

(h) $CH_2 = CH - CH = CH_2$ $CH_3 - C \equiv C - CH_3$

$$\begin{array}{ccc} HC & \!\!\!=\!\!\! & CH \\ & | \quad\quad | & \\ H_2C & \!\!\!-\!\!\! & CH_2 \end{array}$$

$CH_3 - CH = C = CH_2$ $CH_3 - CH_2 - C \equiv CH$

$$\begin{array}{ccc} HC & \!\!\!=\!\!\! & CH \\ & \diagdown \;\; \diagup & \\ & CH & \\ & | & \\ & CH_3 & \end{array}$$

$$\begin{array}{ccc} HC & \!\!\!=\!\!\! & C-CH_3 \\ & \diagdown \;\; \diagup & \\ & CH_2 & \end{array}$$

$$\begin{array}{ccc} H_2C & \!\!\!-\!\!\! & CH_2 \\ & \diagdown \;\; \diagup & \\ & C & \\ & \| & \\ & CH_2 & \end{array}$$

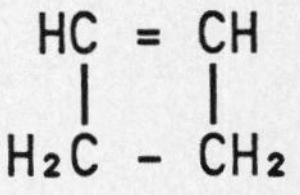

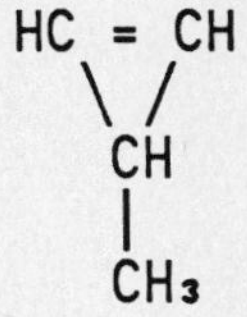

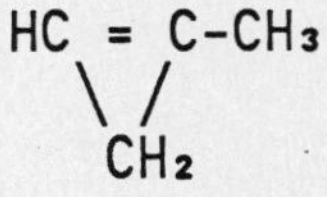

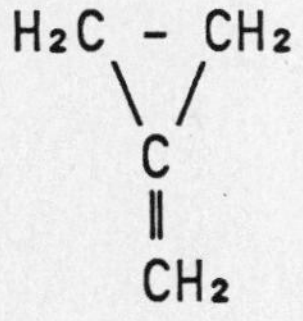

17. (a) $CH_3 - CH_2 - CH_2 - CH_2 - CH_3$

(b) $\begin{array}{c} H_2C - CH - CH_3 \\ | | \\ H_2C - CH_3 \end{array}$

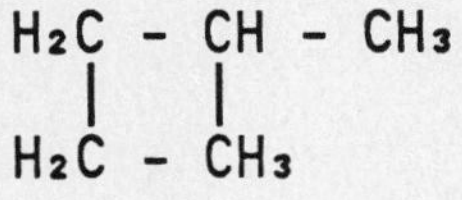

(c) $CH_3 - CH = CH - CH = CH - CH_3$

(d) $CH_3 - CHBr - CHBr - CH_3$

(e)

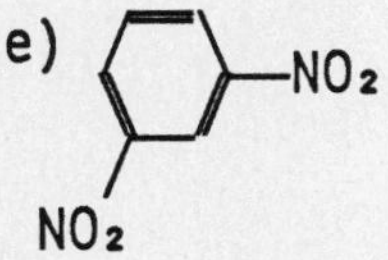

(f) $C \equiv CH$

(g) $CH_3 - H_2C \quad\quad CH_2 - CH_3$

$CH_3 - H_2C \quad\quad CH_2 - CH_3$

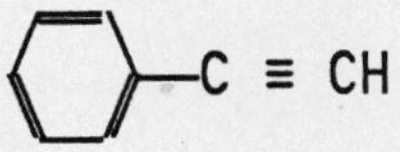

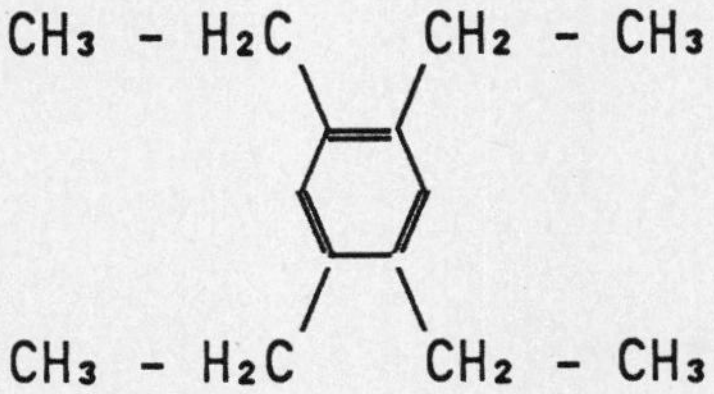

18. (a) propane, (b) 2,2-dimethylpropane, (c) pentane, (d) methylpropane, (e) cyclopropane, (f) propene, (g) 2,3-dimethylbutane, (h) methylbutane.

19. (a) 2-butene, (b) 1-butene, (c) 1-butyne, (d) 1,2-dimethylcyclopropane, (e) 1,3-pentadiene, (f) methyl-1,4-butadiene, (g) 1-chloropropene, (h) 1,2-dichlorocyclobutane.

20. (a) ethyl, (b) isopropyl, (c) phenyl, (d) *tert*-butyl, (e) cyclohexyl, (f) isobutyl, (g) *sec*-butyl, (h) cyclopropyl.

21. (a) Br   Br

(b) $CH_3 - CH - CH_3$
           |
          SO_3H

(c) $CH_3 - CBr_2 - CH_3$

(d)   O   O
           || ||
    $HO - C - CH_2 - CH_2 - CH_2 - CH_2 - C - OH$

(e)   O   O
           || ||
    $HO - C$—⬡—$C - OH$

(f) ⬡—CH_2Cl

(g) ⬡—CH_3 ⬡—CH_3 Cl—⬡—CH_3
       | |
       Cl Cl

(h) CH_3
                 |
          ⬡—CH
                 |
                 CH_3

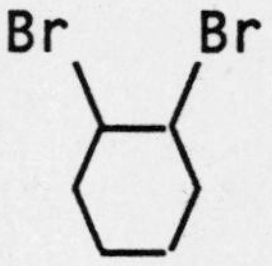

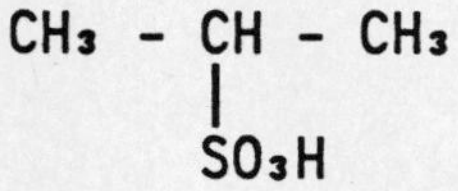

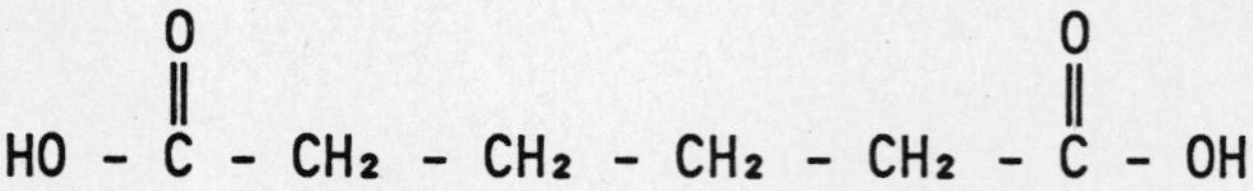

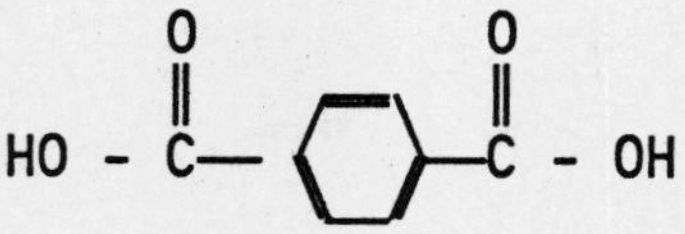

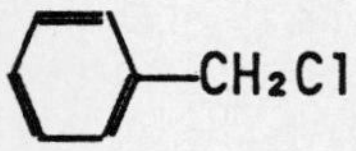

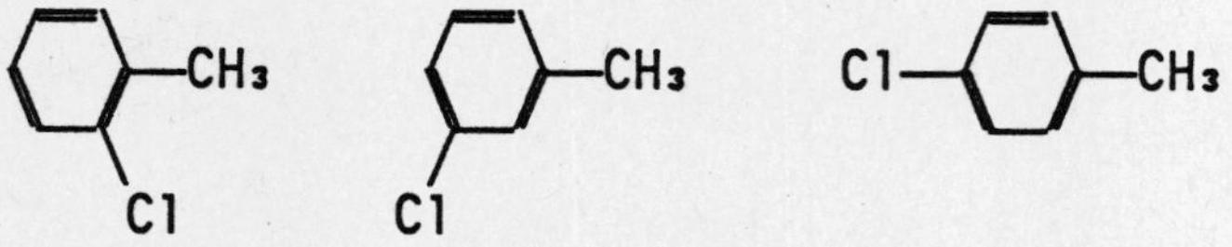

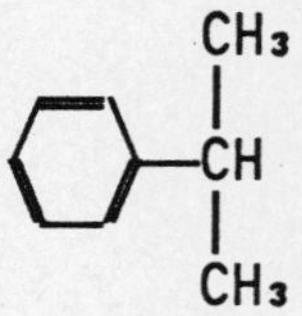

22. $CH_3 - CH_2 - CH_2 - CH_2 - CH_2 - CH_2 - CH_3$ $CH_3 - CH - CH_2 - CH_2 - CH_2 - CH_3$
                    heptane |
                                                                  CH_3 2-methylhexane

$CH_3 - CH_2 - CH_2 - CH_2 - CH_2 - CH_3$ CH_3
                    | |
                    CH_3 3-methylhexane $CH_3 - C - CH_2 - CH_2 - CH_3$
                                                               |
                                                              CH_3 2,2-dimethylpentane

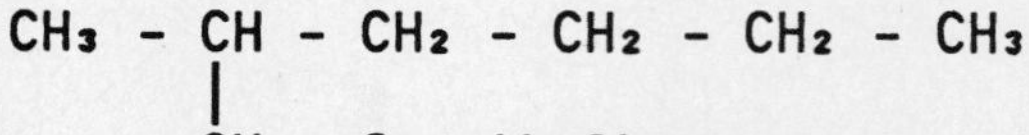

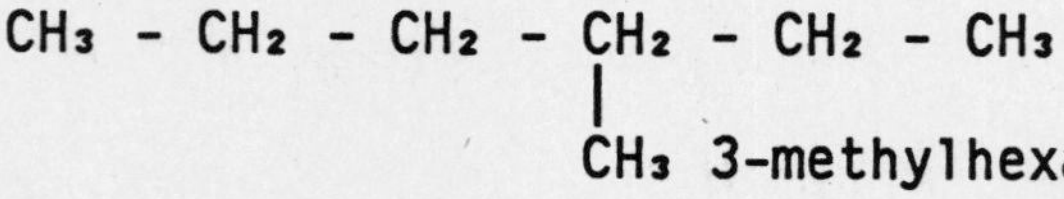

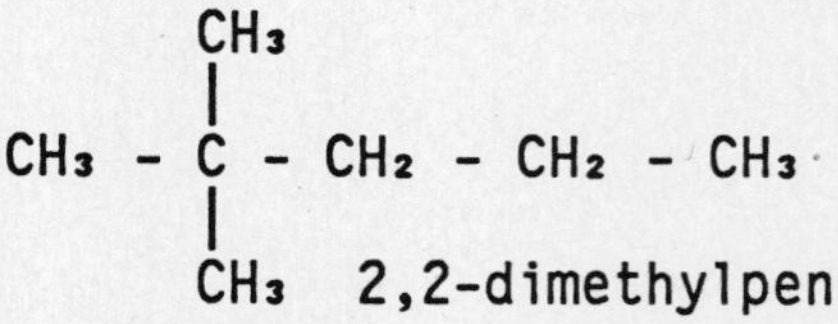

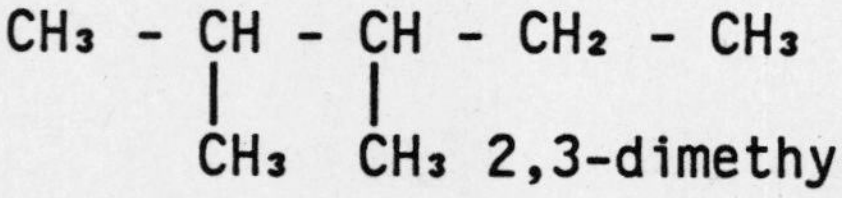

$CH_3 - CH - CH - CH_2 - CH_3$ $CH_3 - CH - CH_2 - CH - CH_3$
        | | | |
       CH_3 CH_3 2,3-dimethylpentane CH_3 CH_3 2,4-dimethylpentane

              CH_3 CH_3
              | |
$CH_3 - CH_2 - C - CH_2 - CH_3$ $CH_3 - C - CH - CH_3$
              | | |
              CH_3 3,3-dimethylpentane CH_3 CH_3 2,2,3-trimethylbutane

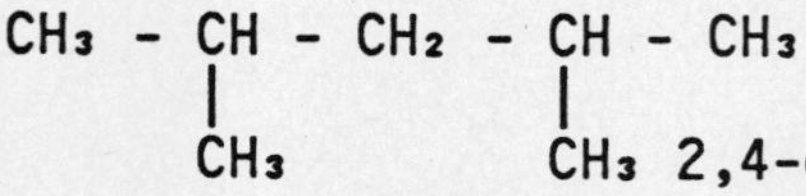

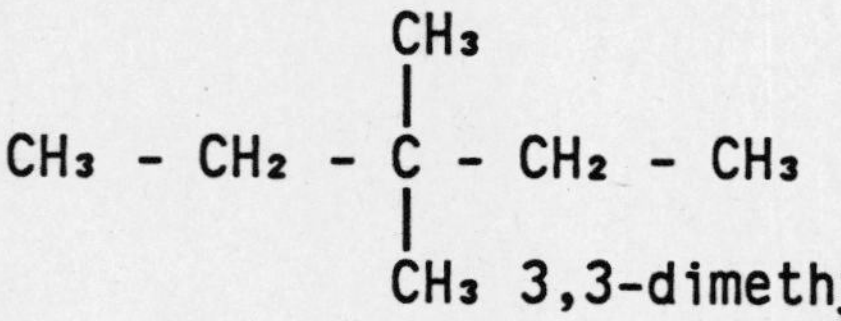

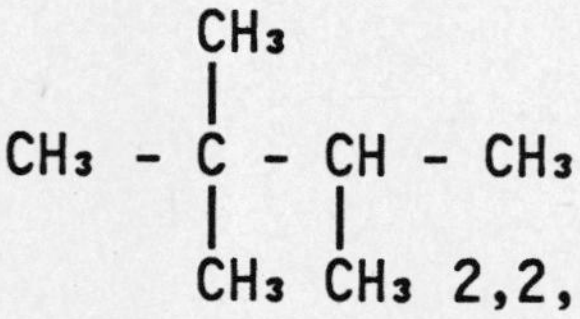

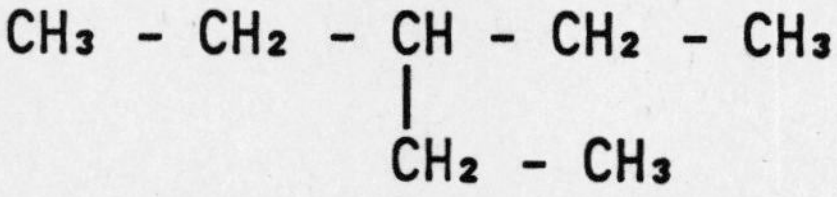

$CH_3 - CH_2 - CH - CH_2 - CH_3$
                    |
                    $CH_2 - CH_3$ 3-ethylpentane

$CH_3 - CH_2 - CH_2 - CH_2 - CH_2 - CH_2 - CH_2 - CH_3$
              octane

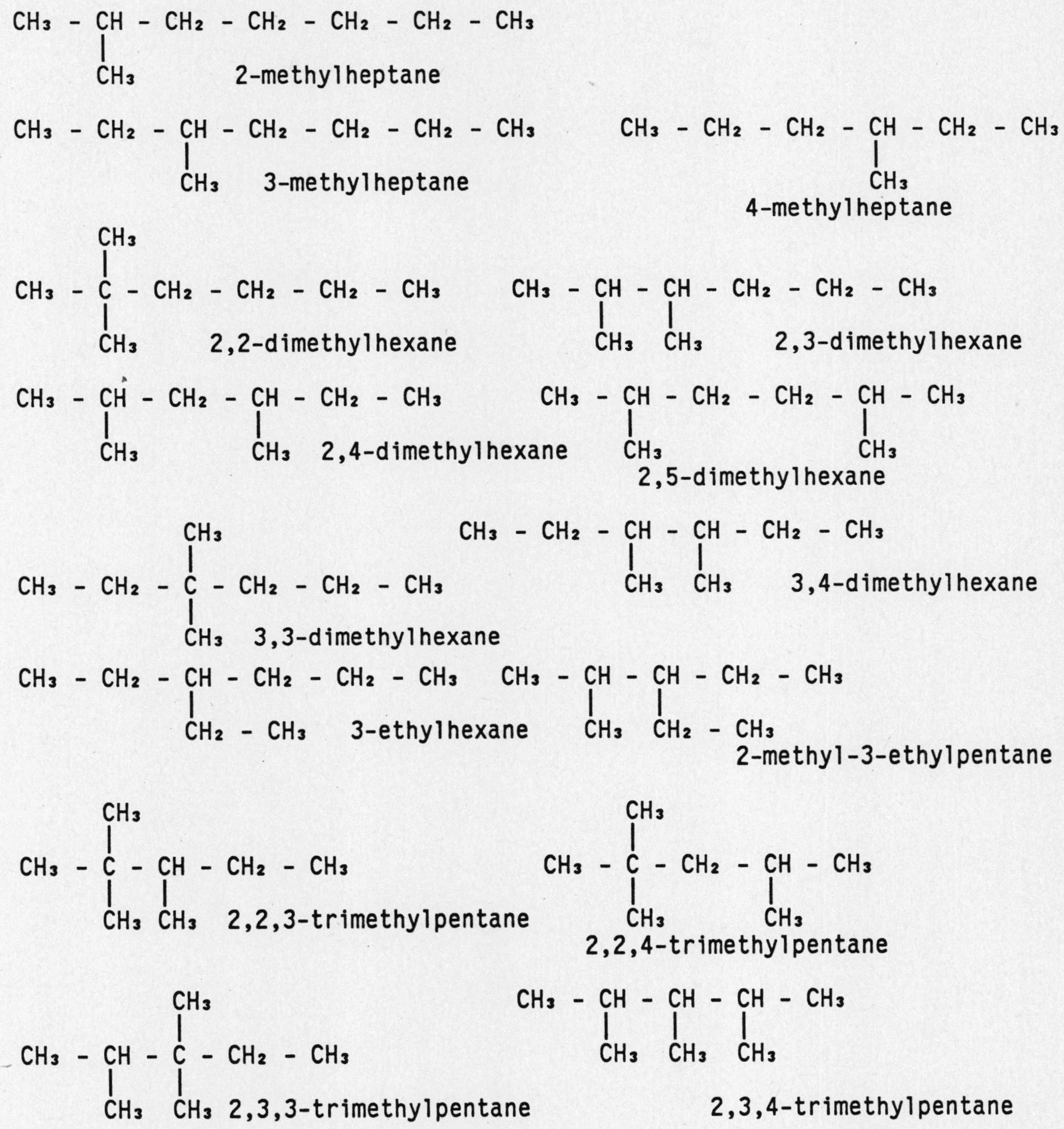
2-methylheptane
3-methylheptane
4-methylheptane
2,2-dimethylhexane
2,3-dimethylhexane
2,4-dimethylhexane
2,5-dimethylhexane
3,3-dimethylhexane
3,4-dimethylhexane
3-ethylhexane
2-methyl-3-ethylpentane
2,2,3-trimethylpentane
2,2,4-trimethylpentane
2,3,3-trimethylpentane
2,3,4-trimethylpentane

23. Water gas is a mixture of H_2 and CO. The equation for the preparation is $C(s) + H_2O(g) \xrightarrow{\Delta} H_2(g) + CO(g)$ and the equations for the combustion are $2CO(g) + O_2(g) \rightarrow 2CO_2(g)$ and $2H_2(g) + O_2(g) \rightarrow 2H_2O(g)$.

24. Combustion means the vigorous reaction with O_2 to produce flame and heat. The products are CO_2 and H_2O.

25. (a) $CH_4(g) + 2O_2(g) \rightarrow CO_2(g) + 2H_2O(l)$, (b) $CH_3CH_2OH(l) + 2O_2(g) \rightarrow CO_2(g) + 3H_2O(l)$; (c) $C_{12}H_{22}O_{11}(s) + 12O_2(g) \rightarrow 12CO_2(g) + 11H_2O(l)$.

26. (i) ethyl chloride, (ii) ethylene, (iii) diethyl ether, (iv) ethyl alcohol, (v) ethyl acetate, (vi) acetaldehyde, (vii) acetic acid.

27. (a) ketone, carbonyl group; (b) tertiary alcohol, hydroxyl group; (c) polyfunctional (aldehyde, acid anhydride, acyl halide), aldehyde, anhydride, and carbonyl halide groups; (d) alkyl dihalide, halo groups; (e) amide, amido group; (f) polyfunctional (primary amine, amide, ester), amino, amido, and ester groups.

28. for example

$$\begin{array}{c} CH_2-CH_3 \\ | \\ CH_3-C-CH_2-CH_2-CH_3 \\ | \\ OH \end{array}$$

29. K_b for $(C_2H_5)_3N$ is greater than for $(CH_3)_3N$.

30. (a) $CH_3-CH_2-OH + NaI$

(b) 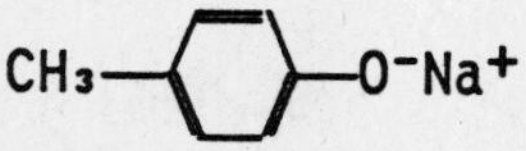 $+ H_2O$

(c) 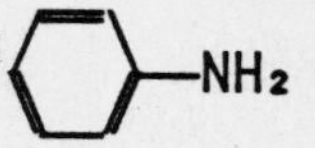 $+ H_2O + NaCl$

(d) 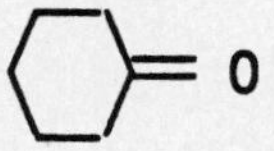

(e) $Ba(CH_3COO)_2 + 2H_2O$

31. (a) $CH_3-CH_2-NH_3^+Br^-$

(b) $CH_3-CH_2-NH_2 + HBr$

(c) $2CH_3-\underset{\displaystyle O^-Na^+}{\underset{|}{CH}}-CH_2-CH_3 + H_2$

32. (a)

$$CH_3-\overset{\displaystyle OH}{\overset{|}{C}}-CH_3 + HBr \rightarrow CH_3-\overset{\displaystyle Br}{\overset{|}{CH}}-CH_3 + H_2O$$

(b) $CH_3-CH_2-CH_2-Br \xrightarrow[ROH\ \Delta]{NaOH} CH_3-CH=CH_2 \xrightarrow{Br_2} CH_3-\overset{\displaystyle Br}{\overset{|}{CH}}-CH_2-Br$

(c) $CH_3-CH_2-OH + HI \rightarrow CH_3-CH_2-I + H_2O$

(d) $CH_3-CH_2-OH \xrightarrow{HBr} CH_3-CH_2-Br \xrightarrow{NH_3} (CH_3-CH_2)_4N^+Br^-$

33. $HC\equiv C-C\equiv C-CH=C=CH-CH=CH-CH=CH-CH_2-\overset{\displaystyle O}{\overset{\|}{C}}-OH$

34. $CH_3-CH_2-O-\overset{\displaystyle O}{\overset{\|}{C}}-H + NaOH \rightarrow CH_3-CH_2-OH + H\overset{\displaystyle O}{\overset{\|}{C}}-O^-Na^+$

35. (a) $-\overset{\displaystyle O}{\overset{\|}{C}}-$ (b) $-\overset{\displaystyle O}{\overset{\|}{C}}-OH$ (c) $R-\overset{\displaystyle O}{\overset{\|}{C}}-$ (d) $-\overset{\displaystyle O}{\overset{\|}{C}}-O^-$

36. (a) $CH_3 - CH_2 - C(=O) - H$ (b)  (c) $CH_3 - C(CH_3)_2 - C(=O) - OH$

(d) 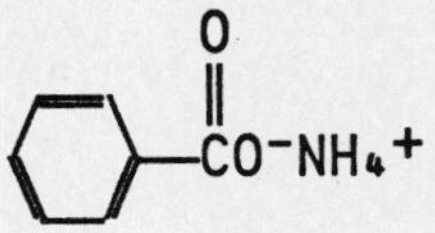 (e) $CH_3 - C(=O) - O - C(=O) - CH_2 - CH_3$ (f) $CH_3 - CH(CH_3) - O - C(=O) - H$

37. (a) propanal, (b) phenyl ethanoate, (c) ethyl phenyl ketone, (d) propanamide, (e) 2-methylpropanoic acid, (f) N-methylbenzamide.

38. (a) $CH_3 - CH_2 - Br$ (b) $CH_3 - CH_2 - O - CH_3$ (c) $CH_3 - CH_2 - C(=O) - O - CH_3$

(d) $CH_3 - CH_2 - C(=O) - H$ (e) $CH_3 - C(=O) - CH_2 - CH_3$ (f) $CH_3 - CH_2 - C(=O) - OH$

(g) $CH_3 - CH_2 - CH_2 - OH$ (h) $CH_3 - CH_2 - C(=O) - NH_2$ (i) $CH_3 - CH_2 - C(=O) - Cl$

(j) $CH_3 - CH_2 - C(=O) - O - C(=O) - CH_2 - CH_3$ (k) $CH_3 - CH(NH_2) - CH_3$

(l) 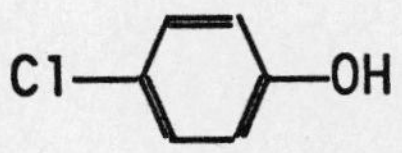 (m) $CH_3 - CH_2 - CH_2 - O^-Na^+$

39. (a) alcohol, (b) carboxylic acid, (c) aldehyde, (d) ketone, (e) ester or carboxylic acid anion.

40. (a) two $-C(=O) - H$ (b) two $- OH$ (c) two $- O - CH_2 - CH_3$ (d) three $- C(=O) -$

(e) two $- C(=O) - OH$ (f) $C \equiv C$ and $- OH$

SKILLS

Skill 32.1 Naming hydrocarbons

Organic compounds are named by common names, by historical names, and by using a systematic method known as the IUPAC system. Some of the rules of the IUPAC system are summarized below.

The names of the first four unbranched alkanes are methane, ethane, propane, and butane.

CH_4	$CH_3 - CH_3$	$CH_3 - CH_2 - CH_3$	$CH_3 - CH_2 - CH_2 - CH_3$
methane	ethane	propane	butane

Names of the higher members consist of a numerical stem and the suffix "ane." For example

$CH_3 - CH_2 - CH_2 - CH_2 - CH_3$	$CH_3 - CH_2 - CH_2 - CH_2 - CH_2 - CH_2 - CH_3$
pentane	heptane

The alkyl group formed by removing a hydrogen from a terminal carbon atom is named by replacing "ane" by "yl." For example

$$CH_3 - CH_2 - CH_2 - CH_2 - CH_2 -$$

pentyl

A saturated branched alkane is named by prefixing the name of the side chains (named as alkyl groups) to the name of the alkane corresponding to the longest continuous carbon atom chain in the molecule. The carbon atoms in the longest chain are numbered to give the lowest possible "locants" to the side chains. The locant precedes the prefix naming the side chain and is separated from it by a hyphen. If two or more side chains are present, there will be two or more prefixes, each with its appropriate locant. Prefixes are cited in alphabetical order or in order of complexity. If there are two or more identical side chains, prefixes such as di, tri, etc., should be used and a locant must be given for each side chain.

```
 4     3     2    1
CH3 - CH2 - CH - CH3
             |
            CH3
   2-methylbutane
```

```
 4     3    2    1
CH3 - CH - CH - CH3
       |    |
      CH3  CH3
 2,3-dimethylbutane
```

```
 5     4     3    2    1
CH3 - CH2 - CH - CH - CH3
             |    |
            CH2  CH3
             |
            CH3
   3-ethyl-2-methylpentane
```

Foreign atoms such as chlorine (called "chloro" when used as a prefix) may be treated in the same way as alkyl side chains, with appropriate locants and multipliers. The alkyl group formed by a branched alkane is named as above except that the carbon atom with the missing H atom is numbered "1."

```
 4     3    2     1
CH3 - CH - CH2 - CH2 -
       |
      CH3
   3-methylbutyl
```

Unbranched alkenes having one double bond are named by replacing the "ane" of the corresponding alkane with "ene." If there are two or more double bonds, the ending will have a prefix identifying the number, for example, diene represents two double bonds. The chain is numbered so as to give the lowest possible numbers to the double bonds.

$$\overset{5}{C}H_3 - \overset{4}{C}H_2 - \overset{3}{C}H = \overset{2}{C}H - \overset{1}{C}H_3$$

2-pentene

$$\overset{6}{C}H_3 - \overset{5}{C}H = \overset{4}{C}H - \overset{3}{C}H_2 - \overset{2}{C}H = \overset{1}{C}H_2$$

1,4-hexadiene

$$\begin{array}{c} CH_3 \\ | \\ \overset{5}{C}H_3 - \overset{4}{C} - \overset{3}{C}H_2 - \overset{2}{C}H = \overset{1}{C}H_2 \\ | \\ CH_3 \end{array}$$

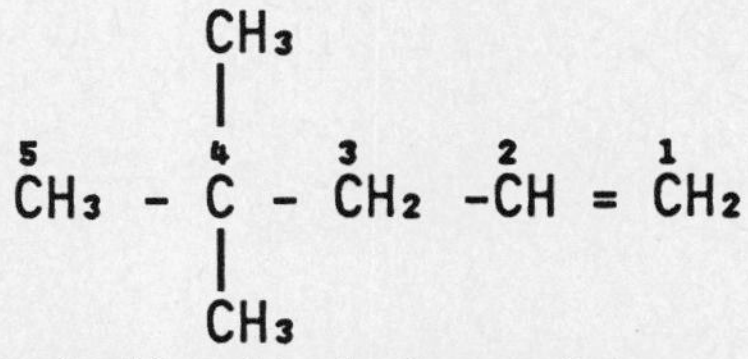

4,4-dimethyl-1-pentene

Unbranched alkynes are named similarly, except that the ending "yne" is used. Branched alkenes and alkynes are named as derivatives of the unbranched chain containing the maximum number of multiple bonds.

The cycloalkanes (with no side chains) have a ring of carbon atoms. The number of carbons in the ring is indicated by the name of the corresponding alkane and the fact that it is a ring is indicated by the prefix "cyclo." Univalent groups for these cycloalkanes are named by replacing the ending "ane" by "yl" with the carbon atom with the free valence being numbered as 1.

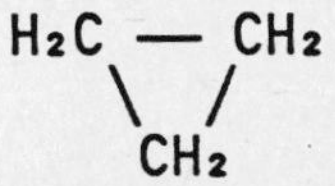

$$\begin{array}{c} H_2C - CH_2 \\ \backslash\ \ / \\ CH_2 \end{array}$$

cyclopropane

$$\begin{array}{c} H_2C - CH- \\ \backslash\ \ / \\ CH_2 \end{array}$$

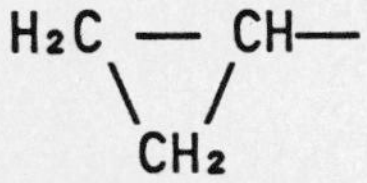

cyclopropyl

The names of unsaturated monocyclic hydrocarbons (with no side chains) are named by replacing the "ane" by "ene," "yne," "adiene," etc., in the name of the corresponding cycloalkane. The position of a substituent such as an alkyl group on the ring is indicated by a prefix to the cycloalkane name. The numbering for the locants is chosen for the carbons in the ring to make the numbers of the locants as small as possible. The common locants "ortho," meta," and "para" are used in place of 1,2-, 1,3-, and 1,4- on disubstituted benzene rings.

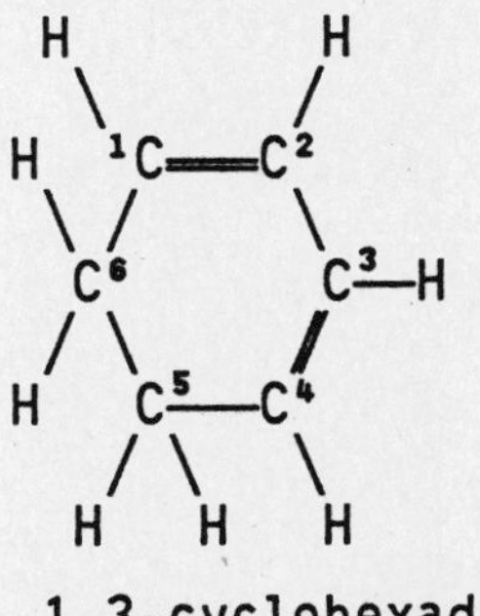

1,3-cyclohexadiene

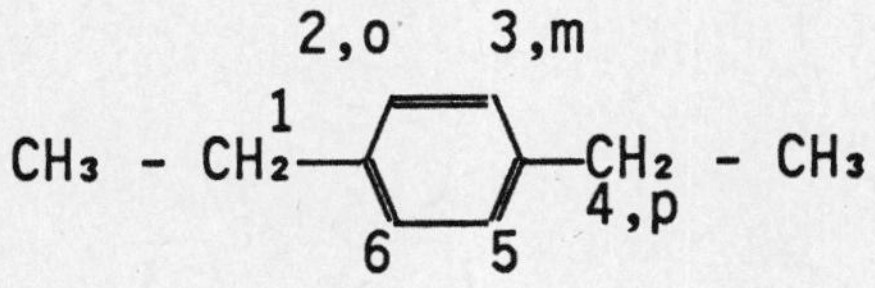

1,4-diethylbenzene or *p*-diethylbenzene

Skill Exercise: Name the following compounds or groups:

(a) $CH_3 - CH_2 - CH - CH_3$
$\qquad\qquad\qquad\quad |$
$\qquad\qquad\qquad\ CH_3$

(b) $CH_3 - CH - CH_2 - CH - CH_3$
$\qquad\qquad\quad |\qquad\qquad\quad |$
$\qquad\qquad CH_3 - CH_2 \qquad CH_2 - CH_3$

(c) $CH_3 - CH_2 -$

(d) $H_2C - CH_2$
$\qquad\ \ |\qquad\ \ |$
$\quad\ H_2C - CH_2$

(e) $H_3C\quad CH_3$
$\qquad\quad \diagdown \diagup$
$\qquad\qquad C$
$\qquad\quad \diagup \diagdown$
$\qquad H_2C\quad\ CH_2$
$\qquad\quad |\qquad\ \ |$
$\qquad H_2C——CH_2$

(f) $H_2C = CH - CH = CH_2$

(g) $CH_3 - CH - CH = CH - CH_3$
$\qquad\qquad\quad |$
$\qquad\qquad\ CH_3$

(h) $CH_3 - CH = C = CH_2$

(i) $CH_2 = C - CH = CH_2$
$\qquad\qquad\quad |$
$\qquad\qquad\ CH_3$

(j) $CH_3 - C \equiv CH$

(k)
$\qquad\qquad\quad CH = CH$
$\qquad\qquad \diagup\qquad\qquad \diagdown$
$\qquad\quad HC\qquad\qquad\qquad CH$
$\qquad\quad \|\qquad\qquad\qquad\qquad \|$
$\qquad\quad HC\qquad\qquad\qquad CH$
$\qquad\qquad \diagdown\qquad\qquad \diagup$
$\qquad\qquad\quad CH = CH$

(l)
$\qquad\qquad\qquad —CH_3$
$\qquad CH_2 - CH_3$

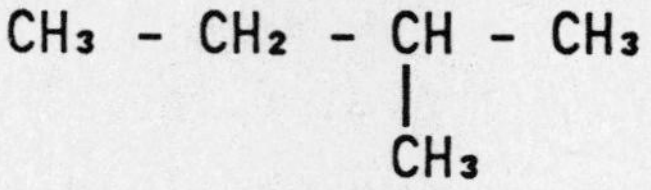

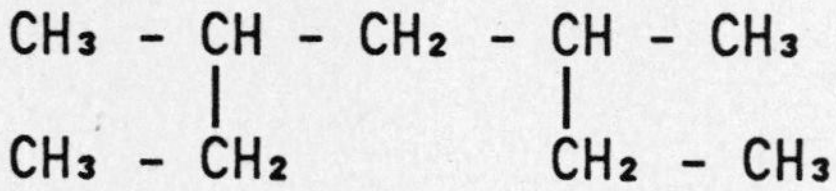

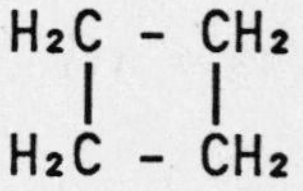

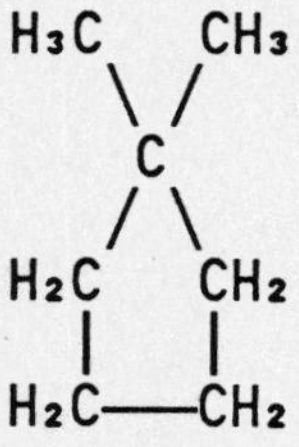

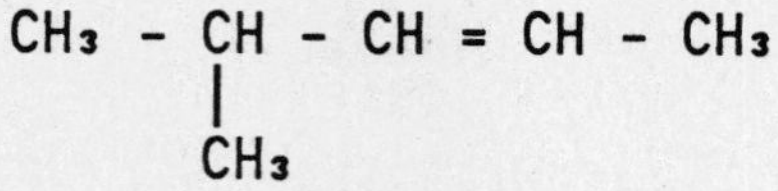

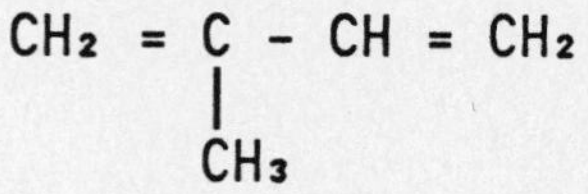

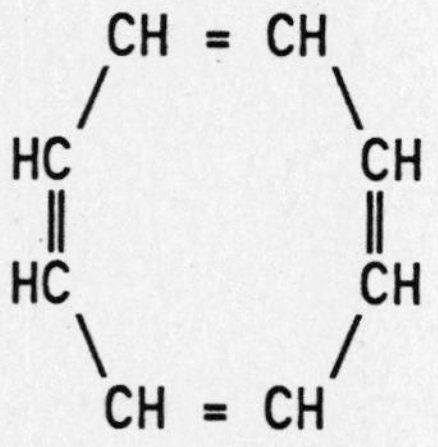

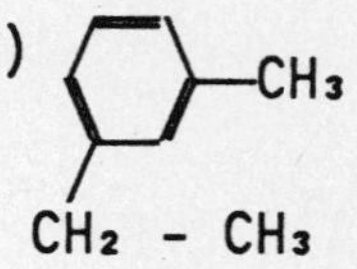

Answers: (a) 2-methylbutane, (b) 3,5-dimethylheptane, (c) ethyl group, (d) cyclobutane, (e) 1,1-dimethylcyclopentane, (f) 1,3-butadiene, (g) 4-methyl-2-pentene, (h) 1,2-butadiene, (i) 2-methyl-1,3-butadiene, (j) propyne, (k) 1,3,5,7-cyclooctatetrene, (l) 1-methyl-3-ethylbenzene or m-methylethylbenzene.

Skill 32.2 Understanding structural and optical isomerism

Structural isomers are compounds having the same molecular formula, but with different arrangements of atoms. For example, the following structural isomers can be written for the molecular formula C_4H_6:

$\qquad CH_2 = CH - CH = CH_2$

$\qquad CH_2 = C = CH - CH_3$

$\qquad\quad HC = CH$
$\qquad\quad\ |\qquad\ |$
$\qquad\ H_2C - CH_2$

$\qquad CH \equiv C - CH_2 - CH_3$

$\qquad CH_3 - C \equiv C - CH_3$

$\qquad\quad HC = CH$
$\qquad\qquad \diagdown \diagup$
$\qquad\qquad CH - CH_3$

$\qquad\quad HC = C - CH_3$
$\qquad\qquad \diagdown \diagup$
$\qquad\qquad\ CH_2$

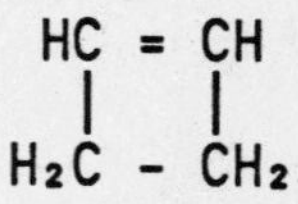

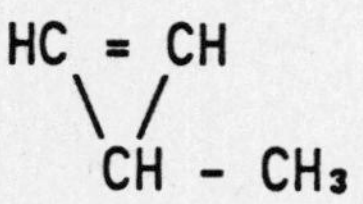

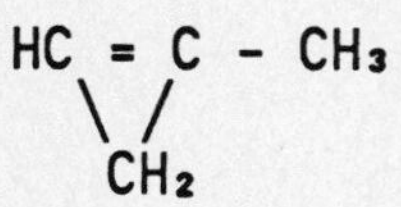

Each of the above compounds has its own set of properties.

Skill Exercise: Write as many of the structural isomers as you can for the molecular formula C_4H_8. *Answer*: 1-butene, 2-butene (*cis* and *trans* isomers), 2-methyl-1-propene, cyclobutane, 1-methylcyclopropane.

Molecules containing an asymmetric carbon atom (an atom that is bonded to four different atoms or groups) will be optically active. Likewise, other molecules that are asymmetric (such as derivatives of propadiene and certain substituted ring structures) or contain other asymmetric atoms are also optically active. For example, the following compounds are optically active:

```
        H                              H   O              ⎡      R'     ⎤+
        |                              |   ‖              ⎢      |      ⎥
CH3 - C - CH2 - CH3        CH3 - CH2 - C - C - OH         ⎢  R - N - R" ⎥
        |                              |                  ⎢      |      ⎥
        OH                             CH3                ⎣      R"'    ⎦
```

Skill Exercise: Identify the optically active species from the following compounds:

```
(a)        O                      (b)     OH          (c)              H
           |                              |                            |
     CH3 - N - CH2 - CH2 - CH3     CH3 - C - CH2 - CH3    CH3 - CH2 - CH - C - OH
           |                              |                         |
           CH2 - CH3                      CH3                       CH3
```

Answers: a, c.

Skill 32.3 Naming organic compounds with functional groups

The nomenclature of alkyl halides is easily understood by referring to the rules of nomenclature of hydrocarbons given in *Skill 32.1 Naming hydrocarbons*. The name of the halogen functional group is "halo." For example

```
CH3 - CH - CH3                 Cl
      |                        |
      Cl                 CH3 - C - CH2 - Cl
2-chloropropane                |
                               F
                   1,2-dichloro-2-fluoropropane
```

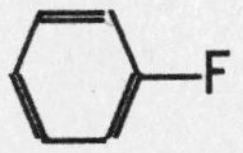
fluorobenzene

Skill Exercise: Name the following compounds:

```
(a) CH3 - CH2 - CH - CH3     (b) CCl4     (c) CH3 = CHCl2     (d) CH2Cl - CH2Cl
                |
                Cl
```

and write structural formulas for (e) 4-bromo-1-butene, (f) 1-chlorobutane, (g) *o*-dichlorobenzene, and (h) hexachlorobenzene. *Answers*: (a) 2-chlorobutane, (b) tetrachloromethane, (c) 1,1-dichloroethane, (d) 1,2-dichloroethane,

(e) $CH_2 = CH - CH_2 - CH_2 - Br$ (f) $CH_3 - CH_2 - CH_2 - CH_2 - Cl$

(g) 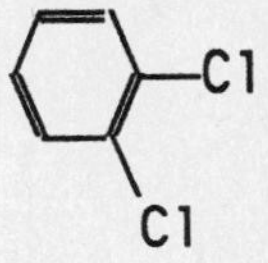

(h) 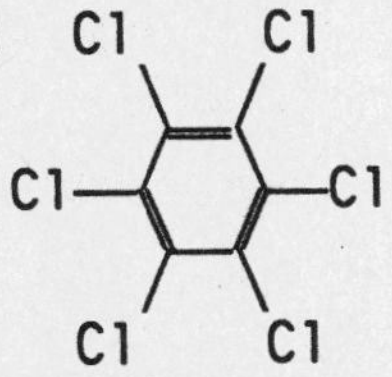

When naming nonaromatic alcohols, the name is derived from the largest hydrocarbon chain that includes the OH group by dropping the final "e" and adding "ol." When necessary, a number is used to show the position of the hydroxyl group. Aromatic alcohols are named as substituted benzene compounds or as derivatives of phenol. For example

$CH_3 - CH_2 - CH_2 - OH$
1-propanol
(*n*-propyl alcohol)

OH
|
$CH_3 - C - CH_3$
|
CH_3
2-methyl-2-propanal
(*tert*-butyl alcohol)

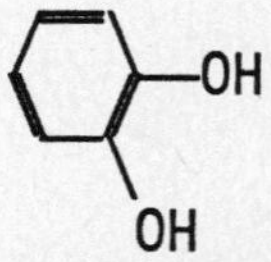

1,2,4-trihydroxybenzene

Skill Exercise: Name the following compounds:

(a) 

(b) $CH_3 - CH - CH_2 - OH$
|
CH_3

(c) $CH_3 - OH$

and write the structural formulas for (d) phenol, (e) 2-propanol, and (g) ethanol.
Answers: (a) 4-chlorophenol, (b) 2-methyl-1-propanol or isobutyl alcohol, (c) methanol or methyl alcohol,

(d) 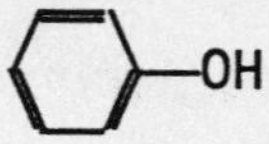

(e) $CH_3 - CH - CH_3$
|
OH

(f) $CH_3 - CH_2 - OH$

Ethers are named as an alkoxy derivative of the longest-chain hydrocarbon to which the alkoxy group is attached. The position of attachment is given by a number starting at the end of the chain nearest the alkoxy group. For example

$CH_3 - O - CH_2 - CH_3$
methoxyethane
(methyl ethyl ether)

CH_3
|
$CH_3 - O - C - CH_3$
|
CH_3
2-methoxy-2-methylpropane
(methyl *tert*-butyl ether)

Skill Exercise: Name the following compounds:

(a) $CH_3 - CH_2 - O - CH - CH_3$
|
CH_3

(b) $CH_3 - CH_2 - O - CH_2 - CH_3$

and write the structural formulas for (c) methoxymethane and (d) 2-ethoxypropane.
Answers: (a) 2-ethoxypropane or ethyl isopropyl ether, (b) methoxyethane or diethyl ether,

```
(c) CH3 - O - CH3            (d) CH3 - O - CH - CH3
                                           |
                                           CH3
```

Amines are named as derivatives of ammonia. For example

```
CH3 - CH2 - NH2                   CH3
ethanamine                         |
(ethyl amine)               CH3 -  C - CH3
                                   |
                                  NH2
                            2-methyl-2-propanamine
                            (tert-butylamine)
```

Skill Exercise: Name the following compounds:

```
(a) CH3 - CH - CH3           (b) (CH3)3N
          |
          NH2
```

and write structural formulas for (c) diethanamine and (d) ethan-2-butanamine.

Answers: (a) 2-propanamine or isopropylamine, (b) trimethanamine or trimethylamine,

```
(c) CH3 - CH2 - N - CH2 - CH3        (d)          CH3   H
                |                                  |    |
                H                     CH3 - CH2 - CH - N - CH2 - CH3
```

Aldehydes are named by using the ending "al" and ketones by using the ending "one." For ketones, a number is used to represent the location of the carbonyl group. For example

```
    O                                                    O
    ||                                                   ||
H - C - H                         CH3 - CH2 - CH2 - C - CH3
methanol                          2-pentanone
(formaldehyde)                    (methyl n-propyl ketone)
```

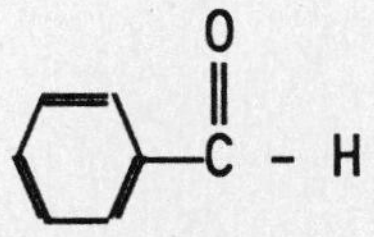

benzaldehyde

Skill Exercise: Name the following compounds:

```
(a)      CH3  O               (b)            O
          |   ||                             ||
    CH3 - CH - C - H           CH3 - CH2 - C - CH2 - CH3
```

and write structural formulas for (c) ethanal and (d) 2-methyl-3-pentanone.

Answers: (a) 2-methyl propanal or isobutyraldehyde, (b) 3-pentanone or diethyl ketone,

```
(c)     O             (d)      CH3  O
        ||                      |   ||
  CH3 - C - H           CH3 - CH - C - CH2 - CH3
```

Organic acids are named by replacing the final "e" of the hydrocarbon name by "oic." For example

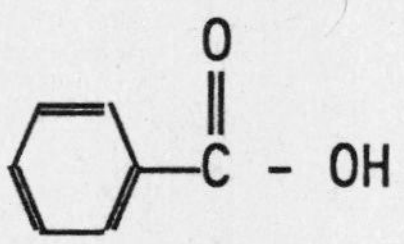

benzoic acid

$$CH_3 - \overset{\overset{O}{\|}}{C} - OH$$

ethanoic acid
(acetic acid)

Skill Exercise: Name the following compounds:

(a) $CH_3 - CH_3 - CH_3 - \overset{\overset{O}{\|}}{C} - OH$ (b) $CH_3 - \overset{\overset{Br}{|}}{CH} - \overset{\overset{Br}{|}}{CH} - \overset{\overset{O}{\|}}{C} - OH$

and write the structural formulas for (c) 2-methylpropanoic acid and (d) methanoic acid. *Answers*: (a) butanoic acid or *n*-butyric acid, (b) 2,3-dibromobutanoic acid,

(c) $CH_3 - \overset{\overset{CH_3}{|}}{CH} - \overset{\overset{O}{\|}}{C} - OH$ (d) $H - \overset{\overset{O}{\|}}{C} - OH$

The names of esters consist of two parts. The first part is derived from the name of the alcohol and the second part is derived from the name of the carboxylic acid with the "ic" replaced by "ate." For example

$$CH_3 - \overset{\overset{O}{\|}}{C} - O - CH_3$$

methyl ethanoate
(methyl acetate)

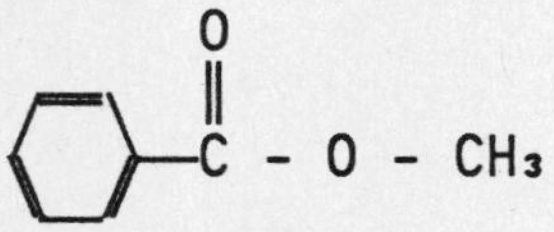

methyl benzoate

Skill Exercise: Name the following compounds:

(a) $H - \overset{\overset{O}{\|}}{C} - O - CH_3$ (b) $CH_3 - CH_2 - \overset{\overset{O}{\|}}{C} - O - CH_3$

and write structural formulas for (c) ethyl ethanoate and (d) 2-propyl 2-methyl propionate. *Answers*: (a) methyl methanoate or methyl formate, (b) methyl propionoate or methyl propionate

(c) $CH_3 - \overset{\overset{O}{\|}}{C} - O - CH_2 - CH_3$ (d) $CH_3 - \overset{\overset{CH_3}{|}}{CH} - \overset{\overset{O}{\|}}{C} - O - \overset{\overset{CH_3}{|}}{CH} - CH_3$

PRACTICE TEST

1 (7 points). Classify each of the following compounds as (i) an alkane, (ii) an alkene, (iii) an alkyne, (iv) aromatic, (v) having conjugated bonds, (vi)

a cycloalkane, (vii) having a normal hydrocarbon chain, (viii) being a saturated hydrocarbon, and/or (ix) being an unsaturated hydrocarbon:

(a) ```
(b)   CH2   (c) CH3 - C = CH2   (d) H - C ≡ C - H
         / \ |
      H2C - CH2   CH3
```

(e) $CH_3 - CH_2 - CH_3$  (f) $CH_2 = C = CH - CH_3$  (g) $CH_2 = CH - CH = CH_2$

2 (8 points). Identify which pair(s) of substances represent (a) optical isomers, (b) geometric isomers, and (c) structural isomers.

```
(i)        OH                                 OH
           |                                  |
   CH3 - C - CH2 - CH3          CH3 - CH2 - C - CH3
           |                                  |
           CH = CH2                           CH = CH2

(ii)       OH
           |
   CH3 - C - CH2 - CH3          CH3 - CH - O - CH2 - CH3
           |                          |
           CH = CH2                   CH = CH2

(iii) H3C        CH2 - CH3          H       CH2 - CH3
         \      /                    \     /
          C = C                       C = C
         /      \                    /     \
        H        CH2 - OH          H3C      CH2 - OH

(iv) H        CH2 - CH3          HO - CH2 - H2C       CH2 - CH3
      \      /                                \     /
       C = C                                   C = C
      /      \                                /     \
     H        CH2 - CH2 - OH                 H       H
```

3 (12 points). Match the name of the type of compound to the formula given.

(a) R-X	(i) acid anhydride
(b) R-OH	(ii) acyl halide
(c) R-O-R	(iii) alcohol
(d) $R\text{-}NH_2$	(iv) aldehyde
(e) RCHO	(v) alkoxide
(f) R_2CO	(vi) alkyl halide
(g) RCOOH	(vii) amide
(h) RCOOR'	(viii) amine
(i) RCO-X	(ix) carboxylic acid
(j) $RCONH_2$	(x) ester
(k) RO^-M^+	(xi) ether
(l) RCOO-OCR'	(xii) ketone
```

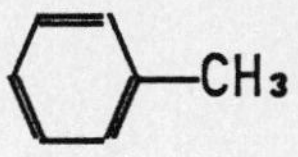

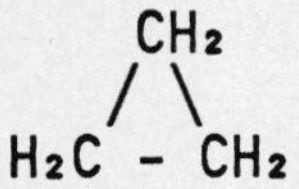

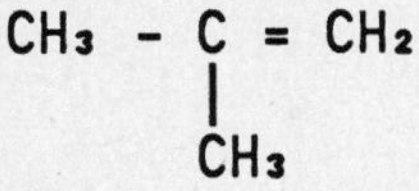

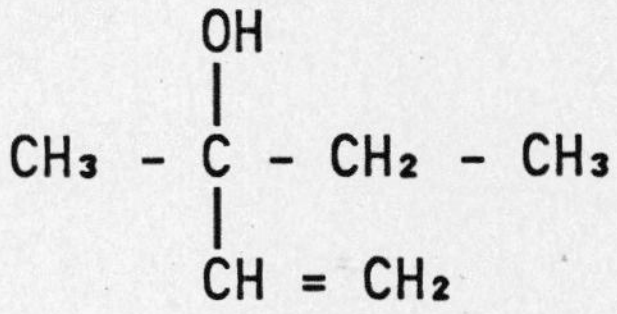

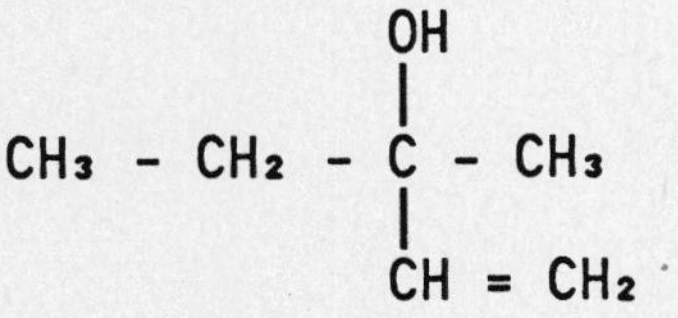

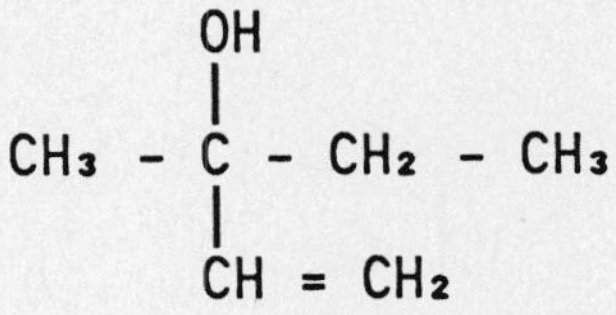

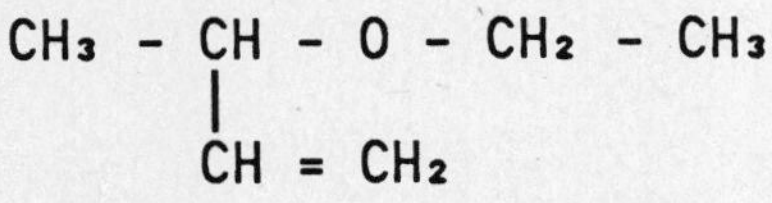

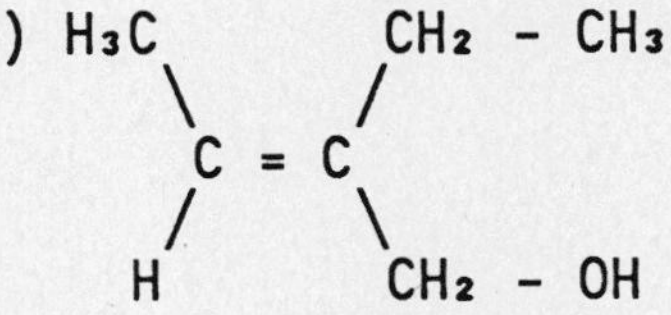

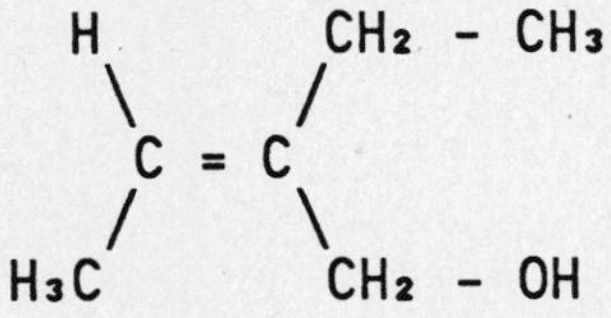

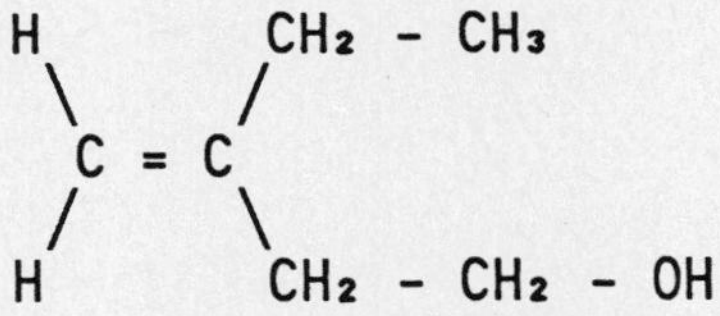

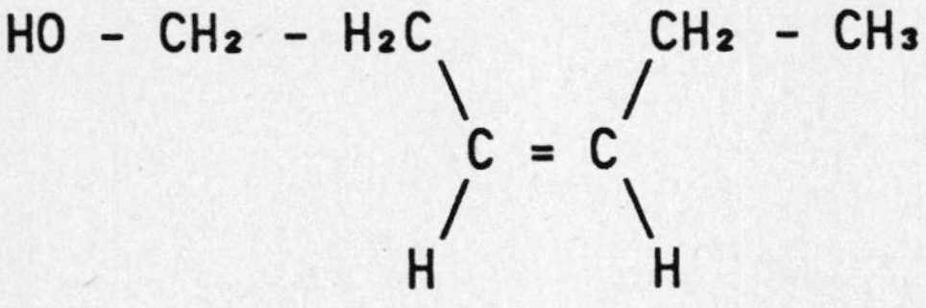

4 (34 points). Choose the response that best completes each statement.

(a) The products of the complete combustion of a hydrocarbon are

(i) CO_2 and H_2O (ii) CO_2 and H_2O_2

(iii) CO_2, O_3, and H_2O (iv) $C_{12}H_{22}O_{11}$

(b) Paraffin is a long-chain hydrocarbon. The solubility of paraffin would be expected to be highest in

(i) water (ii) ammonia

(iii) gasoline (iv) alcohol

(c) Which of the following is an alkene?

(i) $HC \equiv CH_2 - CH_3$ (ii) $CH_3 - CH_2 - CH_3$

(iii) CH_3— 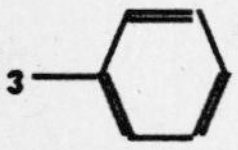 (iv) 

(d) Catalytic addition of hydrogen to an alkene produces

(i) an alkane (ii) an aromatic ring

(iii) a carbohydrate (iv) an alkyne

(e) The carbon-halogen bond in an alkyl halide is

(i) acidic

(ii) polar

(iii) easily removed to form the free halogen

(iv) all of these answers

(f) Methanol is commercially prepared by

(i) $CO(g) + 2H_2(g) \xrightarrow{\Delta} CH_3OH(g)$

(ii) $H_2C{=}CH_2 + H_2O \rightarrow CH_3CH_2OH$

(iii) fermentation of grains, fruits, etc.

(iv) none of these answers

(g) Which is *not* a reaction of alcohol?

(i) as a weak acid, e.g., $2ROH + 2Na \rightarrow 2RO^-Na^+ + H_2$

(ii) as a weak base, e.g., $ROH + HBr \rightarrow RBr + H_2O$

(iii) dehydration, e.g., $2ROH \rightarrow R{-}O{-}R + H_2O$

(iv) $ROH + NH_3 \rightarrow RNH_2 + H_2O$

(h) Aromatic amines

(i) are very weak bases because the unshared electron pair on the nitrogen is delocalized around the ring

(ii) readily lose one of the protons on the amine group and act as strong acids

(iii) can be reduced to the corresponding nitro group

(iv) none of these answers

(i) The hybridization of the carbon atom in a carbonyl group is

(i) *sp* (ii) sp^2

(iii) sp^3 (iv) sp^3d^2

(j) Aldehydes and ketones are prepared by

(i) oxidation of alcohols (ii) hydrolysis of alkyl halides

(iii) reduction of phenols (iv) saponification

(k) Unsubstituted monocarboxylic acids

(i) are prepared easily by oxidation of ketones

(ii) form "carbonite" salts

(iii) have pK_a values in the range of 4 to 5

(iv) are stronger bases than acids

(l) Esterification is the reaction between

(i) two alcohols (ii) a carboxylic acid and an alcohol

(iii) two carboxylic acids (iv) an ester and a hot aqueous base

(m) The correct name for the compound

```
                CH₃
                 |
CH₃ - CH₂ - C - Cl
                 |
                CH₃
```

is:

(i) chloropentane (ii) chlorodimethylpropane

(iii) 1-chloro-1,1-dimethylpropane (iv) 2-chloro-2-methylbutane

(n) An organic molecule has the empirical formula C_2H_4O. The molar mass is about 88 g. The molecular formula for this compound is

(i) $C_3H_4O_3$ (ii) C_2H_4O

(iii) $C_4H_8O_2$ (iv) $C_6H_{12}O_3$

(o) Which is *not* true?

(i) Oxidation of primary alcohols gives aldehydes.

(ii) Oxidation of secondary alcohols gives ketones.

(iii) Ketones are easily oxidized.

(iv) Aldehydes and ketones are commercially prepared by dehydrogenation reactions.

(p) Which of the following compounds will react with bromine?

(i)
```
        O
        ‖
CH₃ - C - OH
```

(ii)
```
H       H
 \     /
  C = C
 /     \
H       H
```

(iii) 

(iv) CCl_4

(q) The reaction between a carboxylic acid and an alcohol in the presence of a dehydrating agent produces an

    (i) aldehyde   (ii) ether

    (iii) anhydride   (iv) ester

5 (10 points). A hydrocarbon was separated from a mixture and analyzed. What is the empirical formula if it contained 85.7 mass % C? The approximate molecular mass was 55 u. What is the molecular formula?

6 (10 points). Write the various structural formulas and give the names for the six isomers having the molecular formula C_4H_8.

7 (9 points). Write chemical equations representing (a) the hydrolysis of methyl chloride to methyl alcohol, (b) the oxidation of methyl alcohol to formaldehyde, (c) the oxidation of formaldehyde to formic acid, and (d) the esterification of formic acid with methyl alcohol to give methyl formate.

8 (10 points). Calculate the mass of alcohol produced by the two-step process

$$H_2C=CH_2 + H_2SO_4 \rightarrow CH_3CH_2OSO_3H$$

$$CH_3CH_2OSO_3H + H_2O \rightarrow CH_3CH_2OH + H_2SO_4$$

from 10.0 g of $H_2C=CH_2$. Assume that the first process is 65 % efficient and the second reaction is 83 % efficient.

Answers to practice test

1. (a) iv, v, ix; (b) vi, viii; (c) ii, ix; (d) iii, vii, ix; (e) i, vii, viii, (f) ii, vii, ix; (g) ii, v, vii, ix.

2. (a) i; (b) iii; (c) ii, iv.

3. (a) vi, (b) iii, (c) xi, (d) viii, (e) iv, (f) xii, (g) ix, (h) x, (i) ii, (j) vii, (k) v, (1) i.

4. (a) i, (b) iii, (c) iv, (d) i, (e) ii, (f) i, (g) iv, (h) i, (i) ii, (j) i, (k) iii, (1) ii, (m) iv, (n) iii, (o) iii, (p) ii, (q) iv.

5. CH_2, C_4H_8.

6. $CH_3 - CH_2 - CH = CH_2$
    1-butene

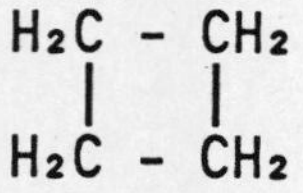

$$\begin{array}{cc} H_3C & H \\ \diagdown & \diagup \\ C = C \\ \diagup & \diagdown \\ H & CH_3 \end{array}$$
trans-2-butene

$$\begin{array}{cc} H_2C - CH_2 \\ | \quad\quad | \\ H_2C - CH_2 \end{array} \quad cyclobutane$$

$$H_2C - CH - CH_3$$
$$\diagdown \diagup$$
$$CH_2 \quad\quad methylcyclopropane$$

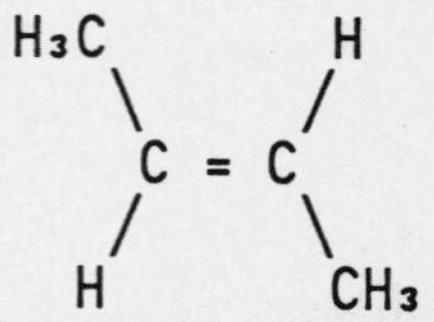

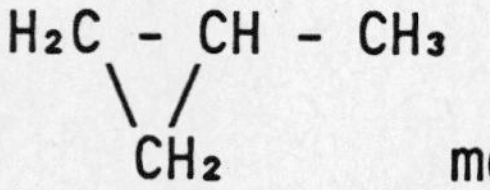

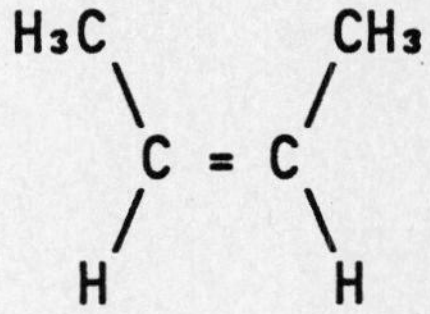

$$H_3C \qquad CH_3$$
$$\backslash \qquad /$$
$$C = C \qquad\qquad CH_3 - \underset{\underset{CH_3}{|}}{C} = CH_2 \quad \text{methylpropene}$$
$$/ \qquad \backslash$$
$$H \qquad H \quad cis\text{-2-butene}$$

7. (a) $CH_3Cl + H_2O \rightarrow CH_3OH + HCl$, (b) $CH_3OH \xrightarrow{\text{oxidation}} H_2CO$,

   (c) $H_2CO \xrightarrow{\text{oxidation}} HCOOH$, (d) $HCOOH + CH_3OH \rightarrow HCOOCH_3 + H_2O$.

8. 8.9 g.

ADDITIONAL NOTES

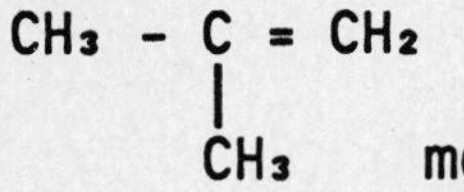

CHAPTER 33

ORGANIC POLYMERS AND BIOCHEMISTRY

CHAPTER OVERVIEW

The first part of the chapter discusses the structure and properties of polymers. The second part of the chapter is an introduction to some of the macromolecules important in biochemistry--carbohydrates, lipids, proteins, and nucleic acids.

COMPETENCIES

Definitions for the following terms should be learned:

- active site
- adenosine diphosphate (ADP)
- adenosine triphosphate (ATP)
- amino acid
- anabolism
- carbohydrate
- catabolism
- chain reaction (addition) polymerization
- copolymer
- denaturation
- deoxyribonucleic acid (DNA)
- elastomers
- enzyme
- glass transition temperature
- homopolymer
- hydrophilic
- hydrophobic
- lipid
- macromolecules
- metabolism
- monomers
- nucleic acid
- nucleotide
- peptide bond
- polypeptide
- protein
- ribonucleic acid (RNA)
- step reaction (condensation) polymerization
- substrate
- thermoplastic
- thermoset polymer

General concepts and skills that should be learned include

- the formation and properties of polymers
- the classes of molecules important to biochemistry and some of their descriptive chemistry

QUESTIONS

1. Complete the following sentences by filling in the various blanks with these terms: adenosine diphosphate, adenosine triphosphate, amino acids, anabolism, carbohydrates, catabolism, enzymes, lipids, metabolism, nucleic acids, nucleotides, peptide bonds, and proteins.

   (a)______ are simple sugars; (b)______ are nonpolymeric hydrophobic substances; (c)______ are long, unbranched polymer chains of (d)______ linked by (e)______; and (f)______ are polymers of (g)______--each of which consists of a five-carbon sugar, a nitrogen-containing base, and a phosphate group. The breakdown of large molecules ((h)______) and the synthesis of large molecules from simpler components ((i)______) are the two processes which make up the (j)______ of an organism. Organisms use the energy released by catabolic reactions to make (k)______ from (l)______ and phosphate, and (m)______ couple the reverse of this reaction to anabolic reactions so that the energy is available for the reactions to occur.

2. Complete the following sentences by filling in the various blanks with these terms: copolymer, glass transition temperature, homopolymer, macromolecules, monomers, thermoplastic, and thermoset.

   Polymers, or (a)______, are produced by bonding together large numbers of smaller molecules known as (b)______. A (c)______ is formed by the polymerization of a single type of monomer and a (d)______ by the polymerization of two or more different polymers. A (e)______ polymer softens when heated and resolidifies when cooled. A (f)______ plastic does not melt when heated--it is permanently rigid. At the (g)______, the flexibility of the polymer increases markedly.

3. Match the class of compound to the use or example given.

(a) sucrose	(i) amine
(b) steroid	(ii) carbohydrate
(c) alkaloid	(iii) enzyme
(d) catalyst	(iv) lipid

4. Choose the response that best completes each statement.

(a) A biochemical molecule containing an amine group and a carboxylic acid group is a

(i) protein (ii) carbohydrate
(iii) lipid (iv) nucleic acid

(b) The loss of water between a carboxylic acid group and an amine group

(i) gives a hemiacetal (ii) is a peptide bond
(iii) gives a glycosidic linkage (iv) is an ester linkage

(c) Carbohydrates containing many sugar monomers are

(i) polyhydroxyl aldehydes (ii) monosaccharides
(iii) disaccharides (iv) polysaccharides

(d) Most fats and oils are

(i) proteins (ii) carbohydrates
(iii) lipids (iv) nucleic acids

(e) Lipids that are esters of long-chain fatty acids and long-chain monohydroxyl alcohols are

(i) waxes (ii) steroids
(iii) terpenes (iv) alkaloids

(f) Which is *not* a part of a nucleotide?

(i) heterocyclic base that contains either a purine or a pyrimidine ring
(ii) a carboxylic acid group
(iii) a sugar
(iv) one or more phosphate groups

(g) The repeating unit in a nucleic acid is

(i) an amino acid (ii) a monosaccharide
(iii) a glyceride (iv) a nucleotide

(h) The type of linkage between the third carbon atom of the sugar of one nucleotide and the phosphate group of another nucleotide in a nucleic acid is classified as

(i) a peptide (ii) a glycoside
(iii) an ester (iv) none of these answers

(i) Deoxyribonucleic acid (DNA) is thought to be the chemical compound responsible for the process of heredity. The density of DNA is 1.1 g/cm^3 and the molar mass is estimated to be 6.0×10^8 g. What is the average volume occupied by one DNA molecule?

(i) 1.1×10^{-15} cm³ (ii) 9.1×10^{-16} cm³
(iii) 3.0×10^{-33} cm³ (iv) none of these answers

5. Distinguish clearly among (a) hydrophilic and hydrophobic molecules; (b) aldoses and ketones; (c) monosaccharides, disaccharides, and polysaccharides; (d) starch and cellulose; (e) condensation and hydrolysis; (f) fats and oils; (g) triglycerides and phospholipids; (h) fibrous proteins and globular proteins; (i) 1°, 2°, 3°, and 4° protein structure; (j) DNA and RNA; (k) replication, transcription, and translation; (l) anabolism and catabolism; (m) ADP and ATP.

6. What is the glass transition temperature? What is its significance in determining the applications for which a polymer can be used?

Answers to questions

1. (a) Carbohydrates, (b) lipids, (c) proteins, (d) amino acids, (e) peptide bonds, (f) nucleic acids, (g) nucleotides, (h) catabolism, (i) anabolism, (j) metabolism, (k) adenosine triphosphate, (l) adenosine diphosphate, (m) enzymes.
2. (a) macromolecules, (b) monomers, (c) homopolymer, (d) copolymer, (e) thermoplastic, (f) thermoset, (g) glass transition temperature.
3. (a) ii, (b) iv, (c) i, (d) iii.
4. (a) i, (b) ii, (c) iv, (d) iii, (e) i, (f) ii, (g) iv, (h) iii, (i) ii.
5. (a) *Hydrophilic molecules* are soluble in water; *hydrophobic molecules* are not. (b) *Aldoses* are polyhydroxyl aldehydes; *ketoses* are polyhydroxyl ketones. (c) *Monosaccharides* are carbohydrates consisting of single sugar monomers; *disaccharides* are composed of two simple sugars; and *polysaccharides* are polymers of many sugar monomers. (d) *Starch* is a hydrophilic polymer of α-D-glucose that is used for energy storage; *cellulose* is a hydrophobic polymer of β-D-glucose that is used for structural support. (e) *Condensation* is a reaction in which monomers are joined by elimination of a water molecule; *hydrolysis* is the reverse reaction, in which a bond between two monomers is broken by addition of the elements of water. (f) *Fats* are glycerides at room temperature; *oils* are liquids because of unsaturated fatty acid chains. (g) *Triglycerides* are esters formed by the condensation of the 3-carbon alcohol glycerol with three fatty acids; *phospholipids* have two of the glycerol hydroxyl

groups of glycerol esterified with fatty acids and the third esterified with phosphoric acid or a derivative of phosphoric acid. (h) *Fibrous proteins* consist of relatively extended polypeptide chains, arranged so that they intertwine with or lie parallel to neighboring chains; *globular proteins* consist of a polypeptide chain wound and folded into a compact but irregular shape. (i) *Primary* (1°) *protein structure* refers to the sequence of amino acids in the polypeptide chain; *secondary* (2°) *protein structure* refers to regular coiling or zigzagging of the polypeptide chains; *tertiary* (3°) *protein structure* refers to the three-dimensional folding of the polypeptide chain in a globular protein; and *quaternary* (4°) *protein structure* refers to the way in which several polypeptide chains are arranged and joined to one another. (j) *Deoxyribonucleic acid* (DNA) is a double-stranded nucleic acid which carries genetic information; *ribonucleic acid* (RNA) is involved in putting this information to work in the cell. (k) *Replication* is the process of each DNA strand having a new, complementary "strand" synthesized, resulting in two DNA molecules; *transcription* involves formation of a single-stranded RNA molecule from a DNA gene; and *translation* involves the process of protein synthesis in accordance with a nucleic acid blueprint. (l) *Anabolism* is the process of organisms synthesizing large molecules from simpler components; *catabolism* is the process of organisms breaking down large molecules into smaller, simpler ones. (m) *Adenosine diphosphate* (ADP) is a nucleotide consisting of the sugar ribose, the base adenine, and two phosphate groups; *adenosine triphosphate* (ATP) has the same structure, but with a third phosphate group.

6. At the glass transition temperature there is a dramatic increase in the flexibility of a polymer. Rigid plastics must have glass transition temperatures above the temperature at which they are used; polymers used in applications that require flexibility must have glass transition temperatures well below the temperature at which they will be used.

PRACTICE TEST

1 (15 points). Show the structure formula of the polymer formed from a monomer having the formula $CH_2 = CHX$. Name the polymers in which X is (a) CH_3, (b) Cl, and (c)  .

2 (15 points). Using the letters A and B to represent two different monomers, sketch the following: (a) a branched homopolymer, (b) a linear block copolymer, (c) a cross-linked homopolymer, (d) a branched graft copolymer, and (e) a linear random copolymer.

3 (15 points). Write the structural formula for α-D-glucose and β-D-glucose. A solution of D-glucose contains 36 % of the α form and 64 % of the β form. Calculate the concentration equilibrium constant for the formation of the β form from the α form.

4 (15 points). Write the structural formula for an α-amino acid in (a) a neutral solution, (b) an alkaline solution, and (c) an acidic solution.

5 (10 points). Show the structure of a tripeptide formed from glycine (H_2CNH_2-COOH).

6 (30 points). Choose the response that best completes each statement.

(a) A polymer that softens when it is heated, can be molded or extended, and hardens upon cooling is known as

(i) thermosetting (ii) thermoplastic

(iii) lipid (iv) vulcanized

(b) Which is *not* a reaction step in a chain reaction polymerization reaction?

(i) initiation (ii) elimination

(iii) termination (iv) propagation

(c) Which structure best represents natural rubber?

(i)

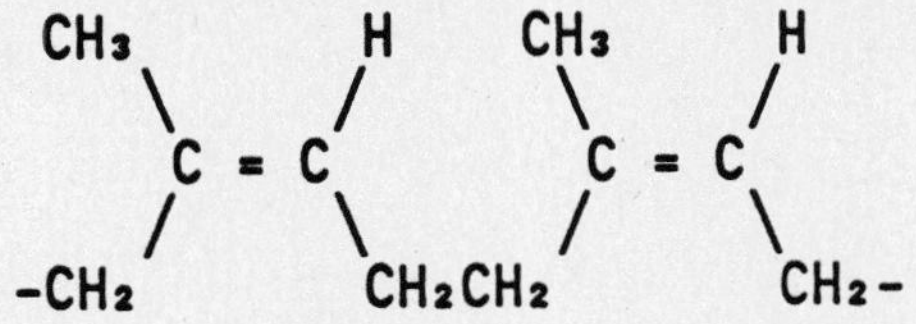

(ii)

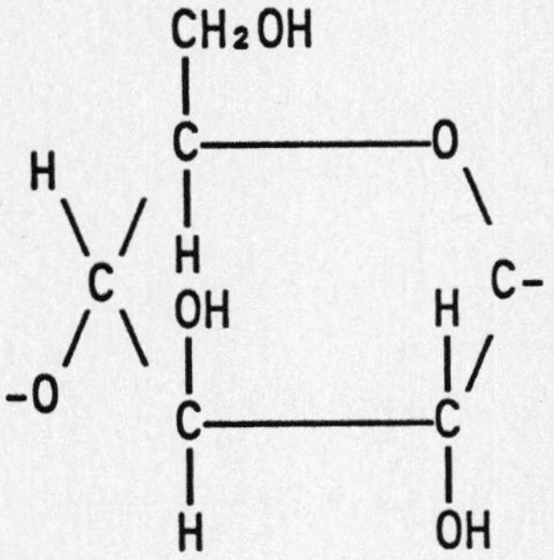

(iii)

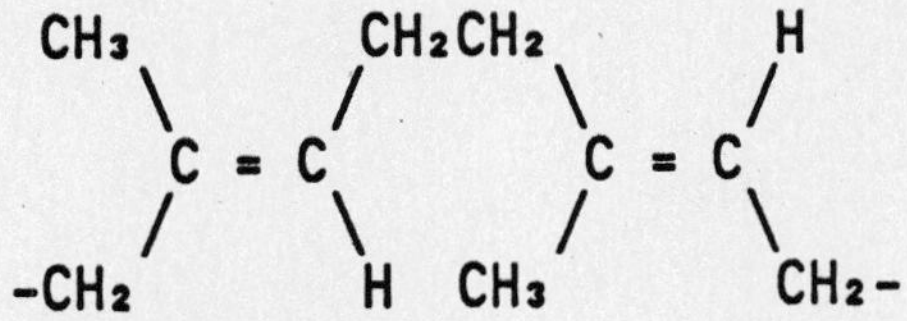

(iv) $-CH_2CH_2-$

(d) monosaccharides are

(i) lipids (ii) nucleic acids

(iii) α amino acids (iv) polyhydroxy aldehydes or ketones

(e) What is a principal function of polysaccharides?

(i) saponification (ii) formation of amino acids

(iii) energy storage (iv) catalyst

(f) Fats and oils are classified as

(i) lipids (ii) proteins

(iii) nucleic acids (iv) natural polymers

(g) Glycerides are

(i) disaccharides (ii) amino acids

(iii) branched copolymers (iv) esters of glycerol and fatty acids

(h) Which formula represents a polyunsaturated oil?

(i) $H_2C - O - CO - C_{14}H_{29}$
$HC - O - CO - C_{14}H_{29}$
$H_2C - O - PO_3-(CH_2)_2-N(CN_3)_3$

(ii) $H_2C - O - CO - C_{14}H_{29}$
$HC - O - CO - C_{12}H_{25}$
$H_2C - O - CO - C_{14}H_{27}$

(iii) $H_2C - O - CO - C_{12}H_{25}$
$HC - O - CO - C_{13}H_{27}$
$H_2C - O - CO - C_{14}H_{29}$

(iv) $H_2C - O - CO - C_{14}H_{27}$
$HC - O - CO - C_{14}H_{25}$
$H_2C - O - CO - C_{14}H_{25}$

(i) A molecule with the formula shown to the right would be classified as

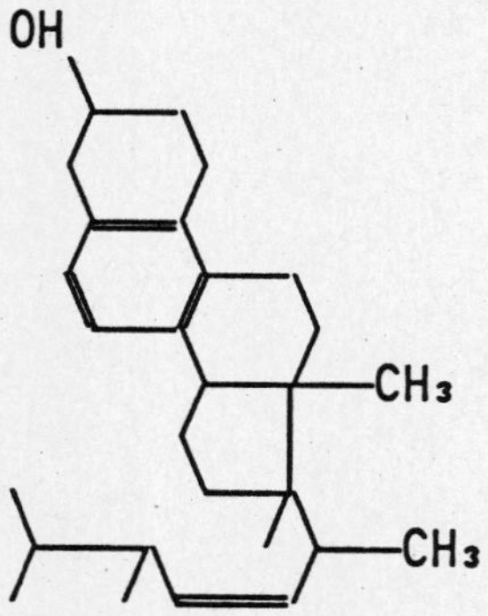

(i) a steroid

(iii) a phospholipid

(ii) a glyceride

(iv) a polysunsaturated oil

(j) The molecule or molecules with which an enzyme interacts is called a

(i) phospholipid (ii) steroid

(iii) substrate (iv) glyceride

(k) The repeating units of nucleic acids are

(i) α-amino acids (ii) nucleotides

(iii) monosaccharides (iv) codons

(l) Which nucleic acid carries genetic information?

(i) DNA (ii) RNA

(iii) ADP (iv) ATP

(m) The synthesizing of large molecules from simpler molecules by a living organism is

(i) catabolism (ii) anabolism
(iii) esterification (iv) glycerolisis

(n) Which is true?
- (i) ADP contains ribose, adenine, and three phosphate groups.
- (ii) ATP contains sucrose, alanine, and three phosphate groups.
- (iii) ADP has a higher energy content than ATP.
- (iv) ATP has a higher energy content than ADP.

(o) Which reaction is important in metabolism?
- (i) $6CO_2 + 6H_2O \rightarrow C_6H_{12}O_6 + 6O_2$
- (ii) $x\ H_2CCH_2 \rightarrow -(CH_2\text{-}CH_2)_{\overline{x}}$
- (iii) $ATP + H_2O \rightarrow ADP + P_i + H^+$
- (iv)

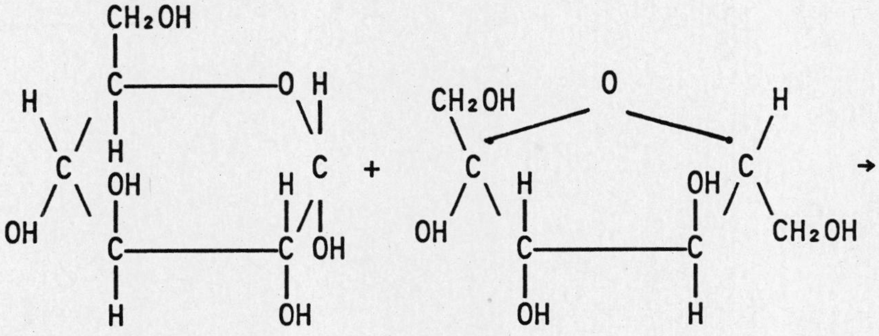

$\rightarrow$ sucrose + H_2O

Answers to practice test

1. $-\!\!\left(CH_2\underset{\displaystyle X}{\underset{|}{C}}H \right)\!\!-_n$; (a) polypropylene, (b) poly(vinyl chloride), (c) polystyrene.

2. (a)
```
A - A - A - A - A - A - A
        |       |
        A       A
        |
        A
```
(b) A-A-A-A-B-B-B-B-B-B-A-A-A-B-B-B-A-A-A

(c)
```
A - A - A - A - A - A - A - A
        |           |
        A           A
        |           |
A - A - A - A - A - A - A - A
```
(d)
```
A - A - A - A - A - A - A
    |               |
    B               B
    |               |
    B               B
                    |
                    B
```
(e) A-A-A-B-A-A-B-B-A-B-A-B-B-A

3.

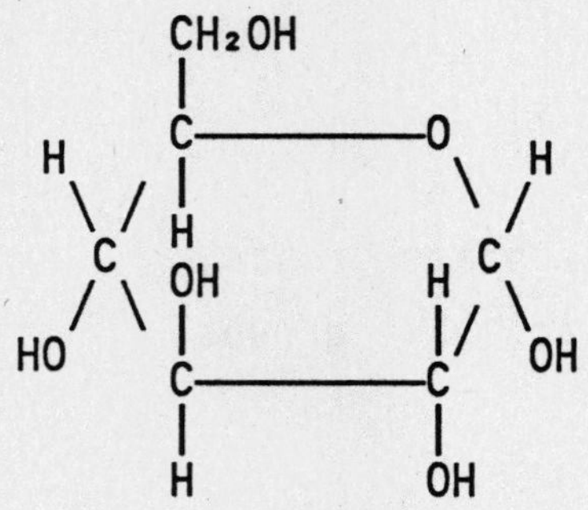

α form   β form

$K = 1.8$

4. (a)

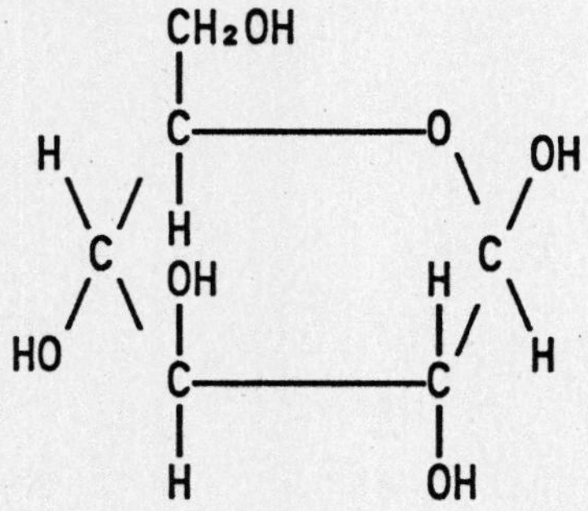

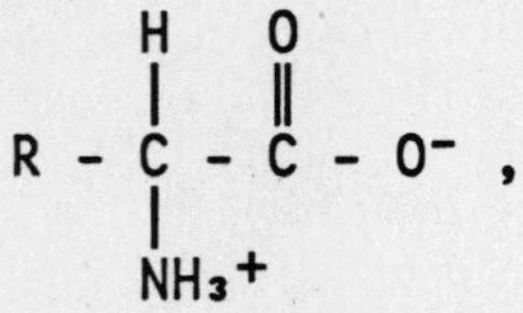

$R - \underset{\underset{NH_3^+}{|}}{\overset{\overset{H}{|}}{C}} - \underset{}{\overset{\overset{O}{\parallel}}{C}} - O^-$,

(b)

$R - \underset{\underset{NH_2}{|}}{\overset{\overset{H}{|}}{C}} - \underset{}{\overset{\overset{O}{\parallel}}{C}} - O^-$,

(c)

$R - \underset{\underset{NH_3}{|}}{\overset{\overset{H}{|}}{C}} - \underset{}{\overset{\overset{O}{\parallel}}{C}} - OH$

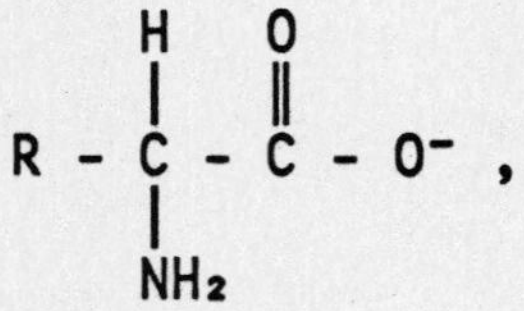

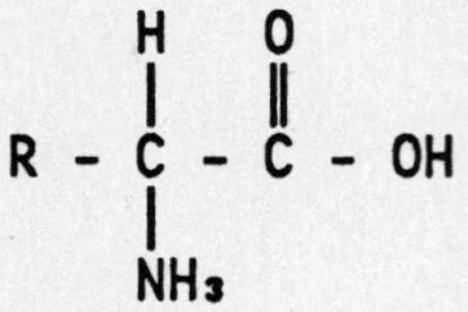

5.

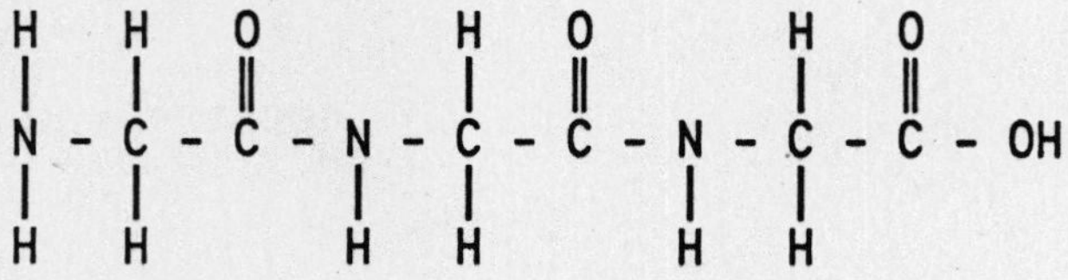

6. (a) ii, (b) ii, (c) i, (d) iv, (e) iii, (f) i, (g) iv, (h) iv, (i) i, (j) iii, (k) ii, (l) i, (m) ii, (n) iv, (o) iii.

ADDITIONAL NOTES

CHAPTER 34

INORGANIC QUALITATIVE ANALYSIS--CHEMICAL PRINCIPLES REVIEWED

CHAPTER OVERVIEW

This chapter opens with a brief introduction to anion and cation analysis. The remainder of the chapter is devoted to a review of equilibrium reactions--for acids and bases, redox reactions, complex formation, and the precipitation and dissolution of solids. This review concentrates on how the different types of equilibrium reactions are utilized in the qualitative analysis for cations.

COMPETENCIES

Definitions for the following terms should be learned:

- centrifugate
- decanting
- filtrate
- group reagent
- reagent
- residue
- supernatant solution
- unknown

General concepts and skills that should be learned include

- using equilibrium conditions in inorganic qualitative analysis

QUESTIONS

1. Complete the following sentences given by filling in the various with these terms: centrifugate, decanted, dissolution, excess, filtrate, group reagent, precipitate, reagent, residue, supernatant solution, and unknown.

The following steps were carried out on a cation (a)______ during a laboratory period. Dilute HCl was added as a (b)______ to precipitate the cations as a group. To separate the solid, the reaction mixture was either centrifuged and the (c)______ was poured off ((d)______) or the mixture was filtered and the (e)______ passed through the filter. After the (f)______ was washed, part of the precipitate was subjected to (g)______ by heating the solid in contact with fresh water. The (h)______ was shown to contain Pb^{2+} by using CrO_4^{2-} as a (i) ________ to form $PbCrO_4$. The remaining (j) ________ dissolved in (k) ________ aqueous ammonia and was shown to contain Ag^+.

2. Match the correct equation to each process.

(a) dissolution by complexation	(i) $H_2S(aq) \rightleftharpoons 2H^+ + S^{2-}$
(b) controlled hydroxide precipitation	(ii) $Fe(CN)_2(s) + 4CN^- \rightarrow [Fe(CN)_6]^{4-}$
(c) dissolution by the formation of a weak acid	(iii) $NH_4^+ + CO_3^{2-} \rightleftharpoons NH_3(aq) + HCO_3^-$
	(iv) $SnS_2(s) + 4H^+ \rightarrow Sn^{4+} + 2H_2S(aq)$
(d) controlled carbonate precipitation	(v) $NH_3(aq) + H_2O(l) \rightleftharpoons NH_4^+ + OH^-$
(e) dissolution by redox	(vi) $3CuS(s) + 2NO_3^- + 8H^+ \rightarrow$
(f) controlled sulfide precipitation	$3Cu^{2+} + 3S(s) + 2NO(g) + 4H_2O(l)$

3. What is the general principle that governs competing equilibria?

Answers to questions

1. (a) unknown, (b) group reagent, (c) centrifugate or supernatant solution, (d) decanted, (e) filtrate, (f) precipitate, (g) dissolution, (h) centrifugate or supernatant solution, (i) reagent, (j) residue, (k) excess.
2. (a) ii, (b) v, (c) iv, (d) iii, (e) vi, (f) i.
3. Competing equilibria are governed by a very important general principle: An ion will form the precipitate, complex ion, or other species that is in equilibrium with the smallest concentration of that ion in solution.

PRACTICE TEST

1 (10 points). Write net ionic equations for
  (a) the precipitation of Hg^{2+} by $H_2S(g)$
  (b) the dissolution of AgCl by aqueous ammonia
  (c) water reacting as an acid with CO_3^{2-}
  (d) the amphoteric behavior of $Al(OH)_3(s)$ in the presence of H^+
  (e) the amphoteric behavior of $Al(OH)_3(s)$ in the presence of OH^-
  (f) the reduction of NO_3^- to NO(g), the oxidation of $H_2S(aq)$ to S, and the overall redox equation
  (g) the production of H_2S from thioacetamide in acidic solution
  (h) the controlled precipitation of carbonates by using NH_3, NH_4^+ and CO_3^{2-}

2 (10 points). What is the concentration of Fe^{2+} in equilibrium with a 0.10 M solution of CN^-? Assume that $[Fe(CN)_6^{4-}]$ = 0.010 mol/L. $K_d = 1.3 \times 10^{-37}$ for the complex.

3 (20 points). What value of $[CN^-]$ is needed to form a precipitate with Ag^+? Assume that $[Ag^+] = 1.7 \times 10^{-4}$ mol/L. $K_{sp} = 2.3 \times 10^{-16}$ for AgCN(s). What value of $[CN^-]$ is needed to dissolve the AgCN as $[Ag(CN)_2]^-$? $K = 4 \times 10^{-7}$ for the equation

$$[Ag(CN)_2]^- \rightleftharpoons AgCN(s) + CN^-$$

4 (20 points). What pH would serve to separate Cu^{2+} from Pb^{2+}? Assume that both ions are at a concentration of 0.010 mol/L. $K_{sp} = 1.3 \times 10^{-20}$ for $Cu(OH)_2$ and 1.2×10^{-15} for $Pb(OH)_2$.

5 (20 points). The equilibrium constant for the equation

$$NH_4^+ + CO_3^{2-} \rightleftharpoons NH_3(aq) + HCO_3^-$$

is 11.7. Will $BaCO_3$ precipitate from a 0.0010 M solution of Ba^{2+} that also contains $[NH_4^+] = 0.5$ mol/L, $[NH_3] = 0.5$ mol/L, and $[HCO_3^-] = 0.003$ mol/L? $K_{sp} = 2.0 \times 10^{-9}$ for $BaCO_3$.

6 (20 points). Would adding Sn^{4+} to an aqueous solution of H_2S cause the H_2S to be oxidized to S? $E° = 0.15$ V for Sn^{4+}/Sn^{2+} and $E° = 0.142$ V for S/H_2S. Would S^{2-} be oxidized to S? $E° = -0.447$ V for S/S^{2-}.

Answers to practice test

1. (a) $Hg^{2+} + H_2S(g) \rightarrow HgS(s) + 2H^+$; (b) $AgCl(s) + 2NH_3(aq) \rightarrow [Ag(NH_3)_2]^+ + Cl^-$; (c) $CO_3^{2-} + H_2O(l) \rightarrow HCO_3^- + OH^-$; (d) $Al(OH)_3 + 3H^+ \rightarrow Al^{3+} + 3H_2O(l)$; (e) $Al(OH)_3(s) + OH^- \rightarrow [Al(OH)_4]^-$; (f) $NO_3^- + 4H^+ + 3e^- \rightarrow NO(aq) + 2H_2O(l)$, $H_2S(aq) \rightarrow S(s) + 2H^+ + 2e^-$, $3H_2S(aq) + 2NO_3^- + 2H^+ \rightarrow 2NO(aq) + 4H_2O(l) + 3S(s)$; (g) $CH_3CSNH_2(aq) + H_3O^+ + H_2O(l) \rightarrow CH_3COOH(aq) + NH_4^+ + H_2S(aq)$; (h) $NH_4^+ + CO_3^{2-} \rightarrow NH_3(aq) + HCO_3^-$.
2. 1.3×10^{-33} mol/L.
3. 1.4×10^{-12} mol/L, 7×10^{-11} mol/L.
4. between 5.04 and 7.54.
5. $[Ba^{2+}][CO_3^{2-}] = 3 \times 10^{-7} > K_{sp}$, yes.
6. $E° = 0.01$ V for H_2S oxidation, yes (a little); $E° = 0.60$ V for S^{2-} oxidation, yes.

ADDITIONAL NOTES

CHAPTER 35

INORGANIC QUALITATIVE ANALYSIS--ANIONS, CATIONS, AND THE SCHEME

CHAPTER OVERVIEW

This chapter introduces the analysis scheme and flow charts for the identification of 11 anions and 22 cations. The general properties of the anions, the preliminary anion tests, and the specific anion tests are discussed. The general properties of the cation groups are briefly discussed. In conjunction with flow charts, chemical and practical comments are given about each step in the analysis.

COMPETENCIES

General concepts and skills that should be learned include

- the reactions involved in the preliminary anion tests and the specific anion tests
- the general chemical properties of each cation group that are important for its separation from the other groups
- the reactions involved in cation group separations, intermediate separation steps of cations, and cation confirmatory tests
- using a flow chart

QUESTIONS

1. Match the specific test to the respective anion.

(a) Br^-	(i) add dilute HCl to solid sample; test gas evolved with $Pb(CH_3COO)_2$
(b) CO_3^{2-}	
(c) Cl^-	(ii) add dilute H_2SO_4; test gas evolved with HNO_3, $BaCl_2$, and $KMnO_4$
(d) CrO_4^{2-}	
(e) I^-	(iii) add Zn, dilute H_2O_2, and dilute H_2SO_4; test gas evolved with $Ba(OH)_2$
(f) NO_2^-	
(g) NO_3^-	(iv) add dilute H_2SO_4 and $FeSO_4$

(h) PO_4^{3-} (v) add chlorine water and CCl_4

(i) S^{2-} (vi) add $AgNO_3$; add aqueous NH_3

(j) SO_3^{2-} (vii) add ammonium molybdate - HNO_3

(k) SO_4^{2-} (viii) add acetic acid and barium acetate

(ix) add dilute H_2SO_4 and $FeSO_4$; add concentrated H_2SO_4 to form a second layer

(x) add dilute HCl and $BaCl_2$

2. Match the group reagent to the respective cation group.

(a) Cation Group III (i) $H_2S(aq)$ with dilute acid

(b) Cation Group V (ii) $H_2S(aq)$ with NH_3/NH_4Cl buffer

(c) Cation Group I (iii) HCl(aq)

(d) Cation Group IV (iv) none

(e) Cation Group II (v) $NH_3/(NH_4)_2CO_3$

3. Match the reagent used for confirmatory tests to the respective cation.

(a) Hg^{2+} (i) SO_4^{2-}

(b) Zn^{2+} (ii) S^{2-}

(c) Cu^{2+} (iii) $NaBiO_3$ and H^+

(d) Co^{2+} (iv) Sn^{2+} and Cl^- or Cu

(e) Ni^{2+} (v) K^+ and $[Fe(CN)_6]^{4-}$ or NCS^-

(f) Ba^{2+} (vi) $Ba(CH_3COO)_2$ or H_2O_2 and HNO_3

(g) Pb^{2+} (vii) "HDMG" and CH_3COOH

(h) Fe^{3+} (viii) NH_3

(i) K^+ (ix) $C_2O_4^{2-}$ or flame test

(j) Al^{3+} (x) aluminon and CH_3COOH/CH_3COO^-

(k) Ca^{2+} (xi) NH_3 or $[Fe(CN)_6]^{4-}$ or NCS^- and $C_5H_5N(aq)$

(l) Cr^{3+} (xii) NCS^- or KNO_2

(m) Mn^{2+} (xiii) CrO_4^{2-} or flame test

(n) Hg_2^{2+} (xiv) p-nitrobenzeneazoresorcinol

(o) Mg^{2+} (xv) $[B(C_6H_5)_4]^-$

(p) Na^+ (xvi) Zn^{2+}, UO_2^+, and CH_3COO^-

4. Choose the response that best completes each statement.

(a) The development of a brown to black color in a solution of $MnCl_2$ in concentrated HCl indicates the presence of

(i) oxidizing anions (ii) reducing anions

(iii) unstable oxoacids (iv) anions that form Ag^+ precipitates

(b) The appearance of a dark blue suspension or precipitate in a solution of $FeCl_3$, $K_3[Fe(CN)_6]$, and dilute HCl indicates the presence of

(i) oxidizing anions

(ii) reducing anions

(iii) anions that form Ba^{2+} precipitates

(iv) water-insoluble carbonates

(c) The use of concentrated H_2SO_4 on solid samples during the anion preliminary tests is based on the ability of H_2SO_4 to be

(i) amphoteric

(ii) a strong acid

(iii) a strong oxidizing agent

(iv) a strong oxidizing agent and a strong acid

(d) The characteristic heterogeneous equilibria reactions of the anions are investigated in the preliminary tests by using

(i) H_2SO_4 (ii) $AgNO_3$

(iii) $Ba(NO_3)_2$ (iv) $HClO_4$

(e) A large excess of Cl^- must be avoided during the precipitation of Cation Group I because

(i) $PbCl_2$ is soluble in warm water

(ii) Cl^- is an excellent oxidizing agent

(iii) AgCl and $PbCl_2$ form soluble chloro complexes

(iv) $HgCl_2$ might precipitate

(f) An acidic S^{2-} solution is used to precipitate Cation Group II so that

(i) Cation Group I does not coprecipitate

(ii) only the sulfides having very low K_{sp} values will precipitate

(iii) the more stable β-forms of the sulfides do not form

(iv) $Hg_2{}^{2+}$ does not undergo disproportionation

(g) The alkaline S^{2-} precipitation for Cation Group III results in

(i) the formation of both insoluble hydroxides and sulfides

(ii) oxidation of Cr^{3+} to $CrO_4{}^{2-}$

(iii) the disproportionation of Fe^{2+}

(iv) the formation of polysulfido complexes

(h) The $NH_3/(NH_4)_2CO_3$ buffer used in precipitating Cation Group IV serves to

(i) keep $Mg(OH)_2$ from forming

(ii) eliminate the accumulated S^{2-}

(iii) act also as a confirmatory test
(iv) coprecipitate Cation Groups I to III

5. A sample of Nichrome wire was dissolved in acid and subjected to analysis. The following results were obtained:
   (a) no precipitate was formed in the presence of Cl^-;
   (b) no precipitate was formed with acidic thioacetamide;
   (c) a dark precipitate A formed in alkaline thioacetamide, leaving supernatant Z;
   (d) subsequent treatment of the supernatant Z indicated no other ions were present;
   (e) treatment of A with HCl caused part of the precipitate to dissolve, and the remainder of the precipitate dissolved in HNO_3, forming solution B;
   (f) addition of NaOH and H_2O_2 to B gave a yellow solution C and dark precipitate D;
   (g) precipitate D dissolved in a nitric acid-hydrogen peroxide mixture to give solution E;
   (h) solution E upon treatment with solid potassium chlorate gave no precipitate;
   (i) upon addition of NH_3 to E, a red-brown precipitate F formed, leaving a solution G;
   (j) HCl was added to precipitate F and the resulting solution was divided into two portions--upon addition of KNCS to one portion, a blood-red complex formed and upon addition of $K_4[Fe(CN)_6]$ to the other portion, a dark blue precipitate formed;
   (k) solution G was divided into two portions--upon addition of HDMG, a bright red precipitate formed and upon addition of HCl, NaF, amyl alcohol-ether, and NH_4NCS, no color was observed;
   (l) solution C changed to a red-orange solution Y upon addition of HCl;
   (m) addition of excess NH_3 to Y did not produce a precipitate;
   (n) addition of barium acetate to Y produced a yellow precipitate and a solution H; and
   (o) addition of thioacetamide to H did not produce a precipitate.

   Identify the metals that make up this alloy.

6. A sample of metal used in typesetting was dissolved in acid and subjected to analysis. The following results were obtained:

(a) a white precipitate A formed in the presence of Cl^-;
(b) a precipitate B formed with acidic thioacetamide, leaving supernatant C;
(c) subsequent treatment of C indicated no other ions present;
(d) precipitate A dissolved in hot water, forming solution D;
(e) D was separated into two portions--a white precipitate formed upon addition of H_2SO_4 to one portion and a yellow precipitate formed upon addition of K_2CrO_4 to the second portion;
(f) most of precipitate B dissolved upon treatment with NH_3 and thioacetamide to form a yellow solution E and precipitate F;
(g) solid sulfur was removed from solution E by using acetic acid and hydrochloric acid;
(h) the liquid was divided into two portions--one portion produced a white precipitate upon addition of Al and $HgCl_2$, and the other portion produced an orange precipitate upon addition of NH_3, acetic acid, and $Na_2S_2O_3$;
(i) precipitate F readily dissolved in HNO_3 to form a colorless solution G; and
(j) G formed a white precipitate upon addition of H_2SO_4.

Identify the metals that make up this alloy.

7. A sample of Monel metal was dissolved in acid and subjected to analysis. The following results were obtained:
(a) no precipitate formed upon addition of Cl^-;
(b) a dark precipitate A formed in acidic thioacetamide, leaving supernatant B;
(c) a dark precipitate C formed upon treatment of B with alkaline thioacetamide, leaving supernatant D;
(d) subsequent treatment of D indicated the presence of no additional ions;
(e) precipitate A did not dissolve in ammoniacal thioacetamide, but dissolved in HNO_3 to form a blue solution E;
(f) addition of H_2SO_4 to E did not produce a precipitate, but addition of NH_3 gave a deep blue solution F;
(g) addition of acetic acid and $K_4[Fe(CN)_6]$ to F gave a red precipitate;
(h) precipitate C partially dissolved in HCl and completely dissolved in HNO_3 to give solution G;
(i) addition of NaOH and hydrogen peroxide to G gave a dark precipitate I and a solution H;

(j) further tests on H indicated that no additional ions were present;
(k) precipitate I dissolved in HNO_3 and H_2O_2 to give solution J;
(l) J did not produce a precipitate upon addition of solid $KClO_3$;
(m) addition of NH_3 to J gave a red-brown precipitate K and a blue-green solution L;
(n) after addition of HCl to K and division of the resulting solution into two parts, a blood-red complex was formed upon addition of KNCS to one portion and a dark blue precipitate was formed upon addition of $K_4[Fe(CN)_6]$ to the other portion; and
(o) solution L was divided into two portions and gave a bright red precipitate when acetic acid and HDMG were added to one portion and did not give a colored complex when HCl, NaF, amyl alcohol, ether, and NH_4NCS were added to the other portion.

Identify the metals that make up this alloy.

8. A sample of "hardware bronze" was dissolved in acid and subjected to analysis. The following results were obtained:
(a) a white precipitate A formed upon addition of Cl^-, leaving supernatant B;
(b) a precipitate C formed upon treatment of B in acidic thioacetamide, leaving supernatant D;
(c) a white precipitate E formed upon treatment of D in alkaline thioacetamide, leaving supernatant F;
(d) subsequent treatment of F indicated the presence of no additional ions;
(e) precipitate A dissolved in hot water--the resulting solution was split into two portions and tested with H_2SO_4 and K_2CrO_4, which resulted in a white precipitate and a yellow precipitate, respectively;
(f) precipitate C did not dissolve in ammoniacal thioacetamide;
(g) precipitate C readily dissolved in HNO_3 to form a blue solution G;
(h) addition of H_2SO_4 to G gave a white precipitate and the supernatant remained blue;
(i) addition of NH_3 to the blue supernatant resulted in a deep blue color, and when the supernatant was split into two portions, one of which was treated with $K_4[Fe(CN)_6]$ and C_5H_5N and the other with NH_4NCS, a red precipitate and a green precipitate, respectively, were formed;
(j) precipitate E dissolved in HCl to give solution H;
(k) no precipitate was formed from H upon addition of NaOH and H_2O_2;

(l) the colorless supernatant did not form a precipitate upon addition of HCl or NH_3 or upon addition of barium acetate, but a white precipitate was formed upon the addition of thioacetamide; and
(m) the white precipitate dissolved upon addition of HCl.
Identify the metals that make up this alloy.

9. A sample of silver solder was dissolved in acid and subjected to analysis. The following results were obtained:
(a) a white precipitate A formed upon addition of Cl^-, leaving supernatant B;
(b) a dark precipitate C formed from B in acidic thioacetamide, leaving supernatant D;
(c) a white precipitate E formed from D in alkaline thioacetamide, leaving supernatant F;
(d) further treatment of F indicated that no additional ions were present;
(e) precipitate A did not dissolve in hot water, but readily dissolved in NH_3 to give solution G;
(f) a white precipitate formed from G upon addition of HNO_3;
(g) precipitate C did not dissolve in ammoniacal thioacetamide, but dissolved in HNO_3 to give a blue solution H;
(h) addition of H_2SO_4 to H did not give a precipitate, but addition of NH_3 resulted in a dark blue solution I;
(i) solution I was separated into two portions after the addition of acetic acid--one portion gave a red precipitate upon addition of $K_4[Fe(CN)_6]$ and the other portion gave a green precipitate upon the addition of NH_4NCS;
(j) precipitate E dissolved in HCl, giving solution J;
(k) no precipitate formed from J upon addition of NaOH and H_2O_2;
(l) no precipitate formed from J upon addition of HCl or NH_3;
(m) no precipitate formed from J upon addition of barium acetate, but a white precipitate formed upon addition of thioacetamide; and
(n) the white precipitate dissolved upon addition of HCl.
Identify the metals that make up the alloy.

Answers to questions

1. (a) v, (b) iii, (c) vi, (d) viii, (e) v, (f) iv, (g) ix, (h) vii, (i) i, (j) ii, (k) x.
2. (a) ii, (b) iv, (c) iii, (d) v, (e) i.

3. (a) iv, (b) ii, (c) xi, (d) xii, (e) vii, (f) xiii, (g) i, (h) v, (i) xv, (j) x, (k) ix, (l) vi, (m) iii, (n) viii, (o) xiv, (p) xvi.
4. (a) i, (b) ii, (c) iv, (d) ii, (e) iii, (f) ii, (g) i, (h) i.
5. Fe, Cr, Ni.
6. Pb, Sb, Sn.
7. Ni, Cu, Fe.
8. Cu, Zn, Pb.
9. Ag, Cu, Zn.

PRACTICE TEST

1 (10 points). Choose the anions that would form a precipitate with Ba^{2+} in a slightly acidic solution: (i) SO_3^{2-}, (ii) CO_3^{2-}, (iii) CrO_4^{2-}, (iv) PO_4^{3-}, and (v) SO_4^{2-}. Write net ionic equations illustrating how each of these anions undergo reaction in the test solution.

2 (10 points). Choose the anions that would form a brown to black color when added to a solution of $MnCl_2$ dissolved in concentrated HCl: (i) NO_2^-, (ii) NO_3^-, (iii) S^{2-}, and (iv) I^-. Write reasonable chemical equations representing the formation of the dark color.

3 (15 points). An unknown was known to contain a combination of CrO_4^{2-}, NO_3^-, Cl^-, and/or Br^- ion(s). Describe the specific tests that could be used to determine which anion(s) could be present. Write chemical equations for the various reactions.

4 (15 points). Write chemical equations for the reactions described below:
(a) disproportionation of Hg_2^{2+}
(b) the dissolution of AgCl(s) by NH_3 and confirmatory test that uses H^+
(c) the dissolution of PbS(s) by oxidation of S^{2-} with HNO_3(aq)
(d) the dissolution of $PbSO_4$(s) by OH^- as a complex
(e) the confirmatory test for Mn^{2+} that uses $NaBiO_3$ in acidic media
(f) the confirmatory tests for Fe^{3+} that uses $K_4[Fe(CN)_6]$ and NCS^-
(g) the reduction of $Co(OH)_3$ in acidic H_2O_2
(h) the confirmatory test for Na^+ that uses CH_3COO^-, Zn^{2+}, and UO_2^+

5 (10 points). Match the group precipitating reagent to the respective cation group.

(a) Zn^{2+}, Mn^{2+}, Fe^{2+} or Fe^{3+} Co^{2+}, Ni^{2+}, Al^{3+}, Cr^{3+}	(i) none
	(ii) $NH_3/(NH_4)_2CO_3$
(b) Hg_2^{2+}, Pb^{2+}, Ag^+	(iii) HCl(aq)
(c) Mg^{2+}, Na^+, K^+, NH_4^+	(iv) acidic H_2S(aq)
(d) Ca^{2+}, Ba^{2+}	(v) H_2S(aq) with NH_3/NH_4Cl buffer
(e) Hg^{2+}, Pb^{2+}, Cu^{2+}, Sn^{2+} or Sn^{4+}, Sb^{3+} or SbO^+	

6 (20 points). Choose the response that best completes each statement.

(a) The confirmatory test for Hg_2^{2+} that uses NH_3 is based on
(i) oxidation-reduction (ii) complexation of Hg_2^{2+}
(iii) a buffered precipitation (iv) acid-base neutralization

(b) The insoluble hydroxide of lead(II)
(i) is formed during the precipitation of Cation Group I
(ii) dissolves in strongly alkaline solutions
(iii) serves as the confirmatory test for Pb^{2+}
(iv) is dark brown

(c) During the group precipitation of Cation Group III, Fe^{2+} and Fe^{3+} precipitate, respectively, as the
(i) sulfides (ii) hydroxides
(iii) hydroxide and sulfide (iv) carbonates

(d) A blue to blue green complex of NCS^- confirms the presence of
(i) Hg^{2+} (ii) Cu^{2+}
(iii) Fe^{3+} (iv) Co^{2+}

(e) To separate and identify the ions in the mixture Ag^+, Cu^{2+}, and Mg^{2+}, the reagents S^{2-}, Cl^- and OH^- could be used. The correct order for addition is
(i) Cl^-, S^{2-}, OH^- (ii) Cl^-, OH^-, S^{2-}
(iii) S^{2-}, Cl^-, OH^- (iv) OH^-, S^{2-}, Cl^-

(f) An unlabeled reagent bottle in a laboratory contained either aluminum nitrate or lead nitrate. Which chemical test could be used to identify which cation was present?
(i) dilute H_2SO_4; $Al_2(SO_4)_3$ will precipitate while $PbSO_4$ will not
(ii) flame test; Al^{3+} produces a green flame while Pb^{2+} is colorless
(iii) dilute HCl; $PbCl_2$ precipitates while $AlCl_3$ does not
(iv) dilute CH_3COOH; $Pb(CH_3COO)_2$ precipitates while $Al(CH_3COO)_3$ does not

(g) Identify which reagent will separate Hg_2^{2+} from Zn^{2+}.

(i) $HNO_2(aq)$ (ii) $HCl(aq)$

(iii) $NaOH(aq)$ (iv) $NaOH(aq, xs)$

(h) Identify which reagent will separate Al^{3+} from Ba^{2+}.

(i) $HNO_2(aq)$ (ii) $HCl(aq)$

(iii) $NaOH(aq)$ (iv) $NaOH(aq, xs)$

(i) Which of the following statements is *not* true?

(i) $Fe(NO_3)_3$ and $Cr(NO_3)_3$ are water soluble.

(ii) $PbCl_2$ is more soluble in water than $BaCl_2$.

(iii) $CuSO_4$ is more soluble in water than $PbSO_4$.

(iv) ZnS is insoluble in water, whereas NaS is soluble.

(j) Which anion produces a colorless, odoriferous gas upon addition of concentrated H_2SO_4, but does not give a Prussian blue precipitate?

(i) CO_3^{2-} (ii) Cl^-

(iii) SO_3^{2-} (iv) S^{2-}

7 (20 points). The following procedures and observations were made on an unknown sample:

(a) no precipitate formed upon addition of $HCl(aq)$ to a solution of the unknown;

(b) precipitate A was produced when the solution from (a) was made alkaline by using a buffer solution of NH_3/NH_4^+;

(c) A was separated, leaving solution B;

(d) A dissolved completely in a strong alkaline solution, producing solution C;

(e) C was acidified and treated with aluminon by using a buffer solution of CH_3COO^-/CH_3COOH, giving a red solid;

(f) B was divided into two portions, D and E;

(g) D did not produce NH_3 upon addition of OH^- and heat;

(h) E was divided into two portions, F and G; and

(i) F was treated with C_2H_5OH and Na_2HPO_4, giving a white precipitate which dissolved in H^+ and produced a blue lake upon treatment with "S and O" reagent.

Using the flow chart on the next page, identify the cations present in the unknown. What additional tests, if any, would be needed to complete the analysis?

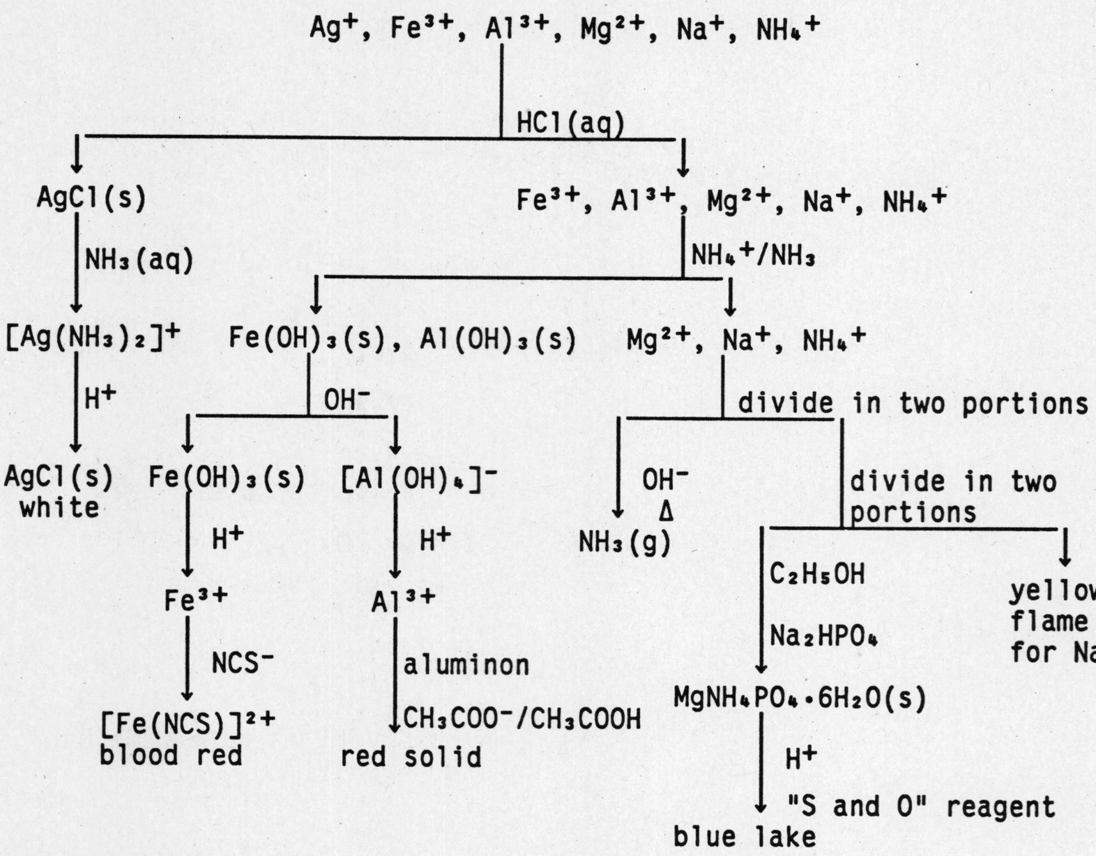

Answers to practice test

1. $SO_4^{2-} + Ba^{2+} \rightarrow BaSO_4(s)$, $SO_3^{2-} + H^+ \rightarrow HSO_3^-$, $CO_3^{2-} + H^+ \rightarrow HCO_3^-$, $CrO_4^{2-} + H^+ \rightarrow HCrO_4^-$, $PO_4^{3-} + H^+ \rightarrow HPO_4^{2-}$.
2. i, ii; $Mn^{2+} \rightarrow Mn^{3+} + e^-$, $Mn^{3+} + xCl^- \rightarrow [MnCl_x]^{(x-3)-}$.
3. CrO_4^{2-} is yellow, add unknown to acetic acid - barium acetate, $CrO_4^{2-} + Ba^{2+} \rightarrow BaCrO_4(s)$(yellow); NO_3^- forms "brown ring" test with H_2SO_4-$FeSO_4$ in contact with concentrated H_2SO_4, Br^- must be removed by using silver acetate, and CrO_4^{2-} must be removed by using barium acetate before testing, $NO_3^- + 3Fe^{2+} + 4H^+ \rightarrow NO(aq) + 3Fe^{3+} + 2H_2O(l)$, $Fe^{2+} + NO(aq) \rightarrow [Fe(NO)]^{2+}$(brown), $Ag^+ + Br^- \rightarrow AgBr(s)$, $Ba^{2+} + CrO_4^{2-} \rightarrow BaCrO_4(s)$; Br^- forms a yellow-brown color in CCl_4 when chlorine water is added, $2Br^- + Cl_2(aq) \rightarrow Br_2(aq) + 2Cl^-$; Cl^- forms a white precipitate with Ag^+ which is soluble in aqueous NH_3, Br^- must be removed by oxidation before testing, $Cl^- + Ag^+ \rightarrow AgCl(s)$, $AgCl(s) + 2NH_3(aq) \rightarrow [Ag(NH_3)_2]^+ + Cl^-$, $2Br^- + S_2O_8^{2-} \rightarrow Br_2(aq) + 2SO_4^{2-}$.
4. (a) $Hg_2^{2+} \rightarrow Hg^{2+} + Hg(l)$; (b) $AgCl(s) + 2NH_3(aq) \rightarrow [Ag(NH_3)_2]^+ + Cl^-$, $[Ag(NH_3)_2]^+ + Cl^- + 2H^+ \rightarrow AgCl(s) + 2NH_4^+$; (c) $3PbS(s) + 8H^+ + 2NO_3^- \rightarrow 3Pb^{2+} + 2NO(g) + 4H_2O(l) + 3S(s)$; (d) $PbSO_4(s) + 4OH^- \rightarrow [Pb(OH)_4]^{2-} +$

SO_4^{2-}; (e) $2Mn^{2+} + 14H^+ + 5NaBiO_3(s) \rightarrow 2MnO_4^- + 5Bi^{3+} + 7H_2O(l) + 5Na^+$; (f) $Fe^{3+} + K^+ + [Fe(CN)_6]^{4-} \rightarrow KFe[Fe(CN)_6](s)$, $Fe^{3+} + NCS^- \rightarrow [Fe(NCS)]^{2+}$; (g) $2Co(OH)_3(s) + 4H^+ + H_2O_2(aq) \rightarrow 2Co^{2+} + 6H_2O(l) + O_2(g)$; (h) $Na^+ + Zn^{2+} + 3UO_2^+ + 9CH_3COO^- + 6H_2O(l) \rightarrow NaZn(UO_2)_3(CH_3COO)_9 \cdot 6H_2O(s)$.

5. (a) v, (b) iii, (c) i, (d) ii, (e) iv.
6. (a) i, (b) ii, (c) i, (d) iv, (e) i, (f) iii, (g) ii, (h) iii, (i) ii, (j) ii.
7. Al^{3+}, Mg^{2+}; run flame test for Na^+ on G.

ADDITIONAL NOTES

A 9
B 0
C 1
D 2
E 3
F 4
G 5
H 6
I 7
J 8